Metabolic Research in Aquatic Animal Nutrition, Physiology and Disease

Metabolic Research in Aquatic Animal Nutrition, Physiology and Disease

Editors

Qingchao Wang
Yan He
Qingsong Tan

MDPI

Basel • Beijing • Wuhan • Barcelona • Belgrade • Novi Sad • Cluj • Manchester

Editors

Qingchao Wang
Huazhong Agricultural University
Wuhan, China

Yan He
Huazhong Agricultural University
Wuhan, China

Qingsong Tan
Huazhong Agricultural University
Wuhan, China

Editorial Office
MDPI
St. Alban-Anlage 66
4052 Basel, Switzerland

This is a reprint of articles from the Special Issue published online in the open access journal *Metabolites* (ISSN 2218-1989) (available at: https://www.mdpi.com/journal/metabolites/special_issues/metabo_aqua_animal).

For citation purposes, cite each article independently as indicated on the article page online and as indicated below:

Lastname, A.A.; Lastname, B.B. Article Title. *Journal Name* **Year**, *Volume Number*, Page Range.

ISBN 978-3-0365-9985-4 (Hbk)
ISBN 978-3-0365-9986-1 (PDF)
doi.org/10.3390/books978-3-0365-9986-1

Contents

metabolites

Editorial

Metabolic Research in Aquatic Animal Nutrition, Physiology and Disease

Yan He [1,2,3], Qingsong Tan [1,2,3] and Qingchao Wang [1,2,3,*]

1 College of Fisheries, Huazhong Agricultural University, Wuhan 430070, China; heyan@mail.hzau.edu.cn (Y.H.); tanqs2000@mail.hzau.edu.cn (Q.T.)
2 Engineering Research Center of Green Development for Conventional Aquatic Biological Industry in the Yangtze River Economic Belt, Ministry of Education, Wuhan 430070, China
3 Key Lab of Freshwater Animal Breeding, Ministry of Agriculture and Rural Affair/Key Laboratory of Agricultural Animal Genetics, Breeding and Reproduction, Ministry of Education, Wuhan 430070, China
* Correspondence: qcwang@mail.hzau.edu.cn

Aquaculture provides a significant amount of high-quality protein for human consumption and is one of the most efficient protein production industries. The nutritional metabolism of aquatic animals plays a crucial role in the flourishing development of aquaculture. Early studies primarily focused on the basic nutritional requirements of aquatic animals, such as protein, fat, and carbohydrates. Protein was determined to be the most important component in aquafeed, having profound effects on the growth and health of aquatic animals. However, in recent years, a shortage of traditional protein sources such as fish meal and soybean meal has hindered the global development of the aquafeed industry [1]. Moreover, the metabolism of aquatic animal proteins, carbohydrates, and lipids is closely linked to their physiological changes and disease occurrence. This Special Issue, "Metabolic Research in Aquatic Animal Nutrition, Physiology, and Disease", includes 15 original research papers from different countries. These papers investigate the effects of alternative protein sources in feed on the metabolism of aquatic animal organisms, the regulation of nutrition metabolism in aquatic animals by feed additives, the metabolic response mechanisms of aquatic animal diseases, and the influence of environmental pressures on the nutritional metabolism of aquatic animals.

Compared to protein sources from terrestrial animals and plants, fish meal is the most important protein source for aquaculture due to its high-quality protein, balanced amino acid profile, and low levels of anti-nutritional factors [2]. However, the rapid development of the aquaculture industry has stimulated a significant demand for fish meal, leading to a continuous shortage in its supply. In fact, from 1950 to 2010, approximately 27% (around 20 million tons per year) of the fish caught by global marine fisheries were utilized to produce products such as fishmeal and fish oil. Notably, 90% of these fish were of food grade or premium quality. This indicates a certain inadequacy in our utilization of fish to sustain ourselves, necessitating a more rational use of fishery resources [3], Therefore, the exploration of alternative protein sources to replace fish meal has become a research hotspot in recent years [4]. Various cottonseed protein products have been investigated as a potential substitute for fish meal in the diets of various fish species [5–7]. Zhang et al. found that enzymatically hydrolyzed cottonseed protein concentrate (ECP), as a cost-effective and easily obtainable plant protein, could be used as a substitute for 15.56% of fish meal without compromising the growth of largemouth bass (*Micropterus salmoides*), and it improved their antioxidant capacity and immune status (contribution 1). Similarly, replacing FM with cottonseed meal protein hydrolysate had similar effects on blunt-snout bream (*Megalobrama amblycephala*) [8]. Cottonseed meal (CSM) has also been successfully used as a replacement for 20% of fish meal without affecting the muscle nutritional deposition in golden pompano (*Trachinotus ovatus*) (contribution 2). Furthermore, Liu et al. incorporated cottonseed protein concentrate (CPC) as a sole protein source in grass carp (*Ctenopharyngodon idellus*) feed

Citation: He, Y.; Tan, Q.; Wang, Q. Metabolic Research in Aquatic Animal Nutrition, Physiology and Disease. *Metabolites* **2024**, *14*, 22. https://doi.org/10.3390/metabo14010022

Received: 4 December 2023
Revised: 21 December 2023
Accepted: 27 December 2023
Published: 28 December 2023

and evaluated its nutritional value and optimal protein level. Their results showed that a protein level of 37.69% in the CPC4 group enhanced the growth performance and health status of grass carp, while high protein levels (CPC5 and CPC6) resulted in relatively poor growth performance (contribution 3). This could be attributed to a decrease in the feeding rate and increased energy expenditure on deamination of excess protein, which increased the nitrogen metabolism burden and negatively impacted fish growth [9]. Wheat gluten can safely replace 100% of fish meal and 61.2% of soy protein concentrate in the diet of red spotted grouper (*Epinephelus akaara*) (contribution 4). Similar findings were also documented in a prior investigation, wherein wheat gluten was partially substituted for dietary fish meal in large yellow croaker [10]. Enzymatic chicken plasma (ECP) derived from enzymatic hydrolysis of poultry by-products is a high-quality animal protein source for most aquatic organisms. Its rich content of amino acids such as arginine, alanine, and taurine can improve the feed acceptance of crustaceans; as such, it is considered a suitable substitute for fish meal [11]. Hlordzi et al. investigated the effects of ECP on the growth performance, digestive enzyme activity, and gene expression of Pacific white shrimp (*Litopenaeus vannamei*), finding that a dietary inclusion level of 32.4% crude protein in ECP1 was more suitable for the growth of juvenile shrimp (contribution 5). Furthermore, Yang et al. explored the utilization of *Clostridium autoethanogenum* protein (CAP), produced through bacterial metabolism of steel mill off-gas, as a replacement for fish meal. CAP successfully replaced 67.1-68.0% of the fish meal in the feed (contribution 6), though the incorporation of CAP into diets necessitates the supplementation of exogenous amino acids, particularly arginine, to maintain an optimal amino acid composition [12–15]. In addition to providing a range of feasible options for substituting protein sources in the aquaculture industry, aiming to improve feed utilization efficiency and reduce production costs, these studies have positive implications for environmental protection and waste utilization.

Another effective way to conserve protein is to fully utilize the energy effects of lipids and carbohydrates in feed [16,17]. However, higher levels of fat or glucose in the feed can lead to abnormal fat deposition in the liver, directly impacting the health of fish as well as their yield and economic benefits [18–20]. Deng et al. found that adding phytosterol glycosides to high-fat feed promoted fat decomposition in the liver of hybrid groupers (♀*Epinephelus fuscoguttatus* × ♂*Epinephelus lanceolatu*), activated the glycolytic pathway, enhanced the utilization of feed energy by fish, and supported the non-specific immune defense mechanisms in fish (contribution 7). Xia et al. demonstrated that regulating endoplasmic reticulum stress in spotted sea bass using 4-PBA, an endoplasmic reticulum stress inhibitor, reduced fat accumulation caused by high-fat feed (contribution 8). Similar results were also reported in some previous studies [21,22]. By combining liver proteomics and lipidomics, Xue et al. investigated the key signaling pathways and biomarkers associated with metabolic fatty liver disease in largemouth bass. They identified FABP1, ABCA1, and VDAC1 as potential biomarkers for the treatment of metabolic fatty liver disease in largemouth bass (contribution 9).

Certain feed additives also have significant benefits for the metabolic health of fish [23–25]. Wu et al. investigated the effects of soybean lecithin (SBL) on the growth, blood parameters, immune capacity, antioxidant capacity, and intestinal barrier function of juvenile largemouth bass. They found that 4% SBL supplementation improved these indicators (contribution 10). Song et al. examined the impact of dietary lactoferrin (LF) supplementation on the growth performance and intestinal health of juvenile orange-spotted grouper (*Epinephelus coioides*) fed a high-soybean meal (SBM) diet. The addition of appropriate LF improved the growth rate of fish, reduced the content of harmful substances in the intestine, and increased the diversity of beneficial bacteria (contribution 11). Similarly, research comparing the intestinal microbiota of grass carp fed different diets showed that diet can modulate the stability of the intestinal microbiota by altering its composition and interspecies interactions, thus regulating the growth metabolism of fish (contribution 12). Broughton et al. focused on the development and addition of a novel oil extracted from genetically modified (GM) oilseeds into Atlantic salmon (*Salmo salar*) feed. Phospholipid profiling analysis revealed that the long-chain polyunsaturated fatty acids (LC-PUFAs), EPA, and DHA derived from GM oilseeds could alter the lipid structure within fish, providing a theoretical basis for supplementing and replacing current terrestrial oilseeds and fish oil (contribution 13).

The water temperature and light intensity are also crucial to the growth and metabolism of fish. Chen et al. have proved that chronic low temperature affects the glucose and lipid metabolism in the liver of freshwater drum (*Aplodinotus grunniens*), with the miR-1/AMPK signaling pathway playing an important role in regulating these processes, providing a theoretical basis for freshwater drum's resistance to low temperature (contribution 14). However, time-restricted feeding (TRF), which has remarkable effects on metabolic regulation caused by circadian rhythm disruption in mammals, failed to mitigate the negative effects of continuous light exposure induced by artificial night light on rainbow trout fry, resulting in increased body fat accumulation and significantly decreased muscle LC-PUFA ratio (contribution 15).

In conclusion, the research papers in the Special Issue "Metabolic Research in Aquatic Animal Nutrition, Physiology and Disease" provide valuable theoretical foundations for protein substitution in the field of aquatic animal nutrition and metabolism. They also delve into the occurrence of metabolic diseases and immune mechanisms in aquatic animals. This in-depth exploration is of great significance in the development of functional feed ingredients or efficient feeds, and in driving the sustainable prosperity of the aquaculture industry.

Funding: This research was funded by Key R & D program of Hubei Province (2021BBA233), Science and Technology Planning Project of the Xinjiang Production and Construction Corps (2023AB079), National Natural Science Foundation of China (Grant No. 32172996), Knowledge Innovation Program of Wuhan-Shuguang Project (2023020201020350), Fundamental Research Funds for the Central Universities (2662023SCPY005).

Conflicts of Interest: The authors declare no conflict of interest.

List of Contributions:

1. Zhang, Q.; Liang, H.; Xu, P.; Xu, G.; Zhang, L.; Wang, Y.; Ren, M.; Chen, X. Effects of Enzymatic Cottonseed Protein Concentrate as a Feed Protein Source on the Growth, Plasma Parameters, Liver Antioxidant Capacity and Immune Status of Largemouth Bass (*Micropterus salmoides*). *Metabolites* **2022**, 12, 1233. https://doi.org/10.3390/metabo12121233.
2. Qin, Y.; He, C.; Geng, H.; Wang, W.; Yang, P.; Mai, K.; Song, F. Muscle Nutritive Metabolism Changes after Dietary Fishmeal Replaced by Cottonseed Meal in Golden Pompano (*Trachinotus ovatus*). *Metabolites* **2022**, *12*, 576. https://doi.org/10.3390/metabo12070576.
3. Liu, G.; Zhou, M.; Wang, X.; Mao, X.; Long, X.; Xie, S.; Han, D.; Tan, Q. Effects of Dietary Cottonseed Protein Concentrate Levels on Growth Performance, Health Status, Flesh Quality and Intestinal Microbiota of Grass Carp (*Ctenopharyngodon idellus*). *Metabolites* **2022**, *12*, 1046. https://doi.org/10.3390/metabo12111046.
4. Cheng, Y.; Wang, Y.; Dong, Z.; Storebakken, T.; Xu, G.; Shi, B.; Zhang, Y. Evaluate of Wheat Gluten as a Protein Alternative for Fish Meal and Soy Protein Concentrate in Red Spotted Grouper *Epinephelus akaara*. *Metabolites* **2023**, *13*, 832. https://doi.org/10.3390/metabo13070832.
5. Hlordzi, V.; Tan, B.; Dong, X.; Zhang, S.; Zhu, L.; Zhang, L.; Hu, X.; Chi, S. Enzymatic Chicken Pulp Promotes Appetite, Digestive Enzyme Activity, and Growth in *Litopenaeus vannamei*. *Metabolites* **2022**, *12*, 698. https://doi.org/10.3390/metabo12080698.
6. Yang, P.; Li, X.; Yao, W.; Li, M.; Wang, Y.; Leng, X. Dietary Effect of *Clostridium autoethanogenum* Protein on Growth, Intestinal Histology and Flesh Lipid Metabolism of Largemouth Bass (*Micropterus salmoides*) Based on Metabolomics. *Metabolites* **2022**, *12*, 1088. https://doi.org/10.3390/metabo12111088.
7. Deng, H.; Zhang, J.; Yang, Q.; Dong, X.; Zhang, S.; Liang, W.; Tan, B.; Chi, S. Effects of Dietary Steroid Saponins on Growth Performance, Serum and Liver Glucose, Lipid Metabolism and Immune Molecules of Hybrid Groupers (♀*Epinephelus fuscoguttatus* × ♂*Epinephelus lanceolatu*) Fed High-Lipid Diets. *Metabolites* **2023**, *13*, 305. https://doi.org/10.3390/metabo13020305.

8. Xia, T.; Liao, Y.; Li, L.; Sun, L.Y.; Ding, N.S.; Wu, Y.L.; Lu, K.L. 4-PBA Attenuates Fat Accumulation in Cultured Spotted Seabass Fed High-Fat-Diet via Regulating Endoplasmic Reticulum Stress. *Metabolites* **2022**, *12*, 1197. https://doi.org/10.3390/metabo12121197.
9. Xue, M.Y.; Yao, T.; Xue, M.; Francis, F.; Qin, Y.C.; Jia, M.; Li, J.; Gu, X. Mechanism Analysis of Metabolic Fatty Liver on Largemouth Bass (*Micropterus salmoides*) Based on Integrated Lipidomics and Proteomics. *Metabolites* **2022**, *12*, 759. https://doi.org/10.3390/metabo12080759.
10. Wu, J.; Yang, W.; Song, R.; Li, Z.; Jia, X.; Zhang, H.; Zhang, P.; Xue, X.; Li, S.; Xie, Y.; et al. Dietary Soybean Lecithin Improves Growth, Immunity, Antioxidant Capability and Intestinal Barrier Functions in Largemouth Bass *Micropterus salmoides* Juveniles. *Metabolites* **2023**, *13*, 512. https://doi.org/10.3390/metabo13040512.
11. Song, T.; Qin, Y.; Ke, L.; Wang, X.; Wang, K.; Sun, Y.; Ye, J. Dietary Lactoferrin Supplementation Improves Growth Performance and Intestinal Health of Juvenile Orange-Spotted Groupers (*Epinephelus coioides*). *Metabolites* **2022**, *12*, 915. https://doi.org/10.3390/metabo12100915.
12. Yang, G.; Xiang, Y.; Wang, S.; Tao, Y.; Xie, L.; Bao, L.; Shen, K.; Li, J.; Hu, B.; Wen, C.; et al. Response of Intestinal Microbiota to the Variation in Diets in Grass Carp (*Ctenopharyngodon idella*). *Metabolites* **2022**, *12*, 1115. https://doi.org/10.3390/metabo12111115.
13. Broughton, R.; Tocher, D.R.; Napier, J.A.; Betancor, M.B. Profiling Phospholipids within Atlantic Salmon *Salmo salar* with Regards to a Novel Terrestrial Omega-3 Oil Source. *Metabolites* **2022**, *12*, 851. https://doi.org/10.3390/metabo12090851.
14. Chen, J.; Song, C.; Wen, H.; Liu, G.; Wu, N.; Li, H.; Xue, M.; Xu, P. miR-1/AMPK-Mediated Glucose and Lipid Metabolism under Chronic Hypothermia in the Liver of Freshwater Drum, *Aplodinotus grunniens*. *Metabolites* **2022**, *12*, 697. https://doi.org/10.3390/metabo12080697.
15. Xu, H.; Shi, C.; Ye, Y.; Song, C.; Mu, C.; Wang, C. Time-Restricted Feeding Could Not Reduce Rainbow Trout Lipid Deposition Induced by Artificial Night Light. *Metabolites* **2022**, *12*, 904. https://doi.org/10.3390/metabo12100904.

References

1. Olsen, R.L.; Hasan, M.R. A limited supply of fishmeal: Impact on future increases in global aquaculture production. *Trends Food Sci. Technol.* **2012**, *27*, 120–128. [CrossRef]
2. Zhao, X.; Wang, Y.; Wang, X.; Ye, J. Growth performance, plasma components, and intestinal barrier in grouper (*Epinephelus coioides*) are altered by dietary fish meal replacement with extruded soybean meal. *Aquac. Rep.* **2021**, *21*, 100863. [CrossRef]
3. Cashion, T.; Frédéric Le Manach, F.L.; Zeller, D.; Pauly, D. Most fish destined for fishmeal production are food-grade fish. *Fish Fish.* **2017**, *18*, 837–844. [CrossRef]
4. Ghosh, K.; Ray, A.K.; Ringø, E. Applications of plant ingredients for tropical and subtropical freshwater finfish: Possibilities and challenges. *Rev. Aquac.* **2019**, *11*, 793–815. [CrossRef]
5. Shen, J.F.; Liu, H.Y.; Tan, B.P.; Dong, X.H.; Yang, Q.H.; Chi, S.Y.; Zhang, S. Effects of replacement of fishmeal with cottonseed protein concentrate on the growth, intestinal microflora, haematological and antioxidant indices of juvenile golden pompano (*Trachinotus ovatus*). *Aquac. Nutr.* **2020**, *26*, 1119–1130. [CrossRef]
6. Ye, G.L.; Dong, X.H.; Yang, Q.H.; Chi, S.Y.; Liu, H.Y.; Zhang, H.T.; Tan, B.P.; Zhang, S. Low-gossypol cottonseed protein concentrate used as a replacement of fish meal for juvenile hybrid grouper (*Epinephelus fuscoguttatus* ♀× *Epinephelus lanceolatus* ♂): Effects on growth performance, immune responses and intestinal microbiota. *Aquaculture* **2020**, *524*, 735309. [CrossRef]
7. He, Y.F.; Guo, X.W.; Tan, B.P.; Dong, X.H.; Yang, Q.H.; Liu, H.; Zhang, S.; Chi, S.Y. Replacing fishmeal with cottonseed protein concentrate in feed for pearl gentian groupers (*Epinephelus fuscoguttatus*♀× *E. lanceolatus*♂): Effects on growth and expressions of key genes involved in appetite and hepatic glucose and lipid metabolism. *Aquac. Rep.* **2021**, *20*, 100710. [CrossRef]
8. Yuan, X.Y.; Jiang, G.Z.; Cheng, H.H.; Cao, X.F.; Shi, H.J.; Liu, W.B. An evaluation of replacing fish meal with cottonseed meal protein hydrolysate in diet for juvenile blunt snout bream (*Megalobrama amblycephala*): Growth, antioxidant, innate immunity and disease resistance. *Aquac. Nutr.* **2019**, *25*, 1334–1344. [CrossRef]
9. Mcgoogan, B.B.; Gatlin Iii, D.M. Dietary manipulations affecting growth and nitrogenous waste production of red drum, *Sciaenops ocellatus* I. Effects of dietary protein and energy levels. *Aquaculture* **1999**, *178*, 333–348. [CrossRef]
10. Wang, P.; Lou, Y.; Feng, J.; He, J.; Zhu, J.; Zhou, Q. Effect of replacing fish meal with wheat gluten meal on growth, serum biochemical indexes and antioxidant enzyme activity of juvenile large yellow croaker (*Larimichthys crocea*). *J. Fish. China* **2018**, *42*, 733–743.
11. Zhuang, Y.; Zhang, W.; Zheng, J.; Tang, Z.; Li, X.; Cao, X.; Zhang, L.; Xu, W.; Mai, K.; Ai, Q. Effects of enzymatic hydrolysis chicken by-product in high plant-based protein diet on growth performance, digestive capacity, antioxidant capacity and non-specific immunity of juvenile turbot (*Scophthalmus maximus* L.). *Aquac. Nutr.* **2021**, *27*, 1578–1589. [CrossRef]
12. Buentello, J.A.; Gatlin, D.M. The dietary arginine requirement of channel cafish (*ctalrus punctaus*) is influenced by endogenous synthesis of arginine from glutamic acid. *Aquaculure* **2000**, *188*, 311–321. [CrossRef]

13. Zhou, F.; Shao, Q.J.; Xiao, J.X.; Peng, X.; Ngandzali, B.O.; Sun, Z.; Wing, K.N. Effects of dietary arginine and lysine levels on growth performance, nutrient utilization and tissue biochemical profile of black sea bream, *Acanthopagrus schlegelii*, fingerlings. *Aquaculture* **2011**, *319*, 72–80. [CrossRef]
14. Murthy, H.S.; Varghese, T.J. Dietary requirement of juveniles of the Indian major carp, *Labeo rohita*, for the essential amino acid lysine. *Isr. J. Aquac.-Bamidgeh* **1997**, *49*, 19–24.
15. Berge, G.E.; Sveier, H.; Lied, E. Effects of feeding Atlantic salmon (*Salmo salar* L.) imbalanced levels of lysine and arginine. *Aquac. Nutr.* **2002**, *8*, 239–248. [CrossRef]
16. Wang, J.; Han, T.; Li, X.; Yang, Y.; Yang, M.; Hu, S.; Jiang, Y.; Harpaz, S. Effects of dietary protein and lipid levels with different protein-to-energy ratios on growth performance, feed utilization and body composition of juvenile, red-spotted grouper (*Epinephelus akaara*). *Aquac. Nutr.* **2017**, *23*, 994–1002. [CrossRef]
17. Li, S.; Mai, K.; Xu, W.; Yuan, Y.; Zhang, Y.; Zhou, H.; Ai, Q. Effects of dietary lipid level on growth, fatty acid composition, digestive enzymes, and expression of some lipid metabolism related genes of orange-spotted grouper larvae (*Epinephelus coioides* H.). *Aquac. Res.* **2016**, *47*, 2481–2495. [CrossRef]
18. Pan, S.; Yan, X.; Dong, X.; Li, T.; Suo, X.; Tan, B.; Zhang, S.; Li, Z.; Yang, Y.; Zhang, H. The positive effects of dietary inositol on juvenile hybrid grouper (♀*Epinephelus fuscoguttatus* ×♂*E. lanceolatu*) fed high-lipid diets: Growth performance, antioxidant capacity and immunity. *Fish Shellfish Immunol.* **2022**, *26*, 84–95. [CrossRef]
19. Fei, S.; Xia, Y.; Chen, Z.; Liu, C.; Liu, H.; Han, D.; Jin, J.; Yang, Y.; Zhu, X.; Xie, S. A high-fat diet alters lipid accumulation and oxidative stress and reduces the disease resistance of overwintering hybrid yellow catfish (*Pelteobagrus fulvidraco*♀× *P. vachelli*♂). *Aquac. Rep.* **2022**, *23*, 101043. [CrossRef]
20. Ding, T.; Xu, N.; Liu, Y.; Du, J.; Xiang, X.; Xu, D.; Liu, Q.; Yin, Z.; Li, J.; Mai, K.; et al. Effect of dietary bile acid (BA) on the growth performance, body composition, antioxidant responses and expression of lipid metabolism-related genes of juvenile large yellow croaker (*Larimichthys crocea*) fed high lipid diets. *Aquaculture* **2020**, *518*, 734768. [CrossRef]
21. Jobling, M.; Koskela, J.; Savolainen, R. Influence of dietary fat level and increased adiposity on growth and fat deposition in rainbow trout, *Oncorhynchus mykiss* (Walbaum). *Aquac. Res.* **1998**, *29*, 601–607. [CrossRef]
22. Zhou, W.; Rahimnejad, S.; Lu, K.; Wang, L.; Liu, W. Effects of berberine on growth, liver histology, and expression of lipid-related genes in blunt snout bream (*Megalobrama amblycephala*) fed high-fat diets. *Fish Physiol. Biochem.* **2018**, *45*, 83–91. [CrossRef] [PubMed]
23. Ciji, A.; Akhtar, M.S.; Tripathi, P.H.; Pandey, A.; Rajesh, M.; Samkamalam, B. Dietary soy lecithin augments antioxidative defense and thermal tolerance but fails to modulate non-specific immune genes in endangered golden mahseer (*Tor putitora*) fry. *Fish Shellfish Immunol.* **2021**, *109*, 34–40. [CrossRef] [PubMed]
24. Siyal, F.A.; El-Hack, M.E.A.; Alagawany, M.; Wang, C.; Wan, X.; He, J.; Wang, M.; Zhang, L. Effect of soy lecithin on growth performance, nutrient digestibility and hepatic antioxidant parameters of broiler chickens. *Int. J. Pharmacol.* **2017**, *13*, 396–402. [CrossRef]
25. Saleh, N.E.; Wassef, E.A.; Kamel, M.A.; El-Haroun, E.R.; El-Tahan, R.A. Beneficial effects of soybean lecithin and vitamin C combination in fingerlings gilthead seabream (*Sparus aurata*) diets on; Fish performance, oxidation status and genes expression responses. *Aquaculture* **2022**, *546*, 737345. [CrossRef]

metabolites MDPI

Article

Effects of Enzymatic Cottonseed Protein Concentrate as a Feed Protein Source on the Growth, Plasma Parameters, Liver Antioxidant Capacity and Immune Status of Largemouth Bass (*Micropterus salmoides*)

Qile Zhang [1], Hualiang Liang [1,2], Pao Xu [1,2], Gangchun Xu [1,2], Lu Zhang [3], Yongli Wang [4], Mingchun Ren [1,2,*] and Xiaoru Chen [4,*]

1 Wuxi Fisheries College, Nanjing Agricultural University, Wuxi 214081, China
2 Key Laboratory of Integrated Rice-Fish Farming Ecology, Ministry of Agriculture and Rural Affairs, Freshwater Fisheries Research Center, Chinese Academy of Fishery Sciences, Wuxi 214081, China
3 Tongwei Co., Ltd., Healthy Aquaculture Key Laboratory of Sichuan Province, Chengdu 610093, China
4 Tongwei Agricultural Development Co., Ltd., Chengdu 610093, China
* Correspondence: renmc@ffrc.cn (M.R.); chenxr@tongwei.com (X.C.)

Abstract: This study appraised the impact of enzymatic cottonseed protein concentrate (ECP) as a fish meal (FM) substitute on the growth and health of largemouth bass (*Micropterus salmoides*) (initial weight 14.99 ± 0.03 g). Five diets with equal nitrogen, fat, and energy were designed to replace 0%, 7.78%, 15.56%, 23.33%, and 31.11% FM by adding 0%, 3.6%, 7.2%, 10.8%, and 14.4% ECP, named ECP0, ECP3.6, ECP7.2, ECP10.8, and ECP14.4, respectively. We fed 300 fish with five experimental diets for 60 days. The results revealed that weight gain rate (WGR) and specific growth rate (SGR) did not notably reduce until the addition of ECP exceeded 7.2%. The proximate composition of fish was not affected by the amount of ECP added in diets. Plasma total protein (TP), albumin (ALB), and high-density lipoprotein (HDL) concentrations increased with the increase of ECP dosage, while the triglyceride (TG) and low-density lipoprotein (LDL) concentrations and alkaline phosphatase (ALP) activity showed an opposite trend. For hepatic antioxidant capacity, the hepatic total superoxide dismutase (T-SOD) and catalase (CAT) activities, glutathione (GSH) content, and the expression of nuclear factor erythroid 2-related factor 2 (Nrf2), superoxide dismutase (SOD), and CAT were increased by ECP, while the hepatic malondialdehyde (MDA) content and the expression of kelch-like-ECH-associated protein 1 (Keap1) were decreased. With regard to inflammation, the expression of nuclear factor-kappa B (NF-κB), interleukin-8 (IL-8), and tumor necrosis factor-α (TNF-α) were inhibited by ECP. In summary, the amount of ECP added to diet can reach 7.2% to replace 15.56% FM without hampering the growth of largemouth bass, and ECP can improve the antioxidant and immune capacity.

Keywords: largemouth bass (*Micropterus salmoides*); enzymatic cottonseed protein concentrate; growth performance; antioxidant capacity; immunity

Citation: Zhang, Q.; Liang, H.; Xu, P.; Xu, G.; Zhang, L.; Wang, Y.; Ren, M.; Chen, X. Effects of Enzymatic Cottonseed Protein Concentrate as a Feed Protein Source on the Growth, Plasma Parameters, Liver Antioxidant Capacity and Immune Status of Largemouth Bass (*Micropterus salmoides*). *Metabolites* **2022**, *12*, 1233. https://doi.org/10.3390/metabo12121233

Academic Editor: Yan He

Received: 12 November 2022
Accepted: 2 December 2022
Published: 8 December 2022

1. Introduction

Largemouth bass (*Micropterus salmoides*) have many advantages, including strong adaptability, fast growth, no intermuscular spines, and good flavor; they are widely liked by humans and have been largely farmed in many countries [1]. The requirement of largemouth bass for dietary protein is high, given that they are carnivorous fish, accounting for approximately 40–50% of dry matter, and the dietary protein is mainly provided by fish meal (FM) [2–5]. In aquatic animals, FM is the best and main source of dietary protein; however, FM resources are limited, the output has difficulty meeting the needs of the rapid development of aquaculture, and the price is gradually rising [6–8]. Therefore, it is necessary to investigate new high-quality and inexpensive alternative protein sources for FM.

Cottonseed meal, the residue of cottonseed after oil extraction, is rich in protein content and yield, and low in price [9]. A variety of cottonseed meal products can be added to aquatic feeds as protein ingredients [10–17]. The application of cottonseed meal not only avoids wasting cottonseed resources but also alleviates the shortage of feed protein resources. As there are some anti-nutritional factors, cottonseed meal needs to be processed to improve the utilization rate. Among various processing methods, enzymatic hydrolysis is an effective and popular method for improving the quality of plant protein because of its non-irritating reaction conditions, less harmful products, lack of effect on the nutritional value of amino acids, and low pollution level on the environment [18]. Previous studies showed that plant proteins with enzymatic hydrolysis treatment have better application effects in animal feed [19,20]. Enzymatic hydrolysis treatment could increase the water-soluble protein, total amino acids, and peptide fraction contents of cottonseed meal [21,22]. Furthermore, various active peptides, such as antioxidant peptides and antimicrobial peptides, could be produced from the enzymatic hydrolysis of cottonseed protein [23–25]. It was reported that crucian carp (*Carassius auratus gibelio*) fed with a diet containing 5% cottonseed meal hydrolysate instead of cottonseed meal had better growth, feed utilization rate, and absorption of zinc and iron [21]. Chinese soft-shelled turtle (*Pelodiscus sinensis*) fed with diets containing cottonseed meal protein hydrolysate had better growth, feed intake, digestive enzymes activity, and intestinal development [26]. In addition, it was reported that adding 3% cottonseed meal protein hydrolysate to the feed of blunt snout bream (*Megalobrama amblycephala*) could reduce FM content from 6% to 2.8%, resulting in better growth performance, antioxidant capacity, and immunity [27].

Up to now, studies on replacing dietary FM with enzymatic cottonseed protein concentrate (ECP) in largemouth bass are still lacking. Therefore, the current study was conducted to investigate the effects of ECP on the growth, proximate composition, plasma biochemical indexes, hepatic antioxidant capacity, and immune status of largemouth bass, and evaluate the potential of ECP as a feed protein material to promote the development of low FM feed for aquatic animals.

2. Results

2.1. Growth Performance

As displayed in Table 1, fish fed with ECP10.8 and ECP14.4 diets were observed to have lower final weight (FW), weight gain rate (WGR), and specific growth rate (SGR) ($p > 0.05$). Compared with the control diet (ECP0), fish fed with ECP3.6, ECP7.2, ECP10.8, and ECP14.4 diets were observed to have similar feed intake (FI), feed coefficient rate (FCR), protein efficiency ratio (PER), and survival rate (SR) ($p > 0.05$).

Table 1. Effect of replacing FM with ECP on growth performance of juvenile largemouth bass (*Micropterus salmoides*) [1].

Diets	IW (g) [2]	FW (g) [3]	WGR (%) [4]	SGR (%/day) [5]	FI (g/fish) [6]	FCR [7]	PER [8]	SR (%) [9]
ECP0	14.88 ± 0.06	72.82 ± 2.66 ^{ab}	389.32 ± 19.31 ^{a}	2.65 ± 0.07 ^{a}	72.09 ± 1.15 ^{ab}	1.26 ± 0.06	1.59 ± 0.08	95.00 ± 5.00
ECP3.6	15.03 ± 0.12	76.77 ± 0.86 ^{a}	411.75 ± 0.88 ^{a}	2.72 ± 0.00 ^{a}	70.95 ± 2.94 ^{ab}	1.25 ± 0.10	1.61 ± 0.12	91.67 ± 5.77
ECP7.2	15.00 ± 0.13	71.11 ± 0.61 ^{b}	374.12 ± 8.08 ^{a}	2.59 ± 0.03 ^{a}	73.41 ± 1.33 ^{a}	1.33 ± 0.05	1.52 ± 0.05	95.00 ± 8.66
ECP10.8	15.08 ± 0.03	60.78 ± 0.75 ^{c}	303.20 ± 5.93 ^{b}	2.33 ± 0.02 ^{b}	66.90 ± 2.21 ^{b}	1.37 ± 0.14	1.46 ± 0.15	98.33 ± 2.89
ECP14.4	14.97 ± 0.18	64.22 ± 2.07 ^{c}	329.22 ± 16.57 ^{b}	2.43 ± 0.07 ^{b}	70.20 ± 2.08 ^{ab}	1.46 ± 0.02	1.37 ± 0.02	88.33 ± 5.77

[1] Data are presented as mean ± SD. [2] Initial body weight (IW). [3] Final body weight (FW). [4] Weight gain rate (WGR, %) = 100 × (final weight (g) − initial weight (g))/initial weight (g). [5] Specific growth rate (SGR, %/day) = 100 × ((Ln (final body weight (g)) − Ln (initial body weight (g)))/days). [6] Feed intake (FI, g/fish) = dry feed fed (g)/fish number. [7] Feed coefficient rate (FCR) = dry feed fed (g)/(final body weight (g) − initial body weight (g)). [8] Protein efficiency ratio (PER) = ((final body weight (g) − initial body weight (g))/protein intake (g). [9] Survival rate (SR, %) = 100 × (final amount of fish)/(initial amount of fish). $^{a–c}$ Values in the same column with different superscripts indicate significant differences ($p < 0.05$), while that with the same letter or no letter superscripts indicate no significant differences ($p > 0.05$).

2.2. Proximate Composition of Whole Fish

As displayed in Table 2, the moisture, crude protein, crude lipid, and ash contents of fish fed with different diets had no significant difference ($p > 0.05$).

Table 2. Effect of replacing FM with ECP on fish whole-body composition of juvenile largemouth bass (*Micropterus salmoides*) [1].

Diets	Moisture (%)	Crude Protein (%)	Crude Lipid (%)	Ash (%)
ECP0	67.53 ± 1.46	17.52 ± 0.01	7.95 ± 0.73	4.62 ± 0.19
ECP3.6	67.40 ± 1.04	17.73 ± 0.47	8.55 ± 0.94	4.55 ± 0.36
ECP7.2	67.74 ± 1.01	17.64 ± 0.26	8.06 ± 0.34	4.55 ± 0.16
ECP10.8	68.59 ± 0.27	17.79 ± 0.06	8.07 ± 0.45	4.95 ± 0.23
ECP14.4	66.33 ± 1.57	18.68 ± 0.18	8.97 ± 0.21	4.89 ± 0.27

[1] Data are presented as mean ± SD.

2.3. Plasma Parameters

As displayed in Table 3, ECP tended to increase the plasma total protein (TP), albumin (ALB), and high-density lipoprotein (HDL) concentrations, while tending to decrease the total cholesterol (TC), triglyceride (TG), and low-density lipoprotein (LDL) concentrations and alkaline phosphatase (ALP) activity. Compared with the control diet, fish fed with ECP7.2, ECP10.8, and ECP14.4 diets had higher TP and ALB concentrations, and fish fed with ECP-containing diets had higher HDL concentrations ($p < 0.05$). Fish fed with ECP14.4 diets had the lowest TC, LDL concentrations, and ALP activity.

Table 3. Effect of replacing FM with ECP on plasma biochemical parameters of juvenile largemouth bass (*Micropterus salmoides*) [1].

Diets	TP (g/L) [2]	ALB (g/L) [2]	TC (mmol/L) [2]	TG (mmol/L) [2]	LDL (mmol/L) [2]	HDL (mmol/L) [2]	ALP (U/L) [2]
ECP0	55.72 ± 7.21 ^{c}	13.61 ± 2.10 ^{c}	12.13 ± 2.24	10.00 ± 1.97 ^{ab}	4.91 ± 1.60 ^{a}	2.37 ± 0.28 ^{c}	194.87 ± 52.78 ^{a}
ECP3.6	61.48 ± 4.68 ^{bc}	15.62 ± 1.81 ^{bc}	12.02 ± 1.45	9.56 ± 1.23 ^{ab}	4.63 ± 0.87 ^{ab}	2.76 ± 0.22 ^{b}	161.16 ± 35.10 ^{abc}
ECP7.2	69.72 ± 4.04 ^{a}	17.34 ± 1.14 ^{a}	12.55 ± 1.06	10.97 ± 1.29 ^{a}	4.78 ± 0.68 ^{a}	3.01 ± 0.13 ^{ab}	169.93 ± 31.01 ^{ab}
ECP10.8	67.74 ± 5.98 ^{ab}	17.42 ± 1.39 ^{ab}	11.63 ± 1.35	8.20 ± 0.92 ^{b}	3.92 ± 0.83 ^{ab}	3.18 ± 0.38 ^{a}	143.72 ± 19.23 ^{bc}
ECP14.4	67.10 ± 4.30 ^{ab}	17.92 ± 1.54 ^{a}	10.66 ± 1.65	6.09 ± 1.20 ^{c}	3.24 ± 1.17 ^{b}	2.96 ± 0.12 ^{ab}	119.13 ± 32.61 ^{c}

[1] Data are presented as mean ± SD. [2] TP, total protein; ALB, albumin; TC, total cholesterol; TG, triglyceride; LDL, low-density lipoprotein; HDL, high-density lipoprotein; ALP, alkaline phosphatase. $^{a–c}$ Values in the same column with different superscripts indicate significant differences ($p < 0.05$), while that with the same letter or no letter superscripts indicate no significant differences ($p > 0.05$).

2.4. Hepatic Antioxidant Parameters and MDA Levels

As shown in Figure 1, ECP tended to improve the hepatic antioxidant capacity of largemouth bass. Compared with the control diet, higher total superoxide dismutase (T-SOD) and catalase (CAT) activities were observed in fish fed with ECP10.8 and ECP14.4 diets, and fish fed with ECP7.2, ECP10.8, and ECP14.4 diets, respectively ($p < 0.05$). Fish fed with ECP-containing diets were observed to have higher glutathione (GSH) content and lower malondialdehyde (MDA) content ($p < 0.05$).

2.5. Expression of Antioxidant Genes mRNA

As presented in Figure 2, compared with the control diet, ECP7.2, ECP10.8, and ECP14.4 diets significantly increased the mRNA expression level of nuclear factor erythroid 2-related factor 2 (Nrf2) and superoxide dismutase (SOD), ECP10.8 and ECP14.4 diets significantly increased the mRNA expression level of CAT, and ECP-containing diets significantly decreased the mRNA expression level of kelch-like-ECH-associated protein 1 (Keap1) ($p < 0.05$).

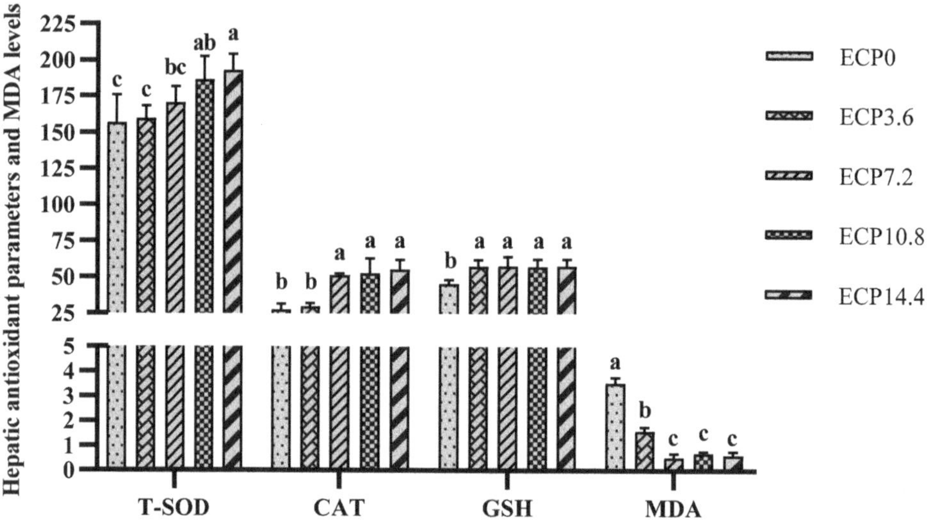

Figure 1. Hepatic antioxidant parameters and MDA levels. T-SOD, total superoxide dismutase; CAT, catalase; GSH, glutathione; MDA, malondialdehyde. Data are presented as mean ± SD. Bars with different superscripts indicate significant differences ($p < 0.05$), while that with the same letter or no letter superscripts indicate no significant differences ($p > 0.05$).

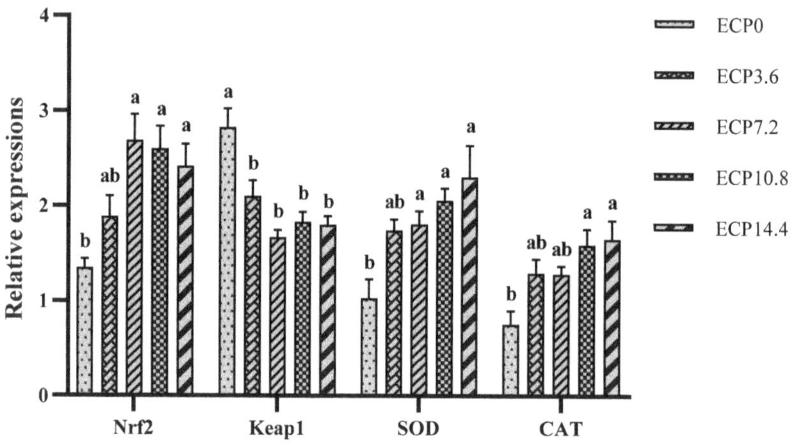

Figure 2. Relative mRNA expression levels of antioxidant genes in liver. Nrf2, nuclear factor erythroid 2-related factor 2; Keap1, kelch-like-ECH-associated protein 1; SOD, superoxide dismutase; CAT, catalase. Data are presented as mean ± SD. Bars with different superscripts indicate significant differences ($p < 0.05$), while that with the same letter or no letter superscripts indicate no significant differences ($p > 0.05$).

2.6. Expression of Immune-Related Genes mRNA

As presented in Figure 3, compared with the control diet, ECP-containing diets significantly decreased the mRNA expression level of nuclear factor-kappa B (NF-κB), interleukin-8 (IL-8), and tumor necrosis factor-α (TNF-α) ($p < 0.05$). ECP-containing diets tended to increase the mRNA expression level of transforming growth factor-β (TGF-β) and interleukin-10 (IL-10), but not significantly ($p > 0.05$).

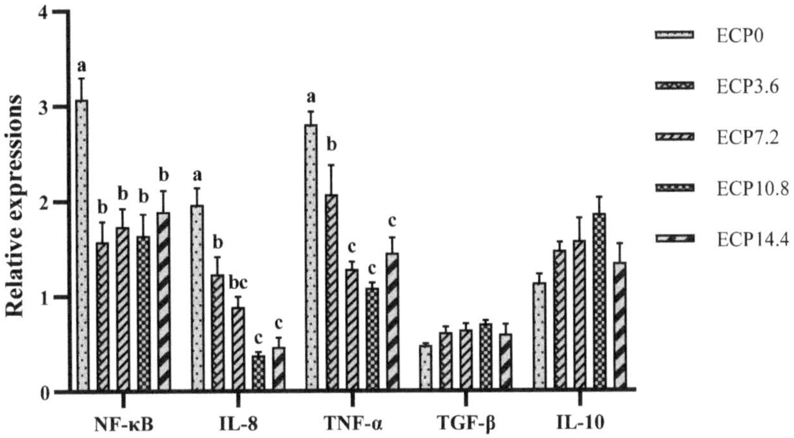

Figure 3. Relative mRNA expression levels of immune-related genes in liver. NF-κB, nuclear factor-kappa B; IL-8, interleukin-8; TNF-α, tumor necrosis factor-α; TGF-β, transforming growth factor-β; IL-10, Interleukin-10. Data are presented as mean ± SD. Bars with different superscripts indicate significant differences ($p < 0.05$), while that with the same letter or no letter superscripts indicate no significant differences ($p > 0.05$).

3. Discussion

As the source of feed protein, various cottonseed protein products have been studied to replace dietary FM in many fish species [10–17]. In the current study, results showed that fish fed with a diet containing 3.6% ECP presented the best growth performance, and 7.2% ECP could be incorporated into the diet to replace 15.56% FM (based on the 45% FM group) without reducing growth performance. However, the growth performance was markedly decreased once the amount of ECP was increased to 10.8% and 14.4%. Likewise, the substitution of cottonseed meal protein hydrolysate for FM had similar effects on blunt snout bream [27]. Similar effects of replacing FM with plant protein hydrolysates have also been found in turbot (*Scophthalmus maximus*), rainbow trout, and largemouth bass [28–30]. The good nutritional value of protein hydrolysates may be the reason for the successful substitution of partial FM. A study on Atlantic cod (*Gadus morhua*) reported that feeding an intact protein diet led to the retaining of more nutrients than feeding an amino acid diet, although the amino acid diet showed faster absorption [31]. The rapid absorption of amino acids and peptides may accelerate the excretion of amino acids or lead to the imbalance of the intake of amino acids, thus reducing the utilization of the amino acid diet. The cottonseed protein concentrate used in the current study was hydrolyzed into amino acids and peptides by enzymes. Once the amount of ECP in the diet was too high, the amino acids and peptides ingested by fish might be supersaturated, leading to a decline in the utilization of ECP, which may be a reason for the decline in growth. Moreover, the addition of ECP notably depressed the activity of plasma ALP in the current study. ALP catalyzes the phosphorylation of many types of molecules under alkaline pH conditions and is involved in membrane transport activities and mineralization of the fish skeleton [32,33]. Therefore, the low activity of ALP possibly contributes to the weak growth performance. The specific reasons for the decline in the growth performance of largemouth bass caused by the addition of ECP to diet need to be confirmed by further research.

In the current study, the proximate composition of fish fed with different diets had no distinct difference, indicating that ECP has little effect on the proximate composition of largemouth bass. Similarly, the substitution of cottonseed meal protein hydrolysate for FM did not affect the proximate composition of blunt snout bream [27]. However, previous studies obtained different results, which indicated that cottonseed protein concentrates in diets affected the proximate composition of largemouth bass [34–37]. The diversity of cottonseed protein products may be the main reason for these different results [38].

Plasma TP and ALB are related to the immune status of fish [39,40]. Results from the current study showed that replacing FM with ECP increased the plasma TP and ALB concentrations, indicating that ECP may enhance the immune response of largemouth bass. Similarly, the substitution of cottonseed meal protein hydrolysate for FM increased the plasma TP and ALB concentrations of blunt snout bream [27]. Diets supplemented with protein hydrolysates also increased the plasma protein level of Olive flounder (*Paralichthys olivaceus*) [41]. To avoid cardiovascular disease caused by lipid accumulation, it is considered to be beneficial when the content of TC, TG, and LDL in plasma are low and the HDL content is high [42,43]. Results from the current study showed that replacing FM with ECP increased plasma HDL concentrations, while decreasing TC, TG, and LDL concentrations, suggesting that ECP might have the effect of lowering plasma lipids and preventing cardiovascular disease. A study on starry flounder (*Platichthys stellatus*) also found that replacing FM with plant protein hydrolysates reduced plasma lipid levels [44].

Antioxidant defense systems, including enzymatic and nonenzymatic systems, play an important role in maintaining fish health [45,46]. MDA, one of the final products of cell membrane lipid peroxidation, is a commonly used indicator to measure the degree of oxidative stress [47]. Results from the current study revealed that the substitution of ECP for FM had a positive effect on increasing hepatic T-SOD and CAT activities and the GSH content while reducing the hepatic MDA content. Correspondingly, results of mRNA expression of antioxidant genes revealed that ECP upregulated the mRNA expression level of Nrf2, SOD, and CAT. These results indicate that ECP could activate the Nrf2 signaling pathway and improve hepatic antioxidant enzyme activity to reduce the oxidative stress of largemouth bass. Similarly, the antioxidant capacities and innate immunity of blunt snout bream hepatocytes were improved by cottonseed meal protein hydrolysate [48]. The antioxidant effect may be due to the fact that ECP contains antioxidant peptides. It was reported that peptide fractions derived from enzymatically hydrolyzed cottonseed protein had obvious effects on inhibiting the formation of MDA in the linoleic acid autoxidation system and scavenging various free radicals [49]. Furthermore, it was reported that the antioxidant capacity of peptides from cottonseed protein hydrolysates remained high or was even enhanced after in vitro digestion [50]. However, there are some different findings, for instance, the hepatic antioxidant capacity of Ussuri catfish (*Pseudobagrus ussuriensis*) was decreased by the substitution of cottonseed meal for dietary FM [51]. Both the serum antioxidant enzyme activities and MDA content in hybrid grouper (♀*Epinephelus fuscoguttatus* × ♂*Epinephelus lanceolatu*) were increased by the substitution of cottonseed protein concentrate for dietary FM [11]. Different results are possibly attributed to the specificity of fish species and tissue, as well as differences in the methods for producing cottonseed protein products [52]. It was reported that, compared with soybean meal or soy protein concentrate, enzyme-treated soybean meal had a more effective role in improving the antioxidant capacity [19].

In addition to the antioxidant system, the liver immune state is intimately related to fish health and growth. It was reported that NF-κB, the critical regulator of proinflammatory gene expression, plays a significant role in inflammation [53]. In the current study, ECP instead of FM markedly inhibited the mRNA expression of NF-κB and proinflammatory cytokines (IL-8 and TNF-α), while it tended to promote the mRNA expression of anti-inflammatory cytokines (TGF-β and IL-10), indicating that ECP has the function of restraining inflammation by inhibiting the NF-κB signal pathway. It was demonstrated that oxidative stress is related to the pathogenesis of various inflammatory diseases, and antioxidant agents can mitigate inflammation [54]. Therefore, the antioxidant peptides produced by the enzymatic hydrolysis of cottonseed protein can not only alleviate oxidative stress but also contribute to suppressing inflammation. Conversely, previous studies reported that cottonseed protein instead of FM induced the expression of proinflammatory genes while suppressing the expression of anti-inflammatory genes in hybrid grouper and silver sillago

(*Sillago sihama Forsskål*) [11,55]. Compared with cottonseed protein concentrate, ECP may better improve antioxidant performance and immunity [34,35]. The diversity between the results of this study and previous studies is likely due to the different cottonseed protein treatment methods. It was reported that, compared with conventional dehulled soybean meal, soybean meal that has undergone enzyme treatment processing effectively improved the nonspecific immunity of largemouth bass [20]. Hence, enzymatic hydrolysis is a great processing method to improve cottonseed protein quality.

4. Materials and Methods

4.1. Experimental Diets

We replaced 0%, 7.78%, 15.56%, 23.33%, and 31.11% FM with 0%, 3.6%, 7.2%, 10.8%, and 14.4% ECP to formulate five iso-nitrogenous (49%) and iso-energetic (19 kJ/g) experimental diets, named ECP0, ECP3.6, ECP7.2, ECP10.8, and ECP14.4, respectively (Table 4). The steps to produce the experimental diets included crushing the ingredients and passing them through 80 mesh sieves, weighing the ingredients according to the formula, fully mixing various ingredients, and granulating the diet using a pelletizer (F-26 (II), South China University of Technology, China). The pellet diets were dried in a ventilated oven at 45 °C, then bagged and placed at −20 °C until use.

Table 4. Ingredients and nutrient composition of experimental diets (% dry basis).

Ingredients (%)	Diets				
	ECP0	ECP3.6	ECP7.2	ECP10.8	ECP14.4
Fish meal [1]	45.00	41.50	38.00	34.50	31.00
Enzymatic cottonseed protein concentrate [1]	0.00	3.60	7.20	10.80	14.40
Blood meal [1]	2.00	2.00	2.00	2.00	2.00
Soybean meal [1]	13.00	13.00	13.00	13.00	13.00
Corn gluten meal[1]	3.00	3.00	3.00	3.00	3.00
Enzymatic hydrolysis of chicken powder [1]	4.00	4.00	4.00	4.00	4.00
Wheat meal	7.00	7.00	7.00	7.00	7.00
Cassava starch	7.00	7.00	7.00	7.00	7.00
Rice bran	6.13	6.13	6.13	6.13	6.13
Microcrystalline cellulose	3.27	2.41	1.58	0.82	0.00
Squid Ointment	2.00	2.00	2.00	2.00	2.00
Fish oil	4.10	4.45	4.75	5.00	5.30
Mineral premix [2]	1.00	1.00	1.00	1.00	1.00
Vitamin premix [2]	1.00	1.00	1.00	1.00	1.00
Monocalcium phosphate	1.00	1.30	1.60	1.90	2.20
Choline chloride	0.50	0.50	0.50	0.50	0.50
L-Lysine [3]	0.00	0.08	0.17	0.25	0.33
L-Methionine [3]	0.00	0.03	0.07	0.10	0.14
Taurine (mg/kg)	0.00	13.30	26.60	39.90	53.20
Proximate Composition (dry basis)					
Crude protein (%)	48.59	48.53	48.61	48.47	48.49
Crude lipid (%)	9.25	9.24	8.98	9.28	9.23
Gross energy (KJ/g)	19.75	19.83	19.87	19.72	19.58

[1] Fish meal, crude protein 66.73%, crude lipid 9.46%; Enzymatic cottonseed protein concentrate, crude protein 64.85%, crude lipid 0.87%; Blood meal, crude protein 90.68%; Soybean meal, crude protein 53.26%, crude lipid 4.25%; Corn gluten meal, crude protein 59.24%, crude lipid 3.30%; Enzymatically hydrolyzed chicken powder, crude protein 84.62%, crude lipid 1.00%. These materials were obtained from Wuxi Tongwei feedstuffs Co., Ltd., Wuxi, China. [2] Mineral premix and vitamins premix were obtained from Wuxi Hanove animal health products Co., Ltd., Wuxi, China. [3] The limiting amino acids (L-*lysine* and L-*methionine*) were supplemented according to the levels of the first group.

4.2. Feeding Trial

The feeding trial was conducted at the Charoen Pokphand Group breeding farm (Huanggang, Hubei, China). Experiment fish were purchased from the Yongda Aquaculture Professional Cooperative (Ezhou, Hubei, China). Firstly, fish were fed with a commercial diet twice a day for two weeks to acclimate to the experimental conditions. We fasted fish for 24 h, then selected 300 lively fish (average initial weight 14.99 ± 0.03 g) and randomly put them into 15 cages (1 m × 1 m × 1 m), with 20 fish per cage and 3 cages per group. The fish were fed with experimental diets to apparent satiety two times (6:30 and 18:30) every day; the feeding trial lasted for 60 days. During the trial period, the water temperature, pH, ammonia nitrogen content, nitrite content, and dissolved oxygen concentration were 28–31 °C, 7.5–8.2, 0–0.2 mg/L, 0.1–0.3 mg/L, and ≥6 mg/L, respectively.

4.3. Sample Collection

The fish were kept in starvation for 24 h before sampling. Then, we collected the quantity and weight of fish in each cage to calculate the growth indicators. Two fish were collected from each cage for general composition analysis, and three fish from each cage were collected for obtaining blood and liver samples. Sample fish were anesthetized with MS-222 before collecting the blood and liver samples. Blood was collected from the tail vein, then centrifuged in a centrifuge at a speed of 3500 rpm for 10 min to collect the upper plasma. The fish was dissected immediately after blood collection to collect liver samples. Serum samples and liver samples were stored at −80 °C.

4.4. Experimental Parameter Detection

The proximate composition analyses of diets and fish were conducted following the methods of AOAC (2003) [56]. We measured the plasma biochemical parameters on an automatic biochemical analyzer with related assay kits. We determined the hepatic antioxidant parameters and MDA levels through the corresponding assay kit. The main methods, assay kits, and testing equipment for index detection are shown in Table 5.

Table 5. The methods and testing equipment of chemical analysis.

Items	Methods, Assay Kits and Testing Equipment
Crude protein	Determined by Hanon K1100 auto kieldahl apparatus (Jinan Hanon Instruments Co., Ltd., Jinan, China).
Crude lipid	Determined by Hanon SOX606 auto fat analy (Jinan Hanon Instruments Co., Ltd., Jinan, China).
Ash	Determined by burning at 550 °C for 5 h in a XL-2A intelligent muffle furnace (Hangzhou Zhuochi Instruments Co., Ltd., Hangzhou, China).
Gross energy	Measured by an oxygen bomb calorimeter IKA C6000 ((IKA Works Guangzhou, Guangzhou, China).
TP [1] ALB [1] TC [1] TG [1] LDL [1] HDL [1] ALP [1]	All plasma parameters were determined by Assay kits (Mindray Bio Medical Co., Ltd., Shenzhen, China) with a Mindray BS-400 automatic biochemical analyzer (Mindray Medical International Ltd., Shenzhen, China).
T-SOD [2] CAT [2] GSH [2] MDA [2]	All hepatic antioxidant parameters and MDA levels were tested according to the instructions of assay kits purchased from Jian Cheng Bioengineering Institute (Nanjing, China).

[1] TP, total protein; ALB, albumin; TC, total cholesterol; TG, triglyceride; LDL, low-density lipoprotein; HDL, high-density lipoprotein; ALP, alkaline phosphatase. [2] T-SOD, total superoxide dismutase; CAT, catalase; GSH, glutathione; MDA, malondialdehyde.

The measurement of relative mRNA expression included extracting total RNA from tissues, detecting the concentration and quality of RNA, and performing quantitative real-time PCR analysis. More detailed information is presented in our previous study [57]. Glyceraldehyde-3-phosphate dehydrogenase (GAPDH) was chosen as the reference gene, and its expression in different groups was shown to be stable [57]. Pfaffl's model was used to analyze the gene expression levels [58]. The specific primer sequences used in this study were designed by reference to previous studies [59–62] and are displayed in Table 6.

Table 6. Primer sequence for RT-qPCR.

Gene [1]	Forward Sequence (5′-3′)	Reverse Sequence (5′-3′)	Source
Nrf2	CTGGTCCGAGACATACGC	CTCAGCAGACGCTCCTTC	Zhao et al. (2021) [59]
Keap1	CGTACGTCCAGGCCTTACTC	TGACGGAAATAACCCCCTGC	Yu et al. (2022) [60]
SOD	TGGCAAGAACAAGAACCACA	CCTCTGATTTCTCCTGTCACC	Gu et al. (2022) [61]
CAT	CTATGGCTCTCACACCTTC	TCCTCTACTGGCAGATTCT	Gu et al. (2022) [61]
NF-κB	CCACTCAGGTGTTGGAGCTT	TCCAGAGCACGACACACTTC	Yu et al. (2022) [60]
IL-8	CGTTGAACAGACTGGGAGAGATG	AGTGGGATGGCTTCATTATCTTGT	Yang et al. (2020) [62]
TNF-α	CTTCGTCTACAGCCAGGCATCG	TTTGGCACACCGACCTCACC	Gu et al. (2022) [61]
TGF-β	GCTCAAAGAGAGCGAGGATG	TCCTCTACCATTCGCAATCC	Gu et al. (2022) [61]
IL-10	CGGCACAGAAATCCCAGAGC	CAGCAGGCTCACAAAATAAACATCT	Gu et al. (2022) [61]
GAPDH	ACTGTCACTCCTCCATCTT	CACGGTTGCTGTATCCAA	Yu et al. (2022) [60]

[1] Nrf2, nuclear factor erythroid 2-related factor 2; Keap1, kelch-like-ECH-associated protein 1; SOD, superoxide dismutase; CAT, catalase; NF-κB, nuclear factor-kappa B; IL-8, interleukin 8; TNF-α, tumor necrosis factor-α; TGF-β, transforming growth factor β; IL-10, interleukin 10; GAPDH, glyceraldehyde-3-phosphate dehydrogenase.

4.5. Statistical Analysis

One-way ANOVA in SPSS 26.0 software was used for the statistical analysis of experimental data. All experimental data were confirmed to conform to normal distribution and homogeneity of variance before any statistical analysis. All data were displayed as mean ± SD. $p < 0.05$ indicated that the variables in different groups were significantly different, and Tukey's multiple comparisons were conducted.

5. Conclusions

The results from the present study showed that 7.2% ECP could be incorporated into the diet to replace 15.56% FM (based on the 45% FM group) without affecting the growth performance of juvenile largemouth bass. ECP-containing diets can improve the hepatic antioxidant capacity and immunity of largemouth bass.

Author Contributions: Conceptualization, M.R.; methodology, Q.Z., H.L., P.X., G.X., L.Z., Y.W. and X.C.; investigation, Q.Z., H.L., P.X., G.X., L.Z., Y.W. and X.C.; data curation, Q.Z., H.L., P.X. and G.X.; writing—original draft preparation, Q.Z.; writing—review and editing, H.L., X.C. and M.R.; supervision, M.R.; project administration, M.R.; funding acquisition, M.R. All authors have read and agreed to the published version of the manuscript.

Funding: This study was financially supported by the National Key Research and Development Program of China (2019YFD0900200), the earmarked fund for CARS (CARS-46).

Institutional Review Board Statement: All animal operations in the present study were carried out in accordance with the animal protection law formulated by the Animal Ethics Committee of Nanjing Agricultural University [Permit number: SYXK (Su) 2011-0036].

Informed Consent Statement: Not applicable.

Data Availability Statement: The authors confirm that the data supporting the findings of this study are available within the manuscript, Tables and Figures.

Conflicts of Interest: Author X.C. and Y.W. are employed by Tongwei Agricultural Development Co., Ltd., L.Z. is employed by Tongwei Co., Ltd. The remaining authors declare that the research was conducted in the absence of any commercial or financial relationships that could be construed as a potential conflict of interest.

References

1. Hussein, G.H.G.; Chen, M.; Qi, P.P.; Cui, Q.K.; Yu, Y.; Hu, W.H.; Tian, Y.; Fan, Q.X.; Gao, Z.X.; Feng, M.W.; et al. Aquaculture industry development, annual price analysis and out-of-season spawning in largemouth bass *Micropterus salmoides*. *Aquaculture* **2020**, *519*, 734901. [CrossRef]
2. Tidwell, J.H.; Webster, C.D.; Coyle, S.D. Effects of dietary protein level on second year growth and water quality for largemouth bass (*Micropterus salmoides*) raised in ponds. *Aquaculture* **1996**, *145*, 213–223. [CrossRef]

3. Huang, D.; Wu, Y.B.; Lin, Y.Y. Dietary protein and lipid requirements for juvenile largemouth bass, *Micropterus salmoides*. *J. World Aquac. Soc.* **2017**, *48*, 782–790. [CrossRef]
4. Cai, Z.N.; Qian, X.Q.; Xie, S.Q. Optimal dietary protein concentrations for largemouth bass (*Micropterus salmoides*) of different sizes (10–500 g). *Aquac. Int.* **2020**, *28*, 831–840. [CrossRef]
5. Li, X.Y.; Zheng, S.X.; Ma, X.K.; Cheng, K.M.; Wu, G.Y. Effects of dietary protein and lipid levels on the growth performance, feed utilization, and liver histology of largemouth bass (*Micropterus salmoides*). *Amino Acids* **2020**, *52*, 1043–1061. [CrossRef]
6. Hardy, R.W. Utilization of plant proteins in fish diets: Effects of global demand and supplies of fishmeal. *Aquac. Res.* **2010**, *41*, 770–776. [CrossRef]
7. Olsen, R.L.; Hasan, M.R. A limited supply of fishmeal: Impact on future increases in global aquaculture production. *Trends Food Sci. Technol.* **2012**, *27*, 12–128. [CrossRef]
8. Jannathulla, R.; Rajaram, V.; Kalanjiam, R.; Ambasankar, K.; Muralidhar, M.; Dayal, J.S. Fishmeal availability in the scenarios of climate change: Inevitability of fishmeal replacement in aquafeeds and approaches for the utilization of plant protein sources. *Aquac. Res.* **2019**, *50*, 3493–3506. [CrossRef]
9. Kumar, M.; Potkule, J.; Patil, S.; Saxena, S.; Patil, P.G.; Mageshwaran, V.; Puniad, S.; Varghesee, E.; Mahapatra, A.; Ashtaputrea, N.; et al. Extraction of ultra-low gossypol protein from cottonseed: Characterization based on antioxidant activity, structural morphology and functional group analysis. *LWT* **2021**, *140*, 110692. [CrossRef]
10. Anderson, A.D.; Alam, M.S.; Watanabe, W.O.; Carroll, P.M.; Wedegaertner, T.C.; Dowd, M.K. Full replacement of menhaden fish meal protein by low-gossypol cottonseed flour protein in the diet of juvenile black sea bass *Cent. Striata. Aquac.* **2016**, *464*, 618–628. [CrossRef]
11. Yin, B.; Liu, H.G.; Tan, B.P.; Dong, X.H.; Chi, S.Y.; Yang, Q.H.; Zhang, S.; Chen, L.Q. Cottonseed protein concentrate (CPC) suppresses immune function in different intestinal segments of hybrid grouper ♀*Epinephelus fuscoguttatus* × ♂*Epinephelus lanceolatu* via TLR-2/MyD88 signaling pathways. *Fish Shellfish Immunol.* **2018**, *81*, 318–328. [CrossRef]
12. Wan, M.G.; Yin, P.; Fang, W.P.; Xie, S.W.; Chen, S.J.; Tian, L.X.; Niu, J. The effect of replacement of fishmeal by concentrated dephenolization cottonseed protein on the growth, body composition, haemolymph indexes and haematological enzyme activities of the Pacific white shrimp (*Litopenaeus vannamei*). *Aquac. Nutr.* **2018**, *24*, 1845–1854. [CrossRef]
13. Shen, J.F.; Liu, H.Y.; Tan, B.P.; Dong, X.H.; Yang, Q.H.; Chi, S.Y.; Zhang, S. Effects of replacement of fishmeal with cottonseed protein concentrate on the growth, intestinal microflora, haematological and antioxidant indices of juvenile golden pompano (*Trachinotus ovatus*). *Aquac. Nutr.* **2020**, *26*, 1119–1130. [CrossRef]
14. Ye, G.L.; Dong, X.H.; Yang, Q.H.; Chi, S.Y.; Liu, H.Y.; Zhang, H.T.; Tan, B.P.; Zhang, S. Low-gossypol cottonseed protein concentrate used as a replacement of fish meal for juvenile hybrid grouper (*Epinephelus fuscoguttatus* ♀× *Epinephelus lanceolatus* ♂): Effects on growth performance, immune responses and intestinal microbiota. *Aquaculture* **2020**, *524*, 735309. [CrossRef]
15. Jiang, S.; Chen, Z.B.; Zhou, F.L.; Yang, Q.B.; Huang, J.H.; Yang, L.S.; Li, Y.D.; Jiang, S.G. Study on partial replacement of fish meal with concentrated dephenolized cottonseed protein in feed of Penaeus monodon. *Aquac. Res.* **2021**, *52*, 3871–3881. [CrossRef]
16. He, Y.F.; Guo, X.W.; Tan, B.P.; Dong, X.H.; Yang, Q.H.; Liu, H.; Zhang, S.; Chi, S.Y. Replacing fishmeal with cottonseed protein concentrate in feed for pearl gentian groupers (*Epinephelus fuscoguttatus*♀× *E. lanceolatus*♂): Effects on growth and expressions of key genes involved in appetite and hepatic glucose and lipid metabolism. *Aquac. Rep.* **2021**, *20*, 100710. [CrossRef]
17. Zhao, W.; Liu, Z.L.; Niu, J. Growth performance, intestinal histomorphology, body composition, hematological and antioxidant parameters of *Oncorhynchus mykiss* were not detrimentally affected by replacement of fish meal with concentrated dephenolization cottonseed protein. *Aquac. Rep.* **2021**, *19*, 100557. [CrossRef]
18. Sun, X.D. Enzymatic hydrolysis of soy proteins and the hydrolysates utilisation. *Int. J. Food Sci. Technol.* **2011**, *46*, 2447–2459. [CrossRef]
19. Ma, X.K.; Shang, Q.H.; Hu, J.X.; Liu, H.S.; Brøkner, C.; Piao, X.S. Effects of replacing soybean meal, soy protein concentrate, fermented soybean meal or fish meal with enzyme-treated soybean meal on growth performance, nutrient digestibility, antioxidant capacity, immunity and intestinal morphology in weaned pigs. *Livest. Sci.* **2019**, *225*, 39–46. [CrossRef]
20. Li, S.L.; Ding, G.T.; Song, F.; Sang, C.Y.; Wang, A.; Chen, N.S. Comparison of dehulled, fermented and enzyme-treated soybean meal in diets for largemouth bass, *Micropterus salmoides*: Effects on growth performance, feed utilization, immune response and intestinal morphology. *Anim. Feed Sci. Technol.* **2020**, *267*, 114548. [CrossRef]
21. Gui, D.; Liu, W.B.; Shao, X.P.; Xu, W.N. Effects of different dietary levels of cottonseed meal protein hydrolysate on growth, digestibility, body composition and serum biochemical indices in crucian carp (*Carassius auratus gibelio*). *Anim. Feed Sci. Technol.* **2010**, *156*, 112–120. [CrossRef]
22. Tang, X.P.; Xiang, R.; Chen, S.J.; Yang, S.F.; Liu, H.; Fang, R.J.; Li, A.K. Effects of Fermented Cottonseed Meal and Enzymatic Hydrolyzed Cottonseed Meal on Amino Acid Digestibility and Metabolic Energy in White Leghorn Rooster. *Pak. J. Zool.* **2018**, *50*, 957–962. [CrossRef]
23. Song, W.G.; Kong, X.Z.; Hua, Y.F.; Chen, Y.M.; Zhang, C.M.; Chen, Y.X. Identification of antibacterial peptides generated from enzymatic hydrolysis of cottonseed proteins. *LWT* **2020**, *125*, 109199. [CrossRef]
24. Wang, L.Y.; Ma, M.G.; Yu, Z.P.; Du, S.K. Preparation and identification of antioxidant peptides from cottonseed proteins. *Food Chem.* **2021**, *352*, 129399. [CrossRef] [PubMed]

25. Filho, J.G.D.O.; Rodrigues, J.M.; Valadares, A.C.F.; Almeida, A.B.D.; Valencia-Mejia, E.; Fernandes, K.F.; Lemes, A.C.; Egea, M.B.; Dyszy, F.H. Bioactive properties of protein hydrolysate of cottonseed byproduct: Antioxidant, antimicrobial, and angiotensin-converting enzyme (ACE) inhibitory activities. *Waste Biomass Valorization* **2021**, *12*, 1395–1404. [CrossRef]
26. Wang, K.Z.; Xu, W.N.; Zhou, M.; Zhang, D.D.; Sun, C.X.; Qian, Y.; Liu, W.B. Effects of fishmeal replacement with cottonseed meal protein hydrolysate on growth, digestion and intestinal histology of juvenile Chinese soft-shelled turtle, *Pelodiscus sinensis*. *Aquac. Nutr.* **2018**, *24*, 1406–1415. [CrossRef]
27. Yuan, X.Y.; Jiang, G.Z.; Cheng, H.H.; Cao, X.F.; Shi, H.J.; Liu, W.B. An evaluation of replacing fish meal with cottonseed meal protein hydrolysate in diet for juvenile blunt snout bream (*Megalobrama amblycephala*): Growth, antioxidant, innate immunity and disease resistance. *Aquac. Nutr.* **2019**, *25*, 1334–1344. [CrossRef]
28. Song, Z.D.; Li, P.Y.; Wang, J.Y.; Sun, Y.Z.; Wang, C.Q. Dietary inclusion of hydrolyzed soybean and cottonseed meals influence digestion, metabolic enzymes, and growth-related hormones and growth of juvenile turbot (*Scophthalmus maximus*). *Aquac. Int.* **2018**, *26*, 1017–1033. [CrossRef]
29. Haghbayan, S.; Shamsaie Mehrgan, M. The effect of replacing fish meal in the diet with enzyme-treated soybean meal (HP310) on growth and body composition of rainbow trout fry. *Molecules* **2015**, *20*, 21058–21066. [CrossRef]
30. Liu, X.; Chi, S.Y.; Li, S.; Cheng, X.L.; Gao, W.H.; Xu, Q.Q.; Zhang, W.B.; Zhou, X.Q. Substitution of fish meal with enzyme-treated soybean in diets for juvenile largemouth bass (*Micropterus salmoides*). *Aquac. Nutr.* **2021**, *27*, 1569–1577. [CrossRef]
31. Berge, G.E.; Lied, E.; Espe, M. Absorption and incorporation of dietary free and protein bound (U14C)-lysine in Atlantic cod (*Gadus morhua*). *Comp. Biochem. Physiol. Part A Physiol.* **1994**, *109*, 681–688. [CrossRef]
32. Öner, M.; Atli, G.; Canli, M. Changes in serum biochemical parameters of freshwater fish *Oreochromis niloticus* following prolonged metal (Ag, Cd, Cr, Cu, Zn) exposures. *Environ. Toxicol. Chem. Int. J.* **2008**, *27*, 360–366. [CrossRef]
33. Ren, M.C.; Liu, B.; Habte-Tsion, H.M.; Ge, X.P.; Xie, J.; Zhou, Q.L.; Liang, H.L.; Zhao, Z.X.; Pan, L.K. Dietary phenylalanine requirement and tyrosine replacement value for phenylalanine of juvenile blunt snout bream, *Megalobrama amblycephala*. *Aquaculture* **2015**, *442*, 51–57. [CrossRef]
34. Liu, Y.L.; Lu, Q.S.; Xi, L.W.; Gong, Y.L.; Su, J.Z.; Han, D.; Zhang, Z.M.; Liu, H.K.; Jin, J.Y.; Yang, Y.X.; et al. Effects of replacement of dietary fishmeal by cottonseed protein concentrate on growth performance, liver health, and intestinal histology of largemouth bass (*Micropterus salmoides*). *Front. Physiol.* **2021**, *12*, 764987. [CrossRef] [PubMed]
35. He, G.Z.; Zhang, T.T.; Zhou, X.M.; Liu, X.P.; Sun, H.; Chen, Y.J.; Tan, B.P.; Lin, S.M. Effects of cottonseed protein concentrate on growth performance, hepatic function and intestinal health in juvenile largemouth bass, *Micropterus salmoides*. *Aquac. Rep.* **2022**, *23*, 101052. [CrossRef]
36. Xu, X.Y.; Yang, H.; Zhang, C.Y.; Bian, Y.H.; Yao, W.X.; Xu, Z.; Wang, Y.Y.; Li, X.Q.; Leng, X.J. Effects of replacing fishmeal with cottonseed protein concentrate on growth performance, flesh quality and gossypol deposition of largemouth bass (*Micropterus salmoides*). *Aquaculture* **2022**, *548*, 737551. [CrossRef]
37. Xie, X.Z.; Wang, J.; Guan, Y.; Xing, S.J.; Liang, X.F.; Xue, M.; Wang, J.J.; Chang, Y.; Leclercq, E. Cottonseed protein concentrate as fishmeal alternative for largemouth bass (*Micropterus salmoides*) supplemented a yeast-based paraprobiotic: Effects on growth performance, gut health and microbiome. *Aquaculture* **2022**, *551*, 737898. [CrossRef]
38. Bian, F.; Zhou, H.; He, G.; Wang, C.; Peng, H.; Pu, X.; Jiang, X.; Wang, X.; Mai, K. Effects of replacing fishmeal with different cottonseed meals on growth, feed utilization, haematological indexes, intestinal and liver morphology of juvenile turbot (*Scophthalmus maximus* L.). *Aquac. Nutr.* **2017**, *23*, 1429–1439. [CrossRef]
39. Soyingbe, A.A.; Ogunyanwo, O.O.; Hammed, T.B.; Adesope, A.O. Effects of sublethal concentrations of diazinon on total protein in tilapia fish (*Oreochromis niloticus*). *IOSR J. Environ. Sci. Toxicol. Food Technol.* **2012**, *1*, 22–25. [CrossRef]
40. Ren, M.C.; Liang, H.L.; He, J.; Masagounder, K.; Yue, Y.; Yang, H.; Ge, X.P.; Xi, B.W. Effects of DL-methionine supplementation on the success of fish meal replacement by plant proteins in practical diets for juvenile gibel carp (*Carassius auratus gibelio*). *Aquac. Nutr.* **2017**, *23*, 934–941. [CrossRef]
41. Khosravi, S.; Bui, H.T.D.; Rahimnejad, S.; Herault, M.; Fournier, V.; Jeong, J.B.; Lee, K.J. Effect of dietary hydrolysate supplementation on growth performance, non-specific immune response and disease resistance of olive flounder (*Paralichthys olivaceus*) challenged with Edwardsiella tarda. *Aquac. Nutr.* **2015**, *21*, 321–331. [CrossRef]
42. Zawistowski, J.; Kopec, A.; Kitts, D.D. Effects of a black rice extract (*Oryza sativa* L. indica) on cholesterol levels and plasma lipid parameters in Wistar Kyoto rats. *J. Funct. Foods* **2009**, *1*, 50–56. [CrossRef]
43. Reinhart, K.M.; Talati, R.; White, C.M.; Coleman, C.I. The impact of garlic on lipid parameters: A systematic review and meta-analysis. *Nutr. Res. Rev.* **2009**, *22*, 39–48. [CrossRef] [PubMed]
44. Song, Z.D.; Li, H.Y.; Wang, J.Y.; Li, P.Y.; Sun, Y.Z.; Zhang, L.M. Effects of fishmeal replacement with soy protein hydrolysates on growth performance, blood biochemistry, gastrointestinal digestion and muscle composition of juvenile starry flounder (*Platichthys stellatus*). *Aquaculture* **2014**, *426*, 96–104. [CrossRef]
45. Ji, K.; Liang, H.L.; Ren, M.C.; Ge, X.P.; Mi, H.F.; Pan, L.K.; Yu, H. The immunoreaction and antioxidant capacity of juvenile blunt snout bream (*Megalobrama amblycephala*) involves the PI3K/Akt/Nrf2 and NF-κB signal pathways in response to dietary methionine levels. *Fish Shellfish Immunol.* **2020**, *105*, 126–134. [CrossRef]
46. Zheng, Q.M.; Wen, X.B.; Han, C.Y.; Li, H.B.; Xie, X.H. Effect of replacing soybean meal with cottonseed meal on growth, hematology, antioxidant enzymes activity and expression for juvenile grass carp, *Ctenopharyngodon idellus*. *Fish Physiol. Biochem.* **2012**, *38*, 1059–1069. [CrossRef]

47. Gaweł, S.; Wardas, M.; Niedworok, E.; Wardas, P. Malondialdehyde (MDA) as a lipid peroxidation marker. *Wiad. Lek.* **2004**, *57*, 453–455.
48. Yuan, X.Y.; Liu, W.B.; Wang, C.C.; Huang, Y.Y.; Dai, Y.J.; Cheng, H.H.; Jiang, G.Z. Evaluation of antioxidant capacity and immunomodulatory effects of cottonseed meal protein hydrolysate and its derivative peptides for hepatocytes of blunt snout bream (*Megalobrama amblycephala*). *Fish Shellfish Immunol.* **2020**, *98*, 10–18. [CrossRef]
49. Gao, D.D.; Cao, Y.S.; Li, H.X. Antioxidant activity of peptide fractions derived from cottonseed protein hydrolysate. *J. Sci. Food Agric.* **2010**, *90*, 1855–1860. [CrossRef]
50. Song, W.G.; Kong, X.Z.; Hua, Y.F.; Li, X.F.; Zhang, C.M.; Chen, Y.M. Antioxidant and antibacterial activity and in vitro digestion stability of cottonseed protein hydrolysates. *LWT* **2020**, *118*, 108724. [CrossRef]
51. Bu, X.Y.; Chen, A.J.; Lian, X.Q.; Chen, F.Y.; Zhang, Y.; Muhammad, I.; Ge, X.P.; Yang, Y.H. An evaluation of replacing fish meal with cottonseed meal in the diet of juvenile Ussuri catfish *Pseudobagrus ussuriensis*: Growth, antioxidant capacity, nonspecific immunity and resistance to *Aeromonas hydrophila*. *Aquaculture* **2017**, *479*, 829–837. [CrossRef]
52. Boboev, A.; Hasanov, A.; Yotova, L.; Hasanov, H. Antioxidant activity of peptides obtained from wheat and cottonseed proteins. *Bulg. J. Agric. Sci.* **2012**, *18*, 103–111.
53. Tak, P.P.; Firestein, G.S. NF-κB: A key role in inflammatory diseases. *J. Clin. Investig.* **2001**, *107*, 7–11. [CrossRef]
54. Li, C.W.; Li, L.L.; Chen, S.; Zhang, J.X.; Lu, W.L. Antioxidant nanotherapies for the treatment of inflammatory diseases. *Front. Bioeng. Biotechnol.* **2020**, *8*, 200. [CrossRef] [PubMed]
55. Liu, H.; Dong, X.H.; Tan, B.P.; Du, T.; Zhang, S.; Yang, Y.Z.; Chi, S.Y.; Yang, Q.H.; Liu, H.Y. Effects of fish meal replacement by low-gossypol cottonseed meal on growth performance, digestive enzyme activity, intestine histology and inflammatory gene expression of silver sillago (*Sillago sihama Forsskál*)(1775). *Aquac. Nutr.* **2020**, *26*, 1724–1735. [CrossRef]
56. Association of Official Analytical Chemists. *Official Methods of Analysis of the Association of Official Analytical Chemists*, 15th ed.; Association of Official Analytical Chemists Inc.: Arlington, TX, USA, 1998.
57. Huang, D.Y.; Liang, H.L.; Ge, X.P.; Zhu, J.; Li, S.L.; Wang, Y.L.; Ren, M.C.; Chen, X.L. Effects of Dietary Lysine Levels on Growth Performance and Glycolipid Metabolism via the AKT/FoxO1 Pathway in Juvenile Largemouth Bass, *Micropterus salmoides*. *Aquac. Nutr.* **2022**, *2022*, 1372819. [CrossRef]
58. Pfaffl, M.W. A new mathematical model for relative quantification in real-time RT–PCR. *Nucleic Acids Res.* **2001**, *29*, e45. [CrossRef] [PubMed]
59. Zhao, L.L.; Liang, J.; Chen, F.K.; Tang, X.H.; Liao, L.; Liu, Q.; Luo, J.; Du, Z.J.; Li, Z.Q.; Luo, W.; et al. High carbohydrate diet induced endoplasmic reticulum stress and oxidative stress, promoted inflammation and apoptosis, impaired intestinal barrier of juvenile largemouth bass (*Micropterus salmoides*). *Fish Shellfish Immunol.* **2021**, *119*, 308–317. [CrossRef] [PubMed]
60. Yu, H.; Liang, H.L.; Ge, X.P.; Zhu, J.; Wang, Y.L.; Ren, M.C.; Chen, X.R. Dietary chlorella (*Chlorella vulgaris*) supplementation effectively improves body color, alleviates muscle inflammation and inhibits apoptosis in largemouth bass (*Micropterus salmoides*). *Fish Shellfish Immunol.* **2022**, *127*, 140–147. [CrossRef]
61. Gu, J.Z.; Liang, H.L.; Ge, X.P.; Xia, D.; Pan, L.K.; Mi, H.F.; Ren, M.C. A study of the potential effect of yellow mealworm (*Tenebrio molitor*) substitution for fish meal on growth, immune and antioxidant capacity in juvenile largemouth bass (*Micropterus salmoides*). *Fish Shellfish Immunol.* **2022**, *120*, 214–221. [CrossRef]
62. Yang, P.; Wang, W.Q.; Chi, S.Y.; Mai, K.S.; Song, F.; Wang, L. Effects of dietary lysine on regulating GH-IGF system, intermediate metabolism and immune response in largemouth bass (*Micropterus salmoides*). *Aquac. Rep.* **2020**, *17*, 100323. [CrossRef]

metabolites

MDPI

Article

Muscle Nutritive Metabolism Changes after Dietary Fishmeal Replaced by Cottonseed Meal in Golden Pompano (*Trachinotus ovatus*)

Yawen Qin [1], Chaoqun He [1], Haoyu Geng [1], Wenqiang Wang [1], Peng Yang [1], Kangsen Mai [1] and Fei Song [1,2,3,*]

[1] Guangzhou Key Laboratory of Subtropical Biodiversity and Biomonitoring, Guangdong Provincial Key Laboratory for Healthy and Safe Aquaculture, Institute of Modern Aquaculture Science and Engineering (IMASE), College of Life Science, South China Normal University, Guangzhou 510631, China; 2020022760@m.scnu.edu.cn (Y.Q.); 2020022843@m.scnu.edu.cn (C.H.); 2021023073@m.scnu.edu.cn (H.G.); 2019022514@m.scnu.edu.cn (W.W.); 2019022432@m.scnu.edu.cn (P.Y.); kmai@ouc.edu.cn (K.M.)

[2] Southern Marine Science and Engineering Guangdong Laboratory (Zhuhai), Zhuhai 519000, China

[3] Southern Marine Science and Engineering Guangdong Laboratory, Zhanjiang 524025, China

* Correspondence: songfei@m.scnu.edu.cn; Tel.: +86-020-85211375

Abstract: Our previous study demonstrated that based on growth performance and feed utilization, cottonseed meal (CSM) could substitute 20% fishmeal (FM) without adverse effect on golden pompano (*Trachinotus ovatus*). Muscle deposition was also an important indicator to evaluate the efficiency of alternative protein sources. Therefore, the present study was conducted to explore the changes of physiobiochemical and nutrient metabolism in muscle after FM replaced by CSM. Four isonitrogenous and isolipidic experimental diets (42.5% crude protein, 14.0% crude lipid) were formulated to replace 0% (CSM0 diet), 20% (CSM20 diet), 40% (CSM40 diet), and 60% (CSM60 diet) of FM with CSM. Juvenile fish (24.8 ± 0.02 g) were fed each diet for 6 weeks. The results presented, which, compared with the CSM0 diet, CSM20 and CSM40 diets, had no effect on changing the muscle proximate composition and free essential amino acid (EAA) concentration. For glycolipid metabolism, the CSM20 diet did not change the mRNA expression of hexokinase (hk), glucose transport protein 4 (glut4), glucagon-like peptide 1 receptor (glp-1r), while over 20% replacement impaired glucose metabolism. However, CSM20 and CSM40 diets had no effect on altering lipid metabolism. Mechanistically, compared with the CSM0 diet, the CSM20 diet did not change muscle nutritive metabolism through keeping the activities of the nutrient sensing signaling pathways stable. Higher replacement would break this balance and lead to muscle nutritive metabolism disorders. Based on the results, CSM could substitute 20–40% FM without affecting the muscle nutritive deposition. All data supplemented the powerful support for our previous conclusion that CSM could successfully replace 20% FM based on growth performance.

Keywords: fish nutrition; marine aquaculture; replacement; cottonseed meal; physiobiochemical; nutrient

Citation: Qin, Y.; He, C.; Geng, H.; Wang, W.; Yang, P.; Mai, K.; Song, F. Muscle Nutritive Metabolism Changes after Dietary Fishmeal Replaced by Cottonseed Meal in Golden Pompano (*Trachinotus ovatus*). *Metabolites* **2022**, *12*, 576. https://doi.org/10.3390/metabo12070576

Academic Editor: Frédérique Courant

Received: 20 May 2022
Accepted: 17 June 2022
Published: 22 June 2022

1. Introduction

Muscle is the largest tissue of teleost fish, accounting for more than 60% of the body weight. It provides premium proteins to meet the requirement for humans [1,2]. Previous studies demonstrated that alternative proteins replace fishmeal (FM) disturbed muscle nutrient metabolism and further resulted in the muscle growth restriction considered as an important limiting factor affecting substitution efficiency [3,4]. Additionally, the main goal of the aquaculture was to improve feed utilization, which in turn efficiently promoted muscle deposition [5–7]. Therefore, muscle nutrient deposition was the important indicator to assess the efficiency of alternative protein sources in substituting FM.

As in mammals, fish muscle growth is dependent on the recruitment of new muscle fiber (hyperplasia) and growth of existing fiber (hypertrophy) [8–10]. The essence of hypertrophy is the deposition of nutrients in muscle [11,12]. These processes are controlled by nutrient sensing signaling pathways at both cellular and systemic levels [13]. Amino acid (AA) transporters are regarded as the first sensor and carrier facilitate AA across the cell membrane, resulting in changes of free AA concentrations in plasma and tissues [14–16]. The target of rapamycin (TOR) and amino acid response (AAR) pathways could sense the nutrient status, especially AA concentrations, at the cellular level [17–19]. Moreover, the insulin/insulin-like growth factor (IGF) system is the main nutrient sensing pathway at a systemic level [20]. As the upstream, IGF through phosphoinositide-3-kinase (PI3K)/protein kinase B (AKT) regulated TOR and AAR signaling to modify nutrient metabolism including glucose, fatty acid and AA [21,22]. Insulin-like growth factor 1 (IGF-1) and insulin-like growth factor 2 (IGF-2) are the major anabolic agents that contribute to tissues, especially muscle nutrient metabolism in teleost fishes [23–25].

Previous studies demonstrated that FM replacement had a profound effect on fish muscle nutrient deposition. Our study on turbot (*Scophthalmus maximus* L.) presented that FM replaced by soybean meal and meat and bone meal at 45% significantly decreased the muscle nutrients' deposition by activating the AAR signaling pathway [26]. Furthermore, a study on the juvenile blunt snout bream (*Megalobrama amblycephala*) found that 5% and 7% cottonseed protein hydrolysate-replaced FM could decrease the muscle glucose, lipid and AA metabolism in muscle through inhibiting the TOR signaling pathway and activating the AMP-activated protein kinase/sirtuin-1 (AMPK/SIRT1) pathway [27]. Mixed plant proteins substituted 40% of FM in turbot and also significantly inhibited the TOR signaling pathway and protein synthesis in the muscle [28]. Moreover, total FM replaced by plant protein blends could suppress the muscle growth of Senegalese sole (*Solea senegalensis*) via altering the expression pattern of genes involved in the GH-IGF signaling pathway [29]. Therefore, the muscle nutritive metabolism response was considered as the crucial index to fully evaluate the optimal alternative ratio for protein sources.

Cottonseed meal (CSM) is known as a non-grain protein source that has cheaper prices, steady supply, relatively higher protein content and a well-balanced AA profile [30,31]. Currently, it has become a popular alternative protein source to replace FM in aquafeeds [32,33]. In the past several years, aquatic nutritionists have carried out research on the application of CSM in different aquafeeds [34,35]. All the studies demonstrated that CSM replacement proportions in feeds were quite different among different species. A study on black seabass (*Centropristis striata*) presented that CSM prepared by a glandless seed could successfully replace dietary 100% FM [36]. However, Bu et al. reported that CSM inclusion over 25.3% could significantly depress the growth performance in the Ussuri catfish *Pseudobagrusus suriensis* [37]. The studies mentioned above only focused on the effect of CSM replacement on growth performance and feed utilization. More attention should also be paid to the changes of muscle nutritive metabolism in order to accurately evaluate the replacement efficiency of CSM.

Golden pompano (*Trachinotus ovatus*) is widely cultivated in Guangdong, Fujian and Hainan provinces. Due to its advantage of growing fast, favorable taste, good nutrient profile and suitability for culture, it has become a commercially important marine fish species in China [38]. In the past decade (2010–2019), golden pompano production in China ranged from 0.75 million tons to 1.68 million tons and was worth 20 billion yuan [39]. Therefore, the industry of golden pompano culture has bright market prospects and huge economic benefits [40]. Previously, our team did some research on the CSM of golden pompano. The results present that, based on growth performance and feed utilization, CSM could substitute 20% FM without adverse effect [41]. As mentioned above, muscle deposition and nutrient metabolism were also the important indicators to evaluate the efficiency of CSM in replacing FM. Therefore, the present study was conducted to explore the physiobiochemical changes and underlying mechanisms of muscle nutrient metabolism after the FM was replaced by CSM. This study will provide a more comprehensive theoretical basis for the rational addition of CSM in the golden pompano diet.

2. Results

2.1. Proximate Composition of Muscle

The muscle proximate compositions of golden pompano after being fed with different diets were shown in Table 1. The values of crude protein (CP) and ash did not significantly differ among fish fed with four different experimental diets ($p > 0.05$). Meanwhile, compared with the control diet, the CSM20 and CSM40 diets also did not significantly change the moisture and crude fat (CF) of muscle ($p > 0.05$). Only the CSM60 diet increased the moisture and decreased CF ($p < 0.05$).

Table 1. Muscle proximate composition of golden pompano fed diets containing various levels of CSM.

	Diet Groups (% Wet Weight)				*p*-Value
	CSM0	CSM20	CSM40	CSM60	
Moisture	69.34 ± 0.080^{b}	69.47 ± 0.32^{ab}	70.24 ± 0.26^{ab}	70.51 ± 0.21^{a}	0.018
CP	20.03 ± 0.17	20.72 ± 0.51	19.84 ± 0.88	20.51 ± 0.23	0.208
CF	8.91 ± 0.65^{a}	7.21 ± 0.026^{ab}	7.21 ± 0.57^{ab}	6.83 ± 0.23^{b}	0.044
Ash	1.26 ± 0.027	1.33 ± 0.026	1.25 ± 0.018	1.31 ± 0.058	0.407

The results were presented as mean ± SEM. Means in the same row with different superscript letters are significantly different ($p < 0.05$). Note: CP, crude protein; CF, crude fat.

2.2. Free AAs Profile in Muscle

Changes of free AAs after fish were fed with different diets in muscle were presented in Table 2. Compared with that of the control diet, the concentration of leucine, threonine, lysine, histidine, arginine, glutamic acid, alanine, aspartic acid, serine and taurine were unaffected by three CSM-containing diets ($p > 0.05$). For other AAs, all the individual essential amino acids (EAAs) and non-essential amino acids (NEAAs) did not present significant differences among CSM0, CSM20 and CSM40 groups ($p > 0.05$), except phenylalanine, proline, tyrosine, NEAA and total amino acid (TAA) demonstrated significantly lower concentrations in CSM40 ($p < 0.05$). The CSM60 diet group presented a decreased free AA profile compared with the control group ($p < 0.05$).

2.3. Gene Expression Profile Related to Glucose Metabolism in Muscle

Figure 1 described the mRNA expression level of genes related to glucose metabolism in muscle after the golden pompano were fed diets containing different CSM levels. Compared with the CSM0 diet, the CSM20 diet had no effect on altering the gene expression of hexokinase (hk), glucose transport protein 4 (glut4) and glucagon like peptide 1 receptor (glp-1r) ($p > 0.05$). In contrast, the CSM20 diet significantly reduced the gene expression level of glucose-6-phosphate (g6pdh), pyruvate kinase (pk), phospho fructokinase-1 (pfk-1), phosphoenolpyruvate carboxykinase (pepck), glucose transport protein 2 (glut2), insulin receptor substrate 1 (irs1) and glucagon-like peptide 1 receptor (igf-ir) in comparison to the CSM0 diet ($p < 0.05$). Moreover, CSM40 and CSM60 diets also inhibited the glucose metabolism level more than the control diet ($p < 0.05$).

2.4. Effects on Muscle Lipid Metabolism by Experimental Diets

Effects of different experimental diets on the mRNA expression levels of lipid anabolism are shown in Figure 2a. ANOVA did not reveal a significant influence of CSM20 on the expression of fatty acid synthetase (FAS), acetyl-CoA carboxylase (ACC), 1-acylglycerol-3-phosphate acyltransferase 3 (AGPAT3), fatty acyl desaturase (FAD) and sterol regulatory element binding protein-1 (SREBP1) in comparison to CSM0 diet ($p > 0.05$). Moreover, the CSM40 diet had no significant effects on altering the expression of FAS, AGPAT3 and SREBP1 ($p > 0.05$). However, compared with the CSM0 diet, three CSM containing diets significantly inhibited the mRNA expression of elovl5 and improved the peroxisome proliferator-activated receptors' alpha (PPARα) and peroxisome proliferator-activated receptors' gamma (PPARγ) expression ($p < 0.05$).

Table 2. Free amino acid profile in muscle of golden pompano fed different experimental diets.

	Diet Groups (ng/mg)				*p*-Value
	CSM0	**CSM20**	**CSM40**	**CSM60**	
Met	43.71 ± 0.93 [a]	44.84 ± 1.13 [ab]	40.89 ± 1.33 [ab]	33.22 ± 1.96 [b]	0.001
Phe	43.62 ± 3.19 [a]	33.38 ± 1.77 [ab]	31.61 ± 0.37 [b]	30.93 ± 2.96 [b]	0.017
Val	167.53 ± 10.73 [a]	139.53 ± 11.59 [ab]	141.54 ± 2.90 [ab]	112.07 ± 10.51 [b]	0.023
Ile	127.65 ± 10.40 [a]	106.72 ± 7.35 [ab]	104.23 ± 1.03 [ab]	83.89 ± 7.86 [b]	0.022
Leu	205.13 ± 18.18	172.39 ± 12.98	166.00 ± 1.41	148.16 ± 13.30	0.07
Thr	178.37 ± 6.70	148.73 ± 16.26	141.79 ± 2.33	137.10 ± 8.44	0.067
Lys	319.69 ± 20.04	368.12 ± 8.13	335.26 ± 27.18	309.99 ± 3.01	0.174
His	63.99 ± 1.81 [ab]	67.40 ± 1.08 [a]	58.14 ± 5.29 [ab]	50.96 ± 2.85 [b]	0.029
Arg	62.77 ± 2.20	66.22 ± 0.27	60.94 ± 5.15	62.51 ± 1.96	0.655
EAA	1212.48 ± 40.76 [a]	1147.33 ± 58.05 [ab]	1080.40 ± 31.39 [ab]	968.84 ± 38.64 [b]	0.021
Glu	400.00 ± 38.53	405.77 ± 30.42	413.25 ± 20.72	446.68 ± 7.68	0.632
Gly	1866.44 ± 86.41 [a]	1854.01 ± 18.75 [a]	1564.49 ± 46.94 [ab]	1349.20 ± 119.88 [b]	0.004
Pro	96.91 ± 7.27 [a]	81.52 ± 0.47 [ab]	76.64 ± 0.59 [b]	71.41 ± 3.19 [b]	0.01
Ala	449.77 ± 11.61	439.88 ± 2.36	429.38 ± 17.31	461.63 ± 4.09	0.251
Asp	15.11 ± 0.84	16.11 ± 1.04	12.72 ± 0.80	12.36 ± 0.86	0.046
Tyr	38.81 ± 2.47 [a]	29.87 ± 0.98 [ab]	29.04 ± 0.12 [b]	28.90 ± 2.80 [b]	0.021
Ser	187.81 ± 3.59	180.68 ± 5.42	177.38 ± 10.78	174.06 ± 8.75	0.64
Tau	1738.72 ± 59.66	1762.94 ± 46.02	1776.66 ± 8.59	1550.36 ± 85.14	0.072
NEAA	4793.57 ± 62.88 [a]	4771.29 ± 67.07 [a]	4479.57 ± 27.84 [b]	4094.60 ± 46.06 [b]	0
Total AA	6006.05 ± 93.11 [a]	5918.42 ± 80.25 [a]	5559.97 ± 24.53 [b]	5063.44 ± 74.82 [c]	0

The results were presented as mean ± SEM. Means in the same row with different superscript letters are significantly different ($p < 0.05$). Note: Met, methionine; Phe, phenylalanine; Val, valine; Ile, isoleucine; Leu, leucine; Thr, threonine; Lys, lysine; His, histidine; Arg, arginine; EAA, essential amino acids; Glu, glutamine; Gly, glycine; Pro, proline; Ala, alanine; Asp, asparagine; Tyr, tyrosine; Ser, serine; Tau, taurine; NEAA, non-essential amino acids; TAA, total amino acids.

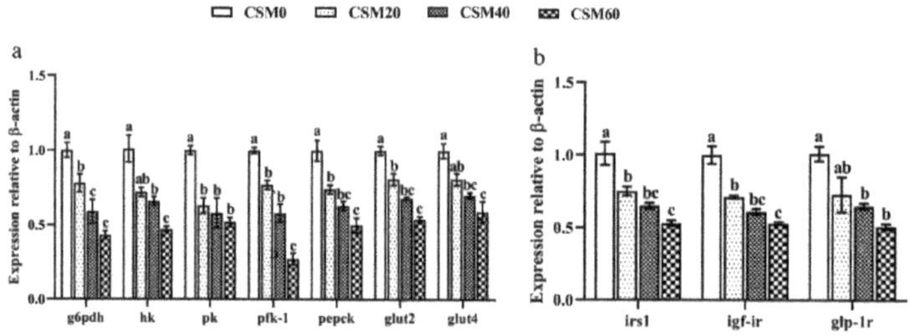

Figure 1. Diagrammatic representation of the gene expression pattern of (**a**) the key enzyme and transporters involved in glucose metabolism and (**b**) hormone receptors involved in insulin signaling pathway and glp-1r signaling in muscle after golden pompano (*Trachinotus ovatus*) were fed CSM0, CSM20, CSM40 and CSN60. Data were represented as mean ± SEM. Different letters above the bars denote significant differences between diet groups at the $p < 0.05$ level. Note: g6pdh, glucose-6-phosphate; hk, hexokinase; pk, pyruvate kinase; pfk-1, phosphofructokinase-1; pepck, phosphoenolpyruvate carboxykinase; glut2, glucose transport protein 2; glut4, glucose transport protein 4; irs1, insulin receptor substrate 1; igf-1r, insulin-like growth factor-1 receptor; glp-1r, glucagon-like peptide 1 receptor.

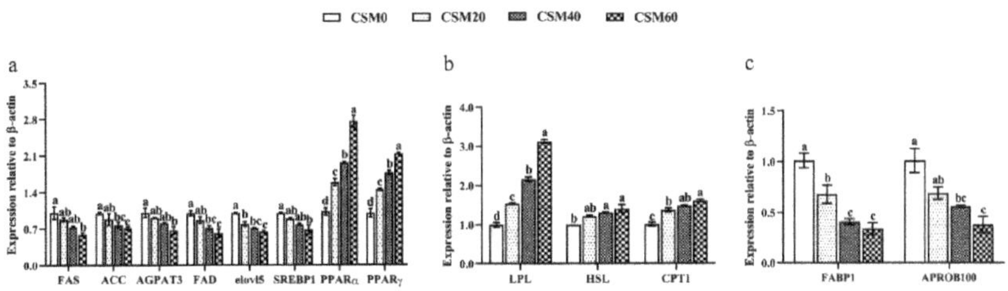

Figure 2. Effects of different experimental diets on the mRNA expression levels of (**a**) lipid anabolism, (**b**) lipid catabolism and (**c**) lipid transporters in muscle after the golden pompano (*Trachinotus ovatus*) were fed different diets. Data were represented as mean ± SEM. Different letters above the bars denote significant differences between diet groups at the $p < 0.05$ level. Note: FAS, fatty acid synthetase; ACC, acetyl-CoA carboxylase; AGPAT3, 1-acylglycerol-3-phosphate acyltransferase 3; FAD, fatty acyl desaturase; elovl5, elongase of very long-chain fatty acids 5; SREBP1, sterol regulatory element binding protein-1; PPARα, peroxisome proliferator activated receptors-alpha; PPARγ, peroxisome proliferator-activated receptors gamma; LPL, lipoprotein lipase; HSL, hormone-sensitive lipase; CPT1, carnitine palmitoyl transferase 1; FABP1, fatty acid binding protein 1; APROB100, apolipoprotein b 100.

The muscle gene expression pattern of the key regulators involved in lipid catabolism after the fish were fed with different diets is presented in Figure 2b. Compared to the control diet, the mRNA expression level of hormone-sensitive lipase (HSL) was not significantly changed by the CSM20 group ($p > 0.05$). Meanwhile, the expression of lipoprotein lipase (LPL) and carnitine palmitoyl transferase 1 (CPT1) of the fish fed CSM20, CSM40 and CSM60 diets were more significantly up-regulated than the control diet ($p < 0.05$).

The effects of dietary CSM substitution on the transcriptional expression level of lipid transporters is illustrated in Figure 2c. No significant differences were detected of the mRNA expression level of apolipoprotein b 100 (APROB100) between the CSM0 and CSM20 groups ($p > 0.05$). However, compared with the CSM0 diet, other diets significantly inhibited the transcriptional expression level of the fatty acid binding protein 1 (FABP1) ($p < 0.05$).

2.5. Gene Expression of AA Transporter and Small Peptide Transporter

The gene expression patterns of L-type amino acid transporter 2 (LAT2), sodium-coupled neutral amino acid transporter 2 (SNAT2) and oligopeptide transporter1 (PEPT1) had a decreasing trend with an increasing CSM replacement level (Figure 3). Compared with the CSM0 diet, the CSM20 diet had no effect on changing the gene expression of LAT2 and PEPT1 ($p > 0.05$), and CSM40 diet did not significantly down-regulate the expression of LAT2 ($p > 0.05$). However, fish fed the CSM20 diet demonstrated significantly lower levels of SNAT2 than the fish fed the CSM0 diet ($p < 0.05$).

2.6. The mRNA Expression of the Genes Related to GH-IGF Axis

The transcription expression levels of key regulators of GH-IGF axis were significantly affected by dietary treatments (Figure 4). No significant differences were found between CSM0 and CSM20 groups in the growth hormone (GH) and IGF-2 ($p > 0.05$). Moreover, the expression of IGF-1 was not significantly down-regulated in fish fed the CSM20 diet and CSM40 diet compared to that of fish fed CSM0 ($p > 0.05$). The CSM60 diet caused a more suppressed GH, IGF-1 and IGF-2 expression pattern than the CSM0 diet ($p < 0.05$).

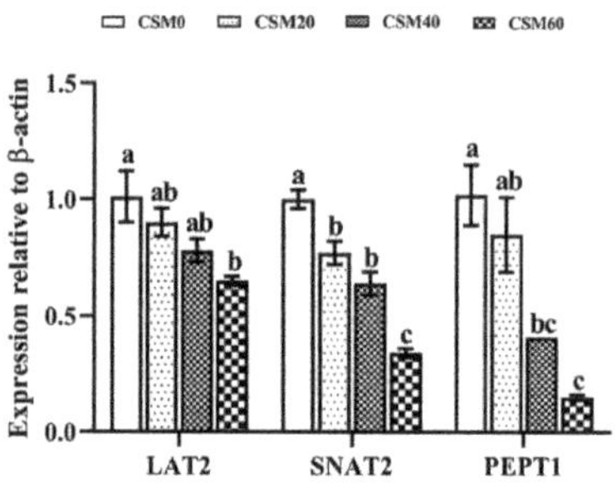

Figure 3. LAT2, SNAT2 and PEPT1 expression in muscle of juvenile golden pompano (*Trachinotus ovatus*) after fed test diets for 6 weeks. Data were represented as mean ± SEM. Different letters above the bars denote significant differences between diet groups at the $p < 0.05$ level. Note: LAT2, L-type amino acid transporter 2; SNAT2, sodium-coupled neutral amino acid transporter 2; PEPT1, oligopeptide transporter1.

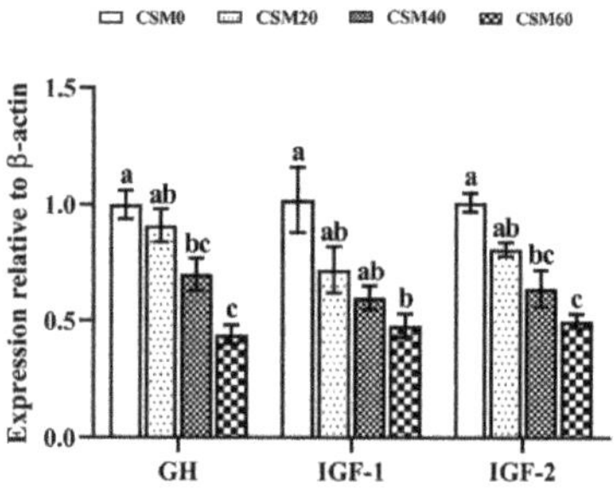

Figure 4. Different diets affected the relative gene expression of the key regulators of GH-IGF-1 axis in muscle of juvenile golden pompano (*Trachinotus ovatus*). Data were represented as mean ± SEM. Different letters above the bars denote significant differences between diet groups at the $p < 0.05$ level. Note: GH, growth hormone; IGF-1, insulin-like growth factor-1; IGF-2, insulin-like growth factor-2.

2.7. Regulations of the TOR and AAR Signaling Pathways

Transcription and protein phosphorylation level of key regulators in TOR and AAR signaling pathways after the feeding trials were presented in Figure 5. In comparison to the control diet, relative mRNA expression of TOR, S6 Ribosomal Protein (S6) and eukaryotic initiation factor 4E-binding protein 1 (4E-BP1) were unaffected by the CSM20 diet ($p > 0.05$). Meanwhile, the CSM40 diet led to a significantly lower transcriptional level of 4E-BP1 ($p < 0.05$) but did not have marked effects on TOR and S6 expression ($p > 0.05$). For the gene expression of detected regulators in the AAR signaling pathway, no significant difference of initiation elongation factor alpha (eIF2α), activating transcription factor 4 (ATF4) and channelopsin2 (CHOP) mRNA expressions were found between fish fed CSM0 and CSM20 diets ($p > 0.05$). In addition, a significant difference of eIF2α and ATF4 mRNA expressions were detected between group CSM40 and group CSM0 ($p > 0.05$).

For the protein phosphorylation analysis, dietary inclusion of CSM did not affect protein phosphorylation level of mTOR after 6 weeks feeding ($p > 0.05$). Meanwhile, CSM20 diet did not had significant effects on altering the protein phosphorylation level of protein kinase B (AKT), S6 and eIF2α ($p > 0.05$) and CSM40 diet did not significantly change the expression pattern of S6 and eIF2α in comparison to CSM0 diet ($p > 0.05$). Meanwhile, the CSM60 diet markedly decreased the phosphorylation level of AKT, S6 and elevated the phosphorylation level of eIF2α ($p < 0.05$).

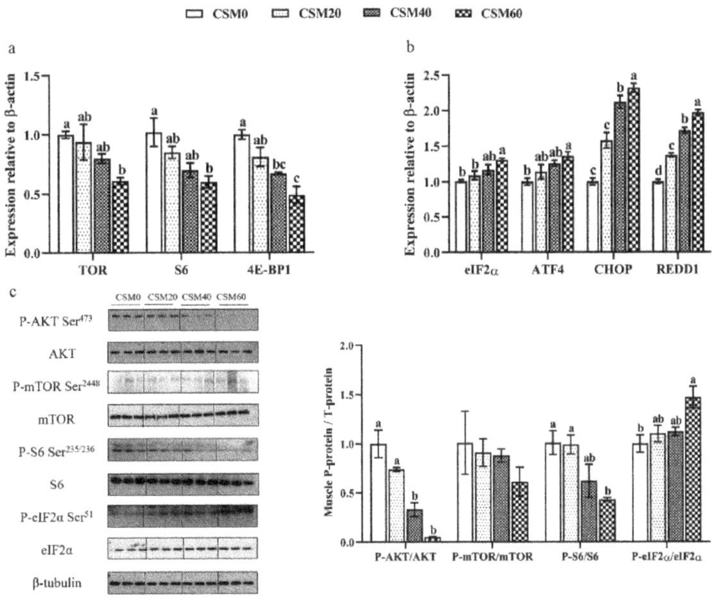

Figure 5. The mRNA expression and protein phosphorylation level of key regulators related to target of rapamycin (TOR) and amino acid response (AAR) signaling pathways in muscle. (**a**) mRNA expression level of TOR signaling pathway; (**b**) mRNA expression of AAR signaling pathway; (**c**) protein phosphorylation level of TOR and AAR signaling pathways. Data were represented as mean ± SEM. Different letters above the bars denote significant differences between diet groups at the $p < 0.05$ level. Note: TOR, target of rapamycin; S6, S6 Ribosomal Protein. 4E-BP1, eukaryotic initiation factor 4E-binding protein 1; eIF2α, initiation elongation factor alpha; ATF4, activating transcription factor 4; CHOP, channelopsin2; REDD1, regulated in development and DNA damage responses 1; AKT, protein kinase B.

3. Discussion

The present study was part of a study exploring the optimum substitution ratio of CSM in golden pompano diets, and it used one growth trial as a research basis that determined the effects of CSM-replaced FM in growth performance and feed efficiency. In fish, the growth is dependent on muscle deposition, which largely related to the muscle nutritive metabolism [42,43]. Therefore, the current study was conducted to evaluate the effects of the CSM substituted FM on muscle nutritive metabolism as well as its potential mechanism.

3.1. CSM Substitution Alter Muscle Nutrient Composition and Free AAs Profile

Muscle is the main site of nutritive deposition for fish species. CP and CF are the important indicators that reflect the nutritive deposition after the fish are fed with different diets [44]. Our present study displayed that compared with the control diet, the CSM20 and CSM40 diets had no effect on changing the moisture, CP, CF and ash content of muscle. Only the fish fed with the CSM 60 diet had significantly increased the moisture and decreased CF of muscle, with the same results for the fermented soybean meal replacement in Japanese seabass (*Lateolabrax japonicus*) [45] and rapeseed meal in *Pseudobagrusus suriensis* [46]. Furthermore, it has been reported that the free AA profile is the key driver to regulate protein and other nutrient metabolisms in muscle [47]. Our current observations indicated that CSM20 and CSM40 diet did not markedly vary the concentration of most individual AA and EAA. Meanwhile, the CSM60 diet reduced the concentration of EAA, NEAA and TAA in muscle. These results were consistent with our previous studies on turbot and shrimp demonstrating the great variation on free AA after a higher percentage of

FM was replaced by alternative proteins [48,49]. We envisaged that the changes of free AA profile after 60% FM replaced by CSM might be partially due to the unbalanced AA composition of higher CSM containing diets [50,51]. A lower level of free AA in muscle not only decreased the bricks for protein synthesis and antioxidative capacity but also disordered the glycolipid metabolism [52,53]. Moreover, the present study also demonstrated that the CSM 60 diet significantly decreased the glycine and tyrosine concentration more than the CSM0 diet. Glycline and glutamic acid play important physiological roles in glutathione (GSH) synthesis, which can protect fish against free radicals by increasing the activity of the antioxidant enzyme of fish [54]. Their results were consistent with our previous study that higher CSM replacement could affect fish immunity [41]. Moreover, data on muscle proximate composition also hinted at lipid deposition changes after fish fed with the CSM60 diet. Thus, we next investigate the changes of the glycolipid metabolism transcriptional level in muscle after golden pompano were fed with gradient CSM addition diets.

3.2. CSM Substitution-Modified Glycolipid Metabolism

It is well known that nutrient compositions are the most important indicators to estimate muscle growth and quality [55,56]. The inhibition of muscle nutrient deposition due to glycolipid metabolism disorder is also the bottleneck of FM replacement [57,58]. Data on the present study illustrated that the mRNA expression of key enzymes and transporters in glycolipid metabolism were extremely sensitive to the dietary protein sources. Only 20% of the CSM replacement could significantly reduce the enzymes' expression of glucose metabolism. However, for lipid metabolism, less than 60% of the CSM replacement did not alter the mRNA expression levels of FAS, AGPAT3 and SREBP1. A similar observation could be found in the study of soy bean protein for cod *Gadus morhua* [59] and rapeseed meal diet for Chinese perch *Siniperca chuatsi* [60]. Furthermore, our recent study on largemouth bass (*Micropterus salmoides*) also presented that the glucose metabolism was more sensitive to feed compositions [61]. Similar to the largemouth bass, the golden pompano was also a carnivorous fish, which had dysglycemia after the higher FM was substituted by alternative proteins. CSM as a plant protein source has a higher content of carbohydrate compared with FM [62]. That might give a reasonable explanation for the dramatic changes of glucose metabolism after the golden pompano were fed with a higher CSM containing diet. Studies on mammal and fish species demonstrated that glycolipid metabolism could be modulated by nutrient sensing signaling pathways [63–65]. Therefore, in order to reveal the mechanisms of the changed nutrient metabolism, we next investigated the response mechanism of nutrient sensing signaling pathways after FM substituted by CSM.

3.3. CSM Substitution through Nutrient Sensing Signaling Pathways to Regulate Glycolipid Metabolism

AA transporters are known to regulate intracellular AA concentrations and TOR activity in lysosomes, which are regarded as the first sensors of dietary nutrients [66]. The present study displayed the variation in the AA profile of muscle after the fish were fed with different CSM containing diets. Therefore, we further measured the mRNA expression level of AA transporters in muscle tissue to explain the underlying mechanism of the changes in the AA profile. Data on the AA transporters' expression demonstrated that CSM40 and CSM60 diets significantly decreased the mRNA expression of LAT2, SNAT2 and PEPT1. Similarly, in blunt snout bream [67] and turbot [68], FM replaced by alternative proteins also decreased the AA transporters' expression. A lower level of AA transporters reduced the AA transmembrane transport efficiency, resulting in an overall decrease in muscle AA levels [69,70]. A restricted AA level in muscle would directly affect the activity of nutrient sensing signaling pathways, and further changed the body growth and metabolism [27,71]. Therefore, we next tried to explore the activity of nutrient sensing signaling pathways after the golden pompano were fed with different diets.

In mammals, it has been well revealed that TOR and AAR were two complementary signalings, which were responsible for sensing the nutrient level to regulate body protein synthesis and nutrient metabolism [72–74]. Moreover, GH and its downstream mediator, IGF-1, constructed a pleotropic axis through the TOR signaling pathway regulating growth, metabolism and organ function [75,76]. Recent studies from our laboratory reported that in largemouth bass, a dietary AA level could through TOR, AAR and GH-IGF signaling to regulate fish growth performance and glycolipid metabolism [53]. Likewise, the present study in golden pompano also demonstrated that dietary protein sources had profound effects on changing nutrient sensing signaling pathways. In detail, compared with the control diet, the CSM60 diet significantly decreased the mRNA expression level and protein phosphorylation level of the key regulators involved in GH-IGF and TOR signaling pathways and increased the mRNA expression level and protein phosphorylation level of the key regulators involved in the AAR signaling pathways in muscle. Our previous study focused on the growth performance after golden pompano fed with the CSM substitution diet also presented a similar changing pattern in the liver [41]. In Olive Flounder (*Paralichthys olivaceus*), the FM replacement modified the growth performance through the IGF system as well [77]. Additionally, the FM substituted with a composite mixture of shrimp hydrolysate and plant proteins in largemouth bass also presented that the appropriate FM replacement with composite mixture could improve the growth performance through activating the TOR signaling pathway [78]. The suppressed TOR signaling pathway would inhibit the whole-body total protein synthesis, especially in muscle. The reduced protein synthesis directly led to decreased muscle deposition. On the other hand, glycolipid metabolism was also inhibited through regulating TOR, AAR and GH-IGF pathways, which resulted in a decrease in the nutrients' deposition of muscle. All the data confirmed our results, in which higher CSM supplementation disturbed nutrient sensing signaling and further inhibited muscle nutrient metabolism. This may explain why CSM could not successfully replace over 40% FM in the golden pompano diet.

4. Materials and Methods

4.1. Experimental Diets

Four isonitrogenous (42.5% crude protein) and isolipidic (14% crude lipid) diets were produced with CSM gradient-replaced FM. Dietary protein was supplied by FM, CSM, corn-gluten meal and poultry byproduct meal. Fish oil, soybean oil and soybean lecithin were used as lipid sources. The control diet (CSM0 diet) was based on FM, and other diets were named the CSM20 diet, CSM40 diet and CSM60 diet, in which CSM was used to replace 20%, 40% and 60% FM, respectively. All dry ingredients were grounded into powder with 320-μm and then mixed thoroughly. An appropriate amount of oil was added in order to meet the requirement of lipid for fish. The resulting doughs were extruded through a pelletizer (F-26, South China University of Technology, Guangzhou, China) after being further homogenized. All the diets were dried at 45 °C to an approximately 10% moisture level and stored in a refrigerator at −20 °C before the feeding trial. The proximate composition of all the experiment diets was given in Table 3. The essential amino acids' composition of each experiment diet was given in Table 4.

4.2. Fish and Feeding Management

The experiment was performed with three hundred and sixty juvenile golden pompanos that were obtained from the Dayawan Fish Farm (Guangdong, China). Before the start of the experiment, all fish were fed with the CSM0 diet for two weeks to adapt to the experimental conditions. Before the experiment, 360 golden pompanos were fasted for 24 h and weighed. Fish with a mean initial weight of 28.42 ± 0.02 g were randomly distributed to 12 sea cages (1.5 m × 1.5 m × 1.5 m). Each experiment diet was randomly assigned to three replicates, and 30 golden pompanos were in each replicate. All treatments were fed their respective feeds twice daily (07:00 and 17:00) to apparent satiation for 6 weeks. The physiobiochemical properties of the water were recorded daily. The temperature and dissolved oxygen of water were maintained at 28–30 °C and 5.0–6.0 mg/L.

Table 3. Experimental diet formulations.

Ingredients	Diets (% Dry Weight)			
	CSM0	CSM20	CSM40	CSM60
Fishmeal	25.00	20.00	15.00	10.00
Cottonseed meal (CSM)	0.00	5.00	10.00	15.00
Corn gluten meal	13.00	13.00	13.00	13.00
Poultry by-product meal	11.00	11.00	11.00	11.00
Soybean meal	8.50	8.50	8.50	8.50
Peanut meal	6.50	6.50	6.50	6.50
Wheat meal	17.50	17.50	17.50	17.50
Fish oil	1.50	2.00	2.40	2.80
Soybean oil	5.00	5.00	5.00	5.00
Soybean lecithin	2.50	2.50	2.50	2.50
Monocalcium phosphate	1.50	1.70	1.90	2.10
Lysine	0.28	0.45	0.60	0.75
Methionine	0.10	0.15	0.20	0.25
Threonine	0.01	0.03	0.05	0.07
Squid paste	1.50	1.50	1.50	1.50
Mineral premix [1]	1.50	1.50	1.50	1.50
Vitamin premix [2]	0.50	0.50	0.50	0.50
Chromium trioxide	0.10	0.10	0.10	0.10
Lutein	0.10	0.10	0.10	0.10
Antioxidant	0.05	0.05	0.05	0.05
Mold inhibitor	0.10	0.10	0.10	0.10
Cellulose	3.76	2.82	2.00	1.18
Proximate composition				
DM (%)	90.24	89.98	89.91	90.12
Crude protein (%)	42.42	42.51	42.58	42.66
Crude lipid (%)	14.00	14.09	14.07	14.05

[1] Mineral premix (mg/kg diet): NaF, 2 mg; KI, 0.8 mg; $CoCl_2 \cdot 6H_2O$ (10 g/kg), 50 mg; $CuSO_4 \cdot 5H_2O$, 10 mg; $FeSO_4 \cdot H_2O$, 80 mg; $ZnSO_4 \cdot H_2O$, 50 mg; $MnSO_4 \cdot H_2O$, 60 mg; $MgSO_4 . 7H_2O$, 1200 mg; $Ca(H_2PO_4)_2 \cdot H_2O$, 3000 mg; NaCl, 100 mg; zeolite, 15,447 mg. [2] Vitamin premix (mg/kg diet): thiamin, 25 mg; riboflavin, 45 mg; pyridoxine HCl, 20 mg; vitamin B12, 0.1 mg; vitamin K3, 10 mg; inositol, 800 mg; pantothenic acid, 60 mg; niacin acid, 200 mg; folic acid, 20 mg; biotin, 1.20 mg; retinal acetate, 32 mg; cholecalciferol, 5 mg; α-to-copherol, 120 mg; ascorbic acid, 2000 mg; choline chloride, 2500 mg; ethoxyquin 150 mg; wheat middling, 14,012 mg.

Table 4. The essential amino acid composition of the experimental diets.

Amino Acids	Diets (% Dry Weight)			
	CSM0	CSM20	CSM40	CSM60
Lys	1.90	1.92	1.92	1.91
Met	0.70	0.70	0.70	0.70
Thr	0.88	0.88	0.88	0.88
Arg	1.29	1.51	1.73	1.95
His	0.47	0.49	0.51	0.53
Ile	0.84	0.81	0.79	0.77
Leu	1.51	1.48	1.45	1.42
Phe	0.89	0.94	0.99	1.04
Val	1.02	1.01	0.99	0.97
Cys	0.15	0.18	0.20	0.23
Tyr	0.66	0.65	0.65	0.64

4.3. Sample Collection

At the end of the feeding trial, all of the golden pompano were fasted for 24 h and weighed. Four fish from each cage were rapidly anesthetized with eugenol (Shanghai Medical Co., Ltd., Shanghai, China), and the dorsal muscle of the fish was cut from the posterior of the anus to the posterior of the dorsal fin, then was quickly collected and immediately frozen. Samples of the muscle molecular analysis were kept in an RNAase tube (Axygen, Union City, CA, USA), snap-frozen in liquid nitrogen and then reversed at −80 °C until gene expression analysis.

4.4. AA Composition of Diet and Proximate Composition; Free AAs of Muscle

Diets were dried by a freezer dryer (ALPHA1-2 LD plus, Christ Co., Ltd., Berlin, Germany) in order to measure the AA compositions. The L-8900 AA analyzer (Hitachi, Tokyo, Japan) was used for analysis after all samples were digested with 6 M HCl for 22 h (Hitachi, Japan).

The proximate compositions of the muscle were determined, followed by the methods reported by Yang [49]. The moisture content was conducted by drying samples to a constant weight in an oven at 105 °C and calculating it as a percentage. The crude lipid was measured by a petroleum ether extraction (B.P. 30–60 °C, 3 h) in a SoxtecTM 2055 extraction. The crude protein was analyzed through a Dumas nitrogen determination apparatus (DT autosampler, Europe Gerhardt Company, Königswinter, Germany). The ash was determined by incinerating samples in a muffle furnace (FO610C, Yamato Scientific Co., Ltd., Tokyo, Japan) at 550 °C for 12 h until the samples were at a constant weight. Moreover, the free AA composition in the muscle was performed by an auto AA analyzer (LA8080; Hitachi, Tokyo, Japan). The muscle sample (0.2 g) and 10% hydrochloric acid were mixed completely and then ground for 5 min. After being centrifuged at 12,000× *g* at 4 °C for 15 min, the supernatants were filtered through 0.22-μm filters for the free AA concentration analysis.

4.5. RNA Extraction and cDNA Synthesis

Total RNA was isolated from the muscle sample using Trizol Reagent (Vazyme Biotech Co., Ltd., Nanjing, China) according to the manufacturer's instructions. After being eluted in diethyl pyrocarbonate (DEPC)-treated water, the RNA samples were subjected to electrophoresis on 1.2% agarose gel to confirm the integrity, and RNA quantity and quality was assessed as a RNA concentration and 260/280 nm absorbance ratio by using a NanoDrop 2000 spectrophotometer (Thermo, NanoDrop Technologies, Wilmington, DE, USA). The cDNA was generated from 1 μg of total RNA by using a Prime Script RT reagent Kit (Vazyme Biotech Co., Ltd., China). Reaction conditions were recommended by the manufacturer's introductions. The cDNA was then diluted with DEPC water to 100 ng/μL and used for real-time qPCR to determine the gene expression levels in muscle.

4.6. mRNA Expression Analysis (RT-qPCR)

The primers used in this study were given in Table 5. Real-time PCR assays were performed on a CFX96 real-time PCR equipment (CFX96, BIO-RAD, Berkeley, CA, USA) in a total volume of 20 μL containing 10 μL of Hiff® qPCR SYBR Green Master Mix (Yeasen, Shanghai, China), 3 μL of diluted cDNA, 0.5 μL of each primer and 6 μL of DEPC water. The amplification reaction was initially conducted at 95 °C for 2 min, followed by 40 cycles of denaturation at 95 °C for 15 s and annealing at 58 °C for 30 s, and finally 72 °C for 20 s. The specificity was confirmed by the melt curve, and the relative mRNA expression levels of target genes were analyzed using the $2^{-\Delta\Delta Ct}$ method. β-actin was used as the internal reference gene to normalize the mRNA expression of genes, and the relative gene expression is relative to the control group.

4.7. Western Blotting

The Western blotting was performed according to Xu's study [48]. The BCA protein assay kit (Beyotime, Biotechnology, Shanghai, China) was used to estimate protein concentrations of samples following the manufacturer's guide. The relative protein expression of total AKT, phosphor-AKT, total TOR, phosphor-TOR, total S6, phosphor-S6, total eIF2α and phosphor- eIF2α were analyzed by sodium dodecyl sulfate-polyacrylamide (SDS-PAGE) and then transferred onto 0.45-μm polyvinylidene fluoride (PVDF) membranes (Millipore, St. Louis, MO, USA) by electrophoretic transfer for 1 h at 100 V. After that, the PVDF membranes were blocked with 5% non-fat milk in Tris-buffered saline in tween 20 (TBST) containing 20 mM Tris-HCl, 500 mM NaCl and 1% Tween-20 at room temperature for 1 h. The membranes were incubated in each primary antibody overnight at 4 °C. After washing the membranes three times with TBST, horseradish peroxidase-labeled secondary antibodies (1:10,000 dilution) was added and incubated for 1 h at room temperature. After washing again with TBST three times, the membranes were then treated with enhanced chemiluminescence (ECL) regents (Beyotime, Biotechnology, China) to visualize the protein bands. Protein expression was quantified with NIH Image 1.63 software, and β-tublin was used as reference.

Table 5. Primers used for determining gene expression.

Target Gene	Forward Sequence (5′–3′)	Reverse Sequence (5′–3′)
g6pdh	CTGTGGCAAAAGTTGGTGTG	CCTGATGATGTGAGGGATGA
hk	CCTTCCTCGTCTTTGTCATTT	TGTCCGTCTCATCCTGGTG
pk	TTTGCCAGTTTCATCCGCT	CCATCACGCCATCGCTCT
pfk-1	TGGGTGGGACCGTGATT	AGGTTGGTGATGCCTTTCTT
pepck	TGGAGTGTTTGTTGGAGCAG	CGAAGTTGTAGCCGAAGAAG
glut2	TCCTGTTTGCTGTGCTGCTT	GTTTTCCGTCCCTTGCG
glut4	AATGGCTGTGGCTGGCTT	AGGTTTTTCCCCGTGTTTCT
irs1	GCTCCACCCCTCTATTATCTCCT	GTACCTCCCACAGTTCCTCAGTC
igf-i r	TTCTGCTGTGCTCTTGTCT	GATGTTTTTGGTGTGGCT
glp-1 r	GGCAATCTCTCCTGTTCCC	AGCCTCTGCTTTTATTCGTG
FAS	GATGGATACAAAGAGCAAGG	GTGGAGCCGATAAGAAGA
ACC	GTTGTCAATCCCAGCCGATC	ATCCACAATGTAGGCCCCAA
AGPAT3	CTTCCTGTTTTGGGCCACTC	GTCGCCATAACTTGAGCCTG
FAD	GAACAATCCCACTTCAACG	AGGAATCCCATACTTCTCACA
elovl5	TACATGGTCACGCTCATTATCC	CCGTTCTGATGCTCCTTCTTTA
SREBP1	GAGCCAAGACAGAGGAGTGT	GTCCTCTTGTCTCCCAGCTT
PPARα	AATCTCAGCGTGTCGTCTT	GGAAATGCTTCGGATACTTG
PPARγ	TCAGGGTTTCACTATGGCGT	CTGGAAGCGACAGTATTGGC
LPL	TTTGTCCTTCCTCGTCACCA	AAGACAGCATCCTCTCCACC
HSL	TCATACCTCCACACCAACCC	GTCTCGCAGTTTCTTGGCAA
CPT1	CTTTAGCCAAGCCCTTCATC	CACGGTTACCTGTTCCCTCT
FABP1	CCAAGGACATCAAGCCAATTAC	TGGTGATTTCAGCCTCCTTAC
APROB100	AAAAGCCACAAGACGAAAGCA	GAAGCAGCAAAAGGCAGAGC
LAT2	CTCCCAGCAGCTTCTCACCAAAC	CTCGTGCCATCTTCATCTCCATC
SNAT2	CTGCTGGCTGCCCTTTTCGGATA	AGGACAGGTGCTGGTTGATGGAG
PEPT1	AACTGGTCTCCTCCAAACGC	GTTGGAGCCATTCCCACTGT
GH	CGGAGCAGTCAGAGTCTTCTACCT	TTCCACAGTAAAACAGTCATCATCAT
IGF-1	CGCAATGGAACAAAGTCGG	AGGAGATACAGCACATCGCACT
IGF-2	GCAAAGACACGGACCCCACT	CGAGGCCATTTCCACAACG
TOR	GGGTCTTATGAGCCAGTGCCAGG	CTTCAGGGTTGTCAGCGGATTGT
S6	GCACTGTCCCTCGCCGTCTT	CTGGGCTTCTTGCCTTCTTT
4E-BP1	ACACCCCAGCAGGAACTTT	GTGACCATCAACGACGCAG
eIF2α	TGTATTCCAGCACCTCAGCC	CGTGGTCGTCATCCGAGTAGA
ATF4	CTGCGTCACCCCTCAACTCC	CATTCGCTCCATCCACAACC
CHOP	CGGAGTTTCTGGATGTTTTGGA	AGGAGGAGGAAGAGGAGGATGA
REDD1	AGCCAAAGACTCAGAATGCG	TGAAAGGTGGGGACAAGGTA
β-actin	TACGAGCTGCCTGACGGACA	GGCTGTGATCTCCTTCTGC

4.8. Statistical Analysis

The statistical analysis was performed using the software SPSS 19.0 (IBM SPSS Statistics, Version 19.0, Armonk, NY, USA), and the analysis results were presented as mean ± standard error of mean (SEM). Prior to analysis, the homogeneity of the variance of measurement data was confirmed. If the data found did not comply with the parametric assumption of normality and homogeneity of variance, data transformations (such as logarithms, square roots and reciprocals) were applied to meet the ANOVA criteria. Significant differences among treatment were analyzed by one-way analysis of variance

(ANOVA), followed by Tukey's multiple range test. For all cases, the level of significance was determined as $p < 0.05$. The figures in this paper were made using GraphPad Prims 8.0 (GraphPad Software Inc., USA).

5. Conclusions

Over all, the present results suggested that CSM could substitute 20–40% FM without affecting the muscle nutritive deposition. All data supplemented powerful support for our previous conclusion that CSM could successfully replace 20% FM based on growth performance. These two studies presented a comprehensive evaluation of the optimal substitution ratio of CSM to replace FM in the golden pompano. Moreover, the underlying mechanism of over 40% FM replaced by CSM that reduced muscle nutrient deposition was also investigated in this study. Inhibited GH-IGF, TOR and activated AAR suppressed glycolipid metabolism and further decreased nutrient deposition in muscle tissue. However, more in-depth mechanisms should be further elucidated in future studies.

Author Contributions: F.S. and K.M. designed the study. Y.Q., C.H. and H.G. performed the study. P.Y. and W.W. analyzed the data. The first draft of the manuscript was written by Y.Q. and F.S.; F.S. revised the paper. All authors have read and agreed to the published version of the manuscript.

Funding: This research was funded by the Foundation and Applied Basic Research Fund project of Guangdong Province (2019A1515110606), Innovation Group Project of Southern Marine Science and Engineering Guangdong Laboratory (Zhuhai) and the Fund of Southern Marine Science and Engineering Guangdong Laboratory (Zhanjiang) (ZJW-2019-06).

Institutional Review Board Statement: The experimental procedures strictly complied with the regulations of the University Animal Care and Use Committee of the South China Normal University (an approval reference number 1002019-02-0016).

Informed Consent Statement: Not applicable.

Data Availability Statement: The data presented in this study are available within the article.

Conflicts of Interest: The authors declare no conflict of interest.

References

1. Periago, M.J.; Ayala, M.D.; López-Albors, O.; Abdel, I.; Martínez, C.; García-Alcázar, A.; Ros, G.; Gil, F. Muscle cellularity and flesh quality of wild and farmed sea bass, *Dicentrarchus labrax* L. *Aquaculture* **2005**, *249*, 175–188. [CrossRef]
2. Nemova, N.N.; Kantserova, N.P.; Lysenko, L.A. The traits of protein metabolism in the skeletal muscle of teleost fish. *J. Evol. Biochem. Physiol.* **2021**, *57*, 626–645. [CrossRef]
3. Willora, F.P.; Nadanasabesan, N.; Knutsen, H.R.; Liu, C.; Sørensen, M.; Hagen, Ø. Growth performance, fast muscle development and chemical composition of juvenile lumpfish (*Cyclopterus lumpus*) fed diets incorporating soy and pea protein concentrates. *Aquac. Rep.* **2020**, *17*, 100352. [CrossRef]
4. Chen, Y.K.; Chi, S.Y.; Zhang, S.; Dong, X.H.; Yang, Q.H.; Liu, H.Y.; Tan, B.P.; Xie, S.W. Evaluation of Methanotroph (*Methylococcus capsulatus*, Bath) bacteria meal on body composition, lipid metabolism, protein synthesis and muscle metabolites of Pacific white shrimp (*Litopenaeus vannamei*). *Aquaculture* **2022**, *547*, 737517. [CrossRef]
5. Quillet, E.; Guillou, S.L.; Aubin, J.; Labbé, L.; Fauconneau, B.; Médale, F. Response of a lean muscle and a fat muscle rainbow trout (*Oncorhynchus mykiss*) line on growth, nutrient utilization, body composition and carcass traits when fed two different diets. *Aquaculture* **2007**, *269*, 220–231. [CrossRef]
6. Oliveira-Júnior, J.C.D.; Aguiar, G.A.C.C.; Carneiro, C.L.D.S.; Ladeira, A.L.F.; Campelo, D.A.V.; Furuya, W.M.; Santos, F.A.C.D.; Zuanon, J.A.S.; Luz, R.K.; Salaro, A.L. Effects of different ratios of crude protein and non-fibrous carbohydrates on growth, metabolism, physiology, nutrient utilization and muscle cellularity of *Lophiosilurus alexandri*, a carnivorous freshwater fish. *Aquaculture* **2021**, *540*, 736685. [CrossRef]
7. Song, D.Y.; Yun, Y.H.; He, Z.J.; Mi, J.L.; Wang, L.M.; Jin, M.; Zhou, Q.C.; Nie, G.X. Fillet texture, physicochemical indexes, muscle cellularity and molecular expression in muscle of Yellow River carp (*Cyprinus carpio haematopterus*) in response to dietary hydroxyproline supplementation. *Aquaculture* **2022**, *549*, 737783. [CrossRef]
8. Silva, P.; Valente, L.M.P.; Olmedo, M.; Galante, M.H.; Monteiro, R.A.F.; Rocha, E. Hyperplastic and hypertrophic growth of lateral muscle in blackspot seabream *Pagellus bogaraveo* from hatching to juvenile. *J. Fish Biol.* **2009**, *74*, 37–53. [CrossRef]
9. Valente, L.M.P.; Moutou, K.A.; Conceição, L.E.C.; Engrola, S.; Fernandes, J.M.O.; Johnston, I.A. What determines growth potential and juvenile quality of farmed fish species? *Rev. Aquac.* **2013**, *5*, S168–S193. [CrossRef]
10. Koganti, P.; Yao, J.B.; Cleveland, B.M. Molecular mechanisms regulating muscle plasticity in fish. *Animals* **2021**, *11*, 61. [CrossRef]

11. Stokes, T.; Hector, A.J.; Morton, R.W.; McGlory, C.; Phillips, S.M. Recent perspectives regarding the role of dietary protein for the promotion of muscle hypertrophy with resistance exercise training. *Nutrients* **2018**, *10*, 180. [CrossRef] [PubMed]
12. Sliva, V.V.; Salomão, R.A.S.; Mareco, E.A.; Pai, M.D.; Santos, V.B. Probiotic additive affects muscle growth of Nile tilapia (*Oreochromis niloticus*). *Aquac. Res.* **2020**, *52*, 2061–2069. [CrossRef]
13. Montesano, A.; Luzi, L.; Senesi, P.; Mazzocchi, N.; Terruzzi, I. Resveratrol promotes myogenesis and hypertrophy in murine myoblasts. *J. Transl. Med.* **2013**, *11*, 310. [CrossRef]
14. Lyndon, A.R.; Davidson, I.; Houlihan, D.F. Changes in tissue and plasma free amino acid concentrations after feeding in Atlantic cod. *Fish Physiol. Biochem.* **1993**, *5*, 365–375. [CrossRef]
15. Larsen, B.K.; Dalsgaard, J.; Pedersen, P.B. Effects of plant proteins on postprandial, free plasma amino acid concentrations in rainbow trout (*Oncorhynchus mykiss*). *Aquaculture* **2012**, *326–329*, 90–98. [CrossRef]
16. Wang, W.Q.; Xu, Y.J.; Chi, S.Y.; Yang, P.; Mai, K.S.; Song, F. Lysine regulates body growth performance via the nutrient-sensing signaling pathways in largemouth bass (*Micropterus salmoides*). *Front. Mar. Sci.* **2020**, *7*, 595682. [CrossRef]
17. Liang, H.L.; Ren, M.C.; Habte-Tsion, H.M.; Ge, X.P.; Xie, J.; Mi, H.F.; Xi, B.W.; Miao, L.H.; Liu, B.; Zhou, Q.L.; et al. Dietary arginine affects growth performance, plasma amino acid contents and gene expressions of the TOR signaling pathway in juvenile blunt snout bream, *Megalobrama amblycephala*. *Aquaculture* **2016**, *461*, 1–8. [CrossRef]
18. Jiang, H.W.; Bian, F.Y.; Zhou, H.H.; Wang, X.; Wang, K.D.; Mai, K.S.; He, G. Nutrient sensing and metabolic changes after methionine deprivation in primary muscle cells of turbot (*Scophthalmus maximus* L.). *J. Nutr. Biochem.* **2017**, *50*, 74–82. [CrossRef]
19. Rabanal-Ruiz, Y.; Otten, E.G.; Korolchuk, V.I. mTORC1 as the main gateway to autophagy. *Essays Biochem.* **2017**, *61*, 565–584. [CrossRef]
20. Mommsen, T.P. Paradigns of growth in fish. *Comp. Biochem. Physiol.* **2001**, *129*, 207–219. [CrossRef]
21. Chi, S.Y.; He, Y.F.; Zhu, Y.; Tan, B.P.; Dong, X.H.; Yang, Q.H.; Liu, H.Y.; Zhang, S. Dietary methionine affects growth and the expression of key genes involved in hepatic lipogenesis and glucose metabolism in cobia (*Rachycentron canadum*). *Aquac. Nutr.* **2020**, *26*, 123–133. [CrossRef]
22. He, Y.F.; Cui, X.; Chi, S.Y.; Tan, B.P.; Dong, X.H.; Yang, Q.H.; Liu, H.Y.; Zhang, S.; Han, F.L. Changes in the PI3K/Akt/TOR signaling pathway after methionine treatment in the primary muscle cells of cobia (*Rachycentron canadum*). *Aquac. Res.* **2021**, *52*, 2783–2790. [CrossRef]
23. Hevrøy, E.M.; El-Mowafi, A.; Taylor, R.G.; Olsvik, P.A.; Norberg, B.; Espe, M. Lysine intake affects gene expression of anabolic hormones in Atlantic salmon, *Salmo salar*. *Gen. Comp. Endocrinol.* **2007**, *152*, 39–46. [CrossRef] [PubMed]
24. Azizi, S.; Nematollahi, M.A.; Amiri, B.M.; Vélez, E.J.; Lutfi, E.; Navarro, I.; Capilla, E.; Gutiérrez, J. Lysine and leucine deficiencies affect myocytes development and IGF signaling in gilthead sea bream (*Sparus aurata*). *PLoS ONE* **2016**, *11*, e0147618. [CrossRef] [PubMed]
25. Kumar, V.; Lee, S.; Cleveland, B.M.; Romano, N.; Lalgudi, R.S.; Benito, M.R.; McGraw, B.; Hardy, R.W. Comparative evaluation of processed soybean meal (EnzoMealTM) vs. regular soybean meal as a fishmeal replacement in diets of rainbow trout (*Oncorhynchus mykiss*): Effects on growth performance and growth-related genes. *Aquaculture* **2020**, *516*, 734652. [CrossRef]
26. Song, F.; Xu, D.D.; Mai, K.S.; Zhou, H.H.; Xu, W.; He, G. Comparative study on the cellular and systemic nutrient sensing and intermediary metabolism after partial replacement of fishmeal by meat and bone meal in the diet of Turbot (*Scophthalmus maximus* L.). *PLoS ONE* **2016**, *11*, e0165708. [CrossRef]
27. Yuan, X.Y.; Liu, M.Y.; Cheng, H.H.; Huang, Y.Y.; Dai, Y.J.; Liu, W.B.; Jiang, G.Z. Replacing fish meal with cottonseed meal protein hydrolysate affects amino acid metabolism via AMPK/SIRT1 and TOR signaling pathway of *Megalobrama amblycephala*. *Aquaculture* **2019**, *510*, 225–233. [CrossRef]
28. Wang, Q.; He, G.; Mai, K.S.; Xu, W.; Zhou, H.H. Fishmeal replacement by mixed plant proteins and maggot meal on growth performance, target of rapamycin signaling and metabolism in juvenile turbot (*Scophthalmus maximus* L.). *Aquac. Nutr.* **2016**, *22*, 752–758. [CrossRef]
29. Valente, L.M.P.; Cabral, E.M.; Sousa, V.; Cunha, L.M.; Fernandes, J.M.O. Plant protein blends in diets for *Senegalese sole* affect skeletal muscle growth, flesh texture and the expression of related genes. *Aquaculture* **2016**, *453*, 77–85. [CrossRef]
30. Lim, C. Substitution of cottonseed meal for marine animal protein in diets for *Penaeus vannamei*. *J. World Aquac. Soc.* **1996**, *27*, 402–409. [CrossRef]
31. Sun, H.; Tang, J.W.; Yao, X.H.; Wu, Y.F.; Wang, X.; Liu, Y.; Lou, B. Partial substitution of fish meal with fermented cottonseed meal in juvenile black sea bream (*Acanthopagrus schlegelii*) diets. *Aquaculture* **2015**, *466*, 30–36. [CrossRef]
32. Huang, Y.J.; Zhang, N.N.; Fan, W.J.; Cui, Y.Y.; Limbu, S.M.; Qiao, F.; Zhao, Y.L.; Chen, L.Q.; Du, Z.Y.; Li, D.L. Soybean and cottonseed meals are good candidates for fishmeal replacement in the diet of juvenile *Macrobrachium nipponense*. *Aquac. Int.* **2018**, *26*, 309–324. [CrossRef]
33. Xie, S.C.; Zhou, Q.C.; Zhang, X.S.; Zhu, T.T.; Guo, C.; Yang, Z.; Luo, J.X.; Yuan, Y.; Hu, X.Y.; Jiao, L.F.; et al. Effect of dietary replacement of fish meal with low-gossypol cottonseed protein concentrate on growth performance and expressions of genes related to protein metabolism for swimming crab (*Portunus trituberculatus*). *Aquaculture* **2022**, *549*, 737820. [CrossRef]
34. Liu, H.K.; Yan, Q.G.; Han, D.; Jin, J.Y.; Zhu, X.M.; Yang, Y.X.; Xie, S.Q. Effect of dietary inclusion of cottonseed meal on growth performance and physiological and immune responses in juvenile grass carp, *Ctenopharyngodon idellus*. *Aquac. Nutr.* **2019**, *25*, 414–426. [CrossRef]

35. Ye, G.L.; Dong, X.H.; Yang, Q.H.; Chi, S.Y.; Liu, H.Y.; Zhang, H.T.; Tan, B.P.; Zhang, S. Low-gossypol cottonseed protein concentrate used as a replacement of fish meal for juvenile hybrid grouper (*Epinephelus fuscoguttatus*♀ × *Epinephelus lanceolatus*♂): Effects on growth performance, immune responses and intestinal microbiota. *Aquaculture* **2020**, *524*, 735309. [CrossRef]
36. Anderson, A.D.; Alam, M.S.; Watanabe, W.O.; Carroll, P.M.; Wedegaertner, T.C.; Dowd, M.K. Full replacement of menhaden fish meal protein by low-gossypol cottonseed flour protein in the diet of juvenile black sea bass *Centropristis striata*. *Aquaculture* **2016**, *464*, 618–628. [CrossRef]
37. Bu, X.Y.; Chen, A.J.; Lian, X.Q.; Chen, F.Y.; Zhang, Y.; Muhammad, I.; Ge, X.P.; Yang, Y.H. An evaluation of replacing fish meal with cottonseed meal in the diet of juvenile Ussuri catfish *Pseudobagrusus suriensis*: Growth, antioxidant capacity, nonspecific immunity and resistance to *Aeromonas hydrophila*. *Aquaculture* **2017**, *479*, 829–837. [CrossRef]
38. Guo, L.; Yang, J.W.; Liu, B.S.; Zhang, N.; Zhu, K.C.; Guo, H.Y.; Ma, Q.W.; Li, Y.L.; Jiang, S.G.; Zhang, D.C. Colinearity based sex-specific marker development in the golden pompano (*Trachinotus ovatus*). *Aquaculture* **2021**, *544*, 737044. [CrossRef]
39. China Golden Pompano Industry Development Report. 2020. Available online: http://www.china-cfa.org/fzjgdt/2020/1222/511.html (accessed on 22 December 2020).
40. Ning, L.J.; Gao, L.L.; Zhou, W.; Liu, S.; Chen, X.Y.; Pan, Q. Beneficial effects of dietary mulberry leaf along with multi-enzyme premixon the growth, immune response and disease resistance of golden pompano *Trachinotus ovatus*. *Aquaculture* **2021**, *535*, 736396. [CrossRef]
41. Qin, Y.W.; He, C.Q.; Wang, W.Q.; Yang, P.; Wang, J.; Qin, Q.B.; Mai, K.S.; Song, F. Changes in growth performance, nutrient metabolism, antioxidant defense and immune response after fishmeal was replaced by low-gossypol cottonseed meal in golden pompano (*Trachinotus ovatus*). *Front. Mar. Sci.* **2021**, *8*, 775575. [CrossRef]
42. Zhou, F.; Song, W.X.; Shao, Q.J.; Peng, X.; Xiao, J.X.; Hua, Y.; Owari, B.N.; Zhang, T.Z.; Ng, W.-K. Partial replacement of fish meal by fermented soybean meal in diets for black sea bream, *Acanthopagrus schlegelii*, juveniles. *J. World Aquac. Soc.* **2011**, *42*, 184–197. [CrossRef]
43. Ma, X.Z.; Feng, L.; Wu, P.; Liu, Y.; Kuang, S.Y.; Tang, L.; Zhou, X.Q.; Jiang, W.D. Enhancement of flavor and healthcare substances, mouthfeel parameters and collagen synthesis in the muscle of on-growing grass carp (*Ctenopharyngodon idella*) fed with graded levels of glutamine. *Aquaculture* **2020**, *528*, 735486. [CrossRef]
44. Zhao, Y.; Li, J.Y.; Yin, L.; Feng, L.; Liu, Y.; Jiang, W.D.; Wu, P.; Zhao, J.; Chen, D.F.; Zhou, X.Q.; et al. Effects of dietary glutamate supplementation on flesh quality, antioxidant defense and gene expression related to lipid metabolism and myogenic regulation in Jian carp (*Cyprinus carpio* var. Jian). *Aquaculture* **2019**, *502*, 212–222. [CrossRef]
45. Liang, X.F.; Hu, L.; Dong, Y.C.; Wu, X.F.; Qin, Y.C.; Zheng, Y.H.; Shi, D.D.; Xue, M.; Liang, X.F. Substitution of fish meal by fermented soybean meal affects the growth performance and flesh quality of Japanese seabass (*Lateolabrax japonicus*). *Anim. Feed Sci. Technol.* **2017**, *229*, 1–12. [CrossRef]
46. Bu, X.Y.; Wang, Y.Y.; Chen, F.Y.; Tang, B.B.; Luo, C.Z.; Wang, Y.; Ge, X.P.; Yang, Y.H. An evaluation of replacing fishmeal with rapeseed meal in the diet of *Pseudobagrusus suriensis*: Growth, feed utilization, nonspecific immunity, and growth-related gene expression. *J. World Aquac. Soc.* **2018**, *49*, 1068–1080. [CrossRef]
47. Alami-Durante, H.; Wrutniak-Cabello, C.; Kaushik, S.J.; Médale, F. Skeletal muscle cellularity and expression of myogenic regulatory factors and myosin heavy chains in rainbow trout (*Oncorhynchus mykiss*): Effects of changes in dietary plant protein sources and amino acid profiles. *Comp. Biochem. Physiol. Part A Mol. Integr. Physiol.* **2010**, *156*, 561–568. [CrossRef]
48. Xu, D.D.; He, G.; Mai, K.S.; Zhou, H.H.; Xu, W.; Song, F. Postprandial nutrient-sensing and metabolic responses after partial dietary fishmeal replacement by soyabean meal in turbot (*Scophthalmus maximus* L.). *Br. J. Nutr.* **2016**, *115*, 379–388. [CrossRef]
49. Yang, P.; He, C.Q.; Qin, Y.W.; Wang, W.Q.; Mai, K.S.; Qin, Q.B.; Wei, Z.H.; Song, F. Evaluation of composite mixture of protein sources in replacing fishmeal for Pacific white shrimp (*Litopenaeus vannamei*): Based on the changing pattern of growth performance, nutrient metabolism and health status. *Aquac. Rep.* **2021**, *21*, 100914. [CrossRef]
50. Villanueva-Gutiérrez, E.; González-Félix, M.L.; Gatlin, D.M.; Perez-Velazquez, M. Use of alternative plant and animal protein blends, in place of fishmeal, in diets for juvenile totoaba, *Totoaba macdonaldi*. *Aquaculture* **2020**, *529*, 735698. [CrossRef]
51. Wang, L.; Yin, N.; Sagada, G.; Hua, Y.; Li, H.; Zhang, J.Z.; Shao, Q.J. Partial replacement of fishmeal with corn gluten meal, pea protein isolate and their mixture in diet of black sea bream (*Acanthopagrus schlegelii*) juveniles: Effects on growth performance, feed utilization and haematological parameters. *Aquac. Res.* **2020**, *51*, 2071–2083. [CrossRef]
52. Zhou, J.S.; Liu, S.S.; Ji, H.; Yu, H.B. Effect of replacing dietary fish meal with black soldier fly larvae meal on growth and fatty acid composition of Jian carp (*Cyprinus carpio var.* Jian). *Aquac. Nutr.* **2018**, *24*, 424–433. [CrossRef]
53. Wang, W.Q.; Yang, P.; He, C.Q.; Chi, S.Y.; Li, S.L.; Mai, K.S.; Song, F. Effects of dietary methionine on growth performance and metabolism through modulating nutrient-related pathways in largemouth bass (*Micropterus salmoides*). *Aquac. Rep.* **2021**, *20*, 100642. [CrossRef]
54. Xue, S.Q.; Chen, S.M.; Ge, Y.X.; Guan, T.; Han, Y. Regulation of glutathione on growth performance, biochemical parameters, non-specific immunity, and related genes of common carp (*Cyprinus carpio*) exposed to ammonia. *Aquaculture* **2022**, *546*, 737241. [CrossRef]
55. Venugopal, V.; Shahidi, F. Structure and composition of fish muscle. *Food Rev. Int.* **1996**, *12*, 175–197. [CrossRef]
56. Knutsen, H.R.; Ottesen, O.H.; Palihawadana, A.M.; Sandaa, W.; Sørensen, M.; Hagen, Ø. Muscle growth and changes in chemical composition of spotted wolffish juveniles (*Anarhichas minor*) fed diets with and without microalgae (*Scenedesmus obliquus*). *Aquac. Res.* **2019**, *13*, 100175. [CrossRef]

57. Herath, S.S.; Haga, Y.; Satoh, S. Effects of long-term feeding of corn co-product-based diets on growth, fillet color, and fatty acid and amino acid composition of Nile tilapia, *Oreochromis niloticus*. *Aquaculture* **2016**, *464*, 205–212. [CrossRef]
58. Lee, S.; Masagounder, K.; Hardy, R.W.; Small, B.C. Effects of lowering dietary fishmeal and crude protein levels on growth performance, body composition, muscle metabolic gene expression, and chronic stress response of rainbow trout (*Oncorhynchus mykiss*). *Aquaculture* **2019**, *513*, 7374435. [CrossRef]
59. Decken, V.D.; Lied, E. Metabolic effects on growth and muscle of soya-bean protein feeding in cod (*Gadus morhua*). *Br. J. Nutr.* **1993**, *69*, 689–697. [CrossRef]
60. Li, J.; Liang, X.F.; Alam, M.S.; Luo, H.C.; Zhang, Y.P.; Peng, B.B.; Xiao, Q.Q.; Zhang, Z.L.; Liu, L.W.; He, S. Adaptation of AMPK-mTOR-signal pathways and lipid metabolism in response to low- and high-level rapeseed meal diet in Chinese perch (*Siniperca chuatsi*). *J. Comp. Physiol. B* **2021**, *191*, 881–894. [CrossRef]
61. Yang, P.; Wang, W.Q.; Chi, S.Y.; Mai, K.S.; Song, F.; Wang, L. Effects of dietary lysine on regulating GH-IGF system, intermediate metabolism and immune response in largemouth bass (*Micropterussalmoides*). *Aquac. Rep.* **2020**, *17*, 100323. [CrossRef]
62. Bellaloui, N.; Turley, R.B. Effects of fuzzless cottonseed pheno-type on cottonseed nutrient composition in near isogenic cotton (*Gossypium hirsutum* L.) mutant lines under well-watered and water stress conditions. *Front. Plant. Sci.* **2013**, *4*, 516–529. [CrossRef] [PubMed]
63. Kuo, Y.H.; Lin, C.H.; Shih, C.C. Ergostatrien-3β-ol from *Antrodia camphorata* inhibits diabetes and hyperlipidemia in high-fat-diet treated mice via regulation of hepatic related genes, glucose transporter 4, and AMP-activated protein kinase phosphorylation. *J. Agric. Food Chem.* **2015**, *63*, 2479–2489. [CrossRef]
64. Xu, C.; Liu, W.B.; Zhang, D.D.; Cao, X.F.; Shi, H.J.; Li, X.F. Interactions between dietary carbohydrate and metformin: Implications on energy sensing, insulin signaling pathway, glycolipid metabolism and glucose tolerance in blunt snout bream *Megalobrama amblycephala*. *Aquaculture* **2018**, *483*, 183–195. [CrossRef]
65. Gu, Z.X.; Mu, H.; Shen, H.H.; Deng, K.Y.; Liu, D.; Yang, M.X.; Zhang, Y.; Zhang, W.B.; Mai, K.S. High level of dietary soybean oil affects the glucose and lipid metabolism in large yellow croaker *Larimichthys crocea* through the insulin-mediated PI3K/AKT signaling pathway. *Comp. Biochem. Physiol. B* **2019**, *231*, 34–41. [CrossRef] [PubMed]
66. Zoncu, R.; Bar-Peled, L.; Efeyan, A.; Wang, S.Y.; Sancak, Y.; Sabatini, D.M. mTORC1 senses lysosomal amino acids through an inside-out mechanism that requires the Vacuolar H^+-ATPase. *Science* **2011**, *334*, 678–683. [CrossRef] [PubMed]
67. Zhou, Q.L.; Habte-Tsion, H.M.; Ge, X.; Xie, J.; Ren, M.; Liu, B.; Miao, L.; Pan, L. Graded replacing fishmeal with canola meal in diets affects growth and target of rapamycin pathway gene expression of juvenile blunt snout bream, *Megalobrama amblycephala*. *Aquac. Nutr.* **2017**, *24*, 300–309. [CrossRef]
68. Song, F.; Xu, D.D.; Zhou, H.H.; Xu, W.; Mai, K.S.; He, G. The differences in postprandial free amino acid concentrations and the gene expression of PepT1 and amino acid transporters after fishmeal partial replacement by meat and bone meal in juvenile turbot (*Scophthalmus maximus* L.). *Aquac. Res.* **2017**, *48*, 3766–3781. [CrossRef]
69. Poncet, N.; Taylor, P.M. The role of amino acid transporters in nutrition. *Curr. Opin. Clin. Nutr. Metab. Care* **2013**, *16*, 57–65. [CrossRef]
70. Yang, X.Y.; Wang, G.H.; Zhao, X.M.; Dong, X.H.; Chi, S.Y.; Tan, B.P. Addition of hydrolysed porcine mucosa to low-fishmeal feed improves intestinal morphology and the expressions of intestinal amino acids and small peptide transporters in hybrid groupers (*Epinephelus fuscoguttatus*♀ × *E. lanceolatus*♂). *Aquaculture* **2021**, *535*, 736389. [CrossRef]
71. Tremblay, F.; Marette, A. Amino acid and insulin signaling via the mTOR/p70 S6 kinase pathway. *J. Biol. Chem.* **2001**, *276*, 38052–38060. [CrossRef]
72. Kim, J.; Guan, K.-L. Amino acid signaling in TOR activation. *Annu. Rev. Biochem.* **2011**, *80*, 1001–1032. [CrossRef] [PubMed]
73. Jansson, T.; Aye, I.L.M.H.; Goberdhan, D.C.I. The emerging role of mTORC1 signaling in placental nutrient-sensing. *Placenta* **2012**, *33*, e23–e29. [CrossRef] [PubMed]
74. Shehab, M.A.; Biggar, K.; Kakadia, J.H.; Dhruv, M.; Jain, B.; Nandi, P.; Nygard, K.; Jansson, T.; Gupta, M.P. Inhibition of decidual IGF-1 signaling in response to hypoxia and leucine deprivation is mediated by mTOR and AAR pathways and increased IGFBP-1 phosphorylation. *Mol. Cell. Endocrinol.* **2020**, *512*, 110865. [CrossRef] [PubMed]
75. Bower, N.I.; Johnston, I.A. Transcriptional regulation of the IGF signaling pathway by amino acids and insulin-like growth factors during myogenesis in Atlantic salmon. *PLoS ONE* **2010**, *5*, e11100. [CrossRef]
76. Asaduzzaman, M.; Sofia, E.; Shakil, A.; Haque, N.F.; Khan, M.N.A.; Ikeda, D.; Kinoshita, S.; Abol-Munafi, A.B. Host gut-derived probiotic bacteria promote hypertrophic muscle progression and upregulate growth-related gene expression of slow-growing Malaysian Mahseer *Tor tambroides*. *Aquac. Rep.* **2018**, *9*, 37–45. [CrossRef]
77. Park, S.-J.; Moon, J.-S.; Seo, J.-S.; Nam, T.-J.; Lee, K.-J.; Lim, S.-G.; Kim, K.-W.; Lee, B.-J.; Hur, S.-W.; Choi, Y.H. Effect of fish meal replacement on insulin-like growth factor-I expression in the liver and muscle and implications for the growth of Olive Flounder *Paralichthys olivaceus*. *Korean J. Fish. Aquat. Sci.* **2019**, *52*, 141–148.
78. Li, S.L.; Dai, M.; Qiu, H.J.; Chen, N.S. Effects of fishmeal replacement with composite mixture of shrimp hydrolysate and plant proteins on growth performance, feed utilization, and target of rapamycin pathway in largemouth bass, *Micropterus salmoides*. *Aquaculture* **2021**, *533*, 736185. [CrossRef]

metabolites

Article

Effects of Dietary Cottonseed Protein Concentrate Levels on Growth Performance, Health Status, Flesh Quality and Intestinal Microbiota of Grass Carp (*Ctenopharyngodon idellus*)

Guoqing Liu [1], Meng Zhou [1], Xiaoyu Wang [1], Xiangjie Mao [1], Xianmei Long [1], Shouqi Xie [2], Dong Han [2] and Qingsong Tan [1,*]

1 Engineering Research Center of Green Development for Conventional Aquatic Biological Industry in the Yangtze River Economic Belt, Ministry of Education/Key Laboratory of Freshwater Animal Breeding, Ministry of Agriculture/Hubei Provincial Engineering Laboratory for Pond Aquaculture, College of Fisheries, Huazhong Agricultural University, Wuhan 430070, China

2 State Key Laboratory of Freshwater Ecology and Biotechnology, Institute of Hydrobiology, Chinese Academy of Sciences, Wuhan 430072, China

* Correspondence: qstan@hotmail.com; Tel.: +86-27-8728-2113

Abstract: The aim of this study was to evaluate the nutritional value of cottonseed protein concentrate (CPC) as a single dietary protein source and the optimal protein level for grass carp (*Ctenopharyngodon idellus*). An 8-week feeding trial was conducted by feeding juvenile grass carp (initial body weight: 4.68 ± 0.01 g) with six experimental diets containing graded levels of protein provided by CPC. The results showed that the optimal CPC level (CPC4) improved the growth performance and health status of grass carp. The optimal dietary protein level was estimated to be 38.61 and 38.66% based on specific growth rate (SGR) and feed efficiency (FE), respectively. The CPC4 group significantly increased the total antioxidant capacity (T-AOC) content and glutathione peroxidase (GSH-Px) activity in the hepatopancreas ($p < 0.05$). In addition, the CPC4 group increased the muscle T-AOC and glutathione (GSH) content and improved muscle hardness, and the gene expression of MRFs, *fgf6a*, *myhc-7*, *myhc-1*, *myhc-4*, *igf-II*, and *tor* was upregulated while *mstn* gene expression was downregulated ($p < 0.05$). Correlation analysis revealed that the optimal dietary CPC level promoted grass carp growth, health, and flesh quality by regulating the relative abundance of intestinal microbes. Furthermore, CPC6 upregulated the ko00480 (Glutathione metabolism) and ko00620 (Pyruvate metabolism) pathways compared to CPC1 ($p < 0.05$), possibly indicating that low dietary CPC levels adversely affected amino acid metabolism in the intestinal microbiota of grass carp, while a high level of CPC will meet the metabolic needs of the body by increasing the utilization of energy.

Keywords: cottonseed protein concentrate; protein requirement; muscle texture; antioxidant capacity; metabolic function

Citation: Liu, G.; Zhou, M.; Wang, X.; Mao, X.; Long, X.; Xie, S.; Han, D.; Tan, Q. Effects of Dietary Cottonseed Protein Concentrate Levels on Growth Performance, Health Status, Flesh Quality and Intestinal Microbiota of Grass Carp (*Ctenopharyngodon idellus*). *Metabolites* **2022**, *12*, 1046. https://doi.org/10.3390/metabo12111046

Academic Editor: Baichuan Deng

Received: 26 September 2022
Accepted: 27 October 2022
Published: 30 October 2022

1. Introduction

Protein is the most expensive ingredient in aquafeed and profoundly impacts the growth and health of aquatic animals. However, the shortage of traditional protein sources, such as fishmeal and soybean meal, hinders the development of the global feed industry [1]. Cottonseed protein plays an essential role in aquaculture as a plant protein source [2]. However, cottonseed contains gossypol and other anti-nutritional factors, limiting its inclusion level in a diet [2]. Through a series of processing, including dehairing, dehulling, low-temperature oil extraction, and solvent extraction, the cottonseed is processed into cottonseed protein concentrate (CPC), which significantly reduce the anti-nutritional factors [3]. Currently, researchers are extensively investigating the potential of CPC as alternative materials in diets [3–5], aiming to alleviate protein shortages. CPC can replace 40–45% of fishmeal in largemouth bass (*Micropterus salmoides*) diets without affecting largemouth bass growth performance [3,4]. Likewise, CPC can replace 24% of the fishmeal in pearl gentian grouper (*Epinephelus fuscoguttatus*♀× *Epinephelus lanceolatu*♂) and promote growth performance [5]. However, nutritional trials using CPC as the sole protein source are uncommon, which is desirable to evaluate the biological value of CPC.

Grass carp (*Ctenopharyngodon idellus*) is herbivorous and feeds on certain water plants in its natural environment [6]. The grass carp culture shares a long history and the largest freshwater aquaculture production in China due to the euryphagic feeding habit and delicious meat of grass carp, which is one of the "four major Chinese carps". Grass carp has been introduced into more than 100 countries around the world for food, weed control, or research [7]. According to FAO statistics, the global aquaculture production in 2020 was 87.5 million tons, among which grass carp production was 5.79 million tons (the top three countries in production are China, Bangladesh, and Iran), accounting for about 6.62% [8]. Nowadays, grass carp culture relies mainly on feeding with compound feed to meet the market demand. Previous studies have determined protein requirements for grass carp based primarily on semi-purified feeds using fishmeal or casein, which indicated that the optimal protein level vary according to the protein materials and fish growth stages [9–12]. Moreover, precise nutrition for aquatic animals requires accurate knowledge regarding the protein sources and their optimal inclusion levels, which has progressively attracted attention from scholars. Therefore, evaluating the optimal dietary protein level in grass carp diets based on CPC as a protein source must be emphasized under the circumstance of precision nutrition.

Flesh quality has always been the most valued part by consumers. The indicators for evaluating flesh quality include cooking loss, texture properties (including hardness, cohesiveness, gumminess, springiness, resilience, and chewiness), pH, and antioxidant capacity [13]. Muscle growth is a dynamic process, including muscle hyperplasia and hypertrophy [14], which is affected by multiple regulatory factors such as muscle regulatory factors (MRFs), insulin-like growth factors (IGFs), fibroblast growth factor 6 (*fgf6*), myostatin (*mstn*), and target of rapamycin (*tor*) [15,16]. There have been many achievements in regulating flesh quality by dietary nutrients in mammals, but the regulation of flesh quality in aquatic animals still needs to be investigated in depth. Existing studies have shown that amino acid supplementation at an appropriate level can improve the flesh quality of grass carp [13] and hybrid bagrid catfish (*Pelteobagrus vachelli*♀× *Leiocassis longirostris*♂) [15]. Grass carp that ate fava beans gained better flesh quality, mainly due to increased muscle hardness [17,18]. Furthermore, a previous study showed that replacing rapeseed meal and cottonseed meal with DDGS induced the myosin isoforms transformation of grass carp, altering the muscle texture properties, histological muscle properties, and gene expression of MRFs [16,19]. A recent study showed that feeding grass carp (6.80 $\pm$ 0.10 g) with graded protein level (soybean meal) diets improved flesh quality at the appropriate protein level [20]. Likewise, in a study of large grass carp (264.11 $\pm$ 0.76 g), optimal protein levels (fish meal, casein and gelatin) also improved muscle texture and antioxidant capacity [12]. Therefore, the effect of varying dietary protein levels from CPC on the muscle histology, myosin heavy chain (*myhc*) gene expression levels, and muscle texture properties deserves investigation, for a more comprehensive nutritional value evaluation of CPC.

The homeostasis of intestinal microbes regulated by nutrients has crucial guiding significance for human health [21]. Rapid, low-cost, and precise DNA sequencing methods, such as those offered by Illumina (San Diego, CA, USA), are becoming increasingly popular and extensively used to examine intestinal microbial composition [21]. Recently, the importance of intestinal microbes in aquatic animals has been paid more attention to by researchers [4,5]. However, previous studies focused on the diversity in intestinal microbial composition after fishmeal replacement or exposure to specific factors [3,22]. Research on the effect of protein level on fish intestinal microbiota has only been reported for Songpu mirror carp (*Cyprinus carpio*) [23]. In addition, the mammalian gut–muscle axis also have received attention [24], but these studies are still in their infancy in aquatic animals. In this regard, a crosstalk analysis between the homeostasis of intestinal microbes and the growth and health status will beneficial, to illustrate the regulative mechanism of the CPC protein level.

To sum up, given the importance of protein resources, the urgent need of farmers for the rapid and healthy growth of fish, and the pursuit of superior flesh quality by consumers, it is necessary to evaluate the nutritional value of CPC as a single protein source and to quantify the optimal protein level for grass carp. Using CPC as a protein source, six treatments were designed containing a gradient of protein levels, subsequently conducting an 8-week feeding trial to investigate the effects of these protein levels on the growth, health, flesh quality, and intestinal microbiota of grass carp.

2. Materials and Methods

2.1. Experimental Diets

Dietary treatments, ingredients, and chemical composition are shown in Table 1. CPC was used as a protein source, starch was used as a carbohydrate source, and soybean oil and fish oil were used as lipid sources to prepare the six experimental diets with different protein contents. The protein contents of the six experimental diets, named CPC1, CPC2, CPC3, CPC4, CPC5, and CPC6, were 24.80%, 30.51%, 33.68%, 37.69%, 41.43%, and 45.61% (expressed in dry matter), respectively; also, 0.1% yttrium oxide (Y_2O_3) was added to the CPC experimental diets for the determination of apparent digestibility coefficients (ADCs) [16]. The CPC was provided by Xinjiang Jinlan Plant Protein Co., Ltd., Shihezi, Xinjiang, China, which contained 61.51% crude protein, 2.36% crude lipid, 5.35% moisture, 5.70% ash, 4.82% crude fiber, 19.55 MJ/kg gross energy, 0.05% tannin, 1.29 mg/kg sterculic acid, and 285 mg/kg free gossypol. According to the instructions provided by the manufacturer, the CPC is made as follows: Briefly, the regular glanded cottonseeds were first crushed and sieved to remove the hulls; then, the kernels were softened under a certain moisture content and rolled into 0.4 mm-thick flakes and dried. The softening, rolling, and drying were processed at a low temperature (60–70 °C). The dried flakes were extracted with mixed solvents of methanol and *n*-hexane in different concentrations for oil extraction and dephenolization, and then subjected to rapid low-temperature drying (not higher than 90 °C) to obtain the final product. The raw materials in the diets were ground and screened through an 80-mesh sieve, then accurately weighed, and finally mixed uniformly using a M-256 mixer (South China University of Technology, Guangzhou. Pellets (diameter: 2.0 mm) were extruded using a laboratory feed extruder (feed moisture: 21%, expansion temperature: 130 °C), air dried, and then stored at −20 °C.

2.2. Fish and Feeding Trial

Juvenile grass carp were purchased from Xinzhou Fisheries Co. Ltd. (Wuhan, China). Prior to the growth trial, the fish were acclimated to the culture conditions and were fed with an even mixture of the six diets. After 2 weeks of acclimation, 540 fish were divided into 18 tanks in a flow-through culture system at random, with 30 fish (initial body weight, 4.68 ± 0.01 g) in each tank (water volume: 300 L, water exchange: 1 L/min). Each diet was delivered randomly to three tanks. The growth trial lasted for 8 weeks. During the growth trial, fish were hand-fed twice a day (08:30 and 15:30). The feed consumed in each tank was recorded daily. Uneaten feed was collected 1 h after feeding and dried in an oven at 60 °C to calibrate grass carp food intake. Throughout the experiment, an air pump was used to ensure that the dissolved oxygen in the water tank was above 6 mg/L; furthermore, the water temperature was maintained at 27 to 28 °C, pH was 7.3 to 7.6, and the natural photoperiod was applied.

Table 1. Ingredients and proximate composition of the experimental diets with different protein levels (expressed as % dry matter).

	CPC1	CPC2	CPC3	CPC4	CPC5	CPC6
Cottonseed protein concentrate (CPC) [1]	38.40	44.50	50.70	56.90	63.10	69.2
Soy oil:Fish oil (1:1)	5.90	5.76	5.62	5.47	5.32	5.14
Corn starch	18.00	16.80	15.60	14.40	13.20	11.98
Microcrystalline cellulose	34.30	29.54	24.68	19.83	14.98	10.29
Vitamin and mineral premix [2]	1.00	1.00	1.00	1.00	1.00	1.00
Calcium biphosphate	1.80	1.80	1.80	1.80	1.80	1.80
Choline chloride (50%)	0.50	0.50	0.50	0.50	0.50	0.50
Yttrium oxide	0.10	0.10	0.10	0.10	0.10	0.10
Proximate composition (%)						
Moisture	6.55	7.14	6.67	7.42	7.42	7.88
Crude protein (dry matter)	24.80	30.51	33.68	37.69	41.43	45.61
Crude lipid (dry matter)	7.92	8.04	7.91	7.95	7.97	8.02
Ash (dry matter)	5.70	6.52	6.67	7.15	7.63	8.29

[1] The CPC was provided by Xinjiang Jinlan Plant Protein Co., Ltd., Shihezi, Xinjiang, China, which contained 61.51% crude protein, 2.36% crude lipid, 5.35% moisture, 5.70% ash, 4.82% crude fiber, 19.55 MJ/kg gross energy, and 285 mg/kg of free gossypol. The essential amino acid contents in the CPC were (g/kg) Lysine, 22.6; Methionine, 9.7; Arginine, 68.6; Histidine, 16.4; Leucine, 30.6; Isoleucine, 16.4; Valine, 23; Tryptophan, 8.1; Phenylalanine, 29.8; and Threonine, 15.9. [2] Each kilogram of the vitamin and mineral premix contained: L-ascorbate-2-monophosphate (35%), 900 mg; vitamin E, 450 mg; inositol, 225 mg; nicotinamide, 120 mg; calcium pantothenate, 60 mg; vitamin A, 30 mg; vitamin K3, 30 mg; vitamin B2, 22.5 mg; vitamin B6, 22.5 mg; vitamin D3, 15 mg; vitamin B1, 15 mg; folic acid, 15 mg; vitamin B12, 120μg; biotin, 3 mg; ferrous sulfate monohydrate, 300 mg; zinc sulfate/sulphate monohydrate, 200 mg; Sodium chloride, 100 mg; manganese sulphate, 25 mg; copper (II) sulfate pentahydrate, 30 mg; cobaltous chloride (10% Co), 5 mg; sodium selenite (10% Se), 5 mg; potassium iodate (2.9%), 3 mg; and magnesium sulphate, 900 mg.

2.3. Sample Collection

At the conclusion of the 8-week growth trial, fish were deprived of food for 24 h. All fish in each tank were counted and collectively weighed. Five fish, randomly selected from each tank, and anesthetized with diluted MS222 (75 mg/L, Aladdin, Shanghai, China), were weighed and their body length measured individually. Then, blood samples were collected from the caudal vein, precipitated at room temperature for 30 min, and serum samples were separated by centrifugation at 4 °C for 10 min at 3000× *g* and stored at −80 °C for analysis. Afterward, the viscera were removed and weighed; then, the hepatopancreas and mesenteric fat were further dissected and weighed. Tissue samples (hepatopancreas and white muscle under the dorsal fin) were collected, fast-frozen in liquid nitrogen, and stored at −80 °C for RNA extraction and enzyme activity evaluation. The intestinal contents of the CPC1, CPC4, and CPC6 groups of fish were collected (in order to explore the effects of low dietary protein level, optimal dietary protein level, and high dietary protein level on the intestinal flora of grass carp), with three biological replicates in each group, and the intestinal contents of six grass carp were collected and mixed as a sample for each biological replicate. The samples were fast-frozen in liquid nitrogen and stored at −80 °C for analysis of intestinal microbial diversity. Furthermore, the hepatopancreas and muscle were sampled and fixed in 4% paraformaldehyde and stored in 70% ethanol for histological observations. Three fish were randomly selected from each tank as samples for muscle texture analysis. In the last 4 weeks of the growth trial, feces were siphoned from each tank twice daily at 2 h after feeding, oven-dried at 60 °C, and then stored at −20 °C for subsequent nutrient apparent digestibility determination. Furthermore, five fish were randomly selected from each tank and stored at −20 °C for proximate analysis of the body. All operating procedures were carried out on ice.

2.4. Sample Analysis

2.4.1. Determination of Serum Biochemical Indices

Serum total protein (TP), glucose (GLU), urea nitrogen (UN), high-density lipoprotein (HDL), low-density lipoprotein (LDL), triglycerides (TG), and cholesterol (CHOL) were detected with an automated biochemical analyzer (Abbott Aeroset®, Abbott Laboratories, Chicago, IL, USA) using commercial test kits. Serum activities of lysozyme (LYS), complement 3 (C3), and the content of immunoglobulin M (IgM) were determined by commercially available kits (the product numbers are ml036413, ml092636, and ml092683, respectively) purchased from Shanghai Enzyme-linked Biotechnology Co. Ltd. (Shanghai, China), according to the manufacturer's instructions.

2.4.2. Proximate Composition

Proximate analysis was conducted for the experimental feed, feces, dorsal muscle, and whole fish according to the standard procedures of AOAC [25]. Briefly, samples were dried to a constant weight at 105 °C to estimate the moisture content. Crude protein was measured using the Kjeldahl method by determining nitrogen ($n \times 6.25$) after acid digestion. Crude lipid was determined by ether extraction using the Soxhlet method. Ash was measured in a muffle furnace at 550 °C for 5 h.

2.4.3. Enzyme Activity of Muscle and Hepatopancreas

The enzyme activities of alanine aminotransferase (ALT), aspartate aminotransferase (AST), superoxide dismutase (SOD), catalase (CAT), total antioxidant capacity (T-AOC), and glutathione peroxidase (GSH-Px), as well as the reduced glutathione (GSH) and malondialdehyde (MDA) contents in the muscle and hepatopancreas were determined using commercial kits (Jian Cheng Bioengineering Institute, Nanjing, China).

2.4.4. Apparent Digestibility

According to a previous study [16], the experimental diets and feces were digested with nitric acid, and the Yttrium content was detected using an IRIS Advantage inductively coupled plasma (ICP) atomic emission spectrophotometer (Thermo Jarrell Ash Corporation, Boston, MA, USA) for the calculation of the apparent digestibility coefficients.

2.4.5. Histological Observation

Paraformaldehyde-fixed muscle and hepatopancreas tissues were dehydrated with gradient grades of ethanol, followed by embedding with paraffin. Thick sections (7 μm) were prepared and stained with hematoxylin–eosin (HE). The stained samples were observed under a light microscope. The number of muscle fibers and diameter were measured using an M-Shot image analysis system (Micro-Shot, Guangzhou, China).

2.4.6. Muscle Textural Properties

Cooking loss of muscle was determined based on a previous study [16]. In short, 2–3 g of muscle was weighed, wrapped in gauze, and then cooked in boiling water (100 °C) for 5 min; then, the muscle was taken out and weighed after removing the surface water with an absorbent paper. Cooking loss was reported in terms of weight loss during heat processing and expressed as % of the initial sample weight. Texture profile analysis (TPA) of the muscle was determined using a TA.XT Plus texture analyzer (Stable Micro Systems, Godalming, UK) equipped with a flat-bottomed cylindrical probe $p/36$ R (20 mm diameter), as previously reported by Kong et al. [19]. The measurements were performed on three fish from each tank.

2.4.7. Gene Expression Quantification

Total RNA was extracted using Trizol™ reagent (Takara, Dalian, China). The RNA integrity and quantity were assessed by Evolution-Capt image analysis software after 1% agarose gel electrophoresis and a 260/280 nm absorbance ratio (NanoDrop® ND-1000, Thermo Fisher Scientific, Waltham, MA, USA). High-quality RNA with a 28 S/18 S ratio $\geq$2 and a 260/280 ratio of 1.9–2.1 was used for further quantification. Subsequently, cDNA was synthesized using a PrimeScript RT reagent kit with a gDNA eraser (Yeasen, Shanghai, China).

Primers for quantitative real-time PCR (qRT-PCR) were designed according to the sequences existing in National Center for Biotechnology Information Search database (Table 2). The qRT-PCR reaction was performed using a Hieff® qPCR SYBR® Green Master Mix (No Rox) (Yeasen, Shanghai, China) on a quantitative thermal cycler (Light Cycler 480 II, Roche). The qRT-PCR reaction procedures were as follows: pre-incubation at 95 °C for 5 min, followed by 40 cycles of 95 °C for 10 s, and annealing temperature (corresponding specific primer pairs) for 20 s and 72 °C for 20 s. Melting curve analysis was performed to ensure that only one fragment was amplified. The relative expression levels of the target genes were calculated by normalization via the $2^{-\Delta\Delta CT}$ method, as described by Pfaffl [26], and using *β-actin* and *ef1α* as the reference genes.

Table 2. Real-time PCR primer sequences.

Genes [1]	Forward Primer (5′-3′)	Reverse Primer (5′-3′)	Amplification Efficiency (%)	Accession Number
myod	ATGGAGTTGTCGGATATTCCCTTC	GCGGTCAGCGTTGGTTGTT	104.45	MG544985
myog	TTACGAAGGCGGCGATAACTT	TGGTGAGGAGACATGGACAGA	101.18	JQ793897
myf5	GTGCCTGTGCCTCATCTCCT	AATGCGTGGTTCACCTTCTTCA	92.41	GU290227
mrf4	TCGCTCCTGTATTGATGTTGATGA	GCTCCTGTCTCGCATTCGTT	107.98	KT899334
fgf6a	CGCATACGAGTCTTCCAT	CCTACGAGAACATCCAACA	102.95	MK050993
fgf6b	TCCAGTCCGCTTCCGAGTA	AGATGAAACCCGATGCCTACA	91.14	MK050992
mstn	CTGACGCCAAGTTCCACATACA	CGACTCTGCTTCAAGTTCTTCTCT	99.15	KP719016
myhc-7	AACTGCGCTGTAACGGTGTA	AGTGTGCCCAAACCTGTACT	101.85	MW113233
myhc-2	ACAGTGGCCAGCATTGATGA	TCCGCAGAGTTCAAACCCAA	101.15	MW113235
myhc-4	ACTCCGCTGACATGCTGAAA	TGTCCAGCACACCAATGAAGA	103.78	MW113236
myhc-1	TTCCGTTGTTGTGTCAGGCT	TACTGGATGACGCGTTTGGT	99.12	MW113234
igf-II	TCTGTGGCAGTCCTCAACAAC	TTCCGCAACTTCTTCGCTCTT	97.78	EF062860
tor	TCCCACTTTCCACCAACT	ACACCTCCACCTTCTCCA	105.68	JX854449
s6k1	ACATAAAGCAGCCTGACG	TGGAGGAGGTAATGGACG	101.51	EF373673
4e-bp1	GCTGGCTGAGTTTGTGGTTG	CGAGTCGTGCTAAAAAGGGTC	99.59	KT757305
β-actin	TATGTTGGTGACGAGGCTCA	GCAGCTCGTTGTAGAAGGTG	98.82	M25013
ef1α	TGACTGTGCCGTGCTGAT	CGCTGACTTCCTTGGTGATT	99.51	GQ266394

[1] *myod*—myogenic differentiation antigen; *myog*—myogenin; *myf5*—myogenic factor 5; *mrf4*—muscle regulatory factor 4; *fgf6a*—fibroblast growth factor 6 a; *fgf6b*—fibroblast growth factor 6 b; *mstn*—myostatin; *myhc-7*—myosin heavy chain 7; *myhc-2*—myosin heavy chain 2; *myhc-1*—myosin heavy chain 1; *myhc-4*—myosin heavy chain 4; *igf-II*—insulin-like growth factor 2; *tor*—target of rapamycin; *s6k1*—ribosomal protein S6 kinase 1; *4e-bp1*—4 e-binding protein 1; *ef1α*—elongation factor 1-alpha.

2.4.8. Intestine Microbiome Analysis

Microbial community DNA was extracted from intestinal contents using the E.Z.N.A.® Soil DNA Kit (Omega Bio-Tek, Norcross, GA, USA). The integrity of the DNA was checked on 1% agarose gel, while the DNA concentration and purity were detected on a NanoDrop 2000 UV–vis spectrophotometer (Thermo Scientific, Wilmington, NC, USA). After amplification of the hypervariable V3–V4 regions of the bacterial 16 S rRNA gene on an ABI GeneAmp® 9700 PCR thermocycler (ABI, Foster, CA, USA) using primer pairs 338 F (5′-ACTCCTACGGGAGGCAGCAG-3′) and 806 R (5′-GGACTACHVGGGTWTCTAAT-3′), the PCR products were sequenced by the Illumina MiSeq PE300 platform (Illumina, San Diego, CA, USA). The raw reads from this study were deposited in the NCBI Sequence Read Archive (SRA) database (Accession Number: PRJNA826686).

The raw reads of 16 S rRNA gene sequencing were demultiplexed, quality filtered, and merged by fastp (version 0.20.0, https://github.com/OpenGene/fastp) and FLASH (version 1.2.7, https://ccb.jhu.edu/software/FLASH/index.shtml). Bioinformatic statistical analysis of the operational taxonomic units (OTUs) at 97% similarity level was performed using Uparse (version 7.0.1090, http://www.drive5.com/uparse/). Compared with the 16 S rRNA database (Silva 138), the RDP classifier Bayesian algorithm was used to perform taxonomic analysis on the representative sequences of OTUs (confidence threshold was 0.7). The Alpha diversity index under different random sampling was calculated using mothur (version 1.30.2, https://www.mothur.org/wiki/Download_mothur) and the difference between groups was compared by Welch's *t*-test. Statistical analysis and graphing of the PCoA (principal co-ordinates analysis) was performed using the R language (version 3.3.1, https://www.r-project.org/); ANOSIM analysis using the weighted UniFrac metric was also performed to judge whether the groupings are meaningful. One-way ANOVA was used to discriminate between-group differences in Firmicutes, Bacteroidetes, and F/B values. Spearman coefficients were used to indicate the correlation between clinical factors and species. In addition, LEfSe (linear discriminant analysis of effect size) performs linear discriminant analysis (LDA) according to the degree of influence of species composition on samples, identifying those species with significant differences. Finally, the 16 S taxonomic lineage based on the Silva database was transformed into the taxonomic lineage of prokaryotes in the KEGG database by Tax4 Fun. KEGG functional annotation was performed on the 16 S RNA gene sequence.

2.5. Statistical Analysis

Data were expressed as the mean ± standard error (SE). The data analysis was carried out with the SPSS computer program, version 26 (IBM, Armonk, NY, USA). After confirmation of the normality and homogeneity of variance (Levene's test) of the results, the effect of diet treatments was identified by one-way analysis of variance (ANOVA), and the difference between groups was further compared by Duncan's test. The difference was considered significant at $p < 0.05$. The data were compared with orthogonal polynomials, and if significance (linear, quadratic, or cubic) was detected, the model was further fitted by regression analysis. A two-slope broken-line linear (2 SBL-LL) model was also tested if the quadratic or cubic regression was significant. The R square (R^2) was used for optimal regression selection to detect the optimal protein level for dependent variables. The dietary protein requirement (provided by CPC) of grass carp was obtained by the broken-line analysis based on SGR and FE.

3. Results

3.1. Growth Performance, Apparent Digestibility, and Morphology Parameters

The effects of graded dietary protein levels on growth performance, apparent digestibility, and morphological parameters are presented in Table 3. After 8 weeks of feeding, the final body weight (FBW) of the grass carp was about three times the initial body weight, and there was no significant difference in survival rate among treatments ($p > 0.05$). As the dietary protein content increased gradually, the FBW, specific growth rate (SGR), and feed efficiency (FE) all exhibited an increasing first and then a decreasing trend in the quadratic model ($p < 0.05$, $R^2 = 0.541$, 0.545, and 0.642), all of which showed the highest values in the CPC4 group. The feeding rate (FR) and protein efficiency ratio (PER) showed a linear downward trend with increasing protein levels ($p < 0.05$, $R^2 = 0.410$ and 0.528). Furthermore, as dietary protein levels increased, the apparent digestibility coefficient of the dry matter (ADC_d) and morphological parameters (including condition factor (CF), hepatosomatic index (HSI), and mesenteric fat index (MFI)) first increased and then decreased in the quadratic model ($p < 0.05$, $R^2 = 0.881$, 0.603, 0.777, and 0.741), while the apparent digestibility coefficient of the protein (ADC_p) exhibited a rising linear model ($p < 0.05$, $R^2 = 0.734$). Based on the broken-line analysis of SGR and FE, the optimal dietary protein (CPC) level of the grass carp was estimated to be 38.61 and 38.66%, respectively (Figure 1).

Table 3. Effects of dietary protein levels on the growth, apparent digestibility, and morphological parameters of grass carp [1].

Items [2]	Diet Treatments						PSE [3]	Orthogonal Contrast [4]			Regression		
	CPC1	CPC2	CPC3	CPC4	CPC5	CPC6		Linear	Quadratic	Cubic	Model [5]	(Pr > F) [6]	R^2
IBW (g)	4.70	4.69	4.70	4.67	4.67	4.67	0.03	0.903	0.447	0.281	Ns	-	-
FBW (g)	11.83 ^{a}	12.39 ^{ab}	14.72 ^{bc}	15.42 ^{c}	14.61 ^{bc}	13.11 ^{abc}	1.23	0.040	0.004	0.234	Qd	0.003	0.541
SR (%)	96.67	97.78	97.78	95.56	97.78	97.78	3.77	0.858	0.870	0.630	Ns	-	-
SGR (%/d)	1.66 ^{a}	1.72 ^{a}	2.04 ^{b}	2.13 ^{b}	2.03 ^{b}	1.84 ^{ab}	0.16	0.030	0.005	0.216	2 SBL-LL	0.003	0.887
FR (%/d)	4.09 ^{b}	3.50 ^{a}	3.63 ^{a}	3.55 ^{a}	3.44 ^{a}	3.44 ^{a}	0.18	0.001	0.060	0.087	Ln	0.004	0.410
FE (%)	38.58 ^{a}	46.81 ^{ab}	51.72 ^{b}	54.78 ^{b}	54.26 ^{b}	50.18 ^{b}	5.03	0.005	0.009	0.872	2 SBL-LL	0.000	0.982
PER	1.56 ^{b}	1.53 ^{b}	1.54 ^{b}	1.45 ^{b}	1.31 ^{ab}	1.10 ^{a}	0.15	0.001	0.078	0.750	Ln	0.000	0.528
ADC_d (%)	66.89 ^{c}	72.02 ^{d}	71.18 ^{d}	71.60 ^{d}	61.51 ^{b}	56.27 ^{a}	1.85	0.000	0.000	0.214	Qd	0.000	0.881
ADC_p (%)	64.43 ^{a}	67.03 ^{a}	69.86 ^{ab}	70.62 ^{ab}	75.99 ^{bc}	79.77 ^{c}	3.54	0.000	0.409	0.698	Ln	0.000	0.734
CF (%)	1.70 ^{a}	1.77 ^{ab}	1.84 ^{bc}	1.89 ^{c}	1.84 ^{bc}	1.79 ^{abc}	0.05	0.032	0.003	0.680	Qd	0.001	0.603
HSI (%)	1.86 ^{a}	2.14 ^{cd}	2.22 ^{d}	2.28 ^{d}	2.02 ^{bc}	1.91 ^{ab}	0.08	0.917	0.000	0.187	Qd	0.000	0.777
MFI (%)	1.29 ^{a}	1.52 ^{b}	1.85 ^{c}	2.25 ^{d}	2.40 ^{e}	1.58 ^{b}	0.04	0.000	0.000	0.000	Qd	0.000	0.741

[1] All data are the means of three parallel tanks (n = 3). Mean values not sharing a common superscript in the same row are significantly different ($p < 0.05$), while mean values in the same row without any superscript are not different. Growth parameters, feed utilization (except PER), and digestibility parameters were calculated with reference to the study of Wang et al. [20], PER was calculated according to the study of Abdel-Tawwab et al. [27] and morphological parameters were calculated according to the study of Abouel Azm et al. [16]. [2] IBW, initial body weight; FBW, final body weight. SR (Survival rate, %) = 100 × (number of survival/total number). SGR (Specific growth rate, %/d) = 100 × (ln (FBW)–ln (IBW))/(experimental period (d)). FE (Feed efficiency, %) = 100 × [FBW (g)–IBW (g)]/dry feed intake (g). FR (Feeding rate, %/d) = 100 × dry feed intake/(experimental days × (FBW + IBW)/2). PER (Protein efficiency ratio) = (FBW (g) – IBW (g))/(dry feed intake (g) × dietary protein content). ADC_d (Apparent digestibility coefficient of dry matter, %) = 100 × (1 – (dietary yttrium content/fecal yttrium content)). ADC_p (Apparent digestibility coefficient of protein, %) = 100 × (1 – (dietary yttrium content/fecal yttrium content) × (fecal protein content/dietary protein content)). HSI (Hepatosomatic index, %) = 100 × (final hepatopancreas weight (g)/final body weight (g)). MFI (Mesenteric fat index, %) = 100 × (mesenteric fat weight (g)/final body weight (g)). CF (Condition factor, %) = fish weight (g) × 100/fish length $(cm)^3$. [3] PSE = Pooled standard error of treatment means (n = 3). [4] If statistical significance ($p < 0.05$) was detected, the model that fits best with the data was selected. [5] Ns = No structure ($p > 0.05$); 2 SBL-LL = Two slope broken line-linear ascending and linear descending; Ln = Linear; Qd = Quadratic. [6] Probability associated with the F-statistic test.

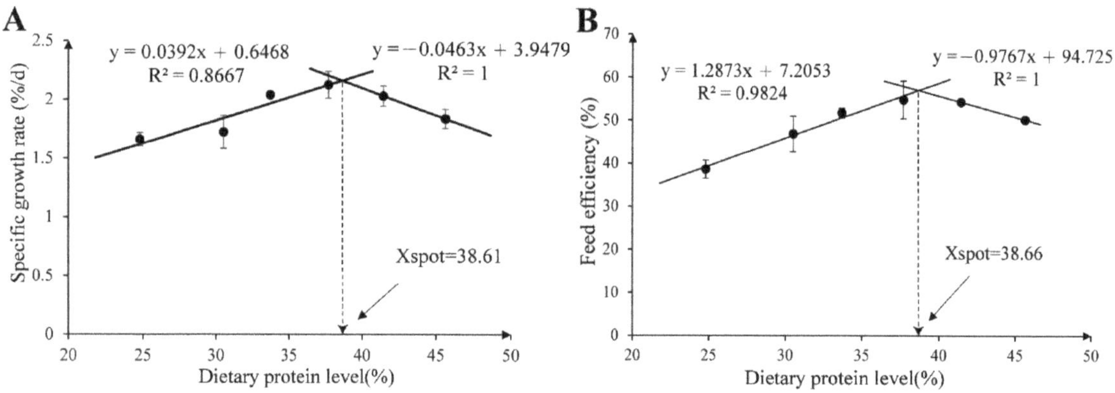

Figure 1. Broken-line analysis of the relationship between the dietary protein level and growth performance of grass carp. (**A**) The regression analysis between the specific growth rate and dietary protein level. (**B**) The regression analysis between the feed efficiency and dietary protein level.

3.2. Body and Dorsal Muscle Proximate Composition

As shown in Table 4, the crude lipid content in the whole body increased linearly (R^2 = 0.375), while the muscle protein content increased first and then decreased in the quadratic model in response to the increasing protein level ($p < 0.05$, R^2 = 0.331). No other significant difference in the proximate composition of the whole body and dorsal muscle was observed between treatments ($p > 0.05$).

Table 4. Effects of dietary protein levels on the body and dorsal muscle proximate composition of grass carp (% fresh weight) [1].

	Diet Treatments						PSE [2]	Orthogonal Contrast [3]			Regression		
	CPC1	CPC2	CPC3	CPC4	CPC5	CPC6		Linear	Quadratic	Cubic	Model [4]	(Pr > F) [5]	R^2
Whole body													
Moisture	72.03	72.54	72.38	71.55	72.38	71.32	0.70	0.180	0.317	0.869	Ns	-	-
Crude Protein	13.65	13.68	13.77	13.77	13.72	13.64	0.24	0.951	0.415	0.863	Ns	-	-
Crude Lipid	9.90	9.95	10.84	11.12	10.85	10.80	0.45	0.005	0.051	0.416	Ln	0.007	0.375
Ash	3.14	3.33	2.66	3.13	2.59	3.09	0.41	0.344	0.360	0.357	Ns	-	-
Dorsal muscle													
Moisture	77.20	76.68	76.19	77.03	77.45	77.13	0.72	0.434	0.249	0.130	Ns	-	-
Crude Protein	18.18	18.69	18.68	18.72	18.35	18.23	0.34	0.649	0.026	0.368	Qd	0.049	0.331
Crude Lipid	3.72	3.89	3.88	3.88	3.77	3.63	0.31	0.610	0.252	0.876	Ns	-	-
Ash	0.44	0.45	0.46	0.39	0.44	0.50	0.05	0.514	0.197	0.181	Ns	-	-

[1] All data are the means of three parallel tanks (n = 3). Mean values not sharing a common superscript in the same row are significantly different ($p < 0.05$), while mean values in the same row without any superscript are not different. [2] PSE = Pooled standard error of treatment means (n = 3). [3] If statistical significance ($p < 0.05$) was detected, the model that fits best with the data was selected. [4] Ns = No structure ($p > 0.05$); Ln = Linear; Qd = Quadratic. [5] Probability associated with the F-statistic test.

3.3. Serum Biochemical Indices and Immune Enzyme Activity

The biochemical indices and immune parameters in the serum of grass carp are presented in Table 5. The serum TP showed a linearly increasing trend as the dietary protein level increased (R^2 = 0.291) but was not significantly different among all treatments. The contents of serum UN and IgM also increased linearly with dietary protein levels ($p < 0.05$, R^2= 0.696 and 0.650). In contrast, the contents of TG, LDL, and C3 in the serum increased first and then decreased, responding to the increasing dietary protein level in the quadratic model ($p < 0.05$, R^2 = 0.710, 0.300, and 0.471, respectively). However, the dietary protein level did not affect the contents of serum GLU, CHOL, HDL, and LYS of grass carp ($p > 0.05$).

Table 5. Effects of dietary protein levels on biochemical indices and immune parameters in the serum of grass carp [1].

	Diet Treatments						PSE [3]	Orthogonal Contrast [4]			Regression		
Indices [2]	CPC1	CPC2	CPC3	CPC4	CPC5	CPC6		Linear	Quadratic	Cubic	Model [5]	(Pr > F) [6]	R^2
TP (g/L)	38.00	38.57	38.17	38.63	40.97	42.00	2.43	0.036	0.322	0.945	Ln	0.021	0.291
GLU (mmol/L)	5.60	5.00	4.70	4.97	5.43	5.67	1.10	0.726	0.236	0.665	Ns	-	-
UN (mmol/L)	0.50 ^{a}	0.53 ^{a}	0.63 ^{a}	0.77 ^{ab}	1.07 ^{bc}	1.10 ^{c}	0.17	0.000	0.391	0.355	Ln	0.000	0.696
TG (mmol/L)	2.50 ^{a}	2.85 ^{a}	3.57 ^{b}	3.99 ^{b}	3.68 ^{b}	3.57 ^{b}	0.35	0.000	0.004	0.448	Qd	0.000	0.710
CHOL (mmol/L)	6.70	6.95	7.17	6.75	6.67	6.60	0.63	0.567	0.424	0.530	Ns	-	-
HDL (mmol/L)	2.69	2.70	2.73	2.75	2.78	2.77	0.32	0.202	0.778	0.776	Ns	-	-
LDL (mmol/L)	4.22 ^{a}	4.37 ^{ab}	4.61 ^{b}	4.44 ^{ab}	4.33 ^{ab}	4.30 ^{ab}	0.18	0.884	0.027	0.334	Qd	0.069	0.300
C3 (μg/mL)	122.62 ^{a}	125.61 ^{ab}	127.58 ^{ab}	130.02 ^{bc}	133.96 ^{c}	124.87 ^{ab}	2.87	0.016	0.005	0.025	Qd	0.008	0.471
IgM (μg/mL)	32.20 ^{a}	32.29 ^{a}	38.00 ^{b}	38.83 ^{b}	38.42 ^{b}	40.42 ^{b}	2.08	0.000	0.199	0.751	Ln	0.000	0.650
LYS (ng/mL)	181.00	187.22	178.87	179.95	180.99	176.60	6.50	0.230	0.660	0.737	Ns	-	-

[1] All data are the means of three parallel tanks (n = 3). Mean values not sharing a common superscript in the same row are significantly different ($p < 0.05$), while mean values in the same row without any superscript are not different. [2] TP, total protein; GLU, glucose; UN, urea nitrogen; TG, triglyceride; CHOL, cholesterol; HDL, high-density lipoprotein; LDL, low-density lipoprotein; C3, complement 3; IgM, immunoglobulin M; LYS, lysozyme. [3] PSE = Pooled standard error of treatment means (n = 3). [4] If statistical significance ($p < 0.05$) was detected, the model that fits best with the data was selected. [5] Ns = No structure ($p > 0.05$); Ln = Linear; Qd = Quadratic. [6] Probability associated with the F-statistic test.

3.4. Antioxidative Capacity, Metabolic Enzymes, and Histological Observation of Hepatopancreas

As shown in Figure 2, hepatopancreatic T-AOC, GSH-Px, AST, and ALT activities significantly increased first and then decreased in response to the increasing dietary protein level in the quadratic model (R^2 = 0.666, 0.461, 0.558, and 0.386, respectively). The activities of T-AOC and ALT peaked in the CPC3 group ($p < 0.05$), while GSH-Px and AST showed the highest values in the CPC4 group ($p < 0.05$). The hepatopancreatic MDA content significantly decreased first and then increased with dietary protein level and was the lowest in the CPC4 group ($p < 0.05$, $R^2 = 0.623$). In contrast, SOD and CAT activities were not significantly different among all treatments ($p > 0.05$).

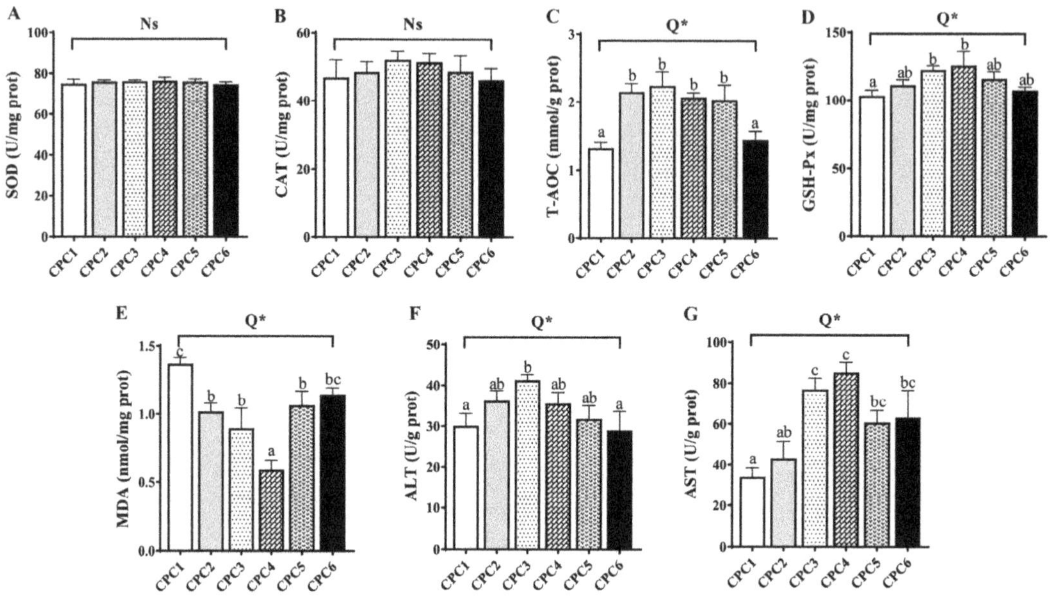

Figure 2. Effects of the dietary protein levels on antioxidative capacity and transaminase activities of grass carp hepatopancreas. (**A–G**) SOD: superoxide dismutase; CAT: catalase; T-AOC: total antioxidant capacity; GSH-Px: glutathione peroxidase; MDA: malondialdehyde; ALT: alanine aminotransferase; AST: aspartate aminotransferase. All data were the means of three parallel tanks ($n = 3$). Mean values not sharing a common superscript in the same row are significantly different ($p < 0.05$), while mean values in the same row without any superscript are not different. If statistical significance ($p < 0.05$) was detected, the model that fits best with the data was selected. Ns = No structure ($p > 0.05$); Q = Quadratic; * $p < 0.05$.

As shown in Figure 3, the dietary protein level significantly affected the hepatopancreatic microstructure of grass carp. CPC3 and CPC4 groups showed more normal hepatocytes without noticeable swelling and atrophy. However, CPC1 and CPC6 groups showed more abnormal hepatocytes, nuclear migration, cellularity vacuum, and lipid deposition. Some hepatocyte nuclear migration and cell vacuums were observed in the CPC2 and CPC5 groups.

3.5. Muscle Texture Analysis

As shown in Table 6, the cooking loss of muscle increased linearly with the dietary protein level ($p < 0.05$, $R^2 = 0.611$). The muscle texture parameters, including hardness, cohesiveness, gumminess, and chewiness in the cooked meat, increased first and then decreased in response to the increasing dietary protein level in the quadratic model ($p < 0.05$, $R^2 = 0.687$, 0.480, 0.637, and 0.645, respectively). Muscle springiness was not affected by the dietary protein level ($p > 0.05$). Muscle resilience first decreased and then increased in the quadratic model ($p < 0.05$, $R^2 = 0.717$), while pH showed a linear downward trend with the dietary protein level ($p < 0.05$, $R^2 = 0.282$).

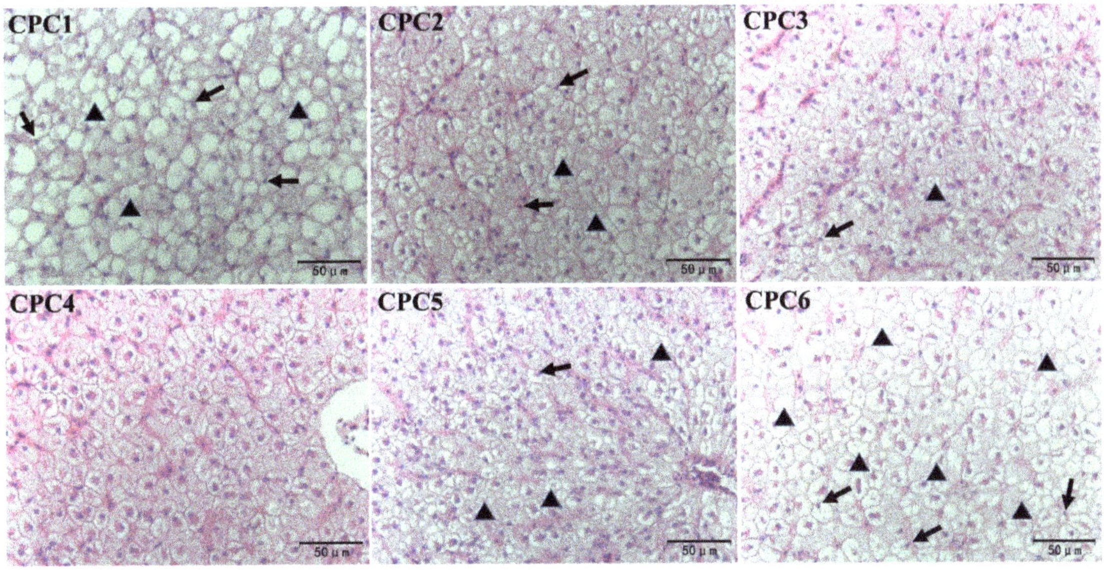

Figure 3. Hepatopancreatic histology of grass carp (hematoxylin and eosin, ×400). Arrows indicate nuclei shifted to the periphery of the hepatocytes; triangles indicate vacuolation.

Table 6. Effects of the dietary protein levels on the muscle texture of grass carp [1].

	Diet Treatments						PSE [2]	Orthogonal Contrast [3]			Regression		
	CPC1	CPC2	CPC3	CPC4	CPC5	CPC6		Linear	Quadratic	Cubic	Model [4]	(Pr > F) [5]	R^2
Cooking loss (%)	29.38 ^{a}	31.17 ^{ab}	33.78 ^{abc}	35.54 ^{bc}	35.58 ^{bc}	38.16 ^{c}	2.71	0.001	0.677	0.781	Ln	0.000	0.611
Hardness (g)	1037.24 ^{a}	1188.03 ^{b}	1353.18 ^{c}	1476.93 ^{d}	1151.90 ^{ab}	1104.84 ^{ab}	65.95	0.289	0.000	0.854	Qd	0.000	0.687
Springiness	0.53	0.51	0.48	0.48	0.52	0.50	0.10	0.448	0.145	0.442	Ns	-	-
Cohesiveness	0.46 ^{a}	0.47 ^{ab}	0.48 ^{abc}	0.49 ^{bc}	0.50 ^{c}	0.47 ^{ab}	0.00	0.020	0.013	0.106	Qd	0.007	0.480
Gumminess	504.58 ^{a}	557.33 ^{ab}	660.12 ^{bc}	741.68 ^{c}	683.29 ^{c}	557.53 ^{ab}	59.94	0.028	0.000	0.065	Qd	0.000	0.637
Chewiness (g)	279.51 ^{a}	325.41 ^{b}	367.64 ^{c}	433.60 ^{d}	313.62 ^{b}	305.25 ^{ab}	15.19	0.051	0.000	0.663	Qd	0.000	0.645
Resilience (g/s)	0.26 ^{c}	0.24 ^{bc}	0.22 ^{ab}	0.21 ^{a}	0.22 ^{ab}	0.23 ^{ab}	0.00	0.004	0.001	0.794	Qd	0.000	0.717
pH	5.40 ^{b}	5.34 ^{ab}	5.31 ^{ab}	5.30 ^{ab}	5.31 ^{ab}	5.22 ^{a}	0.09	0.042	0.918	0.361	Ln	0.023	0.282

[1] All data are the means of three parallel tanks (n = 3). Mean values not sharing a common superscript in the same row are significantly different ($p < 0.05$), while mean values in the same row without any superscript are not different. [2] PSE = Pooled standard error of treatment means (n = 3). [3] If statistical significance ($p < 0.05$) was detected, the model that fits best with the data was selected. [4] Ns = No structure ($p > 0.05$); Ln = Linear; Qd = Quadratic. [5] Probability associated with the F-statistic test.

3.6. Antioxidative Capacity and Histological Observation of Muscle

The antioxidant parameters, including the activities of SOD, T-AOC, and GSH contents in muscle, significantly increased first and then decreased in response to the increasing dietary protein level in the quadratic model ($p < 0.05$, R^2 = 0.442, 0.365, and 0.627, respectively), which were the highest in the CPC2 or CPC4 group, respectively (Figure 4). The MDA content decreased first and then increased in the quadratic model ($p < 0.05$, R^2 = 0.842) responding to the protein level, and was the lowest in CPC3. However, there was no significant difference in muscle CAT activity ($p > 0.05$).

As shown in Figure 5A, dietary protein level affected the microstructure of grass carp dorsal muscle. The frequency distribution of grass carp muscle fibers shows that the CPC4 group was assigned smaller diameters, including class20, class30 ($p < 0.05$, R^2 = 0.624 and 0.384), and class40 (Figure 5B). Large muscle fibers, including class50, class60 ($p < 0.05$, R^2 = 0.346 and 0.580), and class70 diameters, were more assigned to other groups (Figure 5B). Further statistical analysis found that fiber diameter (Figure 5C) and muscle fiber density (Figure 5D) exhibited a quadratic model ($p < 0.05$, R^2 = 0.664 and 0.513), with the largest muscle fiber density and smallest muscle fiber diameter detected in the CPC4 group.

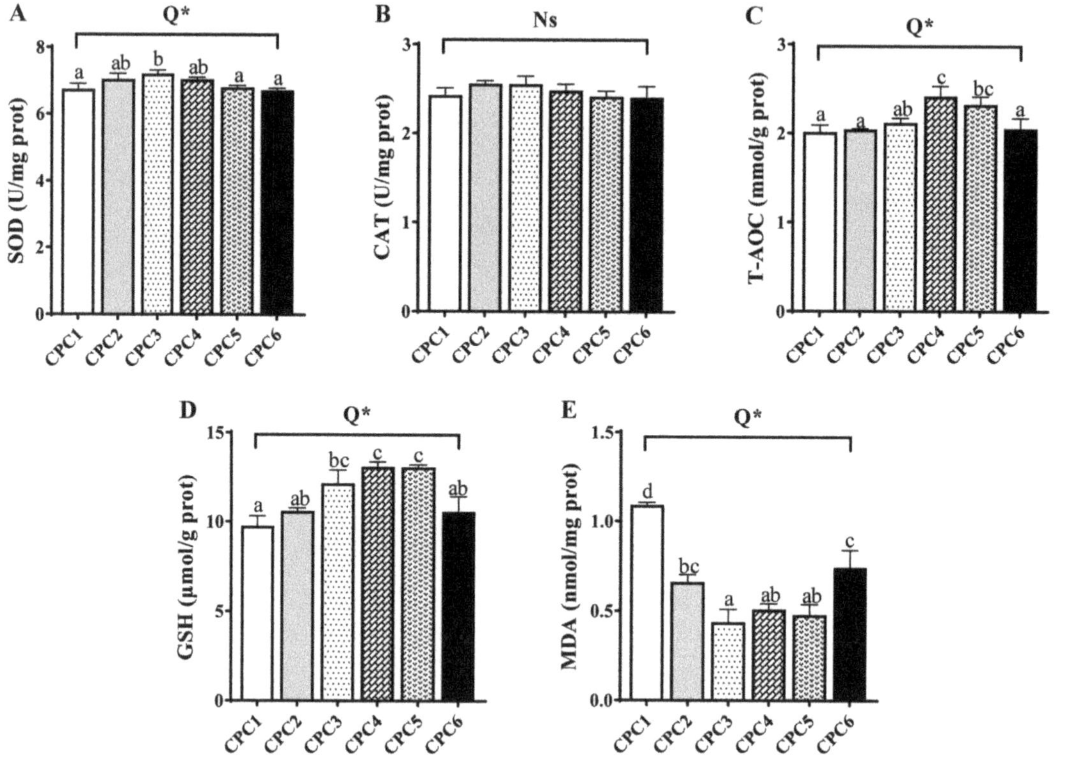

Figure 4. Effects of the dietary protein level on the antioxidant capacity of grass carp muscle. (**A–E**) SOD: superoxide dismutase; CAT: catalase; T-AOC: total antioxidant capacity; GSH: glutathione; MDA: malondialdehyde. All data are means of three parallel tanks ($n = 3$). Mean values not sharing a common superscript in the same row are significantly different ($p < 0.05$), while mean values in the same row without any superscript are not different. If statistical significance ($p < 0.05$) was detected, the model that fits best with the data was selected. Ns = No structure ($p > 0.05$); Q = Quadratic; * $p < 0.05$.

3.7. Muscle-Related Genes Expression

As shown in Figure 6, the gene expression of myogenic differentiation antigen (*myod*), myogenin (*myog*), *mstn*, *fgf6a*, *myhc-7*, *myhc-1*, *myhc-4*, *igf-II*, *tor*, and 4 e-binding protein 1 (*4e-bp1*) in the muscle all varied in the quadratic models ($p < 0.05$, $R^2 = 0.564$, 0.722, 0.443, 0.621, 0.733, 0.631, 0.656, 0.456, 0.785, and 0.567, respectively) as dietary protein level increased. Gene expression of muscle regulatory factor 4 (*mrf4*) varied in the 2 SBL-LL model with increasing dietary protein levels ($p < 0.05$, $R^2 = 0.337$), while myogenic factor 5 (*myf5*) showed a linear upward trend ($p < 0.05$, $R^2 = 0.756$). The maximum gene expression of *myod*, *mrf4*, *fgf6a*, *myhc-1*, *myhc-4*, and *tor* was detected in the CPC4 group, while gene expression of ribosome S6 protein kinase 1 (s6k1) peaked in the CPC5 group. The minimum gene expression values of *mstn*, *4e-bp1*, and *myhc-2* were detected in the CPC4 group. However, dietary protein levels did not significantly affect the *fgf6b* gene expression levels ($p > 0.05$).

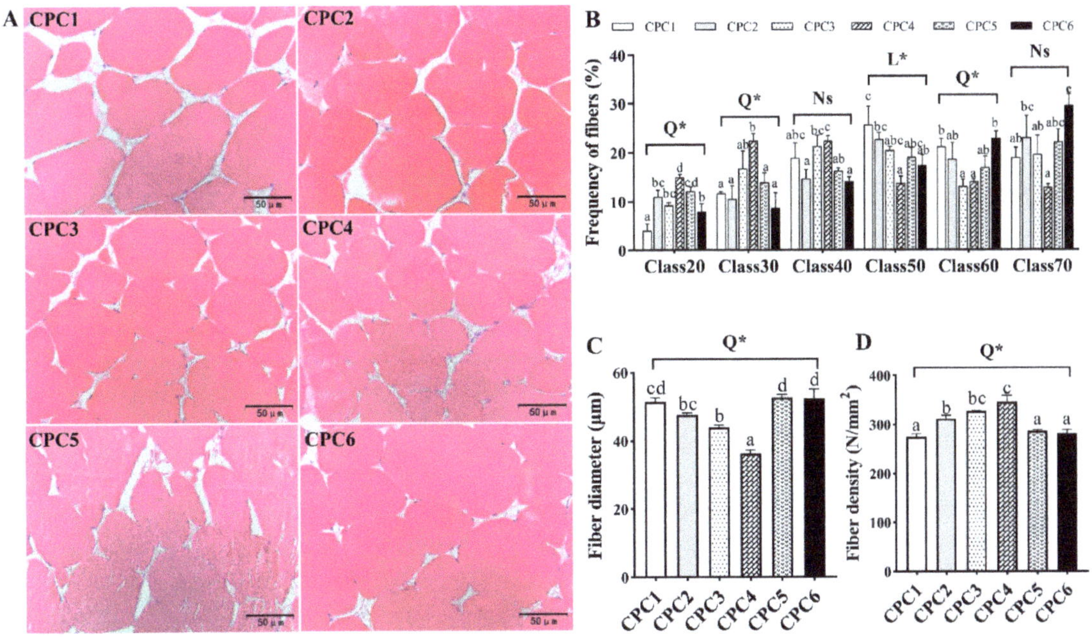

Figure 5. Muscle histology of grass carp fed with different protein levels for eight weeks. (**A**) Observation on muscle fiber of white muscle in grass carp (hematoxylin and eosin, ×400). (**B**) The frequency distribution of the muscle fibers' diameter classes (d, μm): class 20 = d ≤ 20, class 30 = 20 < d ≤ 30, class 40 = 30 < d ≤ 40, class 50 = 40 < d ≤ 50, class 60 = 50 < d ≤ 60, and class 70 = d > 60. (**C**) The muscle fiber diameter (μm). (**D**) The muscle fiber density (n/mm^2). All data are the means of three parallel tanks (n = 3). Mean values not sharing a common superscript in the same row are significantly different ($p < 0.05$), while mean values in the same row without any superscript are not different. If statistical significance ($p < 0.05$) was detected, the model that fits best with the data was selected. Ns = No structure ($p > 0.05$); L = Linear; Q = Quadratic; * $p < 0.05$.

3.8. Intestinal Microbiota of Grass Carp

After pyrosequencing, the samples were leveled to 29,566 sequences according to the minimum sample sequence value among the nine samples. All sequences were delineated as OTUs with 97% sequence similarity values, yielding 485 OTUs. Shannon and rarefaction curves tended to approach saturation plateaus, indicating that the microbial diversity and total OTUs in the different samples did not vary with the apparent change in sequencing depth (Figure 7).

As shown in Figure 8A–C, the Sobs, Chao, and Ace indices were used to assess microbial community richness, which was not significantly affected by dietary protein levels ($p > 0.05$). In addition, alpha diversity revealed a high species coverage index (over 99%), indicating that the sequences were sufficient to capture the species richness of the samples (Figure 8D). However, microbial community diversity, including the Simpson and Shannon indices, responded to dietary protein changes (Figure 8E,F). Compared with the CPC4 group, the Simpson index was significantly lower in the CPC1 ($p < 0.05$) and CPC6 ($p < 0.001$) groups. The PCoA with weighted UniFrac distances (Figure 8G) showed that the samples were separated from each other after eight weeks of feeding with graded dietary protein levels. Further ANOSIM analysis showed that the "Between" box was higher than the boxes of the other three groups (Figure 8H), indicating that the between-group difference was more significant than the within-group difference, which proved that the grouping was statistically significant (p = 0.001, R = 0.8354).

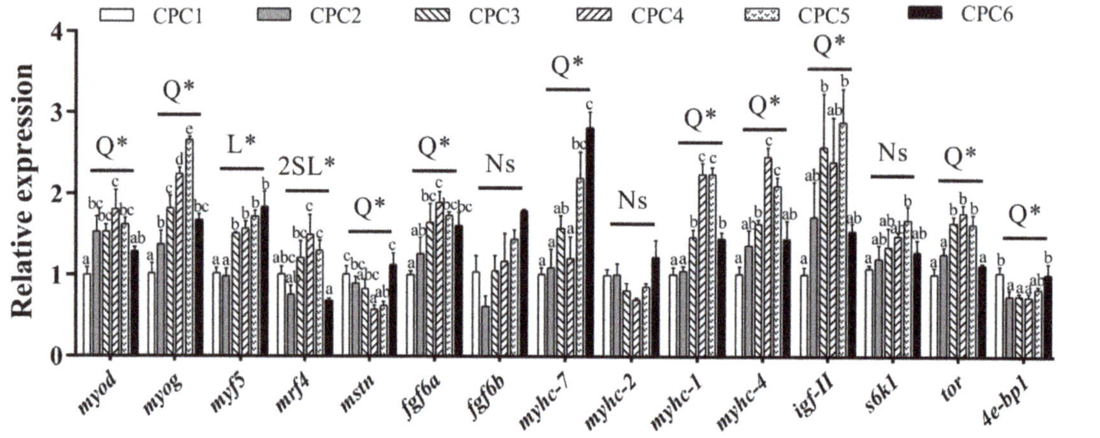

Figure 6. Muscle-related genes expression of grass carp fed with different protein levels for eight weeks. *myod*—myogenic differentiation antigen; *myog*—myogenin; *myf5*—myogenic factor 5; *mrf4*—muscle regulatory factor 4; *fgf6a*—fibroblast growth factor 6 a; *fgf6b*—fibroblast growth factor 6 b; *mstn*—myostatin; *myhc-7*—myosin heavy chain 7; *myhc-2*—myosin heavy chain 2; *myhc-1*—myosin heavy chain 1; *myhc-4*—myosin heavy chain 4; *igf-II*—insulin-like growth factor 2; *tor*—target of rapamycin; *s6k1*—ribosomal protein S6 kinase 1; *4e-bp1*—4 e-binding protein 1; *ef1α*—elongation factor 1-alpha. The cDNAs used to detect *myhc-2* and *myhc-4* were diluted 180 times; others were diluted six times. All data are the means of three parallel tanks ($n = 3$). Mean values not sharing a common superscript in the same row are significantly different ($p < 0.05$), while mean values in the same row without any superscript are not different. If statistical significance ($p < 0.05$) was detected, the model that fits best with the data was selected. Ns = No structure ($p > 0.05$); 2 SL = Two slope broken line-linear ascending and linear descending; L= Linear; Q = Quadratic; * $p < 0.05$.

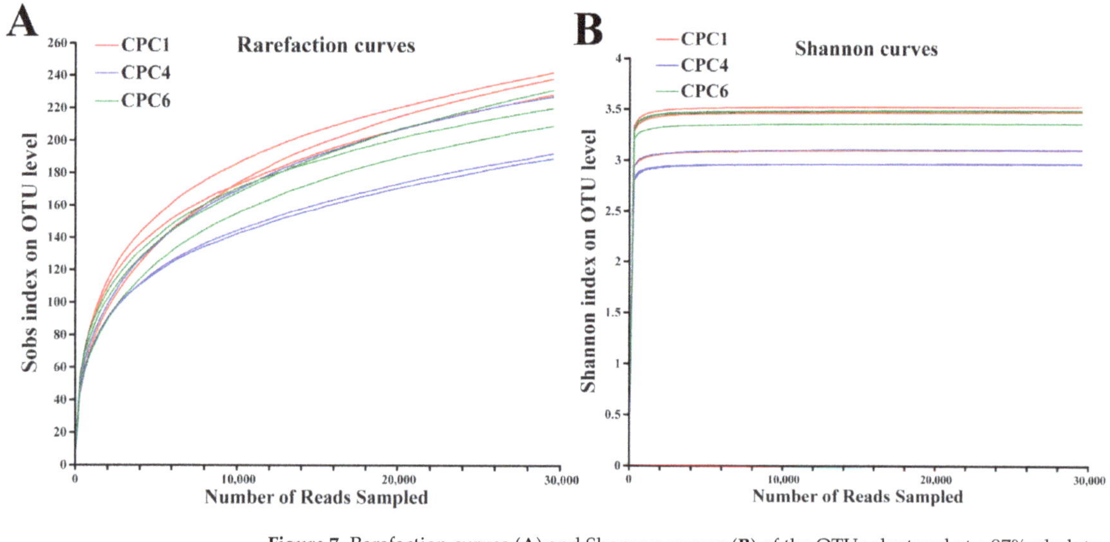

Figure 7. Rarefaction curves (**A**) and Shannon curves (**B**) of the OTUs clustered at a 97% phylotype similarity level of the fish intestinal microbiota from the different dietary protein treatments. The horizontal axis is the number of valid sequences; the vertical axis is the observed number of operational taxonomic units. The Sobs index and Shannon index were used to estimate the richness and diversity of the intestinal microbiota, respectively.

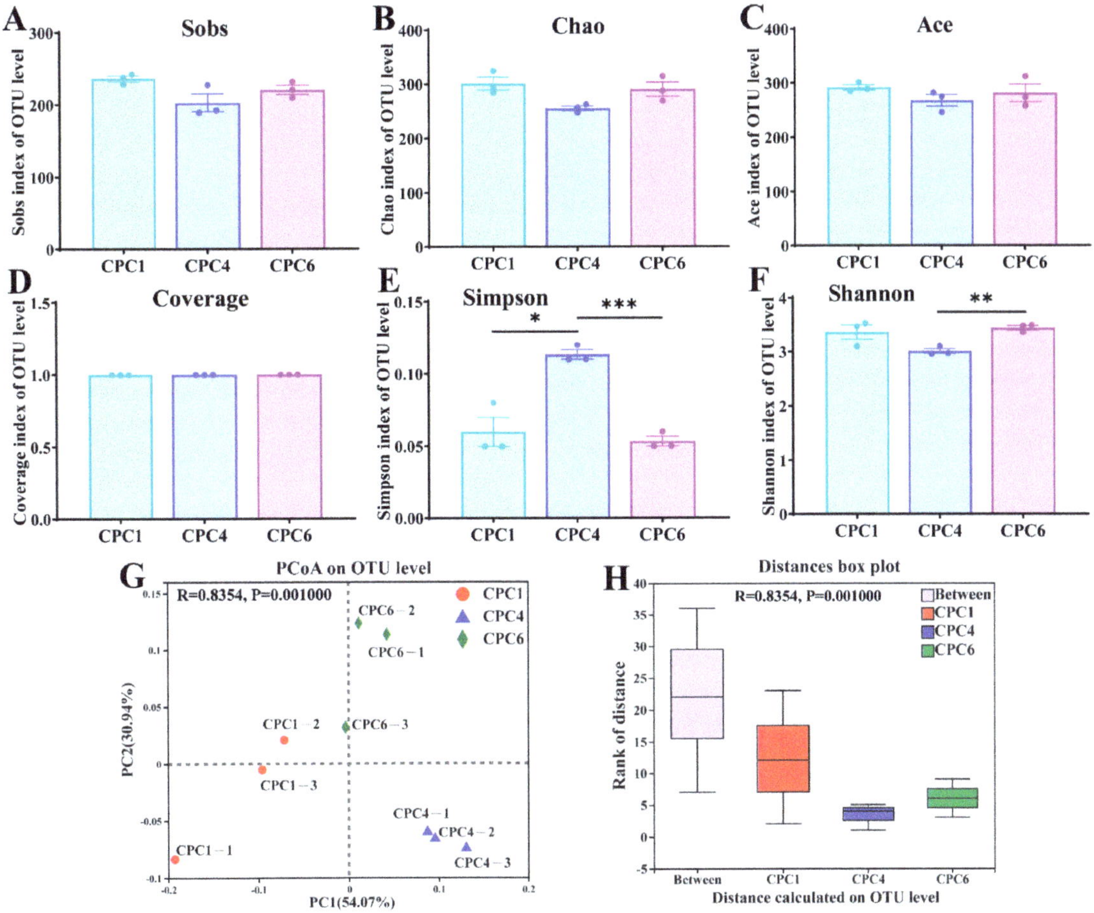

Figure 8. Alpha diversity analysis and Beta diversity analysis of grass carp intestinal microbiota from different dietary protein treatments. The Sobs index (**A**), Chao index (**B**), and Ace index (**C**) were used to reflect the microbial community richness. The coverage index (**D**) represents the community coverage. The Simpson index (**E**) and Shannon index (**F**) were used to reflect the microbial community diversity. Significance analysis was performed using Welch's *t*-test, and two groups with significant differences were marked (* $p \leq 0.05$, ** $p \leq 0.01$, *** $p \leq 0.001$). (**G**) Principal coordinates analysis of the weighted UniFrac scores of the microbial communities. (**H**) ANOSIM analysis (analysis of similarities) is a nonparametric test used to test whether the differences between groups are significantly greater than the differences within groups, where the "Between" box represents the between-group differences, and the other colored boxes represent the within-group differences. ANOSIM analysis showed that R = 0.8354, $p = 0.001$.

At the phylum level, this study found that the dominant phyla in the three groups were *Proteobacteria* (CPC1: 57.26%; CPC4: 36.80%; CPC6: 50.09%), *Fusobacterium* (CPC1: 10.95%; CPC4: 34.59%; CPC6: 13.07%), *Actinobacteria* (CPC1: 19.19%; CPC4: 11.20%; CPC6: 12.13%), *Firmicutes* (CPC1: 6.81%; CPC4: 9.44%; CPC6: 10.98%), and *Bacteroidetes* (CPC1: 1.15%; CPC4: 4.59%; CPC6: 7.88%) (Figure 9A). In addition, the *Bacteroidetes* showed an upward trend, and the relative abundance of the CPC6 group was significantly higher than that of the CPC1 group ($p < 0.05$). The F/B (*Firmicutes* to *Bacteroidetes* ratio) value showed a downward trend with increasing dietary protein level, and the CPC4 and CPC6 groups were significantly lower than the CPC1 group ($p < 0.05$) (Figure 9B). At the genus level, the top five genera with the highest abundance ratio were *Cetobacterium* (CPC1: 10.95%; CPC4: 34.28%; CPC6: 13.05%), *Gemmobacter* (CPC1: 8.95%; CPC4: 9.16%; CPC6: 5.37%), *Aeromonas* (CPC1: 3.89%; CPC4: 6.43%; CPC6: 10.07%), *Demequlna* (CPC1: 6.66%; CPC4: 6.88%; CPC6: 6.29%), and *Micribacterium* (CPC1: 10.67%; CPC4: 2.66%; CPC6: 5.02%) (Figure 9C). Correlation analysis (Figure 10) of the intestinal bacteria genera with grass carp growth and muscle parameters showed that SGR strongly correlates with *Cetobacterium* (r = 0.745), *Akkermansia* (r = 0.837), *Micribacterium* (r = −0.828), *Pseudomonas* (r = −0.678) and *Silanimonas* (r = −0.703). FE showed strong correlation with *Akkermansia* (r = 0.867), *Crenobacter* ,, *Bacteroides* (r = 0.700), *norank_f__Barnesiellaceae* (r = 0.667), *Microbacterium* (r = −0.833), and *Pseudomonas* (r = −0.717). Muscle hardness was negatively correlated with *Microbacterium* (r = −0.717), *Pseudomonas* (r = −0.733), *norank_f__Rhizobiales_Incertae_Sedis* (r = −0.800), *Silanimonas* (r = −0.833), and *Pelomonas* (r = −0.850). Muscle density was also negatively correlated with *norank_f__Rhizobiales_Incertae_Sedis* (r = −0.723). Muscle MDA content was shown to correlate with *Cetobacterium* (r = −0.833), *Akkermansia* (r = −0.800), *Microbacterium* (r = 0.883), *Pseudomonas* (r = 0.767), and *norank_f__Rhizobiales_Incertae_Sedis* (r = 0.667). Muscle GSH content was shown to negatively correlate with *Microbacterium* (r = −0.717), *Pseudomonas* (r = −0.800), *norank_f__Rhizobiales_Incertae_Sedis* (r = −0.800), and *Pelomonas* (r = −0.667).

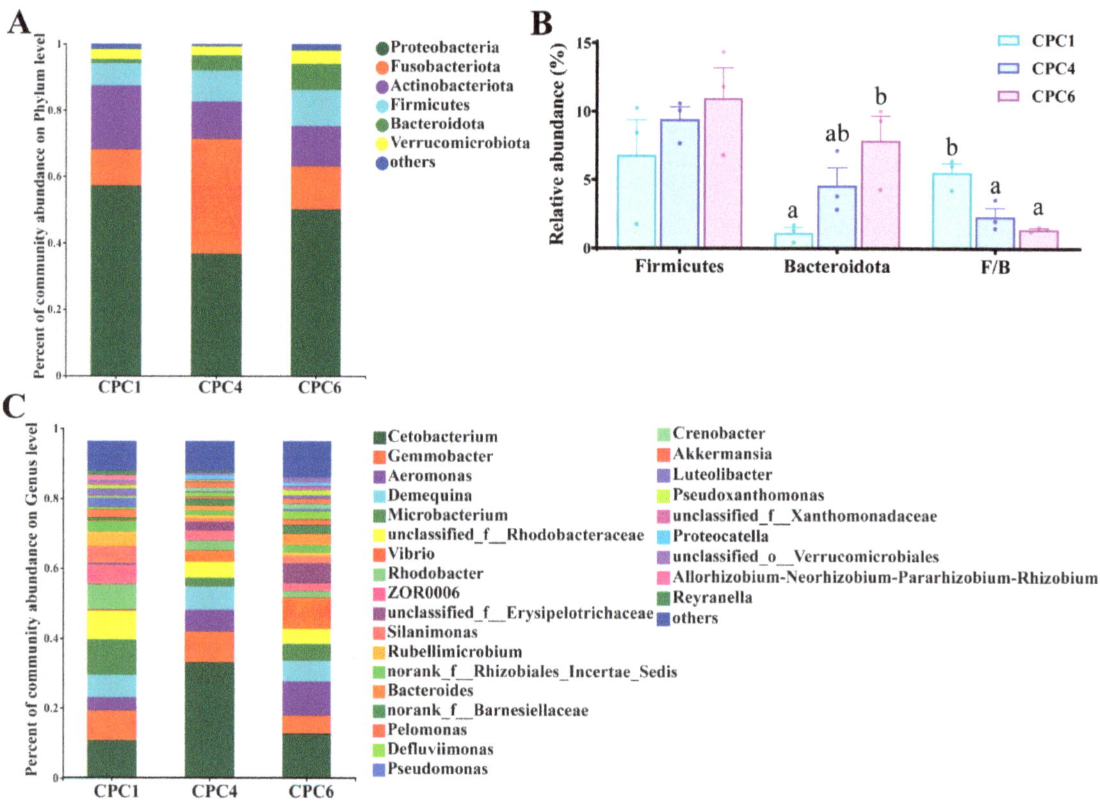

Figure 9. Community composition of intestinal microbiota from different dietary protein treatments. After being fed for eight weeks: the abundance of the intestinal microbial composition at the phylum (**A**) and genus (**C**) level of grass carp in the different protein level groups. Taxa with abundances <1% are included in "Others". (**B**) One-way ANOVA significance tests were performed for Firmicutes, Bacteroides, and F/B (ratio of Firmicutes to Bacteroides). All data are the means of three parallel tanks ($n = 3$). Mean values with different superscripts were significantly different ($p < 0.05$), and the mean values without superscripts were not significantly different.

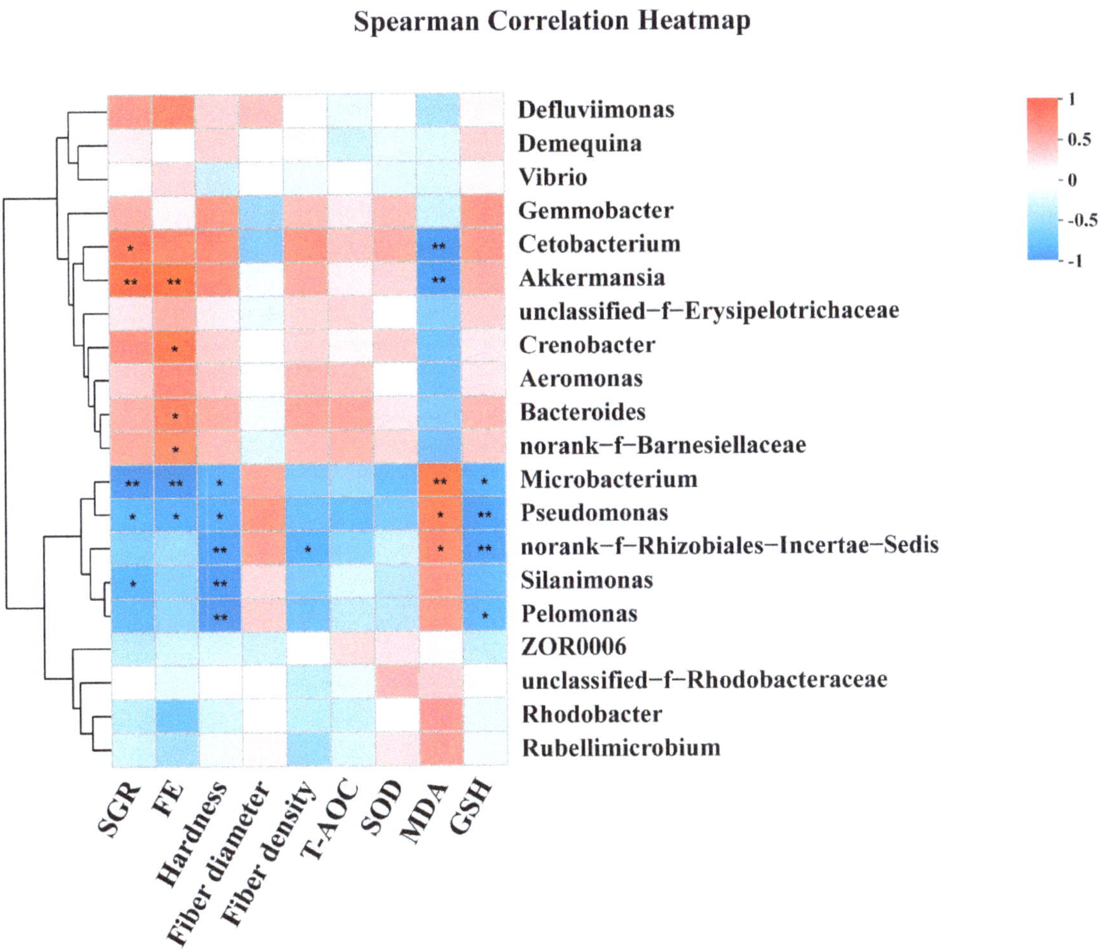

Figure 10. Correlation analysis of the intestinal microbiota and growth, muscle parameters. Spearman test, * $p < 0.05$, ** $p < 0.01$.

LEfSe analysis revealed microbial taxa enriched at different dietary protein levels (Figure 11). CPC1 (red) was mainly enriched in Proteobacteria. CPC4 (blue) was enriched in *Lachnospirales*, *Tessaracoccus*, and *Fusobacterium*. The relative abundance of Bacteroidetes dominates in CPC6 (green).

The predicted results of intestinal microbiota function are presented in Figure 12. KEGG level 1 showed that the three treatments were mainly enriched for metabolism and environmental information processing (Figure 12A). The top three functions predicted by KEGG level 2 (Figure 12B) were carbohydrate metabolism, amino acid metabolism, and membrane transport. However, significant differences were detected only between CPC1 and CPC6 in the predicted pathways of KEGG level 3 in two pathways, ko00480 (Glutathione metabolism) and ko00620 (Pyruvate metabolism) ($p < 0.05$), respectively (Figure 12C–H).

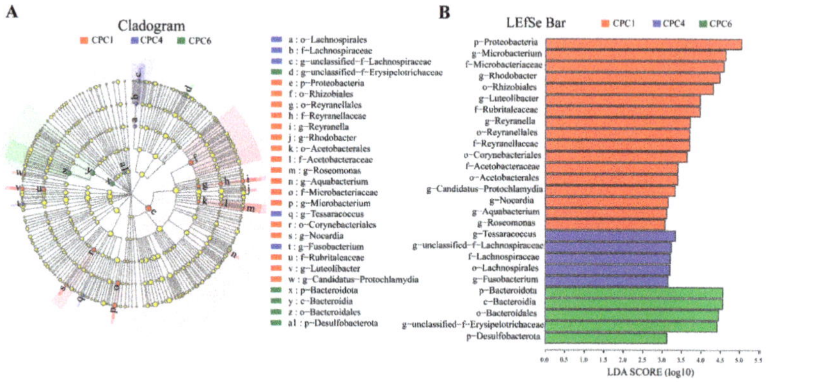

Figure 11. LEfSe analysis identified differential distributions of intestinal microbes among treatment groups with different dietary protein levels. (**A**) Cladogram showing the phylogenetic distribution of intestinal microbes for graded dietary protein level groups generated by LEfSe analysis (layers of the cladogram represent different levels, phyla, classes, orders, families, and genera from the inside out). (**B**) LDA scores showed bacterial differential expression between treatment groups with different dietary protein levels (LDA > 3.0, $p < 0.05$).

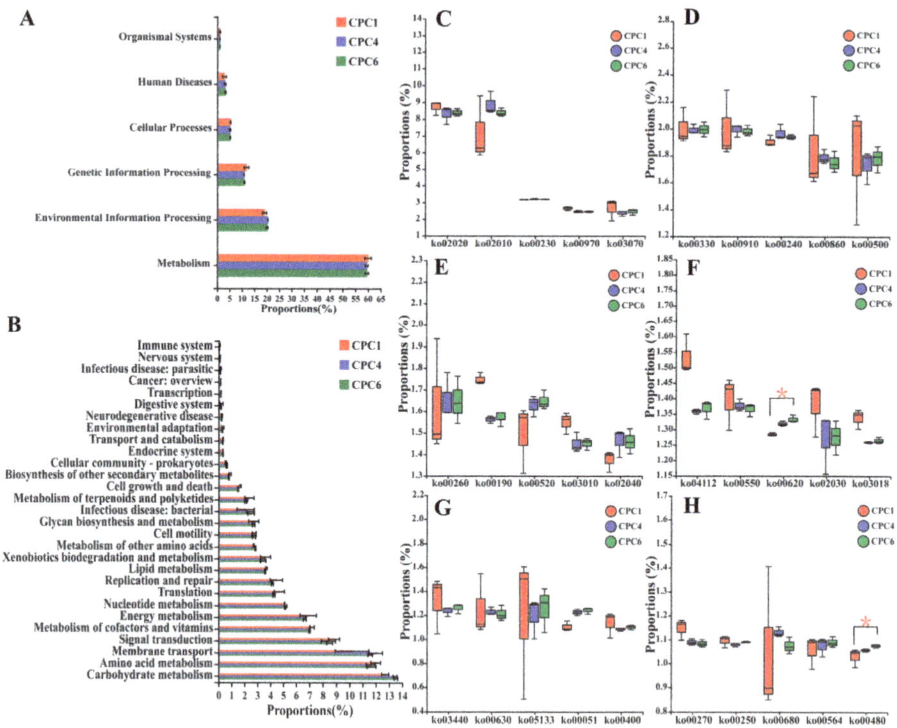

Figure 12. Functional prediction of the intestinal microbiota of grass carp fed with different protein level diets based on Tax4 Fun. (**A**) Functional prediction of the different treatment groups at Level 1. (**B**) Functional prediction of the different treatment groups at Level (top 30 relative abundances). (**C–H**) Differences in KEGG pathway abundance in the different treatment groups at Level 3 (top 30 relative abundances). Kruskal–Wallis rank-sum test was used for significance test (n = 3); * $0.01 < p \leq 0.05$.

4. Discussion

The present study showed that the optimal dietary protein levels improved grass carp SGR and FE, while high protein levels (CPC5 and CPC6) showed relatively poor growth performance. This may be due to the depression in feeding rate and more energy expenditure on processing excess protein for deamination, increasing the nitrogen metabolism burden and affecting fish growth [28]. Based on SGR and FE, the optimal dietary protein levels for juvenile grass carp (4.68 $\pm$ 0.01 g) were estimated to be 38.61 and 38.66% in this study, respectively, which was highly consistent with the 38.63% protein requirement of grass carp (6.80 $\pm$ 0.10 g) [20]. Moreover, the result of this study was close to the 40% requirement for juvenile grass carp (4.27 $\pm$ 0.01 g) reported by Jin et al. [10], but slightly lower than the 41–43% requirement for grass carp fry (0.15–0.20 g) reported by Dabrowski [9]. This variation in dietary protein requirements is strongly associated with different life-history stages, as reported in Abdel-Tawwab et al. [27], as heavier fish reduces the protein requirements. In other words, the present study indicates that CPC is a potential material in juvenile grass carp diet, with similar nutritional value for fish growth compared with soybean meal and other traditional materials.

In this study, PER decreased with increased protein levels, and was significantly lower in the CPC6 group compared with other groups, which was consistent with previous reports in Nile tilapia (*Oreochromis niloticus*) [27] and Songpu mirror carp [23]. Improved CF indicates good health and growth performance [10]. CF peaked in the CPC4 group in the present study, suggesting an optimal dietary protein level may promote grass carp growth by improving morphology. Previous studies have shown that HSI and MFI tend to decrease as dietary protein levels increase [10,27]. In this study, HSI and MFI first showed an increasing and then decreasing trend, which may be due to CPC use. Similar morphological parameters were reported when CPC was used to replace fishmeal in largemouth bass [4] and pearl gentian grouper [5]. The present study showed that high and low protein diets decreased dry matter digestibility, which further explains the depressed growth. In contrast, protein digestibility increased linearly, which indicates that a non-fishmeal diet (based on CPC) is tolerable for grass carp.

There was a significant negative correlation between net protein utilization (NPU) and urea levels [29]. The current study found that urea nitrogen increased linearly with the dietary protein level, suggesting that excess dietary protein can lead to protein waste. Previous studies on rainbow trout (*Oncorhynchus mykiss*) [30] and Nile tilapia [27] have shown that serum lipids tend to increase with dietary protein levels. Consistently, serum biochemical indices of triglyceride and LDL showed a trend of increasing first and then decreasing as dietary protein levels increased in this study. These serum parameters may be elevated due to converting excess protein to lipids and carbohydrates. Complement 3 and immunoglobulin M are important antibacterial compounds related to the immune response of teleost fish [11]. In this study, the content of C3 and IgM in the serum of fish showed quadratic and linear models, respectively, in response to the dietary protein level, which indicated that the CPC4 group had an improved immune ability. Consistently, the optimal protein levels could improve the immune function of the grass carp gut [11]. Similar results were also observed in mirror carp (*Cyprinus carpio*), in that the optimal protein level enhanced the C3, C4, and IgM of mirror carp at different water temperatures [31]. The immune-enhancing effect of appropriate dietary protein can be explained by increasing protein intake and feed utilization with appropriate dietary protein levels [10,27].

Oxidative damage is closely related to antioxidant enzymes and can be expressed using MDA content [32]. The current study demonstrated that optimal dietary protein levels increased the T-AOC content and GSH-Px activity while it reduced the MDA content in the hepatopancreas of grass carp, indicating that optimal dietary protein levels could protect grass carp from oxidative damage. However, the dietary protein levels did not significantly affect the hepatopancreatic SOD and CAT activities. Similarly, Xu et al. [11] reported that the optimal protein levels reduced the MDA contents but did not alter the CAT activities of grass carp in the mid intestine and distal intestine. ALT and AST are essential amino acid metabolizing enzymes and the improvement of ALT and AST reflects the vigorous activity of amino acid metabolism in fish [11,27]. In the current study, hepatopancreatic ALT and AST activities were improved in the CPC4 group, implying that amino acid metabolism was enhanced at the optimal dietary protein levels. These results suggest that the optimal protein levels for improved grass carp growth may be partly due to enhanced amino acid metabolism and improved antioxidant capacity. The hepatopancreas is an essential organ for metabolism in fish, and histological changes are considered a crucial indicator in evaluating nutritional status [19]. In the present study, fish fed CPC3 and CPC4 showed normal polygonal-shaped hepatocytes with large, clear nuclei centrally located without noticeable swelling or atrophy. These results indicate that the appropriate dietary protein level plays an essential role in promoting the antioxidant capacity and maintaining the health of the hepatopancreas. Noticeable, CPC5 and CPC6 diets depressed the hepatopancreatic ALT and AST activities and damaged the microstructure of the hepatopancreas, which indicated a reverse effect on fish metabolism and health, and possibly contributed to the depressed growth.

This study is the first to report the effect of protein levels on flesh quality when CPC was used as the protein source. Cooking loss, texture characteristics (including hardness, springiness, cohesiveness, gumminess, chewiness, resilience), and pH value are crucial parameters for evaluating muscle sensory quality [13,18]. The increasing cooking loss represents a decreased muscle water-holding capacity [12]. In this study, muscle cooking loss increased linearly with protein level, suggesting that a high-protein diet had poor muscle water-holding capacity. In addition, this study showed that the muscle pH value linearly decreased as the protein level increased. The high dietary protein group (CPC6 group) had a reduced pH value compared to the low dietary protein group (CPC1 group), which may be related to the production of more lactic acid in the muscle [12,13]. This study showed that the optimal protein level (CPC4 group) improved the texture characteristics of grass carp, such as hardness, cohesiveness, gumminess, chewiness, and resilience. Recent studies show that muscle shear force or hardness was maximized at the optimal dietary protein levels (fishmeal and casein) [12].

Flesh quality is often closely related to muscle fiber diameter, showing a negative correlation between the muscle fiber diameter and the hardness [14,16,19]. The larger myofiber diameters (class50, 60, and 70) accounted for more of the myofibers of fish from both the low protein level (CPC1 group) and the high protein level (CPC6 group). In contrast, the optimal protein level (CPC4 group) had smaller muscle fibers (class20, 30, and 40). Further statistical analysis showed that CPC4 had a smaller mean diameter of myofibers and a greater density of myofibers. The histological results of this study also confirmed the negative correlation between muscle fiber diameter and hardness.

In the current study, T-AOC was elevated at optimal protein levels, and the MDA content was significantly reduced, suggesting that the optimal protein levels can maintain muscle structural integrity by inhibiting oxidative damage. Further research found that the enhanced antioxidant capacity was partly attributable to the improved GSH content and SOD activity of the muscle. Differences in antioxidant capacity between treatments at different protein levels obtained in a previous study further corroborate the findings of this study [12].

After myoblasts initially form skeletal muscle, satellite cells provide additional nuclei required for skeletal muscle expansion [15]. MRFs regulate satellite cells, *myod* and *myf5* regulate satellite cell activation and proliferation, and *myog* and *mrf4* act on cell differentiation [15]. In addition, *myhc* plays a vital role in fish muscle growth by promoting myofiber proliferation and hypertrophy [16,19]. In this study, MRFs, *myhc-1*, and *myhc-4* were upregulated at the optimal dietary protein levels, indicating that appropriate dietary protein levels can effectively promote grass carp muscle growth. Moreover, optimal protein levels in this study significantly upregulated *fgf6a* and downregulated *mstn*. As reported in grass carp, the *fgf6a* plays a vital role in muscle growth regulation [16]. Besides, *mstn* has been shown to inhibit teleost muscle growth [15]. These results also confirmed that the optimal protein levels in this experiment could promote grass carp muscle growth by regulating MRFs, *fgf6a*, and *mstn*. A previous study showed that transcriptome analysis of broad bean-fed grass carp (higher hardness) muscle detected upregulation of *myog*, which plays a crucial role in promoting the formation of new muscle fibers [17]. Likewise, a recent study found that optimal dietary protein (soybean meal) levels promoted the expression of genes *myhc-1* and *myhc-4* by regulating a family of MRFs and contributed to flesh quality improvement [20]. Optimal dietary protein levels in this investigation may improve flesh quality through the same pattern.

Tor regulates phosphorylation of its downstream effectors, ribosomal S6 kinase 1 (*s6k1*) and eukaryotic translation initiation factor 4 e-binding protein 1 (*4e-bp1*), which ultimately promote protein synthesis in fish and can affect *nrf2* expression as an upstream regulator of antioxidant capacity [13]. Recent studies have shown that dietary protein levels can enhance the antioxidant capacity of grass carp muscle by upregulating *tor* and *s6 k1* [12]. The optimal dietary protein level in this study may promote the muscle antioxidant capacity through the same pattern. In addition, it has been demonstrated that *igf-I* and *igf-II* promote muscle growth in hybrid catfish [15] by binding to *igf1 r*, which may indicate that dietary protein may promote muscle growth through IGFs. Still, the specific mechanism needs to be further studied.

The intestinal microbiome has profound effects on human well-being, including host metabolism, physiology, nutrition, and immune function, and is even referred to as a "metabolic organ" [21]. Recently, studies on the regulation of intestinal microbes by dietary nutrients have received increasing attention [3,23]. In this study, the results of 16 S amplicon sequencing showed that the dietary protein decreased the intestinal microbiota diversity of grass carp, estimated using the Shannon and Simpson indices. The PCoA of the weighted UniFrac distances further revealed that CPC1, CPC4, and CPC6 were separated, and the weighted UniFrac-based ANOSIM revealed significant differences in microbiota structure between different protein levels. These findings suggest that protein levels may alter the intestinal microbiota structure of grass carp. In addition, this study identified the dominant phyla of grass carp were *Proteobacteria*, *Fusobacteriota*, *Actinobacteria*, *Firmicutes*, and *Bacteroidota*, which is consistent with the previous study on cyprinids [6,23]. Abundant *Proteobacteria* can be used to characterize intestinal microbial homeostasis, as dysregulation of homeostasis when the *Proteobacteria* abundance rises often leads to metabolic disturbances or inflammation [21]. A previous study of Songpu mirror carp reported that the abundance of *Proteobacteria* increases with protein levels [23]. This is partly consistent with the result between the CPC4 and CPC6 groups. The abundance of *Proteobacteria* was reduced at the optimal protein level (CPC4), which may benefit the homeostasis of intestinal microbiota.

More and more attention has been paid to studying the abundance ratio (F/B ratio) of *Firmicutes* and *Bacteroides* [33]. In humans and mice, increased F/B ratios are often associated with obesity, diabetes, and metabolic disorders [33]. In this study, the F/B value of the CPC4 group decreased significantly and showed good growth performance and health status, which provided a reference for studying the relationship between F/B and health. A decrease in the F/B ratio with increasing protein levels was consistently observed in Songpu mirror carp [23]. *Firmicutes* can extract energy from food [33]. Increased dietary protein levels in the present study promoted the abundance of *Firmicutes*, suggesting that the better weight gain in CPC4 and CPC6 groups than the low-protein group (CPC1 group) may be attributed to increased energy generation and utilization. The LEfSe analysis showed that the abundance of *Bacteroides* increased in response to the protein levels and contributed significantly to the differences in the CPC6 group. It is reported that *Bacteroidetes* have carbohydrate-related enzymes, and *Bacteroidetes* and *Firmicutes* play an essential role in the energy metabolism and glucose metabolism of organisms [34]. Combined with the apparent digestibility data, these results may imply that more energy is available for growth in the CPC4 group than in the CPC1 group. Still, the increase in ADC_p in the CPC6 group and the decrease in ADC_d indicate that less energy is available for growth in the CPC6 group than the CPC4 group.

At the genus level, *Cetobacterium* was the most abundant in the gut of grass carp through community composition maps, consistent with previous studies on freshwater fish [3,23]. *Cetobacterium* is reported to produce vitamin B12 and can ferment peptides and carbohydrates, and inhibit the growth of harmful bacteria [35]. *Akkermansia* represents a novel biomarker of intestinal metabolic health coupling and is essential for treating metabolic syndrome [36]. Spearman's correlation analysis showed that *Cetobacterium* and *Akkermansia* were strongly associated with growth performance while negatively correlated with muscle MDA content. These results well demonstrate the excellent growth performance of CPC4 in this study and explain its possible assistance to the antioxidant capacity of muscle. In contrast, *Microbacterium* and *Pseudomonas* exhibited a negative correlation with SGR, muscle hardness, and muscle GSH content while they were strongly positively correlated with muscle MDA content. These results confirmed that *Microbacterium* is one of the most closely related cornerstone genera of other bacteria and plays a crucial role in the growth and health of fish [37], and *Pseudomonas* is one of the most important opportunistic pathogens of grass carp [6]. In this study, in addition to being associated with growth and health, these genera also affected muscle hardness and antioxidant capacity. The interrelationship between intestinal microbiota and growth, health, and flesh quality requires more research to elucidate its specific regulatory mechanisms.

Intestinal microbiota affects the host mainly through its metabolites. The phylum *Fusobacterium* can metabolize carbohydrates to butyrate [34,38], which mediates the regulation of intestinal inflammatory processes, atherosclerosis, and immune system maturation [39]. In addition, among *Firmicutes*, the *Lachnospiraceae*, *Lactobacillaceae*, and *Ruminococcaceae* species hydrolyze starch and other sugars to produce butyrate and other short-chain fatty acids (SCFAs) [39], which inhibit the growth of harmful bacteria. In this study, *Fusobacterium* and *Lachnospiraceae* were significantly enriched in the CPC4 group, suggesting that optimal protein levels may further contribute to the health of grass carp by enhancing the abundance of SCFA-producing bacteria. However, the mechanism of their specific interactions needs further study.

The functional prediction of grass carp intestinal microbial communities based on Tax4 Fun revealed that, interestingly, at KEGG pathway level 3, ko00480 (Glutathione metabolism) and ko00620 (Pyruvate metabolism) showed significant differences between the CPC1 and CPC6 groups. Glutathione metabolism is part of amino acid metabolism and contains important antioxidant molecules to protect the body from oxidants [22]. A previous study showed that grass carp fed low protein levels had a lower intestinal GSH-Px and GSH content than the high protein levels [11]. Similarly, ko00480 (Glutathione metabolism) was significantly downregulated in the low protein level treatment group in this study, suggesting that a diet below the optimum protein requirement may impair amino acid metabolism functions and the antioxidant system. Cells convert glucose to pyruvate in the cytoplasmic matrix through glycolysis. Pyruvate can produce a large amount of adenosine triphosphate (ATP) under aerobic conditions. In contrast, pyruvate can produce lactate and a small amount of ATP through anaerobic glycolysis under anoxic conditions [40]. The significant enrichment of the ko00620 (Pyruvate metabolism) pathway in the CPC6 group may indicate that grass carp respond to a high-protein diet by enhancing glucose metabolism to obtain more energy to meet the needs of metabolizing protein. However, the mechanism of action between animal intestinal microbiota and metabolic function requires more studies.

Microcrystalline cellulose is the most commonly used filler and binder in fish feed [41]. Previous studies on the dietary protein requirements of red drum (*Sciaenops ocellatus*) [42] and dietary carbohydrate-to-lipid ratios of channel catfish (*Ictalurus punctatus*) [43] reached 15.56 and 40.61% microcrystalline cellulose use, respectively, which suggest that the use of a varying amount of cellulose in protein requirement studies is acceptable. In addition, a recent study also showed that using 1.84–31.84% microcrystalline cellulose in grass carp diets did not produce adverse effects [41]. This study was conducted to meet the demand for extruded feed in the aquatic feed market by producing extruded feed, and varying starch levels may disable the production of extruded diets at high starch levels, so microcrystalline cellulose was used as a filler to make the extruded feed. This might result in a relatively high optimal protein level. However, when comparing the optimal protein level with other results in grass carp, this study is acceptable.

5. Conclusions

In short, the optimal dietary protein level for juvenile grass carp provided by CPC was estimated to be 38.61–38.66%. Optimal dietary CPC levels promote grass carp performance, health, and flesh quality. In addition, optimal dietary CPC levels help maintain intestinal microbiota homeostasis. The results of this study demonstrate that CPC is a protein source with potentially significant application given its similar nutritional value compared to traditional protein sources. Altogether, the results of this study provide a theoretical basis for the development of sustainable and high-performance aquafeeds for valuable aquaculture species. However, based on the growth results in this study, we should pay attention to the amino acid composition of the CPC raw materials and fish carcasses, and explore the amino acid balance technology to better the CPC utilization in the future. In addition, the specific components of CPC that can play a role in affecting intestinal flora, and their specific mechanisms of regulating intestinal flora growth and health, are also worth exploring.

Author Contributions: G.L. carried out the experiment and drafted the manuscript; Q.T. designed the experiment and revised the manuscript; M.Z. and X.W. helped to culture the fish; X.M. analyzed the data; X.L. helped to carry out the sample analysis; S.X. and D.H. assisted with the experimental design and manuscript revision. All authors have read and agreed to the published version of the manuscript.

Funding: This work was financially supported by the National Key R&D Program of China (2019YFD0900200) and the National Natural Science Foundation of China (Grant No. 32072950).

Institutional Review Board Statement: This study was approved by the Ethical Committee for Laboratory Animals Care and Use of Huazhong Agricultural University. All procedures were carried out following the Guide for the Care and Use of Laboratory Animals.

Informed Consent Statement: Not applicable.

Data Availability Statement: Data is contained within the article.

Conflicts of Interest: The authors declare that there are no competing interest.

References

1. Olsen, R.L.; Hasan, M.R. A limited supply of fishmeal: Impact on future increases in global aquaculture production. *Trends Food Sci. Technol.* **2012**, *27*, 120–128. [CrossRef]
2. Montoya-Camacho, N.; Marquez-Ríos, E.; Castillo-Yáñez, F.J.; Cárdenas López, J.L.; López-Elías, J.A.; Ruíz-Cruz, S.; Jiménez-Ruíz, E.I.; Rivas-Vega, M.E.; Ocaño-Higuera, V.M. Advances in the use of alternative protein sources for tilapia feeding. *Rev. Aquac.* **2019**, *11*, 515–526. [CrossRef]
3. Xie, X.; Wang, J.; Guan, Y.; Xing, S.; Liang, X.; Xue, M.; Wang, J.; Chang, Y.; Leclercq, E. Cottonseed protein concentrate as fishmeal alternative for largemouth bass (*Micropterus salmoides*) supplemented a yeast-based paraprobiotic: Effects on growth performance, gut health and microbiome. *Aquaculture* **2022**, *551*, 737898. [CrossRef]
4. Xu, X.; Yang, H.; Zhang, C.; Bian, Y.; Yao, W.; Xu, Z.; Wang, Y.; Li, X.; Leng, X. Effects of replacing fishmeal with cottonseed protein concentrate on growth performance, flesh quality and gossypol deposition of largemouth bass (*Micropterus salmoides*). *Aquaculture* **2022**, *548*, 737551. [CrossRef]
5. Chen, G.; Yin, B.; Liu, H.; Tan, B.; Dong, X.; Yang, Q.; Chi, S.; Zhang, S. Effects of fishmeal replacement with cottonseed protein concentrate on growth, digestive proteinase, intestinal morphology and microflora in pearl gentian grouper (♀*Epinephelus fuscoguttatus* × ♂*Epinephelus lanceolatu*). *Aquac. Res.* **2020**, *51*, 2870–2884. [CrossRef]
6. Wu, S.; Wang, G.; Angert, E.R.; Wang, W.; Li, W.; Zou, H. Composition, diversity, and origin of the bacterial community in grass carp intestine. *PLoS ONE* **2012**, *7*, e30440. [CrossRef]

7. Lin, S.; Milardi, M.; Gao, Y.; Wong, M.H. Sustainable management of non-native grass carp as a protein source, weed-control agent and sport fish. *Aquac. Res.* **2022**. [CrossRef]
8. Food and Agriculture Organization of the United States (FAO). Global Aquaculture Production Quantity. 1950–2020. Available online: https://www.fao.org/fishery/statistics-query/en/aquaculture/aquaculture_quantity (accessed on 24 October 2022).
9. Dabrowski, K. Protein requirements of grass carp fry (*Ctenopharyngodon idella* Val.). *Aquaculture* **1977**, *12*, 63–73. [CrossRef]
10. Jin, Y.; Tian, L.X.; Xie, S.W.; Guo, D.Q.; Yang, H.J.; Liang, G.Y.; Liu, Y.J. Interactions between dietary protein levels, growth performance, feed utilization, gene expression and metabolic products in juvenile grass carp (*Ctenopharyngodon idella*). *Aquaculture* **2015**, *437*, 75–83. [CrossRef]
11. Xu, J.; Wu, P.; Jiang, W.D.; Liu, Y.; Jiang, J.; Kuang, S.Y.; Tang, L.; Tang, W.N.; Zhang, Y.A.; Zhou, X.Q.; et al. Optimal dietary protein level improved growth, disease resistance, intestinal immune and physical barrier function of young grass carp (*Ctenopharyngodon idella*). *Fish Shellfish. Immunol.* **2016**, *55*, 64–87. [CrossRef]
12. Xu, J.; Feng, L.; Jiang, W.D.; Wu, P.; Liu, Y.; Jiang, J.; Kuang, S.Y.; Tang, L.; Zhou, X.Q. Different dietary protein levels affect flesh quality, fatty acids and alter gene expression of Nrf2-mediated antioxidant enzymes in the muscle of grass carp (*Ctenopharyngodon idella*). *Aquaculture* **2018**, *493*, 272–282. [CrossRef]
13. Wang, B.; Liu, Y.; Feng, L.; Jiang, W.D.; Kuang, S.Y.; Jiang, J.; Li, S.H.; Tang, L.; Zhou, X.Q. Effects of dietary arginine supplementation on growth performance, flesh quality, muscle antioxidant capacity and antioxidant-related signalling molecule expression in young grass carp (*Ctenopharyngodon idella*). *Food Chem.* **2015**, *167*, 91–99. [CrossRef] [PubMed]
14. Johnston, I.A. Muscle development and growth: Potential implications for flesh quality in fish. *Aquaculture* **1999**, *177*, 99–115. [CrossRef]
15. Zhao, Y.; Li, J.Y.; Jiang, Q.; Zhou, X.Q.; Feng, L.; Liu, Y.; Jiang, W.D.; Wu, P.; Zhou, J.; Zhao, J. Leucine improved growth performance, muscle growth, and muscle protein deposition through AKT/TOR and AKT/FOXO3 a signaling pathways in hybrid catfish *Pelteobagrus vachelli* × *Leiocassis longirostris*. *Cells* **2020**, *9*, 327. [CrossRef]
16. Abouel Azm, F.R.; Kong, F.; Tan, Q.; Zhu, Y.; Yu, H.; Yao, J.; Luo, Z. Effects of replacement of dietary rapeseed meal by distiller's dried grains with solubles (DDGS) on growth performance, muscle texture, health and expression of muscle-related genes in grass carp (*Ctenopharyngodon idellus*). *Aquaculture* **2021**, *533*, 736169. [CrossRef]
17. Xu, W.H.; Guo, H.H.; Chen, S.J.; Wang, Y.Z.; Lin, Z.H.; Huang, X.D.; Tang, H.J.; He, Y.H.; Sun, J.J.; Gan, L. Transcriptome analysis revealed changes of multiple genes involved in muscle hardness in grass carp (*Ctenopharyngodon idellus*) fed with faba bean meal. *Food Chem.* **2020**, *314*, 126205. [CrossRef] [PubMed]
18. Lin, W.L.; Zeng, Q.X.; Zhu, Z.W. Different changes in mastication between crisp grass carp (*Ctenopharyngodon idellus* C. et V) and grass carp (*Ctenopharyngodon idellus*) after heating: The relationship between texture and ultrastructure in muscle tissue. *Food Res. Int.* **2009**, *42*, 271–278. [CrossRef]
19. Kong, F.; Abouel Azm, F.R.; Wang, X.; Zhu, Y.; Yu, H.; Yao, J.; Luo, Z.; Tan, Q. Effects of replacement of dietary cottonseed meal by distiller's dried grains with solubles on growth performance, muscle texture, health and expression of muscle-related genes in grass carp (*Ctenopharyngodon idellus*). *Aquac. Nutr.* **2021**, *27*, 1255–1266. [CrossRef]
20. Wang, X.Y.; Liu, G.Q.; Xie, S.Q.; Pan, L.; Tan, Q.S. Growth and meat quality of grass carp (*Ctenopharyngodon idellus*) responded to dietary protein (soybean meal) level through the muscle metabolism and gene expression of myosin heavy chains. *Front. Nutr.* **2022**, *9*, 833924. [CrossRef]
21. Guinane, C.M.; Cotter, P.D. Role of the gut microbiota in health and chronic gastrointestinal disease: Understanding a hidden metabolic organ. *Ther. Adv. Gastroenterol.* **2013**, *6*, 295–308. [CrossRef]
22. Eroglu, A.; Dogan, Z.; Kanak, E.; Atli, G.; Canli, M. Effects of heavy metals (Cd, Cu, Cr, Pb, Zn) on fish glutathione metabolism. *Environ. Sci. Pollut. Res.* **2015**, *22*, 3229–3237. [CrossRef] [PubMed]
23. Fan, Z.; Wu, D.; Zhang, Y.; Li, J.; Xu, Q.; Wang, L. Carbonate alkalinity and dietary protein levels affected growth performance, intestinal immune responses and intestinal microflora in Songpu mirror carp (*Cyprinus carpio* Songpu). *Aquaculture* **2021**, *545*, 737135. [CrossRef]
24. Przewłócka, K.; Folwarski, M.; Kaźmierczak-Siedlecka, K.; Skonieczna-Żydecka, K.; Kaczor, J.J. Gut-muscle axis exists and may affect skeletal muscle adaptation to training. *Nutrients* **2020**, *12*, 1451. [CrossRef] [PubMed]
25. AOAC. *Official Methods of Analysis of the Association of Official Analytical Chemists International*, 16th ed.; AOAC: Arlington, VA, USA, 1995.
26. Pfaffl, M.W. A new mathematical model for relative quantification in real-time RT-PCR. *Nucleic Acids Res.* **2001**, *29*, e45. [CrossRef]
27. Abdel-Tawwab, M.; Ahmad, M.H.; Khattab, Y.A.; Shalaby, A.M. Effect of dietary protein level, initial body weight, and their interaction on the growth, feed utilization, and physiological alterations of Nile tilapia, *Oreochromis niloticus* (L.). *Aquaculture* **2010**, *298*, 267–274. [CrossRef]
28. Mcgoogan, B.B.; Gatlin Iii, D.M. Dietary manipulations affecting growth and nitrogenous waste production of red drum, *Sciaenops ocellatus* I. Effects of dietary protein and energy levels. *Aquaculture* **1999**, *178*, 333–348. [CrossRef]
29. Bassily, N.; Michael, K.; Said, A. Blood urea content for evaluating dietary protein quality. *Mol. Nutr. Food Res.* **1982**, *26*, 759–764. [CrossRef]
30. Lone, K.; Inch, B.; Matty, A. Changes in the blood chemistry of rainbow trout, *Salmo gairdneri* Rich, in relation to dietary protein level, and an anabolic steroid hormone, ethylestrenol. *J. Fish Biol.* **1982**, *20*, 597–606. [CrossRef]

31. Huang, J.F.; Xu, Q.Y.; Chang, Y.M. Effects of temperature and dietary protein on gene expression of *Hsp70* and *Wap65* and immunity of juvenile mirror carp (*Cyprinus carpio*). *Aquac. Res.* **2015**, *46*, 2776–2788. [CrossRef]
32. Martínez-Álvarez, R.M.; Morales, A.E.; Sanz, A. Antioxidant defenses in fish: Biotic and abiotic factors. *Rev. Fish Biol. Fish.* **2005**, *15*, 75–88. [CrossRef]
33. Stojanov, S.; Berlec, A.; Štrukelj, B. The influence of probiotics on the firmicutes/bacteroidetes ratio in the treatment of obesity and inflammatory bowel disease. *Microorganisms* **2020**, *8*, 1715. [CrossRef] [PubMed]
34. Zhang, J.; Meng, H.; Kong, X.; Cheng, X.; Ma, T.; He, H.; Du, W.; Yang, S.; Li, S.; Zhang, L. Combined effects of polyethylene and organic contaminant on zebrafish (*Danio rerio*): Accumulation of 9-Nitroanthracene, biomarkers and intestinal microbiota. *Environ. Pollut.* **2021**, *277*, 116767. [CrossRef] [PubMed]
35. Tsuchiya, C.; Sakata, T.; Sugita, H. Novel ecological niche of *Cetobacterium somerae*, an anaerobic bacterium in the intestinal tracts of freshwater fish. *Lett. Appl. Microbiol.* **2008**, *46*, 43–48. [CrossRef] [PubMed]
36. Anhê, F.F.; Pilon, G.; Roy, D.; Desjardins, Y.; Levy, E.; Marette, A. Triggering *Akkermansia* with dietary polyphenols: A new weapon to combat the metabolic syndrome? *Gut Microbes* **2016**, *7*, 146–153. [CrossRef]
37. Kan, H.; Zhao, F.; Zhang, X.X.; Ren, H.; Gao, S. Correlations of gut microbial community shift with hepatic damage and growth inhibition of *Carassius auratus* induced by pentachlorophenol exposure. *Environ. Sci. Technol.* **2015**, *49*, 11894–11902. [CrossRef]
38. Rimoldi, S.; Gliozheni, E.; Ascione, C.; Gini, E.; Terova, G. Effect of a specific composition of short-and medium-chain fatty acid 1-Monoglycerides on growth performances and gut microbiota of gilthead sea bream (*Sparus aurata*). *PeerJ* **2018**, *6*, e5355. [CrossRef]
39. Vacca, M.; Celano, G.; Calabrese, F.M.; Portincasa, P.; Gobbetti, M.; De Angelis, M. The controversial role of human gut lachnospiraceae. *Microorganisms* **2020**, *8*, 573. [CrossRef]
40. Abdel-Wahab, A.F.; Mahmoud, W.; Al-Harizy, R.M. Targeting glucose metabolism to suppress cancer progression: Prospective of anti-glycolytic cancer therapy. *Pharmacol. Res.* **2019**, *150*, 104511. [CrossRef]
41. Cai, W.; Liang, X.F.; Yuan, X.; Liu, L.; He, S.; Li, J.; Li, B.; Xue, M. Different strategies of grass carp (*Ctenopharyngodon idella*) responding to insufficient or excessive dietary carbohydrate. *Aquaculture* **2018**, *497*, 292–298. [CrossRef]
42. Serrano, J.A.; Nematipour, G.R.; Gatlin Iii, D.M. Dietary protein requirement of the red drum (*Sciaenops ocellatus*) and relative use of dietary carbohydrate and lipid. *Aquaculture* **1992**, *101*, 283–291. [CrossRef]
43. Garling, D.L., Jr.; Wilson, R.P. Effects of dietary carbohydrate-to-lipid ratios on growth and body composition of fingerling channel catfish. *Prog. Fish Cult.* **1977**, *39*, 43–47. [CrossRef]

metabolites

MDPI

Article

Evaluate of Wheat Gluten as a Protein Alternative for Fish Meal and Soy Protein Concentrate in Red Spotted Grouper *Epinephelus akaara*

Yanbo Cheng [1,2,3], Yongchao Wang [1], Zhiyong Dong [1,2], Trond Storebakken [2], Guohuan Xu [3], Bo Shi [1,*] and Yuexing Zhang [1,*]

1 National Engineering Research Center for Marine Facilities Aquaculture, Marine Science and Technology College, Zhejiang Ocean University, Zhoushan 316022, China; chengyanbo@gdim.cn (Y.C.); waych24678@163.com (Y.W.); zhiyong0310@126.com (Z.D.)

2 Department of Animal and Aquacultural Sciences, Faculty of Biosciences, Norwegian University of Life Sciences, NO-1432 Ås, Norway; trond.storebakken@nmbu.no

3 Guangdong Provincial Key Laboratory of Microbial Culture Collection and Application, State Key Laboratory of Applied Microbiology Southern China, Institute of Microbiology, Guangdong Academy of Sciences, Guangzhou 510070, China; xuguohuan@gdim.cn

* Correspondence: shibo@zjou.edu.cn (B.S.); yuexing.zhang@zjou.edu.cn (Y.Z.)

Abstract: The aim of this study was to evaluate the effects of wheat gluten as a substitute for fish meal (FM) and soy protein concentrate (SPC) in the low-fishmeal-based extruded diet in red spotted grouper *Epinephelus akaara*. Eight isonitrogenous (441–456 g kg^{-1}) and isocaloric (21.5–22.0 MJ kg^{-1}) diets were produced, including the control diet (R0), three diets with 33.3, 66.7, and 100% FM being replaced by a mixture of wheat gluten, wheat, and taurine (GWT) (RF1, RF2, RF3), three diets with 33.3, 66.7, and 100% SPC replaced by GWT (RS2, RS2, RS3) and one diet with 50% FM and 50% SPC replaced by GWT (RFS). Results showed that feed intake (FI), weight gain (WG), protein retention efficiency, and liver superoxide dismutase activity increased linearly, while feed conversion ratio (FCR) decreased linearly with the decrease of dietary FM. Additionally, FI, WG, and FCR significantly increased with decreasing dietary SPC. Overall, 100% FM or 61.2% SPC can be safely replaced by wheat gluten in the red-spotted grouper diet containing 20.0% FM and 21.4% SPC.

Keywords: *Epinephelus akaara*; fish meal; wheat gluten; soy protein concentrate

Citation: Cheng, Y.; Wang, Y.; Dong, Z.; Storebakken, T.; Xu, G.; Shi, B.; Zhang, Y. Evaluate of Wheat Gluten as a Protein Alternative for Fish Meal and Soy Protein Concentrate in Red Spotted Grouper *Epinephelus akaara*. *Metabolites* **2023**, *13*, 832. https://doi.org/10.3390/metabo13070832

Academic Editors: Qingchao Wang, Yan He and Qingsong Tan

Received: 8 June 2023
Revised: 4 July 2023
Accepted: 6 July 2023
Published: 10 July 2023

1. Introduction

With the development of the aquaculture industry, aquaculture by-products such as fish waste, shrimp, molluscs, crustaceans, and sand silkworms are no longer sustainable for the development of the grouper culture industry. Therefore, high-quality formula feeds are urgently needed. Fish meal (FM) is the main protein source in aquatic feed, especially for carnivorous fish. However, the limited supply of FM and increasing demand make it necessary to investigate the substitution of FM with plant-based proteins that are highly nutritious, widely available, inexpensive, and easily stored [1]. Among all plant proteins, soy protein concentrate (SPC) has been widely used due to its nutritional characteristics (high protein and balanced amino acid profile) and relatively low content of anti-nutritional factors compared with soybean or other soybean derivatives [2]. Previous research indicated that SPC had become a widely recognised FM alternative in marine fish diets, such as rainbow trout *Oncorhynchus mykiss* [3], Japanese flounder *Paralichthys olivaceus* [4], red sea bream *Pagrus major* [5–7], and hybrid grouper *Epinephelus fuscoguttatus* ♀ × *E. lanceolatus* ♂ [8–10]. However, the price of SPC often fluctuates owing to the unstable price of soybeans; thus, finding other plant protein sources is still necessary.

Vital wheat gluten is another promising plant protein source in aquafeeds due to its high protein content and digestion, balanced amino acid composition, low content of non-starch polysaccharides, and anti-nutritional factors. Moreover, a small amount of starch can be used to accelerate ripening and as a binder during processing. Research showed that wheat gluten has similar protein digestibility to FM and SPC [11]. Currently, wheat gluten has been used as an FM substitute in rainbow trout [12,13], Atlantic halibut *Hippoglossus hippoglossus* [14], European sea bass *Dicentrarchus labrax* L. [15], large yellow croaker *Larimichthys crocea* [16], and did not cause adverse effects on growth or health.

Red spotted grouper *Epinephelus akaara*, known as red grouper or Hong Kong grouper, is commonly cultured in Southeast Asia and China. The optimum dietary protein, carbohydrate, and lipid levels for red spotted grouper were 508.3 g kg^{-1} (initial body weight = 7.88 $\pm$ 0.04 g), 76.4 g kg^{-1} (IBW = 7.79 $\pm$ 0.01 g), and less than 91.1 g kg^{-1} (IBW = 2.51 g), respectively [17–19]. Moreover, the optimal dietary protein requirements for orange-spotted grouper *Epinephelus coioides* were 480 (IBW = 10.7 $\pm$ 0.2 g), 521.84 (IBW = 10.02 $\pm$ 0.22 g) and 466.65 (IBW = 102.8 $\pm$ 1.02 g) g kg^{-1} reported by Luo et al. (2004), Yan et al. (2021) and Yan et al. (2020), respectively [20–22]. Groupers need a high protein level (about 500 g kg^{-1}) for optimum growth. However, due to protein is the most expensive ingredient in the diet, some other protein sources have been used to replace fishmeal in orange-spotted grouper and pearl gentian grouper feeds, such as soybean meal, animal by-product meals, cottonseed protein concentrate, black soldier fly, soy protein concentrate, Tenebrio molitor meal, etc., but less studied in red spotted grouper [23–28]. Therefore, the present study aimed to explore the feasibility of using wheat gluten as a protein source in the low-FM-based diet for red spotted groupers by replacing FM or SPC with a large amount of wheat gluten.

2. Materials and Methods

2.1. Ingredients and Diets

Low-temperature dried FM, SPC, and wheat gluten were used as the main protein sources. The proximate compositions and amino acid profiles of FM, SPC, and wheat gluten are shown in Table 1. GWT is an ingredient blend containing 77.5% wheat gluten, 20.5% wheat flour, and 2.0% taurine. Eight isonitrogenous (441–456 g kg^{-1}) and isocaloric (21.5–22.0 MJ kg^{-1}) diets were formulated, including a control diet (R0, 20% FM, and 21.4% SPC), six diets with 33.3, 66.7, and 100% FM or SPC being gradually replaced by GWT (RF1, RF2, RF3, RS1, RS2, and RS3), and one diet with 50% FM and 50% SPC replaced by GWT (RFS). Diets were produced at the Feed Technology Laboratory of the Feed Research Institute, Chinese Academy of Agricultural Sciences in Beijing. All dry ingredients were ground in a hammer mill through a 0.18 mm screen, mixed, preconditioned, and extruded in a twin-screw extruder (MY56X2A, Muyang, Yangzhou, China) with a 2.0 mm die plate. All extruded pellets were dried to 950 g kg^{-1} dry matter at ambient temperature. Fish oil and soy lecithin were coated into pellets with a vacuum coater (ZJB-100). Feed formulations and chemical compositions of experimental diets are shown in Table 2.

Table 1. Compositions and amino acid profiles of wheat gluten, soy protein concentration, and fish meal in the experiment (DM).

Ingredients	Fish Meal †	Wheat Gluten ‡	Soy Protein Concentration §
Compositions, kg^{-1}			
Dry matter, g	913	935	925
Crude protein, g	748	837	694
Crude fat, g	117	44	47
Starch, g	-	74	-
Ash, g	122	10	58
Gross energy, MJ	21.8	22.0	20.1

Table 1. *Cont.*

Ingredients	Fish Meal †	Wheat Gluten ‡	Soy Protein Concentration §
Essential amino acids (EAA), g (16 g N)$^{-1}$			
Arg	5.5	3.6	7.4
His	1.7	2.0	2.5
Leu	7.0	7.0	8.1
Ile	3.8	3.5	4.6
Lys	7.3	1.8	6.7
Met	2.6	1.4	0.9
Phe	3.8	5.2	5.3
Thr	4.1	2.5	4.1
Tyr	3.2	3.3	3.5
Val	4.5	3.8	4.7
Total EAA ¶	43.3	34.1	47.9
Non-essential amino acids (NEAA), g (16 g N)$^{-1}$			
Ala	6.1	2.7	4.4
Asp	8.4	3.2	11.6
Cys	0.8	1.8	0.3
Glu	13.1	35.2	19.1
Gly	5.8	3.4	4.2
Pro	4.2	12.8	5.0
Ser	4.0	4.6	5.2
Total NEAA	42.3	63.8	49.8
Total AA ¶	85.6	97.9	97.6

† TripleNine®, low-temperature dried fish meal, Esbjerg, Denmark. ‡ AMYGLUTEN 110, Syral Belgium N.V., Aalst, Belgium. § YIHAI®, Wilpromil, Goldensea Grain and Oil Industry Co., Ltd., Wilmar, Qinhuangdao, China. ¶ Trp excluded.

Table 2. Feed formulations and chemical compositions of experimental diets (DM).

Ingredients, g kg^{-1}	R0	RF1	RF2	RF3	RS1	RS2	RS3	RFS
Constant feed ingredients †	221.5	221.5	221.5	221.5	221.5	221.5	221.5	221.5
GWT ‡	-	69.0	138.0	207.0	69.0	138.0	207.0	207.0
Fish meal §	200.0	133.0	66.0	-	200.0	200.0	200.0	100.0
Soy protein concentrate	214.0	214.0	214.0	214.0	142.0	71.0	-	107.0
Wheat flour	266.4	250.4	234.6	217.8	262.9	259.1	255.4	236.3
Fish oil	84.0	89.0	94.0	99.0	84.0	84.0	84.0	92.0
Mono calcium Phosphate ¶	14.0	17.0	20.0	23.0	14.5	14.5	14.5	19.0
L-Lysine ††	-	3.7	7.4	11.0	3.1	6.1	9.1	10.0
DL-Methionine ††	-	0.4	0.8	1.2	-	-	-	0.2
L-Arginine ††	-	1.1	2.1	3.1	2.1	4.2	6.2	4.7
L-Threonine ††	-	0.8	1.5	2.3	0.8	1.5	2.2	2.2
Analysed content, kg^{-1}								
Dry matter, g	951	954	957	956	957	957	957	955
Crude protein, g	441	447	444	449	447	451	456	454
Crude fat, g	135	136	139	142	140	141	131	140
Ash, g	76	69	61	54	72	68	54	59
Gross energy, MJ	21.5	21.6	21.6	22.0	21.5	21.7	21.8	22.0
Essential amino acid, g (16 N)$^{-1}$								

Table 2. *Cont.*

Ingredients, g kg^{-1}	R0	RF1	RF2	RF3	RS1	RS2	RS3	RFS
Arg	6.5	6.5	6.6	6.4	6.5	6.3	6.8	6.2
Ile	4.1	3.9	3.9	3.8	3.9	3.7.	3.7	3.5
Leu	7.2	7.0	7.2	7.0	7.0	6.7	6.9	6.6
Lys	6.1	6.0	6.1	6.1	6.1	5.9	6.2	5.9
Met + Cys	2.1	2.0	2.1	2.1	2.0	2.0	2.5	1.9
Phe + Tyr	7.4	7.3	7.7	7.7	7.1	6.7	7.3	6.8
Thr	3.8	3.8	3.8	3.7	3.8	3.7	3.8	3.5
Val	4.4	4.2	4.1	4.0	4.2	4.0	4.1	3.8
Total essential amino acid	41.7	40.7	41.5	40.7	40.6	39.0	41.4	38.2
Non-essential amino acid, g (16 N)$^{-1}$								
Ala	4.7	4.3	4.0	3.5	4.6	4.2	4.3	3.7
Asp	9.9	8.8	8.6	7.8	8.8	7.8	7.1	6.9
Glu	16.3	18.7	21.0	22.9	18.0	19.1	22.1	21.4
Gly	5.0	4.5	4.3	3.9	4.7	4.5	4.69	4.0
Pro	4.7	5.4	6.1	7.0	5.5	6.0	7.2	7.2
Ser	4.4	4.4	4.6	4.5	4.3	4.2	4.3	4.2
Total non-essential amino acid	45.0	46.0	48.6	49.6	46.0	45.7	49.7	47.3
Total amino acid	86.6	86.8	90.1	90.4	86.5	86.6	90.9	85.6

† Constant feed ingredients, kg^{-1}: Soybean meal 70 g (FENGYUAN®, Goldensea Grain and Oil Industry Co., Ltd., Wilmar, Qinhuangdao, China); Peanut meal 70 g (FENGYUAN®, Goldensea Grain and Oil Industry Co., Ltd., Wilmar, Qinhuangdao, China); Krill meal 50 g (QRILLTM, Antarctic Krill Meal, Aker BioMarine, Oslo, Norway); Soy lecithin 20 g (FENGYUAN®, Goldensea Grain and Oil Industry Co., Ltd., Wilmar, Qinhuangdao, China); Premix 10 g (Vitamin premix (mg kg^{-1} diet): vitamin A, 20; vitamin B_1, 12; vitamin B_2, 10; vitamin B_6, 15; vitamin B_{12}, 8; niacinamide, 100; ascorbic acid, 1000; calcium pantothenate, 40; biotin, 2; folic acid, 10; vitamin E, 400; vitamin K_3, 20; vitamin D_3, 10; inositol, 200; corn protein powder, 150; Mineral premix (mg kg^{-1} diet): $CuSO_4 \cdot 5H_2O$, 10; $FeSO_4 \cdot H_2O$, 300; $ZnSO_4 \cdot H_2O$, 200; $MnSO_4 \cdot H_2O$, 100; KI (10%), 80; Na_2SeO_3 (10% Se), 67; $CoCl_2 \cdot 6H_2O$ (10% Co), 5; NaCl, 100; zeolite, 638; Vitamin premix: mineral premix = 2:1); Choline Cl 1.5 g (Be-long corporation, Nanjing, China). ‡ Mixture of vital wheat gluten, wheat flour, and taurine (ratio: 77.5%, 20.5%, and 2.0%). Wheat flour, BLUEKEY®, Beijing Grain and Oil Industry Co., Ltd., Wilmar, Beijing, China. Taurine-JP8, Qianjiang Yongan Pharmaceutical Co., Ltd., Qianjiang, China. § Aidayufen Co., Ltd., Rongcheng, China. ¶ MCP22, mono calcium phosphate, feed grade, Suntran Industrial Group Ltd., Hefei, China. †† Siwei Development Group Ltd., Hangzhou, China.

2.2. Fish Management

After a 2-week acclimation, a 56-day feeding trial was conducted in the Fish Laboratory of the Sino-European Aquatic Nutrition and Feed Resource Institute, Zhejiang Ocean University, with an indoor seawater recirculation system. Before the feeding trial, 1200 healthy juveniles with a similar size (initial weight 9.55 ± 0.03 g) were selected, batch-weighed and randomly assigned to 24 circular 1000-L tanks (the water volume was approximately 750 L), with 50 fish per tank. Each tank was manually fed four times per day (08:00 a.m., 11:30 a.m., 3:00 p.m., and 6:30 p.m.), with 30 min per meal to ensure apparent satiety. After each feeding, the remaining feed pellets in each tank were counted according to the method of Zhang et al. [29] and siphoned out immediately. The daily feeding rate was tentatively set at 10% in excess based on the average intake over the past three days, and fish could receive more feed if they showed signs of feeding. During the feeding, each tank was supplied with seawater at a flow rate of 8–9 L min^{-1} and additional aeration via a nano-air-stone to ensure the dissolved oxygen was above 6.0 mg L^{-1}. One-third of the seawater in each tank was exchanged daily with an equal amount of new sand-filtered seawater through the filter system. Artificial photoperiod was adopted, average water temperament and salinity were 27.9 °C and 28 ppt throughout the whole trial, and the photoperiod was 12D:12L.

2.3. Sampling

Before the feeding trial, 3 × 12 fish of similar size were selected and stored at −20 °C for whole-body composition analysis. After a 28-day feeding trial, five fish per tank were weighed to calculate weight gain. At the end of the 56-day feeding trial, fish in the same tank were counted and batch-weighed after anaesthesia with MS-222 (90 mg L^{-1}). Five fish per tank were randomly selected to determine body length and weight individually, and then the liver and carcass were weighed to calculate morphologic indexes. Five fish in each tank were taken as whole-body samples. Blood samples were collected from the caudal vein of fish with disposable and disinfected injectors, centrifuged at 4 °C for 10 min (4000× *g*), and then stored at −80 °C until analysis. Liver from six fish per tank were collected for analyses of antioxidant parameters.

2.4. Analysis

Initial and final whole-body samples from the same tank were cut into pieces and grounded with a meat grinder, then autoclaved (YXQ-LS, Xunbo, Shanghai, China) at 120 °C for 30 min, homogenised by a homogeniser (DS-1, Shanghai, China), and oven-dried (Jinghong, China) at 80 °C. Whole-body samples were finely ground with a pestle and mortar until all samples passed through the 0.9 mm screen. Dried whole fish and feed samples were analysed for dry matter (105 °C to constant weight), protein (Kjeldahl's method), lipid (Soxhlet Extraction System, Jingke, Shanghai, China), amino acids (amino acid analyser, L-8900, Hitachi, Tokyo, Japan), ash (550 °C, overnight) and energy (Phillipson Microbomb Calorimeter, Gentry Instruments Inc., Aiken, SC, USA) based on the previous studies [17]. Contents of total protein (TP), triglycerides (TG), cholesterol (TC), glucose (GLU), and activities of glutamate–pyruvate transaminase (ALT), glutamate–oxaloacetate transaminase (AST) and superoxide dismutase (SOD) were determined by the BECKMAN COULTER AU5800 automatic biochemical analyser. In addition, superoxide dismutase (SOD) activity and MDA (malonaldehyde) content in the liver were determined using commercial kits (Nanjing Jiancheng Bioengineering Institute, Nanjing, China) according to the manufacturer's instructions.

2.5. Calculation and Statistical Analysis

The parameters were calculated as follows:

$$\text{Weight gain (WG, g)} = \text{final body weight} - \text{initial body weight};$$

$$\text{Feed conversion ratio (FCR, g DM ingested (g gain)}^{-1}) = \text{feed intake}/(\text{final mean body weight—initial mean body weight});$$

$$\text{Hepatosomatic index (HSI, \%)} = \text{liver weight}/\text{body weight} \times 100;$$

$$\text{Viscerosomatic index (VSI, \%)} = \text{viscera weight}/\text{body weight} \times 100;$$

$$\text{Condition factor (CF, g/cm}^3) = \text{body weight}/\text{body length}^3 \times 100;$$

$$\text{Protein retention efficiency (PRE, \%)} = \text{protein gain}/\text{protein intake} \times 100;$$

$$\text{Energy storage ratio (ERE, \%)} = \text{energy gain}/\text{energy intake} \times 100.$$

All data are presented as means ± S.E.M. (n = 3) and assessed by one-way analysis of variance (ANOVA) using the Origin 8.0 Pro SR4 (Origin Lab. Co., Northampton, MA, USA) software. Results were considered to be significant at $p < 0.05$. Moreover, a follow-up trend analysis using orthogonal polynomial contrasts was performed to determine whether the significances were linear and/or quadratic. Limitations of this study are the post hoc analysis performed after ANOVA and the lower regression index R in several analyses.

3. Results

3.1. Growth Performance, Feed Utilization and Morphologic Indexes

No fish died during the 56-day feeding trial. Furthermore, fish in all groups showed high growth performance and feed utilisation (Table 3). On the 56-day feeding trial, significant differences were recorded in FI, WG and FCR among different groups. Fish fed with the RF3 diet had the highest WG (~240%), whereas fish fed with the diet without wheat gluten supplementation (R0 diet) showed the lowest WG (~150%) among all treatments. Fish fed with RF3, RS2 and RFS diets showed a higher value of FI (24–25 g per fish^{-1}) than those fed the R0 diet (~19 g per fish^{-1}) ($p < 0.05$). The lowest value of FCR was observed in the RF3 and RS2 groups, while the highest value was found in the R0 group. However, no significant differences were recorded in FI, WG, and FCR among all groups on the 28-day feeding trial ($p > 0.05$). Similarly, no significant differences were found in HSI, VSI, and CF among fish fed different experimental diets (Table 4).

Table 3. Growth performance and feed utilisation of red spotted grouper fed diet with different GWT inclusion levels.

Diets	Feed Intake, g DM Fish^{-1}		Weight Gain, g Fish^{-1}		Feed Conversion Ratio, g DM Ingested (g Gain)$^{-1}$	
	0–28 Days	0–56 Days	0–28 Days	0–56 Days	0–28 Days	0–56 Days
R0	11.9 ± 0.28	18.7 ± 0.46 †	12.2 ± 0.36	15.2 ± 0.57 †	0.98 ± 0.01	1.23 ± 0.02 ‡
RF1	12.4 ± 0.16	21.5 ± 1.22 †‡	12.5 ± 0.24	18.6 ± 2.07 †‡	0.99 ± 0.01	1.17 ± 0.06 †‡
RF2	12.0 ± 0.04	22.4 ± 0.94 †‡	12.9 ± 0.14	20.6 ± 1.56 †‡	0.93 ± 0.01	1.09 ± 0.05 †‡
RF3	12.1 ± 0.25	24.8 ± 1.10 ‡	12.5 ± 0.35	24.1 ± 1.04 ‡	0.96 ± 0.02	1.03 ± 0.01 †
RS1	12.3 ± 0.52	21.1 ± 0.92 †	12.1 ± 0.73	18.2 ± 1.22 †	1.02 ± 0.03	1.17 ± 0.03 †‡
RS2	13.1 ± 0.17	25.0 ± 1.17 ‡	14.2 ± 0.33	23.6 ± 1.83 ‡	0.92 ± 0.01	1.07 ± 0.03 †‡
RS3	12.4 ± 0.61	21.4 ± 0.67 †‡	12.7 ± 1.01	18.3 ± 0.80 †‡	0.98 ± 0.03	1.17 ± 0.02 †‡
RFS	12.7 ± 0.20	23.9 ± 1.59 ‡	13.4 ± 0.29	22.0 ± 2.28 †‡	0.95 ± 0.02	1.10 ± 0.05 †‡
ANOVA *p*	0.27	0.01	0.15	0.01	0.05	0.03

Data were represented as means ± S.E.M. Different superscript symbols indicate significant differences ($p < 0.05$) among treatments.

Table 4. Morphologic indexes of red spotted grouper fed diet with different GWT inclusion levels.

Parameters	R0	RF1	RF2	RF3	RS1	RS2	RS3	RFS	*p*
HSI, %	2.08 ± 0.25	1.76 ± 0.25	2.16 ± 0.16	2.03 ± 0.20	1.76 ± 0.22	1.97 ± 0.20	2.08 ± 0.15	1.81 ± 0.26	0.61
VSI, %	13.8 ± 0.23	13.8 ± 0.28	13.4 ± 0.46	14.3 ± 0.35	13.5 ± 0.48	13.4 ± 0.30	13.2 ± 0.19	13.2 ± 0.39	0.31
CF, g/cm^3	2.72 ± 0.04	2.84 ± 0.08	2.89 ± 0.07	2.93 ± 0.06	2.80 ± 0.06	2.80 ± 0.04	2.96 ± 0.09	2.81 ± 0.10	0.27

Data were represented as means ± S.E.M.

To further investigate the effects of wheat gluten replacing FM and SPC in red spotted grouper, the regression method was used to analyse the correlation between wheat gluten supplementation level and WG, FI, and FCR. The results showed that the above indices varied linearly with decreasing dietary FM level but quadratic polynomially with reducing dietary SCP level (Figure 1). Specifically, based on the first-order fitting equation, the level of dietary GWT and SPC were estimated to be 207 g kg^{-1} and 214 g kg^{-1}, respectively (Figure 1A,C,E). The optimum levels of wheat gluten according to FI, WG, and FCR were 128 g kg^{-1} (GWT inclusion level was ~61.7%), 131 g kg^{-1} (~63.7%) and 127 g kg^{-1} (~61.2%) based on the second-order regression equation with 200 g kg^{-1} FM in the feed formula, respectively (Figure 1B,D,F).

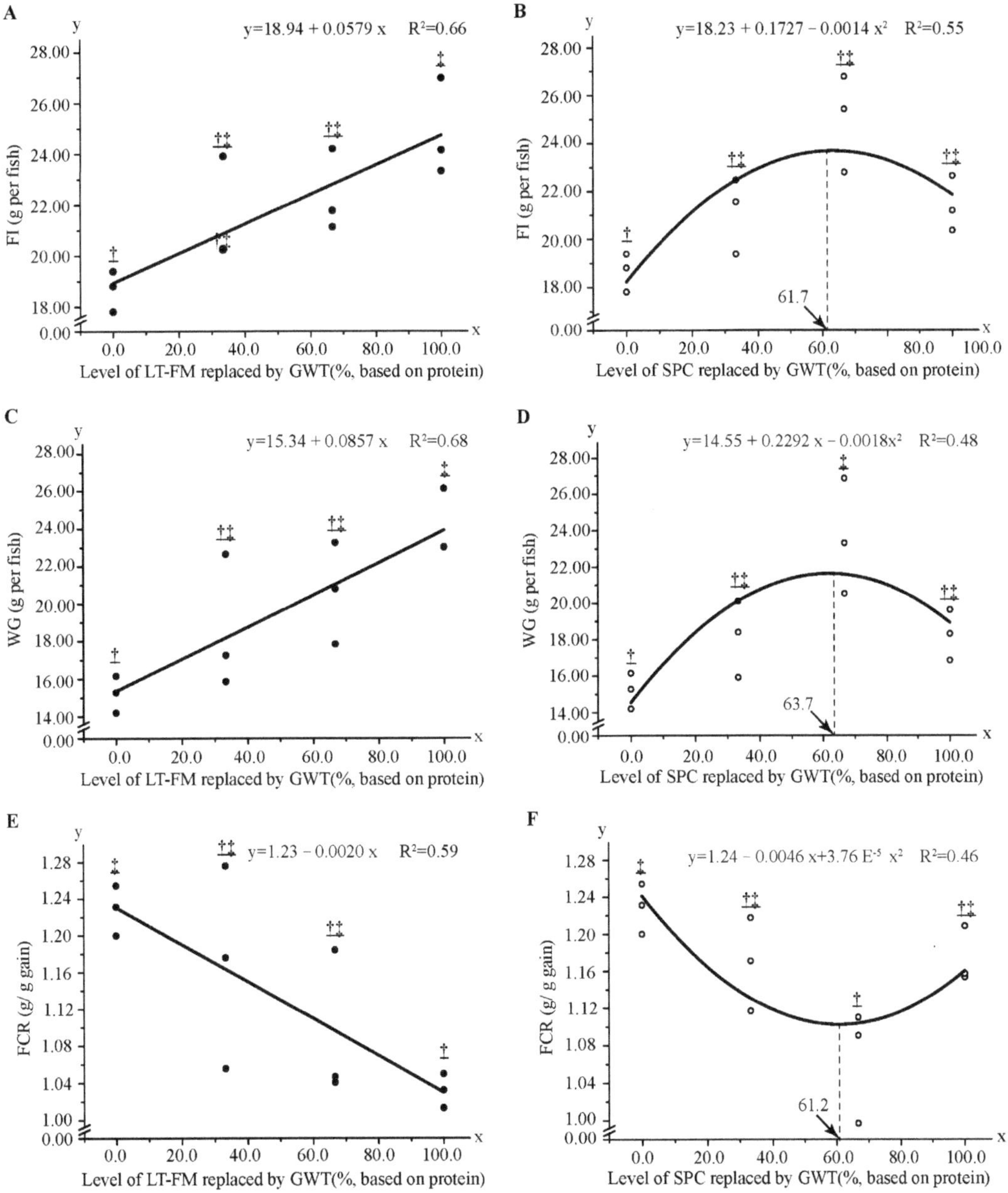

Figure 1. Regression analysis of growth performance and feed utilisation in red spotted grouper fed diet with different GWT inclusion levels. Data were analysed by one-way ANOVA followed by Tukey multiple comparisons, with different underlined symbols indicating significant differences ($p < 0.05$) among treatments. (**A**,**C**,**E**) show the regression analysis of replacing LT-FM with GWT on growth performance and feed utilisation. (**B**,**D**,**F**) represent the regression analysis of replacing SPC with GWT on growth performance and feed utilisation. GWT, a mixture of wheat gluten, wheat, and taurine; LT-FM, low-temperature dried fish meal; SPC, soy protein concentrate.

3.2. Body Compositions and Nutrient Retentions

The proximate compositions in whole-body and nutrient retentions are shown in Table 5. Contents of protein, fat and ash, and energy level and ERE were not significantly different among all treatments ($p > 0.05$). Fish fed with the diets RF2 and RF3 had the highest PRE, whereas the R0 diet showed the lowest value. PRE increased linearly with the increase of dietary FM substitution level (Figure 2A), while ERE increased quadratically with increasing dietary SPC replacement level (Figure 2B). The maximum ERE was obtained when 54.3% SPC was replaced by wheat gluten based on the second-order regression equation.

Table 5. Body compositions and nutrition retentions of red spotted grouper fed diet with different GWT inclusion levels.

Diets	Dry Matter, g kg^{-1}	Proximate Composition, g kg^{-1}			Energy, KJ kg^{-1}	Retention, %	
		Protein	Lipid	Ash		Protein (PRE)	Energy (ERE)
R0	331 ± 5.78	176 ± 1.53	106 ± 2.85	49.7 ± 2.24	8.06 ± 0.12	32.7 ± 0.28 †	34.3 ± 0.71
RF1	326 ± 5.29	174 ± 5.24	103 ± 3.84	47.9 ± 2.37	8.43 ± 0.34	33.9 ± 1.66 †‡	34.7 ± 2.33
RF2	333 ± 9.45	178 ± 2.73	111 ± 4.37	44.3 ± 2.78	8.26 ± 0.20	38.1 ± 1.17 ‡	38.6 ± 1.18
RF3	324 ± 5.78	173 ± 2.40	110 ± 2.19	40.5 ± 1.50	8.24 ± 0.14	37.6 ± 0.68 ‡	39.4 ± 1.19
RS1	335 ± 0.33	179 ± 1.67	110 ± 1.86	46.7 ± 0.35	8.30 ± 0.06	35.9 ± 1.20 †‡	36.8 ± 0.45
RS2	327 ± 3.18	175 ± 0.88	108 ± 2.73	44.5 ± 1.37	8.19 ± 0.13	36.4 ± 1.04 †‡	38.5 ± 1.03
RS3	333 ± 1.00	182 ± 0.88	105 ± 1.00	46.1 ± 1.13	8.17 ± 0.02	35.7 ± 0.74 †‡	35.4 ± 0.26
RFS	330 ± 4.73	178 ± 4.33	109 ± 4.70	43.5 ± 2.75	8.22 ± 0.11	36.5 ± 0.64 †‡	37.3 ± 1.69
ANOVA *p*	0.79	0.53	0.64	0.11	0.88	0.03	0.08

Data were represented as means ± S.E.M. Different superscript symbols indicate significant differences ($p < 0.05$) among treatments.

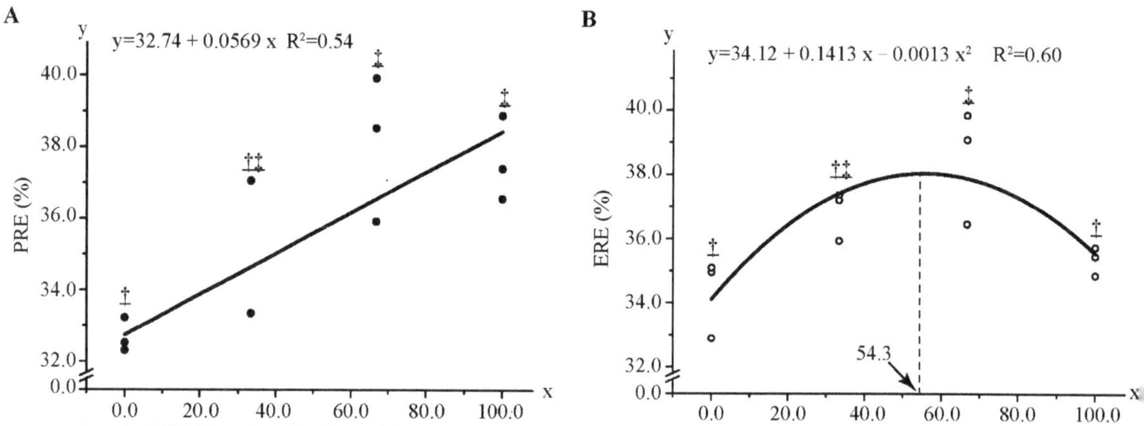

Figure 2. Regression analysis of protein retention efficiency (**A**) and energy retention efficiency (**B**) in red spotted grouper fed diet with different GWT inclusion levels. Data were analysed by one-way ANOVA followed by Tukey multiple comparisons, with different underlined symbols indicating significant differences ($p < 0.05$) among treatments. ERE, energy retention efficiency; GWT, a mixture of wheat gluten, wheat and taurine; LT-FM, low-temperature dried fish meal; PER, protein retention efficiency; SPC, soy protein concentrate.

3.3. Serum and Liver Physiological and Antioxidant Parameters

After the 56-day feeding trial, no significant differences were found in serum TP, TC, TG, GLU, ALT, AST, SOD, and MDA among fish fed diets with FM and/or SPC gradually replaced by GWT (Table 6). However, according to a linear regression model ($y = 249.1 + 0.9798x$, $R^2 = 0.51$), liver SOD activity markedly increased among fish fed diets, with FM gradually replaced by wheat gluten. When SPC was gradually replaced by wheat gluten, liver SOD activity decreased in a quadratic polynomial model ($y = 273.2 - 1.372x + 0.0071x^2$, $R^2 = 0.92$) (Figure 3).

Table 6. Biochemical indices in plasma and liver of red spotted grouper fed diet with different GWT inclusion levels.

Parameters	Diets								*p*
	R0	RF1	RF2	RF3	RS1	RS2	RS3	RFS	
Plasma									
TP, g/L	32.5 ± 1.36	28.7 ± 5.14	34.3 ± 2.17	30.7 ± 1.03	34.4 ± 1.60	33.2 ± 0.76	34.2 ± 0.27	32.7 ± 1.89	0.61
TG, mmol/L	4.20 ± 1.60	2.73 ± 1.10	2.43 ± 0.42	1.76 ± 0.34	5.04 ± 2.03	4.20 ± 3.03	4.01 ± 0.22	1.88 ± 0.88	0.70
CHOL, mmol/L	3.94 ± 0.72	2.61 ± 0.69	2.41 ± 0.26	1.69 ± 0.02	4.00 ± 0.52	2.72 ± 0.75	3.14 ± 0.11	2.60 ± 0.31	0.07
GLU, mmol/L	7.96 ± 0.32	6.42 ± 0.92	6.01 ± 0.85	5.51 ± 0.56	7.52 ± 1.92	6.22 ± 1.68	6.31 ± 0.48	6.39 ± 1.09	0.81
ALT, U/L	1.83 ± 0.88	7.50 ± 4.54	2.17 ± 0.17	3.17 ± 0.33	2.00 ± 0.29	3.17 ± 0.60	2.67 ± 0.17	2.67 ± 0.93	0.37
AST, U/L	122 ± 11.1	102 ± 32.5	142 ± 14.1	128 ± 15.4	117 ± 35.0	119 ± 15.2	132 ± 10.8	94.5 ± 22.3	0.80
SOD, U/mL	21.0 ± 0.42	22.4 ± 0.19	22.0 ± 0.44	22.3 ± 0.66	20.5 ± 0.19	21.7 ± 0.39	20.8 ± 0.22	21.5 ± 0.16	0.12
Liver									
SOD, U/mg prot	273 ± 3.38 ‡	233 ± 4.95 †	342 ± 13.2 §	345 ± 9.40 §	237 ± 4.36 †	212 ± 7.80 †	208 ± 1.70 †	283 ± 3.65 ‡	<0.01
MDA, nmol/mg prot	1.74 ± 0.04	1.52 ± 0.02	2.17 ± 0.01	2.08 ± 0.38	2.77 ± 0.66	1.58 ± 0.01	1.54 ± 0.01	1.40 ± 0.03	0.05

Data were represented as means ± S.E.M. Different superscript symbols indicate significant differences ($p < 0.05$) among treatments.

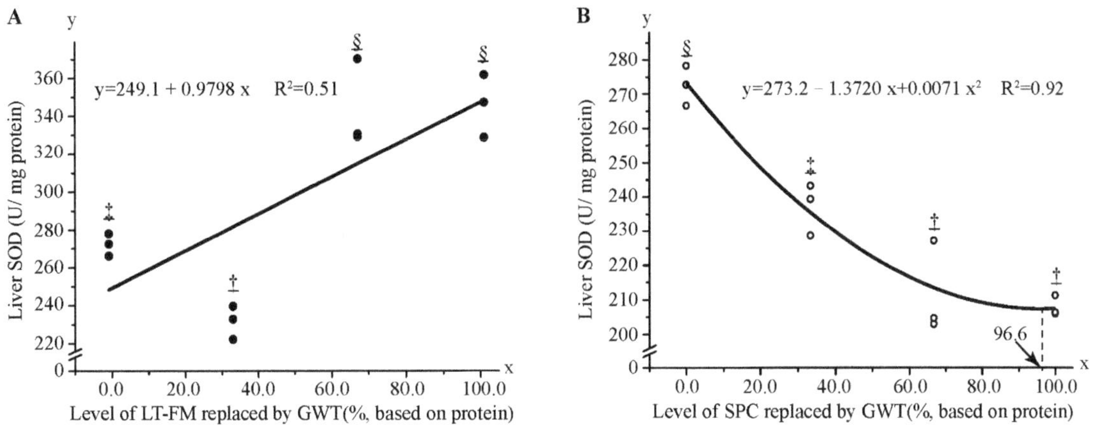

Figure 3. Regression analysis of liver superoxide dismutase (SOD) activity in red spotted grouper fed diet with different GWT inclusion levels. Data were analysed by one-way ANOVA followed by Tukey multiple comparisons, with different underlined symbols indicating significant differences ($p < 0.05$) among treatments. (**A**,**B**) show the regression analysis of replacing LT-FM and SPC with GWT on liver SOD activity, respectively. GWT, a mixture of wheat gluten, wheat and taurine; LT-FM, low-temperature dried fish meal; SPC, soy protein concentrate.

4. Discussion

The present study clearly showed the feasibility of wheat gluten as the main protein source in low-FM diets (20%) for red spotted grouper. Fish fed a diet with GWT supplementation showed a WG value of 190–250%, while the WG in fish fed diet R0 was around 159%. This growth rate was slightly lower than the study by Wang et al. [18], who selected similar size fish (~7.88 g) and a fish-meal-casein-based diet. Furthermore, the effect of diet on the growth of aquatic animals was time-accumulating. The growth performance of fish was not significantly different on the 28th feeding day, but it was markedly different on the 56th feeding day. At the end of the feeding, an increase of WG was observed in fish fed GWT supplementation diets compared with the R0 diet, including that wheat gluten could improve the growth of red spotted grouper juveniles. Similar results were also reported in a previous study, in which wheat gluten partly replaced dietary FM in large yellow croaker [16]. However, different results were reported for European sea bass [15] and Atlantic halibut [14], with their growth not significantly affected by dietary wheat gluten levels. Increased growth due to the supplementation of GWT could be attributed to the higher feed intake and utilisation. Some supplements were used in this study, such as krill meal, taurine and lecithin, which can ensure the attractivity of feed. Krill meal is often used as a feed stimulant due to its high palatability to marine fish, especially the high plant protein content for carnivorous fish. Plant proteins are deficient in taurine, which is necessary for groupers, especially in the high plant protein diet [30]. Previous studies have demonstrated that taurine is a critical nutrient in plant-based diets for Japanese flounder [31], Cobia Rachycentron canadum [32], white seabass *Atractoscion nobilis* [33], sablefish *Anoplopoma fimbria* [34] and meagre *Argyrosomus regius* [35]. The incorporation of lecithin in diets with low or no FM had positive effects on the growth performance of rainbow trout [36] and tinfoil barb *Barbonymus schwanenfeldii* [37]. Secondly, wheat gluten is not only rich in protein but has a much higher digestibility of protein and amino acids than fishmeal and soy protein concentration [11]. Moreover, high levels of anti-nutritional factors due to the increased plant protein level in the diet caused a decrease in feed nutritive value, feed palatable and feed utilisation. However, wheat gluten has fewer antinutritional factors, especially anti-palatability factors, compared with other plant protein sources, such as saponins, alkaloids, fibre, and phytic acid [38]. Thus, all these studies observed good feed intake and high feed utilisation when replacing FM and/or SPC with wheat gluten. Based on the nutritional properties of wheat gluten, balancing the dietary amino acid profile and adding appropriate starch levels may help groupers to adapt to high levels of wheat gluten.

Plant protein sources usually lack one or more essential amino acids. Compared to the FM, wheat gluten is limited in lysine, methionine, arginine, and threonine (Table 1) [39]. Lysine is the most limited amino acid in wheat gluten. The level of lysine in this experiment is approximately 26.0 g kg^{-1} diet (57.7 g kg^{-1} dietary protein), which is higher than the requirement of hybrid grouper (21.6 g kg^{-1} diet, corresponding to 40.5 g kg^{-1} dietary protein) [40], but lower than the *E. coioides* (28.3 g kg^{-1} diet, corresponding to 55.6% of dietary protein) [41]. Methionine is the second most limited amino acid, and its level in this experimental is 27.6–30.3 g kg^{-1} diet (61.3–67.3 g kg^{-1} dietary protein), higher than the optimum requirement of *E. coioides* (13.1 g kg^{-1} diet, corresponding to 27.3 g kg^{-1} dietary protein) [42]. Supplementation of limited amino acids in wheat gluten-supplemented diets is beneficial in improving feed utilisation, which is consistent with previous studies in which crystal amino acids were supplemented in plant protein-based diets [1,43,44]. Moreover, the optimal level of wheat gluten substitution for SPC was estimated to be 61% in this experiment (Figure 1B,D,F). Starch is the main component of wheat flour. Generally, starch is poorly utilised by aquatic animals, and its digestibility decreases with increasing dietary starch levels in almost fish and shrimp species. According to the results of a previous study, the tolerable level of wheat flour in red spotted grouper was 286 g kg^{-1} diet, which is close to 280 g kg^{-1} diet reported in hybrid grouper (*E. lanceolatus* ♂ × *E. fuscoguttatus* ♀) [45], but higher than 72.2 g kg^{-1} diet (incorporation with a cellulose content of ~250 g kg^{-1} diet) and 246 g kg^{-1} diet reported in *E. akaara* [19] and *E. malabaricus* [46] respectively. In this study, higher dietary starch decreased growth performance and feed utilisation in red sported grouper, which was consistent with previous studies on Nile tilapia *Oreochromis niloticus* [47], gibel carp *Carassius auratus* var. *gibelio* [48], grass carp *Ctenopharyngodon idella* [49], hybrid grouper *E. fuscoguttatus* ♀ × *E. lanceolatus* ♂ [50] and largemouth bass *Micropterus salmoides* [51].

Proximate composition and nutrition retention among fish fed diets with or without wheat gluten supplementation are shown in Table 4. The crude protein was about 180 g kg^{-1}, similar to the value reported by Wang et al. [19] in *E. akaara*. Lipid content (~110 g kg^{-1}) in whole-body was slightly higher than the level recorded by Wang et al. [19], mainly due to the higher dietary lipid level. However, no significant differences were observed in fish body compositions (moisture, crude protein, lipid, and ash) in this study, which is consistent with a study in red spotted grouper [18]. However, protein retention was significantly affected by different experimental diets. An increase in protein accumulation was observed in fish fed the diet with FM replaced by wheat gluten. Interestingly, the ERE was not evidently affected in the FM-replaced groups, while in the SPC-replaced groups, there was a secondary increase with increasing dietary wheat gluten levels. Fish, especially carnivorous fish, have limited availability of starch as an energy source. However, moderate wheat gluten could be used as an energy source has been demonstrated in Atlantic salmon *Salmo salar* [39], Atlantic cod *Gadus morhua* [52], and gilthead sea *bream Sparus aurata* [52].

In this study, no significant changes in serum indices were observed when FM or SPC was gradually replaced by wheat gluten. Similar to the results in European sea bass, no significant changes were observed in serum cholesterol when fed a wheat gluten-based diet compared with the FM diet [53]. In this study, liver SOD activity linearly increased with gradually increasing FM replacement levels (0, 1/3, 2/3, and 100%). However, no significant differences were observed in liver MDA content, indicating that wheat gluten has no deleterious effects on oxidative stress products in red spotted grouper. In contrast, a decrease in SOD activity and MDA content was observed in the group in which SPC was replaced by wheat gluten (Figure 3). Glutamate is a non-essential amino acid which is considered to be beneficial for the growth and antioxidant status of aquatic animals. Thus, increased SOD activity may be attributed to the high content of glutamate in wheat gluten, which is higher than FM and SPC [11].

5. Conclusions

The highest growth performance was observed in red spotted grouper fed diets with 100% fishmeal or 66.7% soy protein concentration replaced by wheat gluten. Regression analysis of weight gain and feed conversion efficiency suggested that the optimal levels of wheat gluten substitution for fish meal or soy protein concentration were 20.7% and 13.1%, respectively. The results of this study provide new insights for the development of a fishmeal-free formulation for red spotted grouper.

Author Contributions: Conceptualization, Y.Z.; methodology, Y.C., Y.W., G.X., B.S. and Y.Z.; software, Z.D., Y.W. and T.S.; resources, Y.Z.; writing—original draft preparation, Y.C.; writing—review and editing, B.S., T.S. and Y.Z.; supervision, B.S., Z.D., T.S. and Y.Z.; project administration, B.S., T.S. and Y.Z.; funding acquisition, Y.Z. All authors have read and agreed to the published version of the manuscript.

Funding: This research was funded by the National Key R & D Program of China (2019YFD0900203).

Institutional Review Board Statement: The study was performed in strict accordance with the Laboratory Animal Welfare Guidelines of China (Decree No. 2 of the Ministry of Science and Technology, issued in 1988) and approved by Animal Experimental Ethical Inspection, Institutional Animals Care and Use Committee of Zhejiang Ocean University.

Informed Consent Statement: Not applicable.

Data Availability Statement: Data are available on request due to privacy. The data presented in this study are available on request from the corresponding author.

Conflicts of Interest: The authors declare no conflict of interest.

References

1. Gatlin, D.M.; Barrows, F.T.; Brown, P.; Dabrowski, K.; Gaylord, T.G.; Hardy, R.W.; Herman, E.; Hu, G.; Krogdahl, Å.; Nelson, R.; et al. Expanding the utilization of sustainable plant products in aquafeeds: A review. *Aquac. Res.* **2007**, *38*, 551–579. [CrossRef]
2. Anderson, R.L.; Wolf, W.J. Compositional changes in trypsin inhibitors, phytic acid, saponins and isoflavones related to soybean processing. *J. Nutr.* **1995**, *125*, 581–588.
3. Kaushik, S.J.; Cravedi, J.P.; Lalles, J.P.; Sumpter, J.; Fauconneau, B.; Laroche, M. Partial or total replacement of fish meal by soybean protein on growth, protein utilization, potential estrogenic or antigenic effects, cholesterolemia and flesh quality in rainbow trout, *Oncorhynchus mykiss*. *Aquaculture* **1995**, *133*, 257–274. [CrossRef]
4. Deng, J.; Mai, K.; Ai, Q.; Zhang, W.; Wang, X.; Xu, W.; Liufu, Z. Effects of replacing fish meal with soy protein concentrate on feed intake and growth of juvenile Japanese flounder, *Paralichthys olivaceus*. *Aquaculture* **2006**, *258*, 503–513. [CrossRef]

5. Biswas, A.; Araki, H.; Sakata, T.; Nakamori, T.; Kato, K.; Takii, K. Fish meal replacement by soy protein from soymilk in the diets of red sea bream (*Pagrus major*). *Aquac. Nutr.* **2017**, *23*, 1379–1389. [CrossRef]
6. Biswas, A.; Araki, H.; Sakata, T.; Nakamori, T.; Takii, K. Optimum fish meal replacement by soy protein concentrate from soymilk and phytase supplementation in diet of red sea bream, *Pagrus major*. *Aquaculture* **2019**, *506*, 51–59. [CrossRef]
7. Tola, S.; Fukada, H.; Masumoto, T. Effects of feeding a fish meal-free soy protein concentrate-based diet on the growth performance and nutrient utilization of red sea bream (*Pagrus major*). *Aquac. Res.* **2019**, *50*, 1087–1095. [CrossRef]
8. Mohd Faudzi, N.; Yong, A.S.K.; Shapawi, R.; Senoo, S.; Biswas, A.; Takii, K. Soy protein concentrate as an alternative in replacement of fish meal in the feeds of hybrid grouper, brown-marbled grouper (*Epinephelus fuscoguttatus*) × giant grouper (*E. lanceolatus*) juvenile. *Aquac. Res.* **2018**, *49*, 431–441. [CrossRef]
9. Chen, Y.; Liu, W.; Ma, J.; Wang, Y.; Huang, H. Comprehensive physiological and transcriptomic analysis revealing the responses of hybrid grouper (*Epinephelus fuscoguttatus* ♀ × *E. lanceolatus* ♂) to the replacement of fish meal with soy protein concentrate. *Fish Physiol. Biochem.* **2020**, *46*, 2037–2053. [CrossRef]
10. Wang, J.; Liang, D.; Yang, Q.; Tan, B.; Dong, X.; Chi, S.; Liu, H.; Zhang, S. The effect of partial replacement of fish meal by soy protein concentrate on growth performance, immune responses, gut morphology and intestinal inflammation for juvenile hybrid grouper (*Epinephelus fuscoguttatus* ♀ × *Epinephelus lanceolatus* ♂). *Fish Shellfish Immunol.* **2020**, *98*, 619–631. [CrossRef]
11. Apper-Bossard, E.; Feneuil, A.; Wagner, A.; Respondek, F. Use of vital wheat gluten in aquaculture feeds. *Aquat. Biosyst.* **2013**, *9*, 21. [CrossRef] [PubMed]
12. Pfeffer, E.; Al-Sabty, H.; Haverkamp, R. Studies on lysine requirements of rainbow trout (*Oncorhynchus mykiss*) fed wheat gluten as only source of dietary protein. *J. Anim. Physiol. Anim. Nutr.* **1992**, *67*, 74–82. [CrossRef]
13. Davies, S.J.; Morris, P.C.; Baker, R.T.M. Partial substitution of fish meal and full-fat soya bean meal with wheat gluten and influence of lysine supplementation in diets for rainbow trout, *Oncorhynchus mykiss* (Walbaum). *Aquac. Res.* **1997**, *28*, 317–328. [CrossRef]
14. Helland, S.J.; Grisdale-Helland, B. Replacement of fish meal with wheat gluten in diets for Atlantic halibut (*Hippoglossus hippoglossus*): Effect on whole-body amino acid concentrations. *Aquaculture* **2006**, *261*, 1363–1370. [CrossRef]
15. Messina, M.; Piccolo, G.; Tulli, F.; Messina, C.M.; Cardinaletti, G.; Tibaldi, E. Lipid composition and metabolism of European sea bass (*Dicentrarchus labrax L.*) fed diets containing wheat gluten and legume meals as substitutes for fish meal. *Aquaculture* **2013**, *376–379*, 6–14. [CrossRef]
16. Wang, P.; Lou, Y.; Feng, J.; He, J.; Zhu, J.; Zhou, Q. Effect of replacing fish meal with wheat gluten meal on growth, serum biochemical indexes and antioxidant enzyme activity of juvenile large yellow croaker (*Larimichthys crocea*). *J. Fish. China* **2018**, *42*, 733–743.
17. Jiang, Y.D.; Wang, J.T.; Han, T.; Li, X.Y.; Hu, S.X. Effect of dietary lipid level on growth performance, feed utilization and body composition by juvenile red spotted grouper (*Epinephelus akaara*). *Aquacult. Int.* **2015**, *23*, 99–110. [CrossRef]
18. Wang, J.; Jiang, Y.; Li, X.; Han, T.; Yang, Y.; Hu, S.; Yang, M. Dietary protein requirement of juvenile red spotted grouper (*Epinephelus akaara*). *Aquaculture* **2016**, *450*, 289–294. [CrossRef]
19. Wang, J.; Li, X.; Han, T.; Yang, Y.; Jiang, Y.; Yang, M.; Xu, Y.; Harpaz, S. Effects of different dietary carbohydrate levels on growth, feed utilization and body composition of juvenile grouper *Epinephelus akaara*. *Aquaculture* **2016**, *459*, 143–147. [CrossRef]
20. Luo, Z.; Liu, Y.J.; Mai, K.S.; Tian, L.X.; Liu, D.H.; Tan, X.Y. Optimal dietary protein requirement of grouper *Epinephelus coioides* juveniles fed isoenergetic diets in floating net cages. *Aquac. Nutr.* **2004**, *10*, 247–252. [CrossRef]
21. Yan, X.; Yang, J.; Dong, X.; Tan, B.; Zhang, S.; Chi, S.; Yang, Q.; Liu, H.; Yang, Y. Optimum protein requirement of juvenile orange-spotted grouper (*Epinephelus coioides*). *Sci. Rep.* **2021**, *11*, 6230. [CrossRef]
22. Yan, X.; Yang, J.; Dong, X.; Tan, B.; Zhang, S.; Chi, S.; Yang, Q.; Liu, H.; Yang, Y. The protein requirement of grouper *Epinephelus coioides* at grow-out stage. *Aquac. Nutr.* **2020**, *26*, 1555–1567. [CrossRef]
23. Wang, Y.R.; Wang, L.; Zhang, C.X.; Song, K. Effects of substituting fishmeal with soybean meal on growth performance and intestinal morphology in orange-spotted grouper (*Epinephelus coioides*). *Aquac. Rep.* **2017**, *5*, 52–57. [CrossRef]
24. Millamena, O.M. Replacement of fish meal by animal by-product meals in a practical diet for grow-out culture of grouper *Epinephelus coioides*. *Aquaculture* **2002**, *204*, 75–84. [CrossRef]
25. Chen, G.; Yin, B.; Liu, H.; Tan, B.; Dong, X.; Yang, Q.; Chi, S.; Zhang, S. Effects of fishmeal replacement with cottonseed protein concentrate on growth, digestive proteinase, intestinal morphology and microflora in pearl gentian grouper (♀ *Epinephelus fuscoguttatus* × ♂ *Epinephelus lanceolatu*). *Aquac. Res.* **2020**, *51*, 2870–2884. [CrossRef]
26. Huang, B.; Zhang, S.; Dong, X.; Chi, S.; Yang, Q.; Liu, H.; Tan, B.; Xie, S. Effects of fishmeal replacement by black soldier fly on growth performance, digestive enzyme activity, intestine morphology, intestinal flora and immune response of pearl gentian grouper (*Epinephelus fuscoguttatus* ♀ × *Epinephelus lanceolatus* ♂). *Fish Shellfish Immunol.* **2022**, *120*, 497–506. [CrossRef] [PubMed]
27. Chen, Y.; Ma, J.; Huang, H.; Zhong, H. Effects of the replacement of fishmeal by soy protein concentrate on growth performance, apparent digestibility, and retention of protein and amino acid in juvenile pearl gentian grouper. *PLoS ONE* **2019**, *14*, e0222780. [CrossRef]

28. Song, S.G.; Chi, S.Y.; Tan, B.P.; Liang, G.L.; Lu, B.Q.; Dong, X.H.; Liu, H.; Zhang, S. Effects of fishmeal replacement by Tenebrio molitor meal on growth performance, antioxidant enzyme activities and disease resistance of the juvenile pearl gentian grouper (*Epinephelus lanceolatus* ♂ × *Epinephelus fuscoguttatus* ♀). *Aquac. Res.* **2018**, *49*, 2210–2217. [CrossRef]
29. Zhang, Y.; Guo, M.; Li, N.; Dong, Z.; Cai, L.; Wu, B.; Xie, J.; Liu, L.; Ren, L.; Shi, B. New insights into β-glucan-enhanced immunity in largemouth bass *Micropterus salmoides* by transcriptome and intestinal microbial composition. *Front. Immunol.* **2022**, *13*, 108103. [CrossRef]
30. Lin, Y.H.; Lu, R.M. Dietary taurine supplementation enhances growth and nutrient digestibility in giant grouper *Epinephelus lanceolatus* fed a diet with soybean meal. *Aquac. Rep.* **2020**, *18*, 100464. [CrossRef]
31. Park, G.S.; Takeuchi, T.; Yokoyama, M.; Seikai, T. Optimal dietary taurine level for growth of juvenile Japanese flounder *Paralichthys olivaceus*. *Fish. Sci.* **2002**, *68*, 824–829. [CrossRef]
32. Lunger, A.N.; McLean, E.; Gaylord, T.G.; Kuhn, D.; Craig, S.R. Taurine supplementation to alternative dietary proteins used in fish meal replacement enhances growth of juvenile cobia (*Rachycentron canadum*). *Aquaculture* **2007**, *271*, 401–410. [CrossRef]
33. Jirsa, D.; Davis, D.A.; Salze, G.P.; Rhodes, M.; Drawbridge, M. Taurine requirement for juvenile white seabass (*Atractoscion nobilis*) fed soy-based diets. *Aquaculture* **2014**, *422–423*, 36–41. [CrossRef]
34. Johnson, R.B.; Kim, S.K.; Watson, A.M.; Barrows, F.T.; Kroeger, E.L.; Nicklason, P.M.; Goetz, G.W.; Place, A.R. Effects of dietary taurine supplementation on growth, feed efficiency, and nutrient composition of juvenile sablefish (*Anoplopoma fimbria*) fed plant based feeds. *Aquaculture* **2015**, *445*, 79–85. [CrossRef]
35. de Moura, L.B.; Diógenes, A.F.; Campelo, D.A.V.; de Almeida, F.L.A.; Pousão-Ferreira, P.M.; Furuya, W.M.; Oliva-Teles, A.; Peres, H. Taurine and methionine supplementation as a nutritional strategy for growth promotion of meagre (*Argyrosomus regius*) fed high plant protein diets. *Aquaculture* **2018**, *497*, 389–395. [CrossRef]
36. Iwashita, Y.; Suzuki, N.; Yamamoto, T.; Shibata, J.I.; Isokawa, K.; Soon, A.H.; Ikehata, Y.; Furuita, H.; Sugita, T.; Goto, T. Supplemental effect of cholyltaurine and soybean lecithin to a soybean meal-based fish meal-free diet on hepatic and intestinal morphology of rainbow trout *Oncorhynchus mykiss*. *Fish. Sci.* **2008**, *74*, 1083–1095. [CrossRef]
37. Sohel, M.; Bolong, A.A. Effects of soybean lecithin on growth and survival of tropical tin foil barb. *Ann. Vet. Anim. Sci.* **2020**, *7*, 15–22.
38. Draganovic, V.; van der Goot, A.J.; Boom, R.; Jonkers, J. Assessment of the effects of fish meal, wheat gluten, soy protein concentrate and feed moisture on extruder system parameters and the technical quality of fish feed. *Anim. Feed Sci. Technol.* **2011**, *165*, 238–250. [CrossRef]
39. Storebakken, T.; Shearer, K.D.; Baeverfjord, G.; Nielsen, B.G.; Åsgård, T.; Scott, T.; De Laporte, A. Digestibility of macronutrients, energy and amino acids, absorption of elements and absence of intestinal enteritis in Atlantic salmon, *Salmo salar*, fed diets with wheat gluten. *Aquaculture* **2000**, *184*, 115–132. [CrossRef]
40. Li, X.; Wu, X.; Dong, Y.; Gao, Y.; Yao, W.; Zhou, Z. Effects of dietary lysine levels on growth, feed utilization and related gene expression of juvenile hybrid grouper (*Epinephelus fuscoguttatus* ♀ × *Epinephelus lanceolatus* ♂). *Aquaculture* **2019**, *502*, 153–161. [CrossRef]
41. Luo, Z.; Liu, Y.; Mai, K.; Tian, L.; Tan, X.; Yang, H.; Liang, G.; Liu, D. Quantitative L-lysine requirement of juvenile grouper *Epinephelus coioides*. *Aquac. Nutr.* **2006**, *12*, 165–172. [CrossRef]
42. Luo, Z.; Liu, Y.; Mai, K.; Tian, L.; Yang, H.; Tan, X.; Liu, D. Dietary l-methionine requirement of juvenile grouper *Epinephelus coioides* at a constant dietary cystine level. *Aquaculture* **2005**, *249*, 409–418. [CrossRef]
43. Li, P.; Mai, K.; Trushenski, J.; Wu, G. New developments in fish amino acid nutrition: Towards functional and environmentally oriented aquafeeds. *Amino Acids* **2009**, *37*, 43–53. [CrossRef] [PubMed]
44. Zhang, Y.; Øverland, M.; Xie, S.; Dong, Z.; Lv, Z.; Xu, J.; Storebakken, T. Mixtures of lupin and pea protein concentrates can efficiently replace high-quality fish meal in extruded diets for juvenile black sea bream (*Acanthopagrus schlegeli*). *Aquaculture* **2012**, *354–355*, 68–74. [CrossRef]
45. Luo, Y.; Wu, X.; Li, W.; Jiang, S.; Lu, S.; Wu, M. Effects of different corn starch levels on growth, protein input, and feed utilization of juvenile hybrid grouper (male *Epinephelus lanceolatus* × female *E. fuscoguttatus*). *N. Am. J. Aquac.* **2016**, *78*, 168–173. [CrossRef]
46. Shiau, S.Y.; Lin, Y.H. Carbohydrate utilization and its protein-sparing effect in diets for grouper (*Epinephelus malabaricus*). *Anim. Sci.* **2001**, *73*, 299–304. [CrossRef]
47. Tran-Duy, A.; Smit, B.; van Dam, A.A.; Schrama, J.W. Effects of dietary starch and energy levels on maximum feed intake, growth and metabolism of Nile tilapia, *Oreochromis niloticus*. *Aquaculture* **2008**, *277*, 213–219. [CrossRef]
48. Tan, Q.; Wang, F.; Xie, S.; Zhu, X.; Lei, W.; Shen, J. Effect of high dietary starch levels on the growth performance, blood chemistry and body composition of gibel carp (*Carassius auratus* var. gibelio). *Aquac. Res.* **2009**, *40*, 1011–1018. [CrossRef]
49. Tian, L.X.; Liu, Y.J.; Yang, H.J.; Liang, G.Y.; Niu, J. Effects of different dietary wheat starch levels on growth, feed efficiency and digestibility in grass carp (*Ctenopharyngodon idella*). *Aquac. Int.* **2012**, *20*, 283–293. [CrossRef]
50. Li, S.; Li, Z.; Zhang, J.; Sang, C.; Chen, N. The impacts of dietary carbohydrate levels on growth performance, feed utilization, glycogen accumulation and hepatic glucose metabolism in hybrid grouper (*Epinephelus fuscoguttatus* ♀ × *E. lanceolatus* ♂). *Aquaculture* **2019**, *512*, 734351. [CrossRef]
51. Li, X.; Zheng, S.; Ma, X.; Cheng, K.; Wu, G. Effects of dietary starch and lipid levels on the protein retention and growth of largemouth bass (*Micropterus salmoides*). *Amino Acids* **2020**, *52*, 999–1016. [CrossRef] [PubMed]

52. Bradshaw, C.J.; Higgins, J.; Michael, K.J.; Wotherspoon, S.J.; Hindell, M.A. At-sea distribution of female southern elephant seals relative to variation in ocean surface properties. *ICES J. Mar. Sci.* **2004**, *61*, 1014–1027. [CrossRef]
53. Robaina, L.; Corraze, G.; Aguirre, P.; Blanc, D.; Melcion, J.; Kaushik, S. Digestibility, postprandial ammonia excretion and selected plasma metabolites in European sea bass (*Dicentrarchus labrax*) fed pelleted or extruded diets with or without wheat gluten. *Aquaculture* **1999**, *179*, 45–56. [CrossRef]

metabolites

MDPI

Article

Enzymatic Chicken Pulp Promotes Appetite, Digestive Enzyme Activity, and Growth in *Litopenaeus vannamei*

Vivian Hlordzi [1,2], Beiping Tan [1,2], Xiaohui Dong [1,2], Shuang Zhang [1,2], Lin Zhu [3], Ling Zhang [4,*], Xiangna Hu [4] and Shuyan Chi [1,2,*]

1 Laboratory of Aquatic Nutrition and Feed, College of Fisheries, Guangdong Ocean University, Zhanjiang 524088, China; hlordziviviann@gmail.com (V.H.); bptan@126.com (B.T.); dongxiaohui2003@163.com (X.D.); zhangshuang@163.com (S.Z.)
2 Aquatic Animals Precision Nutrition and High Efficiency Feed Engineering Research Centre of Guangdong Province, Zhanjiang 524088, China
3 Qingdao Bio-Ways Ingredients Bio-Technology Co., Ltd., Qingdao 266071, China; zhu666lin@163.com
4 Shenzhen Institute of Quality & Safety Inspection and Research, Shenzhen 518101, China; huxiangna@163.com
* Correspondence: zhangling923@163.com (L.Z.); chisy@gdou.edu.cn (S.C.)

Abstract: Enzymatic chicken pulp (ECP) is an animal protein source that has been proven to be of excellent nutritional content and good quality for the majority of aquatic organisms because of its quality protein, small peptides, palatability, vitamins, and minerals. An 8-week nutritional trial was conducted to assess the effects of an ECP-based diet on the growth performance, digestive enzyme activity, and gene mRNA expression of Pacific white shrimp (*Litopenaeus vannamei*). Fish soluble pulp (FSP) served as the control group while in the experimental groups, and ECPs with three protein contents were used to replace FSP in equal amounts, named ECP1, ECP2, and ECP3, respectively. No significant difference in weight gain rate, specific growth rate, survival rate, or feed conversion ratio was observed ($p > 0.05$) between the groups. Ash content in the Pacific shrimp's whole body was significantly higher in the ECP1 and ECP3 groups compared to the other groups ($p < 0.05$). Intestinal amylase and protease activities were the highest in the ECP1 and ECP2 groups, respectively ($p < 0.05$). With respect to gene mRNA expression, neuropeptide Y, excitatory amino acid transporter, and fatty acid transport protein 4 were significantly high in the ECP1 group ($p < 0.05$). In conclusion, these three ECPs have their advantages to replace FSP in shrimp feed, but ECP1 is more effective if the effects of digestive enzyme activity, appetite, and expression of growth-related genes are considered.

Keywords: fish soluble pulp; enzymatic chicken pulp; growth performance; appetite; *Litopenaeus vannamei*

Citation: Hlordzi, V.; Tan, B.; Dong, X.; Zhang, S.; Zhu, L.; Zhang, L.; Hu, X.; Chi, S. Enzymatic Chicken Pulp Promotes Appetite, Digestive Enzyme Activity, and Growth in *Litopenaeus vannamei*. *Metabolites* **2022**, *12*, 698. https://doi.org/10.3390/metabo12080698

Academic Editor: Chi Chen

Received: 11 June 2022
Accepted: 20 July 2022
Published: 27 July 2022

1. Introduction

Penaeidae shrimp farming is rapidly growing, with production exceeding 6 million tons in 2018 and more than 75% of harvest being *Litopenaeus vannamei* [1]. Feed is an essential component of cost-effective aquaculture. Aquaculture's successes with respect to nutrition and efficiency heavily relies on supplemental meals [2]. Dietary nutritional value is influenced by the quality of the protein sources employed in the production of the meal [3], particularly for aquaculture animals. Fishmeal (FM), which is the major and high-quality protein source in aquafeed, is very difficult to obtain and very expensive due to high demand [4]. It is very important to reduce the amount of FM used in aquafeed by introducing other protein sources.

Significant volumes of waste are produced during the processing of fish and fowl, including internal organs, fat, skin, feet, skeleton, feathers, and blood. The amount of scrap produced is about 43% (*w*/*w*) of the live weight [5]. Fish soluble pulp (FSP) is produced from fish by-products such as skin, scales, bones, swim bladders, roes, intestines, blood, and liver [4]. These by-products contain large amounts of bioactive rich materials which are mostly underutilized, wasted, or discarded. FSP is rich in protein and amino acids which is suitable for the development of the Pacific white shrimp [6]. Since the early 2000s, global fish catch has reduced significantly, therefore reducing the amount of by-products obtained from this sector [1]. This has led to the study of other ingredients as the primary protein source in the diet of aquatic animals [7,8].

Enzymatic chicken pulp (ECP) is among the animal protein sources proven to be of good quality and high nutrition quality in the diet of most aquatic organisms, particularly for carnivorous species [9,10]. Currently, the poultry by-product is primarily used in pet foods because of its quality protein, palatability, essential fatty acids, vitamins, and minerals [11]. ECP is produced from the enzymatic hydrolysis of poultry by-products such as necks, feet, intestines, blood, and undeveloped eggs, exclusive of feathers. These by-products are first ground, heated to kill microorganisms, and later treated with enzymes. Enzymes are thermally inactivated by heating, and centrifuged to obtain two phases, those being supernatant (water and oil) and solid residue [12]. Most pathogenic microorganisms in ECP are destroyed by the rendering process and conditions (time/temperature) used, with a minimum effect on the digestibility of amino acids [13]. Amino acids such as arginine, alanine, and taurine, which improve feed acceptance in crustaceans, have been observed to be slightly higher in ECP compared to poultry by-product meals [14]. ECP is thought to be a proper substitute protein for FM [15,16] in artificial diets for carnivorous and omnivorous aquaculture species due to consistent availability, relatively low price, nutritional composition, and similar nutritional compositions to FM [17,18].

Due to variability in processing specifications such as time, temperature, and enzymes of poultry by-products into ECP, differences in protein, essential amino acid composition, and lipid contents are observed [19]. In modern rendering facilities, advanced processing technologies are used to counter these challenges [20]. As demand for aquafeed is increasing, several authors confirmed that ECP has considerable potential as an ingredient in fish feed production systems [21].

ECPs with different crude protein levels were incorporated into the diets of Pacific white shrimp and compared with an FSP-based diet to study the effect on growth, survival body composition, mRNA gene expression, and intestinal development.

2. Materials and Methods

2.1. Experimental Diet

In this study, four isonitrogenous (39.7%) and isolipidic (9%) shrimp diets were formulated and prepared. A balanced diet containing FSP was used as a control group (FSP). Three types of ECPs, which contained 32.4%, 33%, and 10.2% crude protein, were added to the diets as ECP1, ECP2, and ECP3 groups, respectively (Tables 1 and 2). The raw materials were crushed and passed through an 80-mesh sieve. After weighing, all ingredients listed in the formula were mixed thoroughly in a Hobart-type mixer to form a moist dough that was placed into the extruder and shaped into strips. The stripes were air-dried, broken into granules, and sealed in bags until the experiment started.

2.2. Experimental Procedure

Healthy Pacific white shrimp larvae were gathered from the southern base of the marine aquaculture seed project of the "863" Program. The purchased shrimps were incubated for two months in an aerated cement pond at the Marine Biological Research Base of Guangdong Ocean University and fed with a commercial diet within this period. Fasting 24 h before the feeding experiment, 720 healthy Pacific white shrimps were batch-weighed to obtain an initial average weight (0.26 ± 0.002 g) and randomly distributed at a stocking density of 30 shrimps per tank in three replicate groups. Indoor fiberglass tanks (0.3 m^3) were used for this study under a natural photoperiod (12 h light/12 h dark) system. The experimental feed was fed manually four times daily (07:00, 11:00, 17:00, and 21:00) to 10% body weight and later regulated to visual apparent satiation for 8 weeks. Oxygen was provided using single-air stones, water temperature ranged from 28–30 °C, pH 8.0–8.2, dissolved oxygen > 7 mg/L, and salinity 28.5–32, respectively. The feed provided was recorded and about 60% of the water in the tanks was changed daily. Mortalities were weighed and recorded daily.

Table 1. Formulation and proximate composition of the experimental diets (% dry matter).

Ingredients	FSP	ECP1	ECP2	ECP3
Brown fish meal	20	20	20	20
Fish soluble pulp	4	0.00	0.00	0.00
ECP 1	0	4	0	0
ECP 2	0	0	4	0
ECP 3	0	0	0	4
Soybean meal	15	15	15	15
Cottonseed protein	11	11	11	12.5
Wheat flour	24	24	24	24
Peanut meal	14	14	14	14
Shrimp shell meal	3	3	3	3
Fish oil	1.6	1.6	1.6	1.6
Soy lecithin	1.5	1.5	1.5	1.5
Soybean oil	1.5	1.5	1.5	1.3
Calcium monophosphate	1	1	1	1
Premix (Vitamin + Mineral)	1	1	1	1
Cholesterol	0.15	0.15	0.15	0.15
Antioxidant	0.45	0.45	0.45	0.45
Microcrystalline cellulose	1.3	1.3	1.3	0
Choline Chloride	0.4	0.4	0.4	0.4
Attractant	0.1	0.1	0.1	0.1
Total %	100	100	100	100
	Nutrient composition (%)			
Moisture	9.34	9.55	9.32	9.21
Crude protein (dry matter)	38.34	38.09	38.29	38.13
Crude lipid (dry matter)	7.14	6.77	6.79	6.55
Ash (dry matter)	10.22	10.24	10.33	10.11

Table 2. Amino acid composition (%).

	FSP	ECP1	ECP2	ECP3	Cottonseed Protein
Aspartic acid	1.99	1.79	1.82	1.25	5.68
Glutamic acid	2.4	2.43	2.48	1.48	12.33
Serine	0.72	0.73	0.74	0.17	2.48
Histidine	1.64	0.82	0.84	0.16	1.73
Arginine	2.35	2.05	2.09	0.52	7.75
Threonine	0.98	0.62	0.63	0.14	1.93
Glycine	1.78	3.28	3.34	0.35	2.45
Alanine	2.94	1.66	1.69	0.71	2.29
Proline	1.96	3.19	3.25	0.41	2.14
Tyrosine	0.78	0.41	0.42	0.21	1.71
Valine	1.04	0.66	0.67	0.29	2.75
Methionine	0.46	0.4	0.41	0.12	0.76
Cystine	0.2	0.39	0.40	0.03	0.76
Isoleucine	0.91	0.67	0.68	0.28	1.87
Leucine	1.72	1.32	1.35	0.29	3.22
Phenylalanine	1.2	0.84	0.86	0.15	3.39
Lysine	2.05	1.23	1.25	0.34	2.62

2.3. Sample Collection

80 shrimps were randomly sampled and stored (−20 °C) for the analysis of the initial proximate composition at the onset of the study. 24 h before the cessation of the feeding trial, shrimps were starved. The weight and number of the shrimps per replicate were

checked and noted to calculate the survival rate (SR), weight gain rate (WGR), specific growth rate (SGR), and feed conversion ratio (FCR). At random, five shrimps per replicate were selected, weighed, and had their length checked, after which their hepatopancreas and intestines were harvested and weighed to denote their condition factor (CF), viscerosomatic index (VSI), and hepatosomatic index (HSI).

$$\text{Weight gain rate }(\%) = 100 \times \frac{(\text{average final individual weight} - \text{average initial individual weight})}{\text{average initial individual weight}}$$

$$\text{Specific growth rate }(\%/\text{d}) = 100 \times \frac{\log_e \text{average final weight} - \log_e \text{average initial weight}}{\text{days of feeding}}$$

$$\text{Feed conversion ratio} = \frac{\text{Feed consumed}}{\text{weight gain}}$$

$$\text{Survival rate }(\%) = 100 \times \frac{\text{Final shrimp number}}{\text{Initial shrimp number}}$$

$$\text{Condition factor }(\text{g/cm}^3) = 100 \times \left(\frac{\text{final weight}}{(\text{shrimp fork length})^3}\right)$$

(from ref [22])

$$\text{Viscerosomatic index }(\%) = 100 \times \left(\frac{\text{visceral weight}}{\text{body weight}}\right)$$

$$\text{Hepatosomatic index }(\%) = 100 \times \left(\frac{\text{hepatopancreas weight}}{\text{body weight}}\right)$$

$$\text{Protein efficiency ratio} = \frac{\text{weight gain}}{\text{protein intake}}$$

$$\text{Protein production value }(\%) = 100 \times \frac{(\text{Final weight} \times \text{Crude protein}) - (\text{Initial weight} \times \text{Crude protein})}{\text{Feed given} \times \text{Crude protein}}$$

2.3.1. Proximate Composition and Analysis

Proximate composition (moisture, crude lipid, crude protein, and ash) of feeds and five shrimps per replicate were analyzed using standard methods obtained from the Association of Official Analytical Chemists [23].

2.3.2. Serum Biochemical Activity

The hemolymph was pooled into 1.5 mL Eppendorf tubes and stored at 4 °C overnight. Stored hemolymph samples were centrifuged (4000 rpm for 10 min at 4 °C) and the serum was harvested for biochemical enzyme activity. Total protein content, aspartate transaminase (AST), alanine transaminase (ALT), catalase (CAT), and lysozyme (LZM) enzyme activities in the serum were evaluated using test kits obtained from Nanjing Jiancheng Institute of Biological Engineering, China. The specific operation methods, absorbance, and calculation formulas were performed according to the test kit instructions. AST and ALT were measured using a full wavelength microplate reader (Thermofisher, Waltham, MA, USA) at 510 nm, and CAT and LZM were measured at 405 nm and 530 nm.

2.3.3. Digestive Enzyme Activity in the Intestine

The intestines of three shrimps per replicate was harvested and frozen in liquid nitrogen and later kept at −80 °C until use. Intestinal amylase, lipase, and protease were determined using the specific operation methods and calculations provided by the commercial test kit (Nanjing Jiancheng Institute of Biological Engineering, Nanjing, China). We weighed the intestine, added 9 times the volume of saline at a ratio of mass (g): volume (mL) of 1:9, homogenized mechanically under ice bath conditions, titled 10%

tissue homogenate, centrifuged at 2500 rpm for 10 min, and used the supernatant as the analysis sample. The absorbance of the amylase, lipase, and protease was measured using the microplate reader (Thermofisher, Waltham, MA, USA) at 660 nm, 550 nm, and 680 nm.

2.3.4. Intestinal Microbiota

DNA library sequencing was performed on the Illumina HiseqTM 2500/4000 by Gene Denovo Biotechnology Co., Ltd., (Guangzhou, China). Bioinformatic analysis was performed using Omicsmart, a real-time interactive online platform for data analysis (http://www.omicsmart.com, accessed on 2 February 2022).

2.3.5. Real-Time Quantitative PCR Analysis

Total RNA from the hepatopancreas of three shrimps per replicate was extracted using Trizol (Invitrogen, Waltham, MA, USA) reagent, and the integrity and quality were verified using 1% agarose gel electrophoresis and spectrophotometer (NanoDrop-2000, ThermoScientific, Waltham, MA, USA). Reverse transcription of RNA extracted was performed using a primeScript TM kit (TaKaRa, Dalian, China). A real-time fluorescent quantitative PCR assay was executed to detect the mRNA expression levels for neuropeptide Y (*npy*), fatty acid transport protein 4 (*fatp*4), excitatory amino acid transporter (*eaat*), growth hormone secretagogue receptor type 1 (*gsh-R1*), cholecystokinin receptor type A-like (*cckar*), and cluster of differentiation 36 (*cd*36), with β-actin gene as the housekeeping gene (Bio-Rad CFX96; Bio-Rad Labs, Hercules, CA, USA) (Table 3). Relative gene expression levels were calculated by $2^{-\triangle\triangle CT}$.

Table 3. Primer sequence used for real-time quantitative PCR analysis.

Gene	Primer	Source
npy	F: GGTGATGTCGAAGTGGCCGGAGTTG R: ACCTCGCCAGGGAGAAGCGGAACCA	GFRP01055388
fatp4	F:CCGACGGGCAAAGCGACTGAACCA R: TCTATTCCACCAGGTATCTTTATCG	KY271629
eaat	F: GTTACAACATCAAACCCGAGACAG R:CCCGAGAAGGTGAAGATGAGGAGC	GGUK01021174
ghs-RI	F: TGCGAAGGAGGAACTCTGAACATT R: CCAAGTAAGTCGCTTCCTGGCTCT	HAAW01018270
cckar	F: ATCGTGTCCCTTGTGCTGTCTGTT R: GTCATCGCCGTCATCTTCTTCGTC	XM_027379811
cd36	F: AACCAAGGTCCTGACCATCAC R: AGGTGAGAGTCGACGAGGAA	[24]
β-actin	F: AAGATGTGTGACGACGAAGTAGC R: AGGATACCTCGCTTGCTCTG	GFRP01025709

npy—neuropeptide Y, *fatp4*—fatty acid transport protein 4, *eaat*—excitatory amino acid transporter, *ghs-RI*—growth hormone secretagogue receptor type 1, *cckar*—cholecystokinin receptor type A-like, *cd36*—cluster of differentiation 36, *β-actin*—beta-actin.

2.4. Statistical Analysis

Data obtained were analyzed using a one-way analysis of variance (ANOVA). Statistical analyses were performed using the SPSS 22.0 for Windows and general differences were significant at $p < 0.05$. Tukey's honestly significant difference (HSD) test was used to compare the mean values between individual treatments. Data are represented as mean values of each group of shrimp ± standard error (SE).

3. Results

3.1. Growth Performance and Survival

The SR of all experimental groups was 80% and above without significant difference ($p > 0.05$), as portrayed in Table 4. No significant difference in IW, WGR, SGR, CF, VSI, or HSI was observed between the experimental groups ($p > 0.05$). FCR was significantly higher in the ECP2 group but was not significantly different from the ECP3 group ($p < 0.05$) (Table 4).

Table 4. Effect of fish soluble pulp meal and enzymatic chicken pulp on the growth and survival of *L. vannamei*.

	FSP	ECP1	ECP2	ECP3
IW	0.26 ± 0.00	0.26 ± 0.00	0.26 ± 0.01	0.26 ± 0.00
WGR/%	2789.92 ± 34.42	2923.21 ± 17.57	2818.48 ± 57.17	2553.41 ± 46.36
SGR (%/d)	5.53 ± 0.08	5.87 ± 0.06	5.52 ± 0.17	5.49 ± 0.22
SR,%	80.00 ± 3.85	88.33.00 ± 2.15	82.22 ± 5.55	86.67 ± 6.67
FCR	1.64 ± 0.02 ^{a}	1.58 ± 0.03 ^{a}	1.84 ± 0.04 ^{b}	1.70 ± 0.5 ^{ab}
CF, g/cm^3	0.81 ± 0.03	0.87 ± 0.01	0.81 ± 0.03	0.81 ± 0.01
VSI,%	5.85 ± 0.09	5.55 ± 0.22	5.11 ± 0.34	5.07 ± 0.32
HSI,%	4.65 ± 0.12	4.11 ± 0.20	4.00 ± 0.28	3.96 ± 0.14

Note: Values are mean values of each group of shrimp (3 replicates) ± SE. Means in each row without superscripts or with the same superscripts do not differ significantly ($p > 0.05$), while those with different superscripts differ significantly ($p < 0.05$). IW-Initial weight, WGR%-Weight gain rate, SGR%/d—Specific growth rate per day, SR%—Survival rate, FCR—Feed conversion ratio, CF%—Condition factor, VSI%—Viscerosomatic index, HSI-Hepatosomatic index.

3.2. Whole Body Composition

As presented in Table 5, no significant difference in moisture and crude protein was observed ($p > 0.05$). Crude lipid was significantly higher in the FSP and ECP3 groups compared to the ECP1 group ($p < 0.05$), but with no significant difference to the ECP2 group ($p > 0.05$). Ash content was significantly higher in the groups fed ECP diets compared to the FSP group ($p < 0.05$), but no significant difference was observed between the FSP and the ECP2 groups ($p > 0.05$). Whole-body PER was significantly higher in the ECP1 group compared to the ECP3 group ($p < 0.05$) with no significant difference to the FSP and ECP2 groups ($p > 0.05$).

Table 5. Effect of fish soluble pulp and enzymatic chicken pulp on the whole-body composition of *L. vannamei*.

	FSP	ECP1	ECP2	ECP3
Moisture	73.44 ± 0.26	74.59 ± 0.57	73.62 ± 0.19	74.05 ± 0.21
Crude protein	69.28 ± 2.11	63.76 ± 3.13	69.23 ± 2.08	64.39 ± 0.63
Crude lipid	6.83 ± 0.78 ^{b}	5.74 ± 0.16 ^{a}	6.41 ± 0.16 ^{ab}	7.47 ± 0.40 ^{b}
Ash	10.43 ± 0.01 ^{a}	11.75 ± 0.04 ^{b}	10.98 ± 0.35 ^{ab}	11.81 ± 0.25 ^{b}
PER	4.67 ± 0.80 ^{ab}	5.41 ± 0.11 ^{b}	4.62 ± 0.10 ^{ab}	4.48 ± 0.27 ^{a}
PPV	40.17 ± 2.09	40.64 ± 6.83	39.57 ± 2.34	32.08 ± 1.00

Note: Values are mean values of each group of shrimps (three replicates) ± SE. Means in each row without superscripts or with the same superscripts do not differ significantly ($p > 0.05$), whiles those with different superscripts differ significantly ($p < 0.05$). PER—Protein efficiency rate, PPV—protein production value.

3.3. Serum Biochemical Indexes

Serum AST and ALT activities in Table 6 were significantly high in ECP 2 compared to FSP and ECP3 groups ($p < 0.05$) but were not significantly different from the ECP1 group ($p > 0.05$). The serum, CAT, LZM activities, and TP content between the groups portrayed no significant difference ($p > 0.05$).

Table 6. Effect of fish soluble pulp and enzymatic chicken pulp in the serum enzyme activities of *L. vannamei*.

	AST, U/L	ALT, U/L	LZM, U/ml	CAT, U/ml	TP, g/ml
FSP	17.94 ± 3.58 ^{a}	46.76 ± 4.17 ^{ab}	275.00 ± 21.43	7.08 ± 3.23	299.67 ± 8.51
ECP1	26.72 ± 4.20 ^{ab}	58.42 ± 4.19 ^{bc}	285.71 ± 19.88	13.28 ± 3.87	297.79 ± 20.72
ECP2	41.76 ± 0.63 ^{b}	59.38 ± 0.37 ^{c}	291.67 ± 5.19	15.86 ± 6.88	324.81 ± 7.92
ECP3	12.07 ± 2.93 ^{a}	40.55 ± 1.12 ^{a}	300.00 ± 5.46	10.45 ± 1.18	289.89 ± 13.86

Note: Values are mean values of each group of shrimps (three replicates) ± SE. Means in each row without superscripts or with the same superscripts do not differ significantly ($p > 0.05$), while those with different superscripts differ significantly ($p < 0.05$).

3.4. Digestive Enzyme Activity in Intestine

In the intestine, no significant difference in lipase activity between the groups was observed ($p > 0.05$). Amylase activity in the intestine was significantly higher in the ECP1 group compared to the other groups ($p < 0.05$). Protease activity was significantly high in the ECP1 and ECP2 groups ($p < 0.05$) but was not significantly different from the ECP3 group ($p > 0.05$). (Table 7).

Table 7. Effect of fish soluble pulp and enzymatic chicken pulp in the intestinal digestive enzyme activities of *L. vannamei*.

	Lipase, U/gprot	Amylase, U/mgprot	Protease, U/mL
FSP	0.81 ± 0.26	3.07 ± 0.27 ^{a}	1352.24 ± 116.5 ^{a}
ECP1	1.06 ± 0.05	5.47 ± 0.03 ^{b}	1944.10 ± 156.08 ^{b}
ECP2	0.52 ± 0.04	3.05 ± 0.03 ^{a}	2322.53 ± 53.75 ^{b}
ECP3	0.60 ± 0.01	3.16 ± 0.01 ^{a}	1828.34 ± 124.36 ^{ab}

Note: Values are mean values of each group of shrimps (three replicates) ± SE. Means in each row without superscripts or with the same superscripts do not differ significantly ($p > 0.05$), while those with different superscripts differ significantly ($p < 0.05$).

3.5. Intestinal Microbiota

3.5.1. Species Composition

In between the groups, most intestinal bacteria observed at the phylum level were classified as Proteobacteria (53%). The 2nd and 3rd predominant phyla were Bacteroidetes (31%) and Perrucomicrobia (10%), respectively. However, the group fed an FSP-based diet had the lowest Proteobacteria level (49.56%), and the highest Verrucomicrobia content at 15.02% (Figure 1A). At the phylum level, the most prevalent bacteria found in the FSP group was Verrucomicrobia. Firmicutes was prevalent in the ECP1 group, Chlamydiae in the ECP2 group, and Patescibacteria in ECP3 as portrayed in the heat map presented in Figure 1B. As displayed in Figure 1C, 229 operational taxonomic units (OTUs) were shared by all the groups. However, the highest unique OTUs were observed in the FSP group (59), followed by the ECP1 group (35), ECP2 group (27), and ECP3 group (15), respectively.

3.5.2. Alpha Diversity

Alpha diversity results indicated the bacterial community were evenly distributed within the groups. The Simpson, Shannon, and Chao1 diversity were enriched in both FSP and ECP groups and did not differ significantly, as demonstrated in Figure 2A–C. As observed in the rarefaction curve, samples were sequenced at a high depth and near saturation to capture enough variety for all groups (Figure 2D).

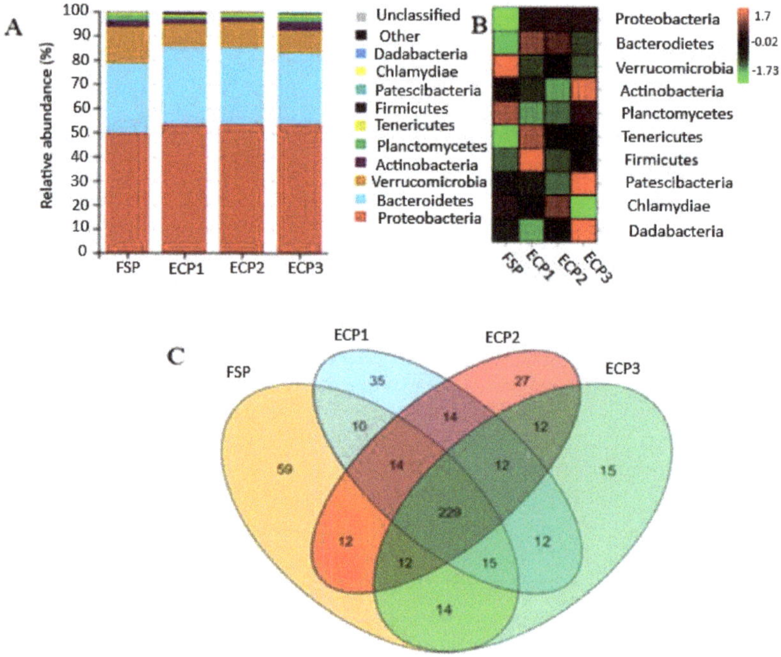

Figure 1. Effect of fish soluble pulp and enzymatic chicken pulp on the bacterial species composition of the intestinal microbiota community of *L. vannamei*. (**A**) Taxonomic distribution; (**B**) Heat map; (**C**) Venn diagram.

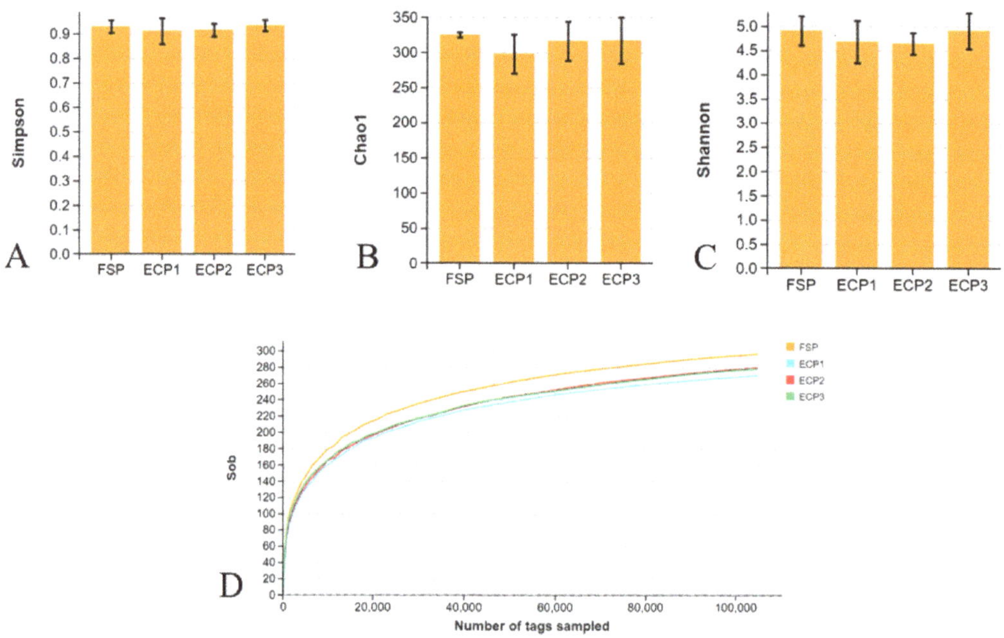

Figure 2. Alpha diversity of present study. (**A**) Simpson diversity; (**B**) Chao1 diversity; (**C**) Shannon diversity; (**D**) Rarefaction.

3.5.3. Beta Diversity

The PCA two-dimensional plot indicated that the gut microbiota of Pacific white shrimp in the FSP group was similar to the ECP3 group, while no association was observed between the FSP group and ECP1 and ECP2 groups (Figure 3).

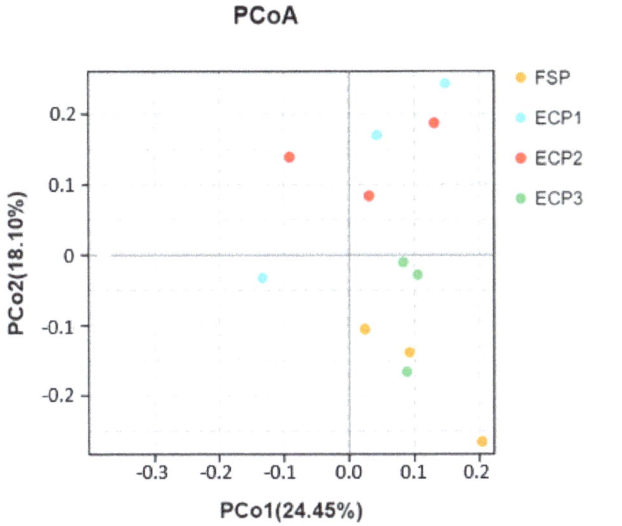

Figure 3. Principal coordinate analysis (PCoA) plot. The scatter plot is of principal coordinate 2 (PCo2) versus principal coordinate 1 (PCo1).

3.6. Gene mRNA Expression in the Hepatopancreas

As demonstrated in Figure 4, the mRNA gene expression of *eaat* was significantly high in the ECP1 group ($p < 0.05$), while *fatp*4 was high in the ECP1 group but was not significantly different from the ECP2 and ECP3 ($p > 0.05$) groups. No significant difference was observed in the gene mRNA expression of *gsh-RI* and *cd*36 ($p > 0.05$) between the groups as demonstrated in Figure 5. The ECP1 group had a significantly high mRNA gene expression of *npy* compared to the other groups ($p < 0.05$), while no significant difference in *cckar* ($p > 0.05$) was observed between the groups (Figure 6).

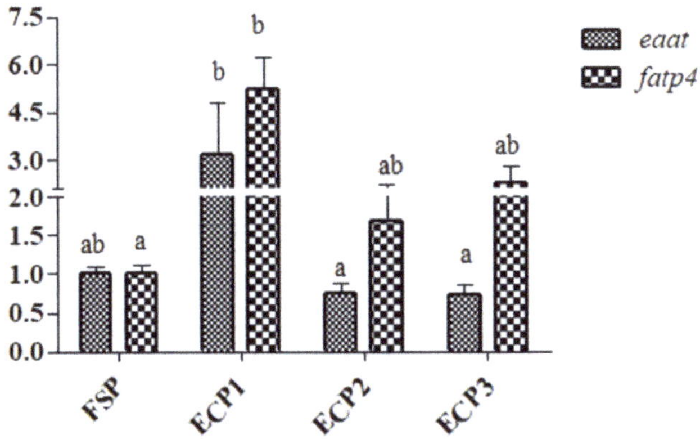

Figure 4. Effect of fish soluble pulp and enzymatic chicken pulp on the expression of *eaat* and *fatp4* on mRNA gene in the hepatopancreas of *L. vannamei*. *eaat*- excitatory amino acid transporter, *fatp4*-fatty acid transport protein 4. Means without superscripts or with the same superscripts do not differ significantly ($p > 0.05$), while those with different superscripts differ significantly ($p < 0.05$).

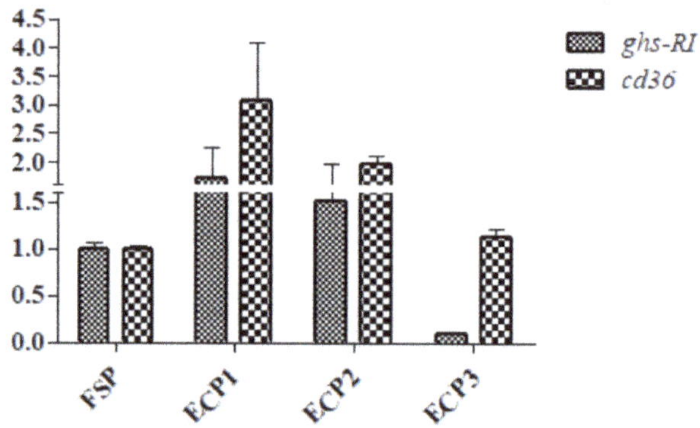

Figure 5. Effect of fish soluble pulp and enzymatic chicken pulp on the expression of *ghs-RI* and *cd36* on mRNA gene in the hepatopancreas of *L. vannamei*. *ghs-RI*—Growth hormone secretagogue receptor type 1, *cd36*—cluster of differentiation 36. Means without superscripts or with the same superscripts do not differ significantly ($p > 0.05$), while those with different superscripts differ significantly ($p < 0.05$).

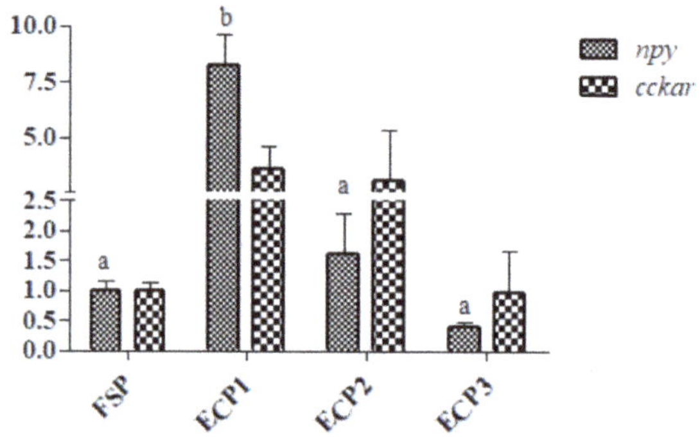

Figure 6. Effect of fish soluble pulp and enzymatic chicken pulp on the expression of *npy* and *cckar* on mRNA gene in the hepatopancreas of *L. vannamei*. *npy*—Neuropeptide Y, *cckar*—cholecystokinin receptor type A-like. Means without superscripts or with the same superscripts do not differ significantly ($p > 0.05$), while those with different superscripts differ significantly ($p < 0.05$).

4. Discussion

In this study, no significant difference in WGR and SGR was observed between the groups fed ECP and FSP diets. This might be due to the high digestibility of ECP and FSP and sufficient amino acids (Table 2) [25]. Better digestive enzymatic activities were found in the ECP1 group which has the better growth WGR and PER. Similar results were observed in juvenile rainbow trout (*Oncorhynchus mykiss*) [26], Pacific white shrimp [27], European seabass (*Dicentrarchus labrax*) [28], and tilapia (*Oreochromis niloticus*) [29] fed enzymatic fish by-products, feather-enzymatic hydrolysates, enzymatically hydrolysed aquaculture by-products, and enzymatic poultry by-products, respectively. Poultry and fish by-products have been considered suitable ingredients in the diet of aquatic animals due to their high nutritional characteristics [30]. After enzymatic treatment, these by-products are digested by enzymes and converted from large molecules to small ones; that is, the protein is enzymatically hydrolyzed to small peptides and amino acids. The nutritional value of these ingredients is enhanced [12]. However, this is not in accordance with studies conducted in genetically improved tilapia (*Oreochromis niloticus*) fed enzymatically hydrolyzed chicken liver compared to the non-enzymatically hydrolyzed chicken liver [29]. The composition of ingredients produced from poultry by-products is dependent on the processing method and quality of by-products used [31]. The growth performance of aquatic organisms was improved by a moderate level of enzymatic protein hydrolysate inclusion, whereas growth performance was impaired by a lower or greater inclusion level [32]. Lower WGR observed in genetically improved tilapia could be related to the high enzymatic chicken liver incorporation in their diet [29,33]. Results obtained in this study could be due to a moderate inclusion level of ECP in the diet of Pacific white shrimps.

VSI, HSI, and CF are good representatives of general wellbeing, health, and feed quality in aquatic organisms [34]. Like juvenile rainbow trout (*Oncorhynchus mykiss*) and genetically improved tilapia fed enzymatic fish by-product and enzymatic chicken liver, respectively [26,29], there was no significant difference observed in VSI, HSI, and CF between the groups in this study. Dietary and environmental factors may have contributed to the variations in whole-body composition [31,35]. The whole-body moisture was not influenced by ECP in this study but an increase in the whole-body ash was observed. This was likely due to the high ash content of ECP, which typically comprises bones and feet [31]. However, with reference to the growth performance data, ECP used in the diet of Pacific white shrimps had no negative effect on growth performance, although whole body ash was higher.

In aquatic organisms, the existence and function of molecules, including LZM, CAT, and immunoglobulins, are strongly connected with the immunological functions of immune organs [7,36]. LZM and CAT activities are significant humoral indications of non-specific immunity in aquatic organisms, in addition to key enzymes in the host's defense system [26,37]. In this study, no significant difference in CAT and LZM was observed in Pacific white shrimps fed diets containing FSP and ECP, but higher levels of CAT and LZM were witnessed in shrimps fed ECP compared to those fed FSP. Similar observations were made by Bae et al. [26] in juvenile rainbow trout fed enzymatically hydrolyzed fish by-products. This could be due to an improvement in the immune component of ECP during enzymatic hydrolysis and the provision of sufficient protein and amino acids in the diets [38]. The total protein level in the serum plays a vital role in the transportation of fatty acids, the maintenance of a steady pH in aquatic organisms [39], and can be used to determine the nutritional status of an organism [40]. There was no significant difference in total protein content of Pacific white shrimps between the groups which was also observed in juvenile rainbow trout [26].

The gut microbiota plays a variety of important roles in the host, notably immune system modification, pathogen defense, anaerobic peptide, and protein metabolism, in addition to the processing of non-digestible food fibers [41,42]. Although no significant difference was observed, Pacific white shrimps fed FSP diets had higher OTUs and alpha diversity indices (Simpson, Chao1, Shannon, and Rarefaction) compared to those fed ECP diets. The nutrient quality in the diet and rearing environment may influence the abundance of Proteobacteria, Bacteroidetes, and Actinobacteria in the gut of Pacific white shrimps [43]. Proteobacteria and Bacteroidetes were the most abundant bacteria found in the gut of the Pacific white shrimps in the groups. Proteobacteria, Bacteroidetes, and Actinobacteria were observed to be higher in Pacific white shrimps fed ECP diets compared to those fed FSP. Proteobacteria are the most common phylum of bacteria in the aquatic systems and aquatic animals' intestines [44]. Bacteroidetes helped to improve the digestion and absorption of nutrients in the feed of Pacific white shrimp [44,45], and in this experiment, more abundant Bacteroidetes were observed in the ECP1 group compared to other groups, which in turn increased WGR, SGR, and CF. Actinobacteria, although a small bacterial group, plays an important part in the formation and maintenance of the immune system and feed metabolism [46]. The Pacific white shrimps fed ECP-based diets had a relatively high Actinobacteria composition and had a higher LZM content in the intestine of Pacific white shrimp. This suggests that ECP could help Pacific white shrimps' immunological responses.

The transport of nutrients is very important for animal health. Amino acid transporter expression is a key measure of aquatic species' ability to absorb amino acids [47]. They are primarily responsible for the transmembrane transport of amino acids in the body, in addition to promoting amino acid assimilation, protein synthesis, and maintaining physiological functioning [48,49]. In this study, ECP1 had a higher *eaat* and *fatp4* mRNA expression level compared to the other groups. Results in this study were similar to results obtained by Zhuang et al. [14] who observed a higher expression level of b^0 neutral amino acid transporter 1 (b^0at1) and neutral and cationic amino acid transporter (y^+lat2) in juvenile turbot fed enzymatic hydrolysis of chicken by-product in high plant-based protein. This could be due to the appropriate amount of free amino acids in ECP1, leading to the stimulation of the transporters of amino acids located in the hepatopancreas [50], which is conducive for growth [51,52].

Glycine, alanine, and proline have strong predatory effects on aquatic animals [53]. The three amino acid contents in the ingredients of FSP, ECP1, ECP2, and ECP3 used in this experiment were 6.68%, 8.13%, 8.28%, and 1.47%, respectively (Table 2). According to the three amino acid content, the appetizing effects of the ECP1 and ECP2 treatment should be better. The gene *npy* expression was upregulated in these two groups, although the expression level in ECP2 was without significant difference compared to the FSP. Another reason could be the different small peptide contents in ECP1 and ECP2 (Figure 7) making the ECP1 feed more attractive, thus significantly increasing the expression of *npy* mRNA level of the shrimps [54].

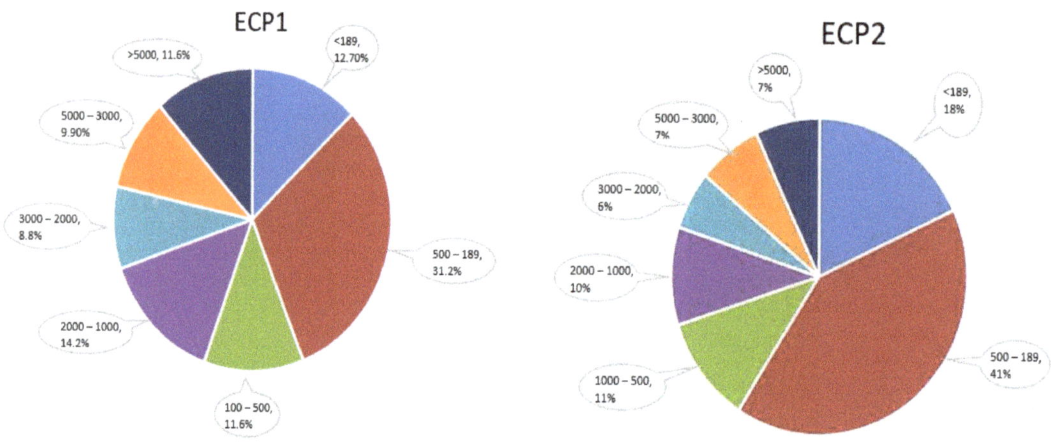

Figure 7. Distribution of molecular weight of small peptides in the raw ingredients of ECP1 and ECP2.

5. Conclusions

ECP1, ECP2, and ECP3 with different protein contents, amino acids, and small peptides, although obtained from the same raw material with different processes, demonstrated different effects. Shrimp that ingested a diet with ECP1 indicated better growth performance, digestive enzyme activity, and upregulated mRNA expression of appetite and growth related genes compared to the other groups. In ECP2, serum AST and ALT were significantly higher than in the other groups. ECP1, which contains a 32.4% crude protein level, is recommended for the diet of Pacific white shrimp larvae.

Author Contributions: V.H. conducted the study, analyzed the data, and drafted the original manuscript. S.C. aided in the experimental design, fund acquisition, supervision of the project, and review and editing of the manuscript. B.T. aided in funding acquisition. X.D., S.Z. and L.Z. (Lin Zhu) aided in the experimental design. L.Z. (Ling Zhang) and X.H. aided in sample analysis. All authors have read and agreed to the published version of the manuscript.

Funding: This study was financially supported by the National Key R&D Program of China (2019YFD0900200) and Special Project on Key Fields of Guangdong Universities, China (Rural Revitalization) (2020ZDZX1034).

Institutional Review Board Statement: The animal study was reviewed and approved by Ethics Review Board of Guangdong Ocean University (GDOU-01/2019).

Informed Consent Statement: Not applicable.

Data Availability Statement: The datasets generated and/or analyzed during the current study are available from the corresponding author on reasonable request.

Conflicts of Interest: The authors declare that they have no known competing financial interests or personal relationships that could have appeared to influence the work reported in this paper.

References

1. FAO. *The State of World Fisheries and Aquaculture 2020: Sustainability in Action*; FAO: Quebec City, QC, Canada, 2020.
2. Omosowone, O.O.; Ogunrinde, A.M. Effect of partial or total supplementation of soybean meal with fluted pumpkin (Telfairia occidentalis) seed meal in the diet of hybrid Catfish (Heteroclarias) fingerlings. *Food Sci. Nutr.* **2018**, *6*, 1190–1195. [CrossRef] [PubMed]
3. Gunben, E.M.; Senoo, S.; Yong, A.; Shapawi, R. High potential of poultry by-product meal as a main protein source in the formulated feeds for a commonly cultured grouper in Malaysia (Epinephelus fuscoguttatus). *Sains Malays.* **2014**, *43*, 399–405.
4. Hlordzi, V.; Wang, J.; Kuebutornye, F.K.A.; Yang, X.; Tan, B.; Li, T.; Cui, Z.; Lv, S.; Lao, T.; Shuyan, C. Hydrolysed fish protein powder is better at the growth performance, hepatopancreas and intestinal development of Pacific white shrimp (Litopenaeus vannamei). *Aquac. Rep.* **2022**, *23*, 101025. [CrossRef]
5. Martínez-Alvarez, O.; Chamorro, S.; Brenes, A. Protein hydrolysates from animal processing by-products as a source of bioactive molecules with interest in animal feeding: A review. *Food Res. Int.* **2015**, *73*, 204–212. [CrossRef]
6. Tlusty, M.F.; Hardy, R.; Cross, S.F. Limiting size of fish fillets at the center of the plate improves the sustainability of aquaculture production. *Sustainability* **2011**, *3*, 957–964. [CrossRef]
7. Chaklader, M.R.; Siddik, M.A.B.; Fotedar, R. Total replacement of fishmeal with poultry by-product meal affected the growth, muscle quality, histological structure, antioxidant capacity and immune response of juvenile barramundi, Lates calcarifer. *PLoS ONE* **2020**, *15*, e0242079. [CrossRef]
8. Ray, G.W.; Liang, D.; Yang, Q.; Tan, B.; Dong, X.; Chi, S.; Liu, H.; Zhang, S.; Rimei, L. Effects of replacing fishmeal with dietary soybean protein concentrate (SPC) on growth, serum biochemical indices, and antioxidative functions for juvenile shrimp Litopenaeus vannamei. *Aquaculture* **2020**, *516*, 734630. [CrossRef]
9. Sugiura, S.H.; Babbitt, J.K.; Dong, F.M.; Hardy, R.W. Utilization of fish and animal by-product meals in low-pollution feeds for rainbow trout Oncorhynchus mykiss (Walbaum). *Aquac. Res.* **2000**, *31*, 585–593. [CrossRef]
10. Zapata, D.B.; Lazo, J.P.; Herzka, S.Z.; Viana, M.T. The effect of substituting fishmeal with poultry by-product meal in diets for Totoaba macdonaldi juveniles. *Aquac. Res.* **2016**, *47*, 1778–1789. [CrossRef]
11. Jayathilakan, K.; Sultana, K.; Radhakrishna, K.; Bawa, A.S. Utilization of byproducts and waste materials from meat, poultry and fish processing industries: A review. *J. Food Sci. Technol.* **2012**, *49*, 278–293. [CrossRef]
12. Wubshet, S.G.; Wold, J.P.; Afseth, N.K.; Böcker, U.; Lindberg, D.; Ihunegbo, F.N.; Måge, I. Feed-forward prediction of product qualities in enzymatic protein hydrolysis of poultry by-products: A spectroscopic approach. *Food Bioprocess Technol.* **2018**, *11*, 2032–2043. [CrossRef]
13. Cruz-Suárez, L.E.; Nieto-López, M.; Guajardo-Barbosa, C.; Tapia-Salazar, M.; Scholz, U.; Ricque-Marie, D. Replacement of fish meal with poultry by-product meal in practical diets for Litopenaeus vannamei, and digestibility of the tested ingredients and diets. *Aquaculture* **2007**, *272*, 466–476. [CrossRef]
14. Zhuang, Y.; Zhang, W.; Zheng, J.; Tang, Z.; Li, X.; Cao, X.; Zhang, L.; Xu, W.; Mai, K.; Ai, Q. Effects of enzymatic hydrolysis chicken by-product in high plant-based protein diet on growth performance, digestive capacity, antioxidant capacity and non-specific immunity of juvenile turbot (Scophthalmus maximus L.). *Aquac. Nutr* **2021**, *27*, 1578–1589. [CrossRef]
15. Thompson, K.R.; Metts, L.S.; Muzinic, L.A.; Dasgupta, S.; Webster, C.D.; Brady, Y.J. Use of turkey meal as a replacement for menhaden fish meal in practical diets for sunshine bass grown in cages. *N. Am. J. Aquac.* **2007**, *69*, 351–359. [CrossRef]
16. Rawles, S.D.; Riche, M.; Gaylord, T.G.; Webb, J.; Freeman, D.W.; Davis, M. Evaluation of poultry by-product meal in commercial diets for hybrid striped bass (Morone chrysops♀ × M. saxatilis♂) in recirculated tank production. *Aquaculture* **2006**, *259*, 377–389. [CrossRef]
17. Yu, Y. Replacement of fishmeal with poultry byproduct meal and meat and bone meal in shrimp, tilapia and trout diets. In *A vances En Nutrición Acuicola. Memorias del Septimo Simposium Internacional de Nutrición Acuícola*; Cruz Suárez, M., Ricque Marie, L.E., Nieto López, D., Villarreal, M.G., Scholz, D., y González, U., Eds.; Nutrición Acuícola: Hermosillo, Sonora, Mexico, 2004; pp. 16–19.
18. González-Rodríguez, Á.; Celada, J.D.; Carral, J.M.; Sáez-Royuela, M.; García, V.; Fuertes, J.B. Evaluation of poultry by-product meal as partial replacement of fish meal in practical diets for juvenile tench (Tinca tinca L.). *Aquac. Res.* **2016**, *47*, 1612–1621. [CrossRef]

19. Dawson, M.R.; Alam, M.S.; Watanabe, W.O.; Carroll, P.M.; Seaton, P.J. Evaluation of poultry by-product meal as an alternative to fish meal in the diet of juvenile Black Sea Bass reared in a recirculating aquaculture system. *N. Am. J. Aquac.* **2018**, *80*, 74–87. [CrossRef]
20. Galkanda-Arachchige, H.S.C.; Wilson, A.E.; Davis, D.A. Success of fishmeal replacement through poultry by-product meal in aquaculture feed formulations: A meta-analysis. *Rev. Aquac.* **2020**, *12*, 1624–1636. [CrossRef]
21. Fasakin, E.A.; Serwata, R.D.; Davies, S.J. Comparative utilization of rendered animal derived products with or without composite mixture of soybean meal in hybrid tilapia (Oreochromis niloticus × Oreochromis mossambicus) diets. *Aquaculture* **2005**, *249*, 329–338. [CrossRef]
22. Udoinyang, E.P.; Amali, O.; Iheukwumere, C.C.; Ukpatu, J.E. Length-weight relationship and condition factor of seven shrimp species in the artisanal shrimp fishery of Iko river estuary, southeastern Nigeria. *Int. J. Fish. Aquat. Stud.* **2016**, *4*, 109–114.
23. Horwitz, W. *Official Methods of Analysis of AOAC International, Volume I, Agricultural Chemicals, Contaminants, Drugs/Edited by William Horwitz*; AOAC International: Gaithersburg, MD, USA, 2010.
24. Shi, B.; Jin, M.; Jiao, L.; Betancor, M.B.; Tocher, D.R.; Zhou, Q. Effects of dietary zinc level on growth performance, lipolysis and expression of genes involved in the calcium/calmodulin-dependent protein kinase kinase-β/AMP-activated protein kinase pathway in juvenile Pacific white shrimp. *Br. J. Nutr.* **2020**, *124*, 773–784. [CrossRef] [PubMed]
25. Yang, Y.; Xie, S.; Lei, W.; Zhu, X.; Yang, Y. Effect of replacement of fish meal by meat and bone meal and poultry by-product meal in diets on the growth and immune response of Macrobrachium nipponense. *Fish Shellfish Immunol.* **2004**, *17*, 105–114. [CrossRef]
26. Bae, J.; Azad, A.K.; Won, S.; Hamidoghli, A.; Seong, M.; Bai, S.C. Effects of enzymatically hydrolyzed fish by-products in diet of juvenile rainbow trout (Oncorhynchus mykiss). *Fish. Aquat. Sci.* **2019**, *22*, 1. [CrossRef]
27. Dios, D. Fishmeal replacement with feather-enzymatic hydrolyzates co-extruded with soya-bean meal in practical diets for the Pacific white shrimp (Litopenaeus vannamei). *Aquac. Nutr.* **2001**, *7*, 143–151.
28. Leduc, A.; Zatylny-Gaudin, C.; Robert, M.; Corre, E.; le Corguille, G.; Castel, H.; Lefevre-Scelles, A.; Fournier, V.; Gisbert, E.; Andree, K.B. Dietary aquaculture by-product hydrolysates: Impact on the transcriptomic response of the intestinal mucosa of European seabass (Dicentrarchus labrax) fed low fish meal diets. *BMC Genom.* **2018**, *19*, 396. [CrossRef]
29. Wu, J.; Liu, W.; Wen, H.; Zhou, Y.; Wu, J. Animal by-products with or without enzymatic hydrolysis completely replacement of fish meal in genetically improved farmed tilapia diets (Oreochromis niloticus). *Aquac. Res.* **2021**, *52*, 291–301. [CrossRef]
30. Barreto-Curiel, F.; Parés-Sierra, G.; Correa-Reyes, G.; Durazo-Beltrán, E.; Viana, M.T. Total and partial fishmeal substitution by poultry by-product meal (petfood grade) and enrichment with acid fish silage in aquafeeds for juveniles of rainbow trout Oncorhynchus mykiss. *Lat. Am. J. Aquat. Res.* **2016**, *44*, 327–335. [CrossRef]
31. Shapawi, R.; Ng, W.-K.; Mustafa, S. Replacement of fish meal with poultry by-product meal in diets formulated for the humpback grouper, Cromileptes altivelis. *Aquaculture* **2007**, *273*, 118–126. [CrossRef]
32. Hatlen, B.; Jakobsen, J.-V.; Crampton, V.; Alm, M.; Langmyhr, E.; Espe, M.; Hevrøy, E.; Torstensen, B.; Liland, N.; Waagbø, R. Growth, feed utilization and endocrine responses in Atlantic salmon (Salmo salar) fed diets added poultry by-product meal and blood meal in combination with poultry oil. *Aquac. Nutr.* **2014**, *21*, 714–725. [CrossRef]
33. Hevroy, E.; Espe, M.; Waagbo, R.; Sandnes, K.; Ruud, M.; Hemre, G.-I. Nutrient utilization in Atlantic salmon (Salmo salar L.) fed increased levels of fish protein hydrolysate during a period of fast growth. *Aquac. Nutr.* **2005**, *11*, 301–313. [CrossRef]
34. Ighwela, K.A.; Ahmad, A.B.; Abol-Munafi, A.B. The selection of viscerosomatic and hepatosomatic indices for the measurement and analysis of Oreochromis niloticus condition fed with varying dietary maltose levels. *Int. J. Fauna Biol. Stud.* **2014**, *1*, 18–20.
35. Yang, Y.; Xie, S.; Cui, Y.; Zhu, X.; Lei, W.; Yang, Y. Partial and total replacement of fishmeal with poultry by-product meal in diets for gibel carp, Carassius auratus gibelio Bloch. *Aquac. Res.* **2006**, *37*, 40–48. [CrossRef]
36. Burgos-Aceves, M.A.; Cohen, A.; Smith, Y.; Faggio, C. Estrogen regulation of gene expression in the teleost fish immune system. *Fish Shellfish Immunol.* **2016**, *58*, 42–49. [CrossRef] [PubMed]
37. Jo, H.; Lee, S.; Yun, H.; Hong, J.; Moniruzzaman, M.; Bai, S.C.; Park, G.; Chee, S.; Jeon, T. Evaluation of dietary fishmeal analogue with addition of shrimp soluble extract on growth and nonspecific immune response of rainbow trout, Oncorhynchus mykiss. *J. World Aquac. Soc.* **2017**, *48*, 583–591. [CrossRef]
38. Sullivan, K.B. Replacement of Fish Meal by Alternative Protein Sources in Diets for Juvenile Black Sea Bass. Master's Thesis, University of North Carolina Wilmington, Wilmington, NC, USA, 2008.
39. Banavreh, A.; Soltani, M.; Kamali, A.; Yazdani-Sadati, M.A.; Shamsaie, M. Immuno-physiological and antioxidant responses of Siberian sturgeon (Acipenser baerii) fed with different levels of olive pomace. *Fish Physiol. Biochem.* **2019**, *45*, 1419–1429. [CrossRef] [PubMed]
40. Acharya, G.; Mohanty, P.K. Comparative haematological and serum biochemical analysis of catfishes Clarias batrachus (Linnaeus, 1758) and Heteropneustes fossilis (Bloch, 1794) with respect to sex. *J. Entomol. Zool. Stud.* **2014**, *2*, 191–197.
41. O'Hara, A.M.; Shanahan, F. The gut flora as a forgotten organ. *EMBO Rep.* **2006**, *7*, 688–693. [CrossRef]
42. Claus, S.P.; Guillou, H.; Ellero-Simatos, S. The gut microbiota: A major player in the toxicity of environmental pollutants? *npj Biofilms Microbiomes* **2016**, *2*, 16003. [CrossRef]
43. Li, E.; Xu, C.; Wang, X.; Wang, S.; Zhao, Q.; Zhang, M.; Qin, J.G.; Chen, L. Gut microbiota and its modulation for healthy farming of Pacific white shrimp Litopenaeus vannamei. *Rev. Fish. Sci. Aquac.* **2018**, *26*, 381–399. [CrossRef]
44. Ayiku, S.; Shen, J.; Tan, B.; Dong, X.; Liu, H. Effects of reducing dietary fishmeal with yeast supplementations on Litopenaeus vannamei growth, immune response and disease resistance against Vibrio harveyi. *Microbiol. Res.* **2020**, *239*, 126554. [CrossRef]

45. Amoah, K.; Huang, Q.; Dong, X.; Tan, B.; Zhang, S.; Chi, S.; Yang, Q.; Liu, H.; Yang, Y. Paenibacillus polymyxa improves the growth, immune and antioxidant activity, intestinal health, and disease resistance in Litopenaeus vannamei challenged with Vibrio parahaemolyticus. *Aquaculture* **2020**, *518*, 734563. [CrossRef]
46. Binda, C.; Lopetuso, L.R.; Rizzatti, G.; Gibiino, G.; Cennamo, V.; Gasbarrini, A. Actinobacteria: A relevant minority for the maintenance of gut homeostasis. *Dig. Liver Dis.* **2018**, *50*, 421–428. [CrossRef]
47. Wei, Y.; Xu, H.; Liang, M. Amino acid absorption and protein synthesis responses of turbot Scophthalmus maximus to lysine and leucine in free, dipeptide and tripeptide forms. *Aquac. Nutr.* **2020**, *26*, 358–367. [CrossRef]
48. Chi, S.Y.; Zhong, J.L.; He, Y.F.; Li, J.M.; Luo, Y. Cloning and tissue expression of amino acid transporter B0AT1 gene in Epinephelus coioides. *J. Guangdong Ocean Univ.* **2017**, *37*, 1–6.
49. Duan, Y.; Li, F.; Liu, H.; Li, Y.; Liu, Y.; Kong, X.; Zhang, Y.; Deng, D.; Tang, Y.; Feng, Z. Nutritional and regulatory roles of leucine in muscle growth and fat reduction. *Front. Biosci.* **2015**, *20*, 796–813.
50. Bakke, S.; Jordal, A.-E.O.; Gómez-Requeni, P.; Verri, T.; Kousoulaki, K.; Aksnes, A.; Rønnestad, I. Dietary protein hydrolysates and free amino acids affect the spatial expression of peptide transporter PepT1 in the digestive tract of Atlantic cod (Gadus morhua). *Comp. Biochem. Physiol. Part B Biochem. Mol. Biol.* **2010**, *156*, 48–55. [CrossRef] [PubMed]
51. Yang, X.; Wang, G.; Zhao, X.; Dong, X.; Chi, S.; Tan, B. Addition of hydrolysed porcine mucosa to low-fishmeal feed improves intestinal morphology and the expressions of intestinal amino acids and small peptide transporters in hybrid groupers (Epinephelus fuscoguttatus♀ × E. lanceolatus♂). *Aquaculture* **2021**, *535*, 736389. [CrossRef]
52. Zhou, P.; Luo, Y.; Zhang, L.; Li, J.; Zhang, B.; Xing, S.; Zhu, Y.; Gao, F.; Zhou, G. Effects of cysteamine supplementation on the intestinal expression of amino acid and peptide transporters and intestinal health in finishing pigs. *Anim. Sci. J.* **2017**, *88*, 314–321. [CrossRef]
53. Carr, W.E.S.; Netherton, I.J.C.; Gleeson, A.R.; Derby, C.D. Stimulants of Feeding Behavior in Fish: Analyses of Tissues of Diverse Marine Organisms. *Biol. Bull.* **1996**, *190*, 149–160. [CrossRef]
54. Basto-Silva, C.; Enes, P.; Oliva-Teles, A.; Balbuena-Pecino, S.; Navarro, I.; Capilla, E.; Guerreiro, I. Dietary protein source and protein/carbohydrate ratio affects appetite regulation-related genes expression in gilthead seabream (Sparus aurata). *Aquaculture* **2021**, *533*, 736142. [CrossRef]

metabolites

MDPI

Article

Dietary Effect of *Clostridium autoethanogenum* Protein on Growth, Intestinal Histology and Flesh Lipid Metabolism of Largemouth Bass (*Micropterus salmoides*) Based on Metabolomics

Pinxian Yang [1,2,3,†], Xiaoqin Li [1,2,3,†], Wenxiang Yao [1,2,3], Menglu Li [1,2,3], Yuanyuan Wang [1,2,3] and Xiangjun Leng [1,2,3,*]

1 National Demonstration Center for Experimental Fisheries Science Education, Shanghai Ocean University, Shanghai 201306, China
2 Centre for Research on Environmental Ecology and Fish Nutrition (CREEFN) of the Ministry of Agriculture and Rural Affairs, Shanghai Ocean University, Shanghai 201306, China
3 Shanghai Collaborative Innovation for Aquatic Animal Genetics and Breeding, Shanghai Ocean University, Shanghai 201306, China
* Correspondence: xjleng@shou.edu.cn
† These authors contributed equally to this work.

Abstract: *Clostridium autoethanogenum* protein (CAP) is a new single-cell protein explored in aquatic feeds in recent years. This study investigated the dietary effects of CAP replacing fishmeal (FM) on the growth, intestinal histology and flesh metabolism of largemouth bass (*Micropterus salmoides*). In a basal diet containing 700 g/kg of FM, CAP was used to substitute 0%, 15%, 30%, 45%, 70% and 100% of dietary FM to form six isonitrogenous diets (Con, CAP-15, CAP-30, CAP-45, CAP-70, CAP-100) to feed largemouth bass (80.0 g) for 12 weeks. Only the CAP-100 group showed significantly lower weight gain (WG) and a higher feed conversion ratio (FCR) than the control ($p < 0.05$). A broken-line analysis based on WG and FCR showed that the suitable replacement of FM with CAP was 67.1–68.0%. The flesh n-3/n-6 polyunsaturated fatty acid, intestinal protease activity, villi width and height in the CAP-100 group were significantly lower than those in the control group ($p < 0.05$). The Kyoto Encyclopedia of Genes and Genomes analysis showed that the metabolic pathway in flesh was mainly enriched in the "lipid metabolic pathway", "amino acid metabolism", "endocrine system" and "carbohydrate metabolism". In conclusion, CAP could successfully replace 67.1–68.0% of dietary FM, while the complete substitution decreased the growth, damaged the intestinal morphology and down-regulated the lipid metabolites.

Keywords: *Clostridium autoethanogenum*; largemouth bass; growth; intestinal healthy; lipid metabolomics

Citation: Yang, P.; Li, X.; Yao, W.; Li, M.; Wang, Y.; Leng, X. Dietary Effect of *Clostridium autoethanogenum* Protein on Growth, Intestinal Histology and Flesh Lipid Metabolism of Largemouth Bass (*Micropterus salmoides*) Based on Metabolomics. *Metabolites* **2022**, *12*, 1088. https://doi.org/10.3390/metabo12111088

Academic Editors: Qingchao Wang, Yan He and Qingsong Tan

Received: 20 September 2022
Accepted: 2 November 2022
Published: 9 November 2022

1. Introduction

In recent years, the largemouth bass (*Micropterus salmoides*) industry has rapidly expanded in many countries, including China. As a typically carnivorous fish, largemouth bass has a high requirement for protein, especially for fish meal. Generally, the commercial feeds of largemouth bass contain high levels of fish meal of up to 40–50%, leading to a high feed cost. Therefore, developing new protein sources to decrease fish meal inclusion has become an urgent task. In largemouth bass, the replacement of fish meal with alternative proteins has been reported on fermented plant protein [1,2], single-cell protein [3] and poultry by-product meal [4]. However, some disadvantages have been found in these alternative proteins, including the imbalanced amino acid composition and high anti-nutritional factors in plant protein sources, and the potential safety problems, such as lipid oxidation and unwarranted source in animal protein sources.

In recent years, bacterial protein has attracted more attention in aquatic feed [5,6]. *Clostridium autoethanogenum* protein (CAP) is a novel bacterial protein produced by the fermentation of *Clostridium autoethanogenum*. In the fermentation process, the bacterial utilizes carbon monoxide from steel-making waste gas and ammonia to form bacterial protein, then the fermentation liquid is centrifuged and dried to obtain CAP [7]. At present, the relatively complete genome sequence of this bacterium has been obtained, and no toxic genes were found [8]. Compared with traditional plant and animal protein sources, CAP contains less anti-nutritional factors and lower Salmonella and biogenic amines content [9]. Several studies have reported the successful inclusion of CAP in aquatic diets. In juvenile largemouth bass (initial body weight of 17.75 g), the substitution of 63% dietary fish meal with CAP showed no adverse impacts on hepatic and hindgut histology [10]. In large-size largemouth bass with a body weight of 224 g, CAP could replace 150 g/kg of fish meal in a diet containing 350 g/kg of fish meal, without adverse effects on the growth performance, feed utilization and intestinal histology [11]. In Jian carp (*Cyprinus carpio var.* Jian), the replacement of soybean meal with CAP promoted the growth performance and antioxidant capacity without an obvious effect on the liver and midgut [12]. In addition, the successful inclusion of CAP in aquatic diets has also been reported in grass carp (*Ctenopharyngodon idellus*) [9], black sea bream (*Acanthopagrus schlegelii*) [13] and tilapia (*Oreochromis niloticus*) [14]. The suitable replacement level of dietary fish meal with CAP was 42.9% for largemouth bass [11] and 45% for white shrimp (*Litopenaeus vannamei*) [15]. However, the excessive inclusion of bacterial protein decreased the growth and feed intake of black sea bream [13] and white shrimp [15]; thus, it seems that CAP could not completely substitute dietary fish meal.

Previous studies have proved the feasibility of replacing fish meal with CAP in commercial diets for largemouth bass [16]. However, in commercial diets, there are various protein ingredients such as chicken meal, soybean protein concentrate, corn gluten, soybean meal besides fish meal, and these ingredients may interfere with the replacement of fish meal with CAP. Therefore, to fully understand the replacement of fish meal with CAP, fish meal was used as the only protein source in the present study, then, different proportions of dietary fish meal were replaced by CAP to investigate the effects on the growth, intestinal histology and flesh lipid metabolism of largemouth bass. The findings will direct the application of CAP in the diets of carnivorous fishes and promote the sustainable development of aquafeeds with less cost and less dependence on marine products.

2. Materials and Methods

2.1. Experimental Design

The basal diet was designed to contain 700 g/kg of fish meal, and then 0%, 15%, 30%, 45%, 70% and 100% of dietary fish meal was replaced by CAP based on iso-protein to form six diets, named as Con (control group), CAP-15, CAP-30, CAP-45, CAP-70 and CAP-100, respectively. The protein ingredients were ground, screened (60-mesh) and mixed with oil and water (27%). Then, the mixture was extruded to form sinking pellets with a diameter of 2.0 mm using a single-screw extruder (SLP–45; Chinese Fishery Machinery and Instrument Research Institute, Shanghai, China) at an extruding temperature of 85–90 °C. All diets were air-dried naturally and preserved at 4 °C until use. The diets' formulation and proximate composition are shown in Table 1.

Table 1. Formulation and proximate composition of experimental diets (g/kg).

Ingredients [1]	CON	CAP-15	CAP-30	CAP-45	CAP-70	CAP-100
Fish meal	700.0	595.0	490.0	385.0	210.0	0.0
CAP	0.0	85.80	171.6	257.3	400.0	571.6
Bone meal	0.0	15.0	30.0	45.0	69.0	98.0
$Ca(H_2PO_3)_2$	19.0	19.0	19.0	19.0	19.0	19.0
Wheat flour	200.0	194.7	189.4	184.2	176.5	166.9
Fish oil	0.0	9.5	19.0	28.5	44.5	63.5
Soybean oil	25.0	25.0	25.0	25.0	25.0	25.0
Soybean lecithin	25.0	25.0	25.0	25.0	25.0	25.0
Choline chloride	1.0	1.0	1.0	1.0	1.0	1.0
Premix [2]	10.0	10.0	10.0	10.0	10.0	10.0
Yeast extract	20.0	20.0	20.0	20.0	20.0	20.0
Total	1000.0	1000.0	1000.0	1000.0	1000.0	1000.0
		Proximate composition				
Crude protein	503.8	498.8	498.9	500.0	498.2	497.8
Crude lipid	117.1	116.7	117.3	115.9	117.1	115.7
Crude ash	121.3	121.2	118.4	116.2	114.4	112.3
Moisture	81.4	81.6	80.9	81.1	79.7	78.1

[1] The ingredients were purchased from the Yuehai Feed Company (Zhejiang, China), and the protein contents of the ingredients are as follows: fish meal (682.0 g/kg), wheat flour (117.0 g/kg), yeast extract (216.0 g/kg). [2] Premix provided the following per kg of diets: Lascorbate-2-mon-phosphate (35%), 600 mg; vitamin E, 300 mg; inositol, 200 mg; niacinamide, 80 mg; calcium pantothenate, 40 mg; vitamin A, 20 mg; vitamin D_3, 10 mg; vitamin K_3, 20 mg; vitamin B_1, 10 mg; vitamin B_2, 15 mg; vitamin B_6, 15 mg; vitamin B_{12}, 8 mg; biotin, 2 mg; wheat middling, 220 mg; zeolite, 332 mg; Fe, 300 mg; Zn, 200 mg; NaCl 100 mg; Cu, 30 mg; Mn, 25 mg; $CoCl_2 \cdot 6H_2O$ (10%Co), 5 mg; Na_2SeO_3 (10%Se), 5 mg; KIO, 3 mg.

The CAP was provided by Beijing Shoulang New Energy Technology Co., Ltd., Beijing, China. The product contains 842.0 g/kg of crude protein, 19.0 g/kg of crude lipid and 32.7 g/kg of crude ash. As a light-yellow powder with a special smell, CAP can be directly added into feed as a protein source. Dietary amino acids and fatty acids compositions of CAP are shown in Tables 2 and 3.

Table 2. Amino acid composition of experimental diets (dry matter basis, g/kg).

Amino Acid	CON	CAP-15	CAP-30	CAP-45	CAP-70	CAP-100
		Essential amino acid				
Lys	37.6	39.5	41.5	43.5	46.7	50.6
Met	14.6	14.4	14.2	14.0	13.7	13.4
Arg	30.3	28.9	27.4	26.0	23.7	20.8
His	15.2	14.5	13.7	13.0	11.7	10.2
Phe	26.2	25.2	24.3	23.3	21.7	19.8
Trp	5.4	5.2	4.96	4.8	4.4	4.0
Val	24.9	26.0	27.1	28.2	30.0	32.2
Ile	20.2	21.8	23.4	25.0	27.7	31.0
Leu	38.6	38.5	38.4	38.3	38.1	38.0
Thr	21.0	21.4	21.8	22.2	22.9	23.7
		Non-essential amino acid				
Cys	21.0	20.1	19.1	18.2	16.6	14.7
Gly	4.6	4.7	4.9	5.0	5.2	5.5
Ser	20.0	21.6	23.3	24.9	27.6	30.8
Pro	43.8	45.6	47.3	49.1	51.9	55.4
Ala	5.4	5.2	5.0	4.8	4.4	4.0
Asp	24.9	26.0	27.1	28.2	30.0	32.2
Tyr	29.9	28.9	27.9	26.8	25.1	23.0
Glu	66.2	59.3	56.9	54.1	46.8	40
Total amino acids	449.8	446.8	448.26	449.4	448.2	449.3

Note: Glu, glutamic acid; Asp, aspartic acid; Leu, leucine; Ile, isoleucine; His, histidine; Gly, glycine; Thr, threonine; Ala, alanine; Arg, arginine; Phe, phenylalanine; Trp, tryptophan; Lys, lysine; Pro, proline; Tyr, tyrosine; Val, valine; Met, methionine; Ser, serine; Cys, cysteine.

Table 3. Fatty acid composition of experimental diets (percentage of fatty acids, %).

Items	Con	CAP-15	CAP-30	CAP-45	CAP-70	CAP-100
C14:0	4.20	4.68	5.23	5.10	5.34	5.61
C15:0	0.46	0.55	0.66	0.75	0.92	1.12
C16:0	15.31	15.20	17.40	19.20	22.68	25.48
C17:0	0.71	0.61	0.52	0.42	0.27	0.08
C18:0	2.83	2.74	2.66	2.57	2.42	2.25
C20:0	0.29	0.25	0.21	0.17	0.11	0.03
SFAs	23.79	24.03	26.67	28.22	31.74	34.57
C16:1	3.90	3.85	3.79	3.74	3.64	3.54
C17:1	0.53	0.54	0.49	0.50	0.49	0.47
C18:1	14.18	15.23	15.28	14.96	15.92	15.86
C20:1	2.64	2.03	1.91	1.84	1.39	1.20
MUFAs	21.25	21.64	21.48	21.04	21.45	21.06
C16:2	0.24	0.25	0.21	0.25	0.19	0.21
C18:2	14.65	13.76	13.08	13.97	13.26	13.06
C20:2	3.01	2.92	2.91	2.66	4.07	3.68
C20:4	2.30	2.21	2.39	1.12	1.09	0.69
n-6 PUFAs	20.21	19.14	18.59	18.01	18.61	17.65
C18:3	2.83	2.58	2.31	2.21	1.99	1.38
C20:5	10.69	10.61	10.56	10.23	8.56	7.61
C22:5	10.83	10.41	10.00	9.78	9.46	9.34
C22:6	8.05	8.35	8.16	7.06	6.86	6.63
n-3 PUFAs	32.40	31.94	31.03	29.29	26.87	24.95
n-3/n-6	1.60	1.67	1.67	1.63	1.44	1.41

2.2. Experimental Fish and Feeding Management

Largemouth bass were supplied by a local aquaculture farm in Qingpu, Shanghai, China. The fish were fed with commercial diets for 10 weeks until the body weight reached about 80 g. Then, a total of 216 fish with an initial body weight of 80.0 ± 0.5 g were randomly allocated into 18 cage s (1.5 × 1.0 × 1.2 m), with 3 replicates (cage) per treatment and 12 fish per cage. The cages were hung in indoor cement pools without direct sunshine. During the feeding period, the six diets were fed to the fish three times daily (08:00, 13:00, 17:00), and the daily feeding rate was 2–2.5% of body weight. The feed intake was adjusted according to the feeding behavior and water temperature to ensure no feed residue was left 10 min after feeding. About one third of the cultured water was renewed with filtrated pond water, and the waste in the pools was siphoned out once a week. Water quality was monitored every day, and the water temperature, dissolved oxygen, pH and ammonia nitrogen levels were 25–30 °C, >5 mg/L, 7.5–8.0 and <0.2 mg/L, respectively. The feeding trial was conducted at the Binhai Aquaculture Station, Shanghai Ocean University, and lasted for 12 weeks.

2.3. Sample Collection

Prior to slaughter, all fish were deprived of diets for 24 h (starvation), then counted and bulk weighed to calculate survival, weight gain (WG), feed intake (FI) and feed conversion ratio (FCR). Five fish per cage were selected randomly to individually measure body weight and body length for the calculation of condition factor (CF). Three fish per cage were euthanized by an overdose of anaesthetic (MS–222), and clinical signs of death were ensured prior to sampling. The blood was drawn from the caudal vein and centrifuged at 4 °C for 10 min (3000× g r/min) to collect the serum, then frozen at −80 °C for further analysis. Then, the three fish were dissected, and visceral, liver and intestinal lipid were weighed to calculate the hepatosomatic index (HSI), viscerosomatic index (VSI) and intestinal fat ratio (IFR). The anterior intestine (1 cm) was sampled for the measurement of morphological structures, and the remaining intestine was stored at −80 °C for the activity measurement of digestive enzyme. About 1 g of dorsal flesh from the left side of two fish per cage was collected and frozen in liquid N_2 for the metabolomic assay, and the remaining flesh was stored at −80 °C until use. In addition, another three fish per cage were stored at −20 °C for the proximate composition analysis of whole fish.

2.4. Analytical Methods

2.4.1. Growth Indicators

WG (%) = 100 × [(final weight (g) − initial weight (g))/initial body weight (g)]

FCR = feed intake (g)/weight gain (g)

CF (g/cm^3) = 100 × final body weight (g)/final body length (cm)^3

Survival (%) = 100 × final number of fish/initial number of fish

VSI (%) = 100 × visceral weight (g)/body weight (g)

HSI (%) = 100 × hepatopancreas weight (g)/body weight (g)

IFR (%) = 100 × (intestinal fat weight [g]/body weight [g])

FI (g/d/fish) = total feed intake (g)/days (d)/number of test fish

2.4.2. Proximate Composition of Diets and Whole Fish

Crude lipid, crude protein, moisture and crude ash contents were analyzed following the method of AOAC (2000a). The moisture and crude ash contents were determined by drying samples to a constant weight at 105 °C in an oven or by combusting samples at 550 °C for 6 h in a muffle furnace. The crude protein content was estimated with the Kjeldahl system method (2300 Auto analyser; FOSS Tecator, AB, Hoganas, Sweden), and the crude lipid was measured gravimetrically after extraction by chloroform-methanol.

2.4.3. Biochemical Analysis

The anterior intestine samples were thawed at 4 °C, homogenized with nine volumes (*w*/*v*) of ice-cold saline (0.86% NaCl) and centrifuged for 15 min (3000× *g* r/min) at 4 °C. The supernatant was collected and preserved at 4 °C for the measurement of digestive enzyme activity and soluble protein in 24 h. Digestive enzyme indexes such as protease and lipase (LPS) were measured by commercial kits (Shanghai Haling Biotechnology Co., Ltd., Shanghai, China), following the protocols provided by the supplier.

The contents of albumin (ALB, bromocresol green method); total protein (TP, biuret method); triglyceride (TG, GPO-PAP method); cholesterol (CHO, CHOD-PAP method); glucose (GLU, glucose oxidase method); malondialdehyde (MDA, TBA method); and the activities of alkaline phosphatase (AKP, AMP method) and superoxide dismutase (SOD, BioTekmethod) were determined using the kits provided by Shanghai Haling Biotechnology Co., Ltd.

2.4.4. The Histology of Anterior Intestine

The anterior intestine (1 cm) was immersed in Bouin's solution for 24 h, and then transferred into 100% ethyl alcohol. The tissue was dehydrated in a series of alcohol solutions and embedded in paraffin. Then, sections (5 µm) were cut and stained with hematoxylin-eosin and sealed with a neutral gum. The morphological structures of the tissues as villus height and width were observed using an imaging microscope (Nikon YS100, Tokyo, Japan). The image was analyzed with the Image J14.0 image analysis software. The selected images contained complete villi and muscle thickness, while the number of complete villi was not less than 1/3 of the intestinal sections, and the goblet cells were counted under 100 times the field of view. Number of goblet cells: number of goblet cells per 200 µm length of mucosa. Villus height, villus width and muscle thickness parameters were measured (10 fields per individual sample) according to the procedures described by Escaffre et al. [17].

2.4.5. Fatty Acid Composition of Flesh

The fatty acid was measured by the method of boron trifluoride methyl esterification. The extracted fat was dissolved in 1 mL of hexane, then 2 mL of 14% boron trifluoride methanol solution was added. After a water bathing of 25 min at 100 °C (the first step of methyl esterification), benzene (2 mL) and methanol solution (2 mL) were added for another water bath (100 °C, 25 min) (the second step of methyl esterification). Then, the samples were mixed with distilled water and n-hexane. After centrifuging at 3000× *g* r/min for 10 min, the supernatant was collected for fatty acid analysis by using an Agilent Technologies 7890B GS Syetem GC/MS (Agilent, Santa Clara, CA, USA).

2.4.6. Non-Targeted Metabolomic Analysis

The control, CAP-30 and CAP-70 groups were used for the metabolomic analysis. Flesh sample was precisely weighed (50 mg), then extracted with a mixture (400 μL) of methanol and water (4:1, *v*/*v*). The sample was treated by a tissue crusher (high-throughput tissue crusher Wonbio-96c, Wanbo Biotechnology Co., Ltd., Shanghai, China) at 50 Hz for 6 min, then subjected to cryogenic sonication treatment for 30 min, kept at −20 °C for 1 h and centrifuged (13,000× *g*) at 4 °C for 15min. The supernatant (20 μL) was collected and transferred for LC-MS/MS analysis by Shanghai Majorbio Bio-Pharm Technology Co., Ltd. (Shanghai, China).

The chromatographic column was ACQUITY UPLC HSS T3 (100 mm × 2.1 mm i.d., 1.8 μm; Waters, Milford, MA, USA). The mobile phases consisted of solvent A (0.1% formic acid) and solvent B (acetonitrile:isopropanol = 1:1 (*v*/*v*) containing 0.1% formic acid). The sample injection volume was 10 μL, and the flowing rate was 0.4 mL/min with a column temperature of 40 °C. Electrospray positive ion (ESI+) mode and electrospray negative ion (ESI−) mode were used to collect the mass spectrum signal (Triple TOFTM5600+, AB Sciex, San Diego, CA, USA). During the period of analysis, all samples were stored at 4 °C, and a quality control (QC) sample was inserted every 5–15 samples to evaluate the stability of the analytical system and assess the reliability of the results.

Metabolism raw data was imported into Progenesis QI (Waters Corporation, Milford, USA) for preprocessing. Statistically significant metabolites among groups were selected with VIP > 1 and $p < 0.05$ for PCA. A partial least squares discriminate analysis (PLS-DA) was used for statistical analysis to determine flesh metabolic changes between comparable groups. The model validity was evaluated from model parameters R2 and Q2, which provided information for the interpretability and predictability to avoid the risk of over-fitting. Differential metabolites between two groups were summarized and mapped into their biochemical pathways through a metabolic enrichment and pathway analysis based on the KEGG (Kyoto Encyclopedia of Genes and Genomes) database search (http://www.kegg.com/ accessed on 3 February 2022) [18].

2.5. Statistical Analysis

The experimental data were presented as the mean and standard deviation (±SD). The statistical analysis was performed using the Statistical Package for the Social Sciences (SPSS) 19.0 for Windows (SPSS, Chicago, IL, USA). A one-way analysis of variance (ANOVA) was combined with the LSD method for multiple comparisons. A follow-up trend analysis was performed using orthogonal polynomial contrast to determine whether significant effects were linear and/or quadratic. The significance level for the differences among treatments was $p < 0.05$. In addition, a nonlinear regression analysis (binomial method) was used, and the CAP protein concentration data were used for binomial transformation and curve estimation.

3. Results

3.1. Growth Performance

In Table 4, the CAP-15, CAP-30, CAP-45 and CAP-70 groups showed the similar WG, FI and FCR to the control ($p > 0.05$), while the WG of CAP-100 group was decreased by 31.8%, and FCR was increased by 0.35 when compared to the control ($p < 0.05$). The VSI of CAP-45, CAP-70 and CAP-100 groups, and the HSI and CF of CAP-100 group were significantly lower than those of the control ($p < 0.05$). The broken-line model based on WG and FCR showed that the proper replacement ratio of dietary fish meal by CAP was 67.1% and 68.0%, equal to 374.6 g/kg and 380.3 g/kg of CAP inclusion (Figure 1a,b), respectively. The binomial regression analysis showed that the CAP protein concentration had a good direct fit with the WG and FCR of largemouth bass, and the optimal CAP concentrations for WG and FCR were 146 g/kg and 181 g/kg, respectively (Figure 2a,b).

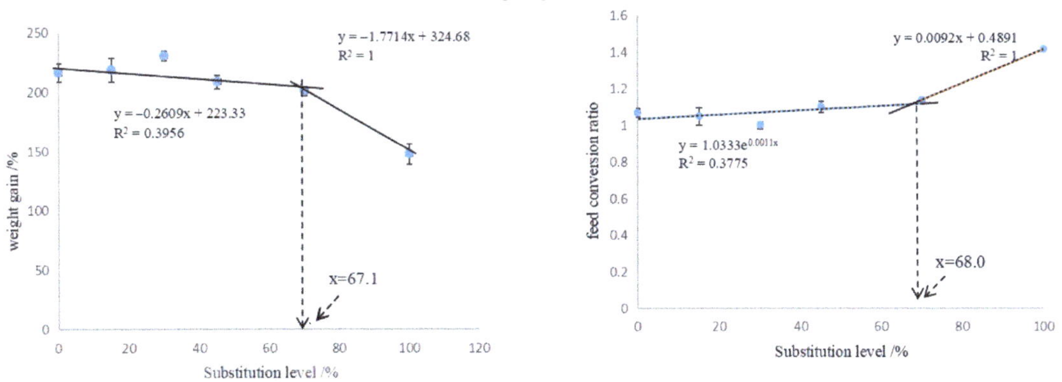

(**a**) (**b**)

Figure 1. Broken-line analysis of weight gain (**a**) and feed conversion ratio (**b**) against substituted ratio of fish meal with CAP.

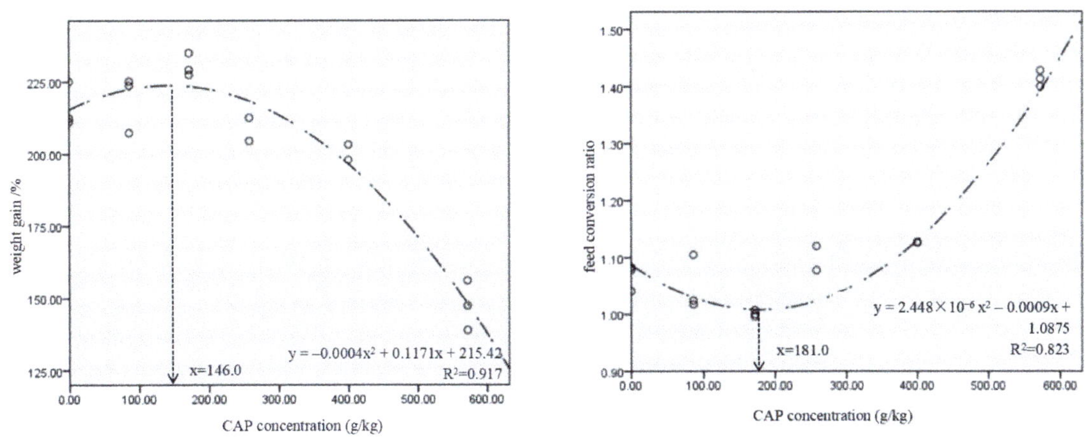

(**a**) (**b**)

Figure 2. Relationship between CAP addition level and weight gain (**a**)/feed conversion ratio (**b**) for largemouth bass. Data are fitted with a nonlinear regression model (binomial regression model).

Table 4. Effects of dietary *Clostridium autoethanogenum* protein on growth and body index of largemouth bass.

Parameters	Diets						Pr > F		
	Con	CAP-15	CAP-30	CAP-45	CAP-70	CAP-100	ANOVA	Linear	Quadratic
IBW/g	80.1 ± 0.2	80.0 ± 0.2	80.2 ± 0.2	80.1 ± 0.2	80.0 ± 0.2	80.2 ± 0.2	-	-	-
FBW/g	253.1 ± 6.2^{ab}	255.1 ± 7.9^{ab}	264.5 ± 3.3^{a}	246.8 ± 4.5^{b}	240.5 ± 3.1^{b}	198.1 ± 6.9^{c}	0.000	0.000	0.000
WG/%	216.3 ± 7.6^{ab}	218.8 ± 9.9^{ab}	230.5 ± 3.9^{a}	208.6 ± 5.6^{b}	200.6 ± 3.7^{b}	147.5 ± 8.6^{c}	0.000	0.000	0.000
FCR	1.06 ± 0.02^{bc}	1.04 ± 0.04^{bc}	1.00 ± 0.01^{c}	1.10 ± 0.03^{b}	1.13 ± 0.00^{b}	1.41 ± 0.01^{a}	0.000	0.000	0.000
FI/g/d/fish	2.20 ± 0.03^{a}	2.18 ± 0.01^{a}	2.18 ± 0.03^{a}	2.15 ± 0.04^{a}	2.19 ± 0.7^{a}	1.99 ± 0.10^{b}	0.002	0.001	0.006
CF/g/cm^3	2.43 ± 0.07^{abc}	2.55 ± 0.05^{a}	2.49 ± 0.15^{abc}	2.35 ± 0.03^{bc}	2.28 ± 0.12^{cd}	2.16 ± 0.02^{d}	0.002	0.000	0.025
VSI/%	8.09 ± 0.17^{a}	8.12 ± 0.18^{a}	7.92 ± 0.10^{ab}	7.65 ± 0.19^{b}	7.70 ± 0.18^{b}	7.78 ± 0.28^{ab}	0.024	0.005	0.173
HSI/%	2.59 ± 0.18^{a}	2.48 ± 0.09^{ab}	2.49 ± 0.09^{ab}	2.35 ± 0.13^{ab}	2.28 ± 0.10^{ab}	2.16 ± 0.17^{b}	0.016	0.001	0.520
MSI/%	1.56 ± 0.13^{a}	1.56 ± 0.07^{ab}	1.52 ± 0.07^{ab}	1.45 ± 0.10^{ab}	1.48 ± 0.07^{ab}	1.39 ± 0.07^{b}	0.202	0.017	0.774
SR/%	100.0	100.0	100.0	100.0	100.0	100.0	-	-	-

Note: Within the same row, values with different superscripts are significantly different ($p < 0.05$). Pr > F: significant probability associated with the F-statistic. The value in the table is the average number of standard deviations ($n = 3$). Abbreviations: IBW, initial body weight; FBW, final body weight; WG, weight gain; FCR, feed conversion ratio; FI, feed intake; CF, condition factor; HSI, hepatosomatic index; VSI, viscerosomatic index.

3.2. Body Composition and Nutrients Retention

In the proximate composition of whole-body, there were no significant differences in crude ash, crude lipid and protein retention among all the groups ($p > 0.05$). The crude protein in the CAP-45, CAP-70 and CAP-100 groups, and the moisture in the CAP-70 and CAP-100 groups were significantly higher, while the lipid retention in the CAP-70 group was significantly lower than that in the control ($p < 0.05$) (Table 5).

Table 5. Effects of dietary *Clostridium autoethanogenum* protein on proximate composition of whole body (fresh weight) and nutrients retention of largemouth bass.

Parameters	Diets						Pr > F		
	Con	CAP-15	CAP-30	CAP-45	CAP-70	CAP-100	ANOVA	Linear	Quadratic
Moisture (g/kg)	686.3 ± 4.1^{b}	693.1 ± 3.8^{ab}	694.3 ± 2.7^{ab}	688.4 ± 5.1^{b}	699.9 ± 4.5^{a}	703.1 ± 2.1^{a}	0.157	0.033	0.230
Crude protein (g/kg)	165.9 ± 6.2^{b}	168.3 ± 3.1^{ab}	172.9 ± 5.3^{a}	171.5 ± 0.9^{a}	172.7 ± 5.9^{a}	170.1 ± 4.1^{ab}	0.170	0.067	0.066
Crude lipid (g/kg)	56.1 ± 1.9	55.7 ± 0.5	50.8 ± 3.6	51.1 ± 2.9	51.2 ± 1.4	55.3 ± 2.7	0.032	0.173	0.027
Crude ash (g/kg)	43.4 ± 2.1	43.3 ± 3.0	47.0 ± 1.4	43.6 ± 1.5	46.4 ± 0.8	44.7 ± 2.5	0.096	0.132	0.322
Protein retention (%)	34.6 ± 0.1	34.6 ± 0.1	36.5 ± 0.1	37.2 ± 0.2	36.4 ± 0.1	35.6 ± 0.2	0.000	0.000	0.000
Lipid retention (%)	51.6 ± 2.2^{a}	50.3 ± 1.4^{ab}	45.3 ± 1.2^{ab}	47.7 ± 0.6^{ab}	43.8 ± 5.4^{b}	51.7 ± 4.5^{a}	0.000	0.067	0.001

Note: Within the same row, values with different superscripts are significantly different ($p < 0.05$).

3.3. Serum Biochemical Indices Analyses

In Table 6, no significant differences were detected in AKP activity, MDA, TG, TP, ALB and GLU contents in the CAP groups ($p > 0.05$) when compared to the control group ($p > 0.05$). CAP-30 and CAP-45 groups showed significantly higher SOD activity than the control ($p < 0.05$).

Table 6. Effects of dietary *Clostridium autoethanogenum* protein on serum biochemical indices of largemouth bass.

Parameters	Diets						Pr > F		
	Con	CAP-15	CAP-30	CAP-45	CAP-70	CAP-100	ANOVA	Linear	Quadratic
MDA (nmol/mL)	161.1 ± 16.9	162.9 ± 1.3	150.1 ± 13.9	158.5 ± 12.0	170.8 ± 12.2	168.6 ± 2.1	0.566	0.313	0.295
SOD (U/mL)	146.9 ± 13.6^{b}	145.1 ± 5.1^{b}	164.0 ± 1.7^{a}	165.0 ± 7.1^{a}	144.1 ± 7.1^{b}	137.8 ± 3.7^{b}	0.016	0.239	0.012
AKP (U/mL)	123.4 ± 2.8^{ab}	129.7 ± 2.5^{a}	128.1 ± 2.5^{a}	125.0 ± 3.6^{ab}	120.0 ± 7.2^{ab}	110.7 ± 5.2^{b}	0.000	0.000	0.000
TP (gprot/L)	52.5 ± 3.8	53.7 ± 6.1	52.8 ± 2.3	53.33 ± 1.5	52.2 ± 4.0	50.5 ± 1.0	0.952	0.500	0.483
ALB (gprot/L)	19.1 ± 2.2^{ab}	20.2 ± 0.4^{a}	19.9 ± 0.9^{a}	20.9 ± 0.9^{a}	17.1 ± 0.9^{b}	17.2 ± 1.5^{b}	0.021	0.025	0.033
GLU (mmol/L)	7.30 ± 0.43^{ab}	8.16 ± 0.15^{a}	8.12 ± 0.72^{a}	7.86 ± 0.61^{ab}	7.13 ± 0.46^{b}	7.30 ± 0.28^{ab}	0.129	0.236	0.067
TG (mmol/L)	17.7 ± 0.6	18.3 ± 0.5	17.0 ± 0.9	18.4 ± 0.5	18.3 ± 0.9	18.3 ± 1.1	0.090	0.169	0.020
CHO (mmol/L)	13.6 ± 1.2^{ab}	14.2 ± 0.2^{a}	13.1 ± 0.9^{ab}	13.0 ± 1.4^{ab}	12.8 ± 0.7^{ab}	11.6 ± 0.8^{b}	0.259	0.028	0.443

Note: Within the same row, values with different superscripts are significantly different ($p < 0.05$). Abbreviations: MDA, malondialdehyde; SOD, superoxide dismutase; AKP, alkaline phosphatase; TP, total protein; ALB, albumin; GLU, glucose; TG, triglyceride; CHO, cholesterol.

3.4. Intestinal Digestive Enzymes

Compared to the control group, the intestinal LPS activity was significantly increased in the CAP-45 group, while the protease activity was significantly decreased in the CAP-100 group ($p < 0.05$) (Table 7).

Table 7. Effects of dietary *Clostridium autoethanogenum* protein on digestive enzymes activity of largemouth bass.

Parameters	Diets						Pr > F		
	Con	CAP-15	CAP-30	CAP-45	CAP-70	CAP-100	ANOVA	Linear	Quadratic
LPS (U/g prot)	356.4 ± 8.8 [bc]	365.5 ± 15.4 [b]	367.0 ± 12.1 [ab]	386.3 ± 5.2 [a]	364.7 ± 8.0 [b]	342.3 ± 6.8 [c]	0.010	0.769	0.000
Protease (U/mg prot)	2074.4 ± 32.5 [a]	2019.7 ± 98.3 [ab]	2006.4 ± 92.8 [ab]	2003.2 ± 76.8 [ab]	1965.1 ± 89.7 [ab]	1909.6 ± 62.1 [b]	0.265	0.023	0.062

Note: Within the same row, values with different superscripts are significantly different ($p < 0.05$).

3.5. The Histology of Anterior Intestine

As shown in Table 8, the villus height of the CAP-100 group and the villus width of the CAP-70 and CAP-100 groups were significantly lower than those of the control group ($p < 0.05$). The muscle thickness did not reveal any significant difference among the treatments ($p > 0.05$).

Table 8. Effects of dietary *Clostridium autoethanogenum* protein on intestinal structure of largemouth bass (μm).

Parameters	Diets						Pr > F		
	Con	CAP-15	CAP-30	CAP-45	CAP-70	CAP-100	ANOVA	Linear	Quadratic
Villus height	1201.5 ± 72.9 [a]	1192.7 ± 36.8 [a]	1256.9 ± 39.8 [a]	1240.2 ± 66.1 [a]	1129.3 ± 37.7 [ab]	939.9 ± 52.3 [b]	0.000	0.001	0.000
Villus width	105.24 ± 5.9 [a]	110.1 ± 4.6 [a]	106.5 ± 3.7 [a]	103.4 ± 2.8 [a]	96.9 ± 3.5 [b]	97.1 ± 5.1 [b]	0.016	0.002	0.132
Muscle thickness	222.2 ± 12.2	191.5 ± 7.6	198.5 ± 5.6	213.5 ± 11.8	211.2 ± 13.4	198.5 ± 8.1	0.050	0.216	0.798

Note: Within the same row, values with different superscripts are significantly different ($p < 0.05$).

When the replacement ratio of fishmeal with CAP reached 70% (CAP-70 and CAP-100 group), some goblet cells were observed to be enlarged, and some intestinal structure was damaged with shed and ruptured villus. The length and width of the intestinal villi tended to decrease with the increasing CAP inclusion (Figure 3).

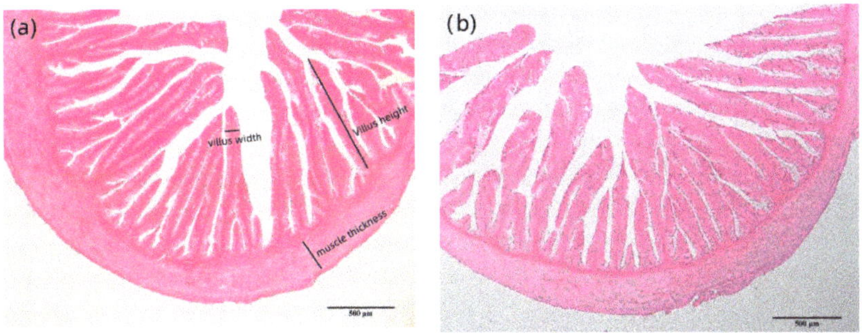

Figure 3. *Cont.*

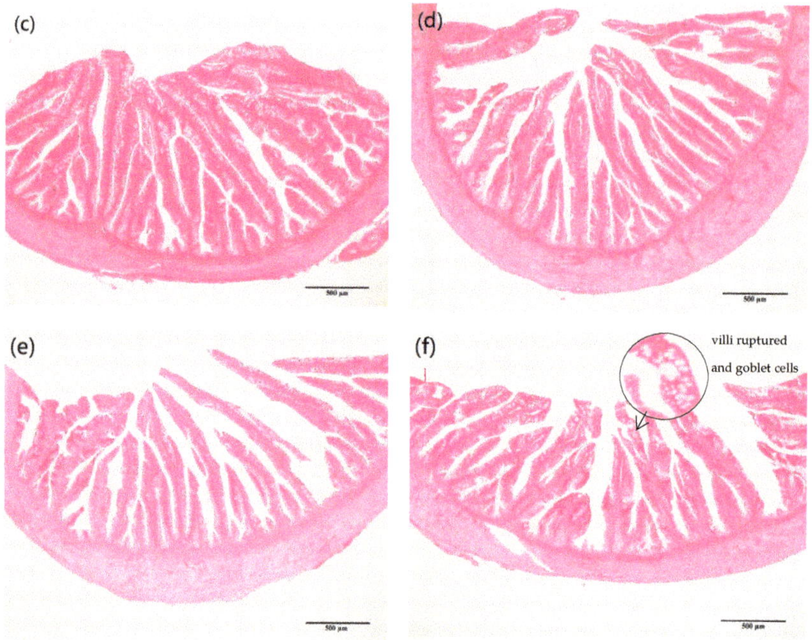

Figure 3. The foregut tissue structure of largemouth bass (40×). (**a**) Foregut tissue of control group; (**b**) foregut tissue of CAP-15 group; (**c**) foregut tissue of CAP-30 group; (**d**) foregut tissue of CAP-45 group; (**e**) foregut tissue of CAP-70 group; (**f**) foregut tissue of CAP-100 group. Note: The scale bar in the figure is the magnification contrast of micrograph.

3.6. Fatty Acid Composition in Flesh

In Table 9, flesh SFAs in the CAP-30, CAP-45 and CAP-70 groups were significantly lower, while C20:4 were higher than those of the control group ($p < 0.05$). CAP-70 and CAP-100 groups showed significantly higher C20:1 than the control ($p < 0.05$). In addition, the n-6PUFAs of the CAP-70 group diet and the n-3/n-6 PUFAs of the CAP-100 group were significantly lower than those of the control ($p < 0.05$).

Table 9. Effects of dietary *Clostridium autoethanogenum* protein on muscle fatty acid composition of largemouth bass (percentage of fatty acids, %).

Parameters	Diets							Pr > F	
	Con	CAP-15	CAP-30	CAP-45	CAP-70	CAP-100	ANOVA	Linear	Quadratic
C14:0	2.03 ± 0.07	1.93 ± 0.06	1.89 ± 0.09	1.90 ± 0.06	1.91 ± 0.07	2.09 ± 0.14	0.268	0.621	0.034
C16:0	12.0 ± 0.9	11.6 ± 0.2	11.1 ± 0.2	10.7 ± 0.6	10.4 ± 0.3	11.9 ± 0.7	0.154	0.290	0.032
C18:0	4.26 ± 0.05	4.11 ± 0.18	4.02 ± 0.24	4.13 ± 0.16	4.08 ± 0.07	4.42 ± 0.10	0.703	0.645	0.251
SFA	18.3 ± 0.9 [a]	17.7 ± 0.1 [ab]	16.9 ± 0.4 [b]	16.7 ± 0.6 [b]	16.7 ± 0.5 [b]	18.3 ± 0.8 [a]	0.135	0.338	0.0.24
C16:1	5.49 ± 0.35	4.82 ± 0.40	5.57 ± 0.40	5.61 ± 0.11	5.18 ± 0.17	5.44 ± 0.34	0.373	0.707	0.968
C18:1	19.1 ± 0.6	19.6 ± 0.6	19.1 ± 0.1	19.5 ± 1.1	19.1 ± 0.9	18.5 ± 0.8	0.877	0.466	0.470
C20:1	0.84 ± 0.05 [b]	0.85 ± 0.03 [b]	1.05 ± 0.4 [ab]	0.99 ± 0.05 [ab]	1.12 ± 0.10 [a]	1.16 ± 0.08 [a]	0.064	0.006	0.806
MUFA	25.5 ± 1.1	25.3 ± 1.3	25.7 ± 0.5	26.1 ± 1.6	25.3 ± 1.2	25.1 ± 1.1	0.950	0.919	0.427
C18:2	16.1 ± 0.4	16.3 ± 0.2	15.5 ± 0.6	15.9 ± 0.3	14.8 ± 0.5	15.9 ± 0.5	0.027	0.044	0246
C20:2	0.63 ± 0.05	0.63 ± 0.3	0.63 ± 0.4	0.60 ± 0.3	0.60 ± 0.10	0.61 ± 0.15	0.688	0.212	0.789
C20:3	1.65 ± 0.2	1.62 ± 0.2	1.63 ± 0.4	1.61 ± 0.02	1.62 ± 0.14	1.61 ± 0.02	0.617	0.164	0.601
C20:4	0.23 ± 0.01 [b]	0.26 ± 0.01 [ab]	0.35 ± 0.05 [a]	0.34 ± 0.03 [a]	0.31 ± 0.06 [a]	0.29 ± 0.02 [ab]	0.133	0.116	0.035
n-6PUFAs	18.6 ± 0.2 [a]	18.8 ± 0.1 [a]	18.1 ± 0.5 [ab]	18.5 ± 0.2 [ab]	17.3 ± 0.4 [b]	18.3 ± 0.1 [ab]	0.017	0.029	0.303
C20:5	3.54 ± 0.15	3.45 ± 0.25	3.22 ± 0.24	3.29 ± 0.21	2.83 ± 0.12	2.70 ± 0.04	0.208	0.023	0.622
C20:5	2.48 ± 0.15	2.36 ± 0.19	2.46 ± 0.20	2.42 ± 0.08	2.48 ± 0.21	2.31 ± 0.08	0.911	0.656	0.745
C22:6	20.1 ± 2.7	20.7 ± 1.5	20.4 ± 1.8	20.3 ± 1.7	19.4 ± 0.4	19.9 ± 0.8	0.975	0.612	0.776
n-3PUFAs	26.2 ± 2.3	26.5 ± 0.9	26.1 ± 1.8	26.0 ± 1.2	24.8 ± 0.3	24.9 ± 0.7	0.728	0.202	00617
n-3/n-6	1.43 ± 0.08 [a]	1.40 ± 0.05 [ab]	1.44 ± 0.06 [a]	1.40 ± 0.05 [ab]	1.43 ± 0.03[a]	1.33 ± 0.04 [b]	0.874	0.554	0.386

Note: Within the same row, values with different superscripts are significantly different ($p < 0.05$). Abbreviations: SFA, saturated fatty acid; MUFA, monounsaturated fatty acid; PUFAs, polyunsaturated fatty acid.

3.7. Flesh Metabolite Profiles

The PCA and PLS-DA models of flesh metabolites were performed to identify changes in metabolites in the control, CAP-30, CAP-70 and CAP-100 groups (Figure 4).

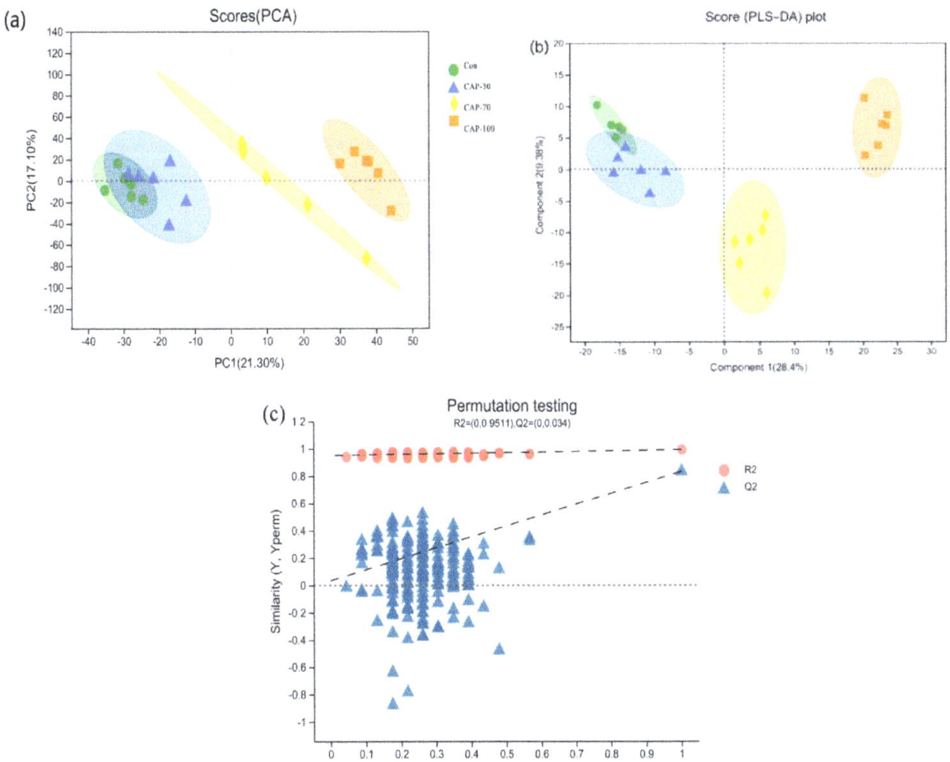

Figure 4. PCA score plots and PLS–DA score plots of muscle metabolomics. (**a**) PCA score plots of muscle metabolomics; (**b**) PLS–DA score plots of ESI+ mode and ESI− mode; (**c**) internal validation of PLS–DA score plots.

The PLS–DA scores plot of the three groups showed strong clustering for flesh metabolism products without any overlap. In the permutation test, the result showed good repeatability and predictability of the model. A total of 507, 1709 and 1838 different metabolites, including 25, 103 and 134 named metabolites, were identified between the CAP-30 group and the control, between the CAP-70 group and the control, and between the CAP-100 group and the control, respectively (Figure 5).

The above identified metabolites were assigned to the KEGG database. In the CAP-30 group, seven KEGG pathways were classified, and the top priority was "lipid metabolism", followed by "amino acid metabolism", "cancer: Overview", "polyketide metabolism", "biosynthesis of other secondary metabolites" and "folding, classification and degradation" (Figure 6a). In the CAP-70 group, 16 KEGG pathways were significantly enriched in "lipid metabolism", "amino acid metabolism", "cancer: Overview", "digestive system", "endocrine system", "carbohydrate metabolism", "other amino acid metabolism" and "nucleotide metabolism" (Figure 6b). In the CAP-100 group, 27 KEGG pathway enrichment were observed as "lipid metabolism", "cancer: Overview", "amino acid metabolism", "digestive system", "endocrine system", "carbohydrate metabolism", "other amino acid metabolism", "metabolism of vitamins", etc. (Figure 6c). Among them, 2, 8 and 10 metabolites in the lipid metabolism pathway were detected in the groups of CAP-30, CAP-70 and CAP-100 when compared with the control group (Table 10).

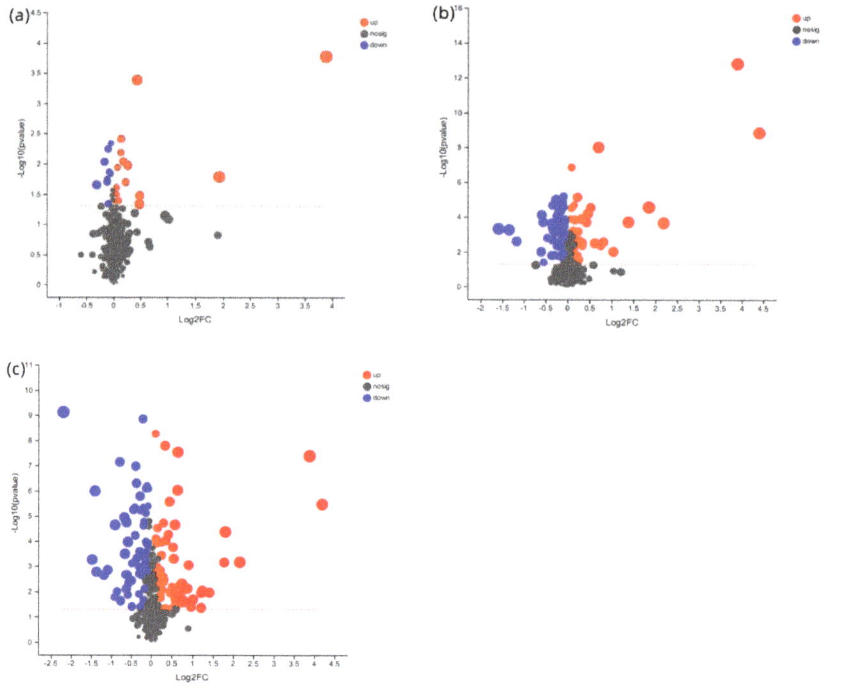

Figure 5. Volcano plots for the potential metabolomic features of muscle samples. (**a**) Volcano plots between CAP-30 and control group; (**b**) CAP-70 and control group; and (**c**) CAP-100 and control group.

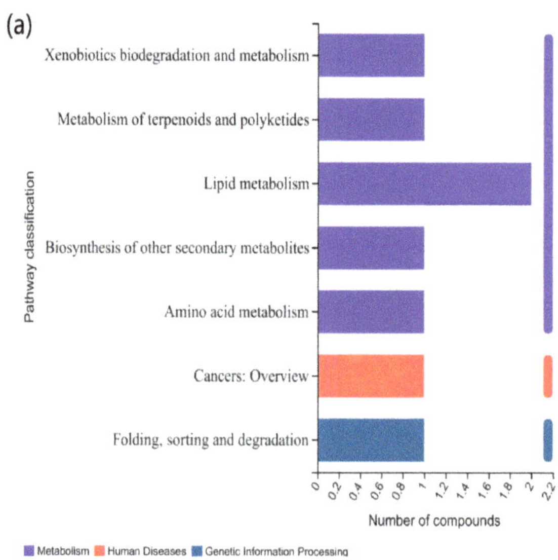

Figure 6. *Cont.*

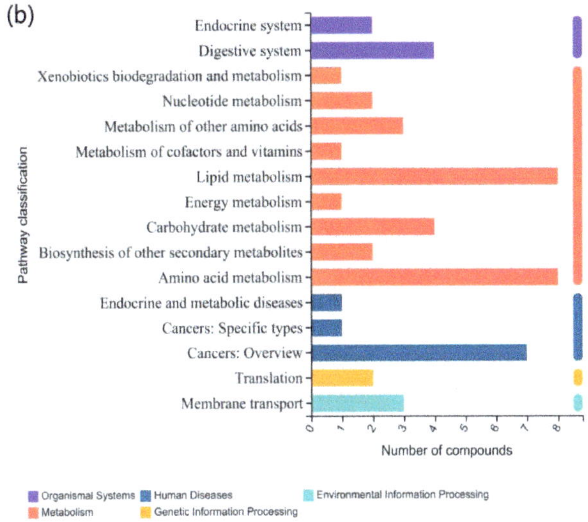

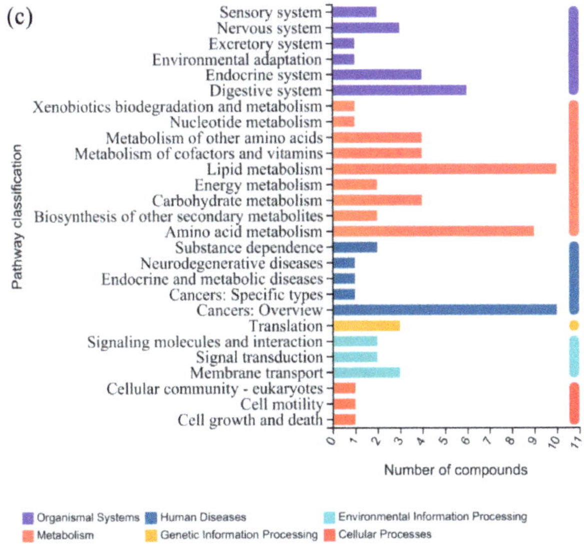

Figure 6. KEGG pathway classification of differently expressed metabolites. (**a**) KEGG pathway classification between CAP-30 and control group; (**b**) CAP-70 and control group; and (**c**) CAP-100 and control group.

Table 10. List of discriminating KEGG metabolites in CAP-30, CAP-70, CAP-100 and control groups.

Metabolite	Formula	FC	p. Value	KEGG Pathway Description
CAP-30 vs Con				
LysoPC(22:1(13Z))	$C_{30}H_{60}NO_7P$	2.59	0.02	Glycerophospholipids
Cortolone	$C_{21}H_{34}O_5$	1.99	0.02	Lipid metabolism; steroid hormone biosynthesis
CAP-70 vs Con				
Uridine diphosphate glucose	$C_{15}H_{24}N_2O_{17}P_2$	1.47	0.01	Glycerolipid metabolism; amino sugar and nucleotide sugar metabolism
9,10-DHOME	$C_{18}H_{34}O_4$	1.19	0.01	Linoleic acid metabolism
Cortolone	$C_{21}H_{34}O_5$	1.16	0.01	Lipid metabolism; steroid hormone biosynthesis
LysoPC(22:1(13Z))	$C_{30}H_{60}NO_7P$	1.29	0.01	Glycerophospholipid metabolism; choline metabolism in cancer
Eicosapentaenoic Acid	$C_{20}H_{30}O_2$	1.03	0.03	Biosynthesis of unsaturated fatty acids
LysoPC(24:0)	$C_{32}H_{66}NO_7P$	0.80	0.03	Glycerophospholipid metabolism; choline metabolism in cancer
LysoPC(20:5(5Z,8Z,11Z,14Z,17Z))	$C_{28}H_{48}NO_7P$	0.95	0.00	Glycerophospholipid metabolism; choline metabolism in cancer
Acetylcholine	$C_7H_{15}NO_2$	0.89	0.04	Glycerophospholipid metabolism; regulation of actin cytoskeleton
CAP-100 vs Con				
9,10-DHOME	$C_{18}H_{34}O_4$	1.19	0.00	Lipid metabolism; linoleic acid metabolism
LysoPC(20:1(11Z))	$C_{28}H_{56}NO_7P$	1.06	0.01	Glycerophospholipid metabolism; choline metabolism in cancer
9(S)-HODE	$C_{18}H_{32}O_3$	1.10	0.00	Linoleic acid metabolism; PPAR signaling pathway
PC(22:6(4Z,7Z,10Z,13Z,16Z,19Z)/18:3 (6Z,9Z,12Z))	$C_{48}H_{78}NO_8P$	1.32	0.01	Glycerophospholipid metabolism; arachidonic acid metabolism; linoleic acid metabolism; α-Linolenic acid metabolism
Galactosylsphingosine	$C_{24}H_{47}NO_7$	0.83	0.01	Sphingolipid metabolism
LysoPC(20:5(5Z,8Z,11Z,14Z,17Z))	$C_{28}H_{48}NO_7P$	0.91	0.00	Lipid metabolism
11b,17a,21-Trihydroxypreg-nenolone	$C_{21}H_{32}O_5$	0.94	0.00	Lipid metabolism; steroid hormone biosynthesis
LysoPC(22:5(7Z,10Z,13Z,16Z,19Z))	$C_{30}H_{52}NO_7P$	0.93	0.01	Glycerophospholipid metabolism; choline metabolism in cancer
Acetylcholine	$C_7H_{15}NO_2$	0.65	0.00	Glycerophospholipid metabolism; regulation of actin cytoskeleton
PC(18:3(6Z,9Z,12Z)/20:5 (5Z,8Z,11Z,14Z,17Z))	$C_{46}H_{76}NO_8P$	0.91	0.00	Glycerophospholipid metabolism; arachidonic acid metabolism; linoleic acid metabolism; α-Linolenic acid metabolism

4. Discussion

4.1. Effect of CAP on the Growth of Largemouth Bass

In this study, CAP successfully replaced dietary fish meal up to 70% without significant effects on the growth performance, and the low inclusion of CAP (146 g/kg) even numerically increased WG and decreased FCR. The fish meal replacement level obtained from the binomial regression analysis was lower than that obtained from the broken-line analysis (Figures 1 and 2). Generally, the aim of replacing dietary fish meal is to produce a growth performance that is no lower than the high fish meal group, rather than better than the high fish meal group. Thus, the proper replacement level of dietary fish meal was suggested to be 67.1–68.0% in the present study. The complete substitution significantly decreased the weight gain and feed intake. Fish meal contains many active substances and unknown growth factors, such as small peptides, taurine, trimethylamine oxide (TMAO), etc., which are deficient in bacterial protein ingredients [19,20]. Small peptides could promote the intestinal tract movement, the transfer and absorption of amino acids, and increase the apparent digestibility [21]. Taurine can improve the utilization of feed and enhance the immune capacity of the fish [22,23]. TMAO can stimulate the fish to feed [24]. In addition, some essential amino acids in CAP are lower than those in fish meal, such as arginine (Table 2). The lack of arginine in diet has been reported to induce the erosion of fin and increase the mortality of channel catfish (*Ictalurus punctatus*) [25]. Meanwhile, the lysine content in CAP diets was relatively higher, which might aggravate the antagonism between lysine and arginine. In black sea bream [26], Indian major carp (*Labeo rohita*) [27] and Atlantic salmon [28], the imbalanced ratio of lysine to arginine in feed was found to decrease the growth. Furthermore, the existence of bacterial cell walls also reduces the nutrients' digestibility. Therefore, the supplementation of exogenous amino acid, especially arginine, should be considered to balance the amino acid composition when CAP is included in diets.

4.2. Effect of CAP on Serum Biochemical Indexes of Largemouth Bass

Serum TP, ALB, TG, CHO and glucose can reflect the metabolism, nutritional status and health of the fish. In the present study, the serum TP content was not affected by dietary CAP, but the serum ALB contents of the CAP-15, CAP-30 and CAP-45 group were significantly higher than that of the control group. In black sea bream [13] and Jian carp [12], replacing dietary fish meal with CAP had no significant effect on serum TP and ALB contents. Usually, the blood glucose content regulated by the nervous and endocrine systems is a dynamic equilibrium, which plays a pivotal role in maintaining the normal activities of fish. No significant difference was observed in serum GLU content among all the groups in this study. Similarly, the replacement of dietary fish meal with CAP (200 g/kg) showed no significant effect on the serum GLU content of Jian carp, but significantly reduced the contents of TG and CHO [12]. In black sea bream, the serum GLU, TG and CHO contents were also not affected by the replacement of fish meal (58.2%) with CAP [13]. Therefore, it seems that the dietary inclusion of CAP has a less negative effect on the nutritional metabolism of largemouth bass.

4.3. Effects of CAP on Digestive Ability and Intestinal Structure of Largemouth Bass

The height and length of intestinal villus can reflect the tissue structure of the intestine, which is positively correlated with the absorption capacity of intestine. In the present study, the complete replacement of fish meal with CAP damaged the intestinal structure with lowered intestinal villus width, height and ruptured villi. Similarly, an intestinal injury was also observed in grass carp, including microvilli shedding, lamina propria loosening and goblet cell reduction, when 10% CAP was used to substitute dietary soybean meal. However, the dietary inclusion of 200 g/kg of CAP showed no pathological damage to the intestine of Jian carp [12], and the replacement of 58.20% of fish meal with CAP did not negatively affect the digestive enzymatic activity of black sea bream [13]. The decreased digestive function and damaged intestinal structure of largemouth by the complete substitution of fish meal bass might be related to the lack of some active substances. The intestinal absorption capacity is related to the intestinal structure, and the decrease in intestinal digestive enzyme activity may be related to the decrease in the height and width of intestinal villi [18]. Some small peptides can stimulate the feed intake and maintain the structure of intestinal epithelium. Taurine can help the digestion and absorption of neutral fats, cholesterol, fat-soluble vitamins and other fat-soluble substances, as well as improve the digestibility of protein [29]. The supplementation of arginine in diets improved the digestion and absorption capacity of red drum (*Sciaenops ocellatus*) [30] and grouper (*Epinephelus coioides*) [31], and dietary arginine also increased the height of intestinal folds and improved the intestinal health of red drum [30].

4.4. Effects of CAP on Flesh Fatty Acid and Lipid Metabolism of Largemouth Bass

In addition to supplying energy, the fatty acids, phospholipids, steroids and their derivatives also play important roles in cell proliferation, apoptosis, immunity and inflammation [32]. Generally, the dietary proportion of fatty acids affects the fatty acid composition in flesh [33,34]. Although fish oil was added in fish meal substituted diets to reach the same lipid level as the control group, the content of n-3 PUFAs and n-3/n-6 PUFAs tended to decrease as the proportion of CAP increased (Table 3). In particular, the highest replacement (CAP-100 group) significantly decreased the n-3/n-6 PUFAs (Table 9). The low PUFAs content in CAP may lead to the decrease in PUFAs content in flesh with the increase in the proportion of CAP replacing fish meal.

Compared to the control group, the most significant metabolic pathway in the CAP-30, CAP-70 and CAP-100 groups was "lipid metabolism" (Table 10). Eight different metabolites related to fat metabolism were found in the comparison of the CAP-70 group and control group, which were involved in the glycerolipid metabolism, linoleic acid metabolism and glycerophospholipid metabolism. Among them, three different metabolites (Acetylcholine, LysoPC (20:5(5Z,8Z,11Z,14Z,17Z)), LysoPC (22:5(7Z,10Z,13Z,16Z,19Z)) were down-regulated. In the CAP-100 vs Con, 10 different metabolites were found to be involved in arachidonic acid, glycerophospholipid metabolism, linoleic acid, linolenic acid and the PPAR signaling pathway, and six different metabolites (Acetylcholine, LysoPC (20:5 (5Z,8Z,11Z,14Z,17Z)), LysoPC (22:5 (7Z,10Z,13Z,16Z,19Z)) were down-regulated. Moreover, six different metabolites (Acetylcholine, Galactosylsphingosine, PC(18:3(6Z,9Z,12Z)/20:5(5Z,8Z,11Z,14Z,17Z)), LysoPC(20:5(5Z,8Z,11Z,14Z,17Z)), LysoPC(24:0),11b,17a,21-Trihydroxypreg-nenolone) were down-regulated and involved in the metabolism of glycerophospholipids, arachidonic acid metabolism, linoleic acid metabolism, α-Linolenic acid metabolism and sphingolipid metabolism (Table 10). Linolenic acid belongs to the n-3 PUFAs family, and linoleic acid and arachidonic acid belong to the n-6 PUFAs family. The n-3 and n-6 PUFAs are necessary for the growth and development of fish, in which they can promote growth, lipid metabolism and immunity. In addition, the elevated n-3 PUFAs and n-3/n-6 PUFAs may play important anti-inflammatory and anti-cancer roles [35,36]. The decreased pathway of phospholipids may affect the function of receptors in the membrane and then impair the normal function of the cell [37]. This might be one reason that the immunity capacity was decreased in fish fed with the high FM substitution levels. Therefore, the decreased growth performance by the complete replacement of fish meal with CAP might be connected with the low polyunsaturated fatty acid content in muscle, the significant changes in lipid metabolism pathways and the down-regulation of some lipid metabolites. Further investigation is needed to focus on the pathways and genes involved in lipid metabolism.

5. Conclusions

In the present study, CAP successfully replaced 70% dietary fish meal without significant effects on the growth and flesh quality of the largemouth bass. The complete replacement of fish meal with CAP reduced the weight gain, digestive ability, damaged the intestinal structure and down-regulated the lipid metabolites of largemouth bass.

Author Contributions: Conceptualization, P.Y. and Y.W.; methodology, M.L.; software, P.Y. and W.Y.; formal analysis, M.L. and X.L. (Xiaoqin Li); investigation, P.Y. and X.L. (Xiangjun Leng); project administration, X.L. (Xiangjun Leng); experimental design and manuscript revision, P.Y. All authors have read and agreed to the published version of the manuscript.

Funding: This work was financially supported by the National Key R&D Program of China (2019YFD0900200).

Institutional Review Board Statement: All animal care and use procedures were approved by the Institutional Animal Care and Use Committee (IACUC) of Shanghai Ocean University (Permit number: SHOU–DW–2021–220), and all authors clearly indicated that such guidelines have been followed.

Informed Consent Statement: Informed consent was obtained from all subjects involved in the study.

Data Availability Statement: All data are contained within the article.

Acknowledgments: The authors would like to thank the team of research assistants who helped with data collection for this study, as well as all the research participants for their time in completing the study.

Conflicts of Interest: The authors declare no conflict of interest.

References

1. Xu, T.; Peng, X.; Chen, Y.; Lin, S.; Huang, X.; Li, Y. Effects of replacing fish meal with fermented mulberry leaves on the growth, lipid metabolism and antioxidant capacity in largemouth bass (*Micropterus salmoides*). *J. Fish. China* **2016**, *40*, 1408–1415. [CrossRef]
2. He, M.; Li, X.Q.; Poolsawa, L.; Guo, Z.H.; Yao, W.X.; Zhang, C.Y.; Leng, X.J. Effects of fish meal replaced by fermented soybean meal on growth performance, intestinal histology and microbiota of largemouth bass (*Micropterus salmoides*). *Aquac. Nutr.* **2020**, *26*, 1058–1071. [CrossRef]

3. Zheng, Y.; Peng, C.; Wu, X.; Han, F.; Xue, M.; Wang, J.; Hu, L. Effects of hydrolyzed yeast on growth performance, lipids metabolism and intestinal structure of largemouth bass (*Micropterus salmoides*). *Chin. J. Anim. Nutr.* **2015**, *27*, 1605–1612. [CrossRef]
4. Ren, X.; Wang, Y.; Chen, J.; Wu, Y.B.; Huang, D.; Jiang, D.L.; Li, P. Replacement of fishmeal with a blend of poultry by-product meal and soybean meal in diets for largemouth bass, *Micropterus salmoides*. *J. World Aquac. Soc.* **2018**, *49*, 155–164. [CrossRef]
5. Aas, T.S.; Grisdale-Helland, B.; Terjesen, B.F.; Helland, S.J. Improved growth and nutrient utilisation in Atlantic salmon (*Salmo salar*) fed diets containing a bacterial protein meal. *Aquaculture* **2006**, *259*, 365–376. [CrossRef]
6. Berge, G.M.; Baeverfjord, G.; Skrede, A.; Storebakken, T. Bacterial protein grown on natural gas as protein source in diets for Atlantic salmon, *Salmo salar*, in saltwater. *Aquaculture* **2005**, *244*, 233–240. [CrossRef]
7. Utturkar, S.M.; Klingeman, D.M.; Bruno-Barcena, J.M.; Chinn, M.S.; Grunden, A.M.; Köpke, M.; Browna, S.D. Sequence data for *Clostridium autoethanogenum* using three generations of sequencing technologies. *Sci. Data* **2015**, *2*, 150014. [CrossRef]
8. Humphreys, C.M.; Mclean, S.; Chatschneider, S.; Millat, T.; Henstra, A.M.; Annan, F.J.; Breitkopf, R.; Pander, B.; Piatek, P.; Rowe, P.; et al. Whole genome sequence and manual annotation of *Clostridium autoethanogenum*, an relevant bacterium. *BMC Genom.* **2015**, *16*, 1085. [CrossRef]
9. Wei, H.C.; Yu, H.H.; Chen, X.M.; Yao, W.; Zou, F.Q.; Chen, P.; Zheng, Y.H.; Wu, X.F.; Liang, X.F.; Xue, M. Effects of soybean meal replaced by *Clostridium autoethanogenum* protein on growth performances, plasma biochemical indexes and hepatopancreas and intestinal histopathology of grass carp (*Ctenopharyngodon idllus*). *Chin. J. Anim. Nutr.* **2018**, *30*, 4190–4201. [CrossRef]
10. Zhu, S.; Gao, W.; Wen, Z.; Chi, S.; Shi, Y.; Hu, W.; Tan, B. Partial substitution of fish meal by *Clostridium autoethanogenum* protein in the diets of juvenile largemouth bass (*Micropterus salmoides*). *Aquac. Rep.* **2022**, *22*, 100938. [CrossRef]
11. Yang, P.X.; Li, X.Q.; Song, B.W.; He, M.; Wu, C.Y.; Leng, X.J. The potential of *Clostridium autoethanogenum*, a new single cell protein, in substituting fish meal in the diet of largemouth bass (*Micropterus salmoides*): Growth, feed utilization and intestinal histology. *Aquac. Fish.* **2021**, *26*, 1–9. [CrossRef]
12. Li, M.Y.; Lian, H.L.; Xie, J.; Chao, W.; Zou, F.; Ge, X.P.; Ren, M.C. Diet supplemented with a novel *Clostridium autoethanogenum* protein have a positive effect on the growth performance, antioxidant status and immunity in juvenile Jian carp (*Cyprinus carpio* var. Jian). *Aquac. Rep.* **2021**, *19*, 100572. [CrossRef]
13. Chen, Y.; Sagada, G.; Xu, B.; Chao, W.; Zou, F.Q.; Ning, W.K.; Sun, Y.; Wang, L.; Zhong, Z.; Shao, Q.J. Partial replacement of fishmeal with *Clostridium autoethanogenum* single-cell protein in the diet for juvenile black sea bream (*Acanthopagrus schlegelii*). *Aquac. Res.* **2020**, *51*, 1000–1011. [CrossRef]
14. Maulu, S.; Liang, H.; Ke, J.; Ren, M.; Ge, X.; Huang, D.Y.; Yu, H. Dietary *clostridium autoethanogenum* protein modulates intestinal absorption, antioxidant status, and immune response in gift (*oreochromis niloticus*) juveniles. *Aquac. Res.* **2021**, *52*, 5787–5799. [CrossRef]
15. Yao, W.X.; Yang, P.X.; Zhang, X.; Xu, X.Y.; Zhang, C.Y.; Li, X.Q.; Leng, X.J. Effects of replacing dietary fish meal with *Clostridium autoethanogenum* protein on growth and flesh quality of Pacific white shrimp (*Litopenaeus vannamei*). *Aquaculture* **2022**, *549*, 737770. [CrossRef]
16. Yang, P.X.; Yao, W.X.; Li, X.Q.; Li, M.L.; Wang, Y.Y.; Wang, P.; Leng, X.J. Dietary effects of fish meal substitution with *Clostridium autoethanogenum* on flesh quality and metabolomics of largemouth bass (*Micropterus salmoides*). *Aquac. Rep.* **2022**, *23*, 101012. [CrossRef]
17. Escaffre, A.M.; Kaushik, S.; Mambrini, M. Morphometric evaluation of changes in the digestive tract of rainbow trout (*Oncorhynchusmykiss*) due to fish meal replacement with soy protein concentrate. *Aquaculture* **2007**, *273*, 127–138. [CrossRef]
18. Song, Y.; Li, R.; Zhang, Y.; Wei, J.; Chen, W.; Chung, C.K.; Cai, Z.W. Mass spectrometry-based metabolomics reveals the mechanism of ambient fine particulate matter and its components on energy metabolic reprogramming in BEAS-2B cells. *Sci. Total Environ.* **2018**, *651*, 3139–3150. [CrossRef]
19. Gause, B.; Trushenski, J. Replacement of fish meal with ethanol yeast in the diets of sunshine bass. *North Am. J. Aquac.* **2011**, *73*, 97–103. [CrossRef]
20. Storebakken, T.; Baeverfjord, G.; Skrede, A.; Olli, J.J.; Marit Berge, G. Bacterial protein grown on natural gas in diets for Atlantic salmon, *Salmo salar*, in freshwater. *Aquaculture* **2004**, *241*, 413–425. [CrossRef]
21. Cheng, Y.B.; Zhang, Y.X.; Dong, Z.Y.; Li, B.; Wang, Y. Effects of dietary replacement of fish meal and soy protein by wheat gluten on plasma biochemical indices and liver anti-oxidative indices of *Nibea japonica* and *Sparus macrocephalus*. *Prog. Fish. Sci.* **2017**, *38*, 106–114.
22. Yun, B.; Ai, Q.; Mai, K.; Xu, W.; Qi, G.S.; Luo, Y.W. Synergistic effects of dietary cholesterol and taurine on growth performance and cholesterol metabolism in juvenile turbot (*Scophthalmus maximus* L.) fed high plant protein diets. *Aquaculture* **2012**, *324–325*, 85–91. [CrossRef]
23. Takagi, S.; Murata, H.; Goto, T.; Endo, M.; Yamashita, H.; Ukawa, M. Taurine is an essential nutrient for yellowtail *Seriola quinqueradiata* fed non-fish meal diets based on soy protein concentrate. *Aquaculture* **2008**, *280*, 198–205. [CrossRef]
24. Zeisel, S.H.; Warrier, M. Trimethylamine N-Oxide, the microbiome, and heart and kidney disease. *Annu. Rev. Nutr.* **2017**, *37*, 157–181. [CrossRef] [PubMed]
25. Buentello, J.A.; Gatlin, D.M. The dietary arginine requirement of channel cafish (*ctalrus punctaus*) is influenced by endogenous synthesis of arginine from glutamic acid. *Aquaculure* **2000**, *188*, 311–321. [CrossRef]

26. Zhou, F.; Shao, Q.J.; Xiao, J.X.; Peng, X.; Ngandzali, B.O.; Sun, Z.; Wing, K.N. Effects of dietary arginine and lysine levels on growth performance, nutrient utilization and tissue biochemical profile of black sea bream, *Acanthopagrus schlegelii*, fingerlings. *Aquaculture* **2011**, *319*, 72–80. [CrossRef]
27. Murthy, H.S.; Varghese, T.J. Dietary requirement of juveniles of the Indian major carp, *Labeo rohita*, for the essential amino acid lysine. *Isr. J. Aquac.-Bamidgeh* **1997**, *49*, 19–24.
28. Berge, G.E.; Sveier, H.; Lied, E. Effects of feeding Atlantic salmon (*Salmo salar L.*) imbalanced levels of lysine and arginine. *Aquac. Nutr.* **2002**, *8*, 239–248. [CrossRef]
29. Li, M.; Lai, H.; Li, Q.; Gong, S.; Wang, R. Effects of dietary taurine on growth, immunity and hyperammonemia in juvenile yellow catfish *Pelteobagrus fulvidraco* fed all-plant protein diets. *Aquaculture* **2016**, *450*, 349–355. [CrossRef]
30. Cheng, Z.; Buentello, A.; Gatlin, D.M. Effects of dietary arginine and glutamine on growth performance, immune responses and intestinal structure of red drum, *Sciaenops ocellatus*. *Aquaculture* **2011**, *319*, 247–252. [CrossRef]
31. Chi, R.S.; Han, F.L.; Tan, B.P.; Dong, X.H.; Yang, Q.H.; Liu, H.Y.; Zhang, S. Effect of dietary arginine level ongrowth performance and intestine morphology of juvenile grouper (*epinephelus coioides*). *Acta Hydrobiol. Sin.* **2016**, *40*, 388–394. [CrossRef]
32. Kohler, B.; Lin, L.; Ferraz-de-Souza, B.; Wieacker, P.; Heidemann, P.; Schroder, V.; Biebermann, H.; Schnabel, D.; Gruters, A.; Achermann, J.C. Five novel mutations in steroidogenic factor 1 (SF1, NR5A1) in 46,XY patients with severe underandrogenization but without adrenal insufficiency. *Hum. Mutat.* **2008**, *29*, 59–64. [CrossRef] [PubMed]
33. Hansen, J.; Berge, G.M.; Hillestad, M.; Krogdahl, S.; Ruyter, B. Apparent digestion and apparent retention of lipid and fattyacids in Atlantic cod (*Gadus morhua*) fed increasing dietary lipid levels. *Aquaculture* **2008**, *284*, 159–166. [CrossRef]
34. Martino, R.C.; Cyrino, J.P.; Portz, L.; Luiz, C.T. Performance and fatty acid composition of surubim (*Pseudoplatystoma coruscans*) fed diets with animal and plant lipids. *Aquaculture* **2002**, *209*, 233–246. [CrossRef]
35. Torstensen, B.E.; Froyland, L.; Lie, O. Replacing dietary fish oil with increasing levels of rapeseed oil and olive oil—effects on Atlantic salmon (*Salmo salar* L.) tissue and lipoprotein lipid composition and lipogenic enzyme activities. *Aquac. Nutr.* **2015**, *10*, 175–192. [CrossRef]
36. Yang, H.; Li, X.Q.; Xu, Z.; Cheng, Z.; Leng, X.J. Effects of three active components in *Eucommia ulmoides* on growth and flesh quality of grass carp (*Ctenopharyngodon idellus*) based on transcriptomics. *Aquac. Nutr.* **2020**, *26*, 1895–1907. [CrossRef]
37. Arts, M.T.; Kohler, C.C. Health and condition in fish: The influence of lipids on membrane competency and immune response. In *Lipids in Aquatic Ecosystems*, 2nd ed.; Kainz, M.T., Brett, M.T., Eds.; Springer: New York, NY, USA, 2009; pp. 237–256.

metabolites

MDPI

Article

Effects of Dietary Steroid Saponins on Growth Performance, Serum and Liver Glucose, Lipid Metabolism and Immune Molecules of Hybrid Groupers (♀*Epinephelus fuscoguttatus* × ♂*Epinephelus lanceolatu*) Fed High-Lipid Diets

Hongjin Deng [1], Jiacheng Zhang [1], Qihui Yang [1,2], Xiaohui Dong [1,2], Shuang Zhang [1,2], Weixing Liang [1,2], Beiping Tan [1,2] and Shuyan Chi [1,2,*]

1 Laboratory of Aquatic Nutrition and Feed, College of Fisheries, Guangdong Ocean University, Zhanjiang 524088, China

2 Guangdong Engineering Technology Research Center of Aquatic Animals Precision Nutrition and High Efficiency Feed, Zhanjiang 524088, China

* Correspondence: chishuyan77@163.com

Abstract: High-lipid diets are attributed to excessive lipid deposition and metabolic disturbances in fish. The aim of this experiment was to investigate the effects of steroidal saponins on growth performance, immune molecules and metabolism of glucose and lipids in hybrid groupers (initial weight 22.71 ± 0.12 g) fed high-lipid diets. steroidal saponins (0%, 0.1% and 0.2%) were added to the basal diet (crude lipid, 14%) to produce three experimental diets, designated S_0, $S_{0.1}$ and $S_{0.2}$, respectively. After an 8-week feeding trial, no significant differences were found between the S_0 and $S_{0.1}$ groups in percent weight gain, specific growth rate, feed conversion ratio, protein efficiency ratio and protein deposition rate ($p > 0.05$). All those in the $S_{0.2}$ group were significantly decreased ($p < 0.05$). Compared to the S_0 group, fish in the $S_{0.1}$ group had lower contents of serum triglyceride and low-density lipoprotein cholesterol and higher high-density lipoprotein cholesterol and glucose ($p < 0.05$). The activities of superoxide dismutase, catalase and glutathione peroxidase were significantly higher, and malondialdehyde contents were significantly lower in the $S_{0.1}$ group than in the S_0 group ($p < 0.05$). Hepatic triglyceride, total cholesterol and glycogen were significantly lower in the $S_{0.1}$ group than in the S_0 group ($p < 0.05$). Activities of lipoprotein lipase, total lipase, glucokinase and pyruvate kinase, and gene expression of lipoprotein lipase, triglyceride lipase and glucokinase, were significantly higher in the $S_{0.1}$ group than in the S_0 group. Interleukin-10 mRNA expression in the $S_{0.1}$ group was significantly higher than that in the S_0 group, while the expression of interleukin-6 and tumor necrosis factor-α genes were significantly lower than those in the S_0 group. In summary, adding 0.1% steroidal saponins to a high-lipid diet not only promoted lipolysis in fish livers, but also activated glycolysis pathways, thus enhancing the utilization of the dietary energy of the groupers, as well as supporting the fish's nonspecial immune-defense mechanism.

Keywords: steroidal saponins; hybrid grouper; high-lipid diet; glucose and lipid metabolism; nonspecial immune

Citation: Deng, H.; Zhang, J.; Yang, Q.; Dong, X.; Zhang, S.; Liang, W.; Tan, B.; Chi, S. Effects of Dietary Steroid Saponins on Growth Performance, Serum and Liver Glucose, Lipid Metabolism and Immune Molecules of Hybrid Groupers (♀*Epinephelus fuscoguttatus* × ♂*Epinephelus lanceolatu*) Fed High-Lipid Diets. *Metabolites* **2023**, *13*, 305. https://doi.org/10.3390/metabo13020305

Academic Editor: Qingsong Tan

Received: 10 January 2023
Revised: 7 February 2023
Accepted: 8 February 2023
Published: 19 February 2023

1. Introduction

Aquaculture has undeniably established its crucial role in global food security and nutrition. The total aquaculture production of the world has reached 122.6 million tons in 2020 [1]. The scale of aquaculture in the world is so large that the more aquatic feeds are required. In recent years, fish-meal and soybean-meal supply have been increasingly tight, which made the prices of some miscellaneous meal fluctuate from time to time, such as cotton meal, rapeseed meal and peanut meal, etc. [2,3]. In view of such a serious situation, full utilization of the energy effect of lipids and carbohydrates in feed to save protein is a more effective way. However, the higher dietary lipids or glucose easily led to the abnormal deposition of lipids in the liver and, directly affecting the health of fish, which would influence the fish yield and economic benefits [4].

Studies in Nile tilapia (*Oreochromis niloticus*), carp (*Cyprinus carpio*) and mice showed that saponins can decrease the total cholesterol (TC) and triglyceride (TG) contents and the activities of glutamic pyruvic transaminase (GPT) and glutamic oxalacetic transaminase (GOT) to protect the body [5–8]. Steroidal saponins, formed by the bonding of steroidal sapogenin and glycosides groups, were widely distributed in roots and rhizomes of plants, mainly found in monocotyledons such as agave, dioscoreaceae and genicaceae [9,10]. Structural diversity of steroidal saponins allows for a series of physiological activities, such as regulating the body's metabolism of carbohydrates and lipids by increasing the activities of lipoprotein lipase and hepatic lipase [7,11]. The right amount of quillaja or soybean saponins have been proven to promote percent weight gain, protein efficiency ratio and protein deposition ratio in juvenile Japanese flounder (*Paralichthys olivaceus*) [12], sea bream (*Sparus aurata*) [13] and carp [14,15], meanwhile enhancing the immune defense by increasing the body's anti-inflammatory and antioxidant capacities.

The grouper, as a common marine fish in China, had a production of 2.04×10^5 tons in 2021, accounting for 11.09% of the marine carnivorous fish cultured in China [16]. However, when groupers consumed diets with higher lipids, the liver showed injury, including larger areas of lipid droplets in liver sections and lower enzyme activities of hepatic CAT and SOD [17,18]. An in vivo experiment of saponin on primary hepatocytes of hybrid groupers fed a high-fat diet showed that saponin had the ability to limit lipid accumulation and improve the oxidative stress in the liver. Groupers that were fed a diet containing 0.4% saikosaponin d displayed the best growth, and fish fed a diet containing 0.8% saikosaponin d had similar weight gain compared to the control group [19]. However, early studies showed that growth performance and enzyme activity of antioxidants were inhibited, and proinflammatory factors were upregulated in sea bream [13], carp [14,15] and juvenile turbot (*Scophthalmus maximus*) [20] fed saponins over 0.2%. Therefore, this experiment was designed to investigate the effects of different levels of steroidal saponins on growth performance, nonspecific immune molecules and glucose and lipid metabolism in hybrid groupers (♀*Epinephelus fuscoguttatus*×♂*E. lanceolatu*) fed high-lipid diets.

2. Materials and Methods

2.1. Experimental Diets

The formula of the basal diet is shown in Table 1. Steroidal saponins (Guangzhou Feixit Biotechnology Co., Ltd., Guangzhou, China) were added to the basal diet at 0%, 0.1% and 0.2% and form three isoproteic (52% crude protein) and isolipidic (14% crude lipid) diets, named as S_0, $S_{0.1}$ and $S_{0.2}$. The ingredients were crushed (HNX-350, Beijing Huanya Tian yuan Machinery Technology Co., Ltd., Beijing, China) and passed through a 60-mesh screen. The solid materials were first mixed (B30 V-mixer, Lifeng Food Machinery Factory, Guangzhou, China) for 15 min, and then oil and water were added and mixed for 10 min to produce a moist dough. The dough was extruded by the twine-screw extruder (F-26, South China University of Technology, Guangzhou, China) through a 2.5 mm die. All pellet diets were air-dried at 25 °C and then stored at −20 °C until used.

Table 1. Ingredients (g/100 g diet) and proximate composition (% dry matter).

Ingredients	S_0
Fish meal	36.00
Poultry by-product meal	10.50
Soybean meal	6.00
Concentrated cottonseed protein	19.00
Wheat flour	16.00
Fish oil	4.25
Soybean oil	4.25

Table 1. *Cont.*

Ingredients	S_0
Choline chloride	0.50
Ca $(H_2PO_4)_2$	1.50
Vitamin C	0.05
Vitamin mix [1]	0.50
Mineral mix [1]	0.50
Betaine	0.50
Ethoxyquin	0.10
Steroid saponins	0.00
Microcrystalline cellulose	0.35
Total	100.00
Proximate analysis (%)	
Moisture	11.50
Crude protein	52.48
Crude lipid	13.93
Crude ash	11.74
Gross energy (KJ g^{-1} DM)	20.52

[1] Vitamin premix and mineral premix were provided by Qingdao Master Biotech Co., Ltd., Qingdao, China.

2.2. Fish and Feeding Trial

The healthy hybrid groupers of consistent genetic background were purchased from a commercial hatchery (Hongxing Hatchery, Zhanjiang, China). The fish were then reared in an outdoor concrete pond (5 m × 4 m × 2 m), fed a commercial diet (50% crude protein, 8% crude lipid) and domesticated for two weeks. All fish were starved for 24 h, and then healthy groupers (initial weight 22.71 ± 0.12 g) were selected randomly and divided into 9 buckets (1 m^3 Fiberglass farming buckets) with 25 fish each. The experiment was conducted in an indoor hydrostatic water culture system at the Marine Biological Research Base of Guangdong Ocean University (Donghai Island, Zhanjiang, China). The fish, reared in three groups with three replicates, were fed the experimental diets to apparent satiation at 8:00 and 16:00 daily for 8 weeks. During the feeding trial, fish were continuously oxygenated every day, kept at a temperature of 30.5 ± 0.8 °C, a salinity of 28–32, dissolved oxygen of 5–6 mg/L and an ammonia content of <0.2 mg/L. The culture water was replaced by 80% each other day.

2.3. Sample Collection and Chemical Analysis

At the end of the feeding trial, all fish in each bucket were made to fast for 24 h. In each replicate, fish were counted and weighed to calculate survival rate (SR), percent weight gain (PGR), specific growth rate (SGR), protein efficiency ratio (PER) and feed conversion ratio (FCR). Two fish were randomly selected in each replicate for whole-fish composition analysis. Three groupers were randomly selected from each replicate and measured for body length and weight, and their dissected livers and viscera were weighed to evaluate the condition factor (CF), hepatopancreas index (HSI) and viscerosomatic index (VSI).

Six fish were randomly selected from each replicate after weighing, and blood was collected from the tail vein. The blood samples were left at 4 °C for 12 h and centrifuged at 4 °C and 3500 rpm for 15 min. The serum obtained with centrifugation was for the analysis of biochemical indexes and antioxidative enzyme activities. The soybean-sized liver was cut in each replicate and washed with saline, and then preserved in 4% formaldehyde solution for making the histological section stained with PAS. Glycogen was quantified with software IPwin32 (6.0.0.260). Four livers in each replicate were obtained, two for measuring the activity of enzymes and another two for analyzing gene expression. The analysis methods for the index are listed in Table 2.

Table 2. The chemical analysis used in the experiment.

Items	Methods	Item Code	Reference/Assay Kits/Section
Composition of diets and whole body			
Moisture	Drying at 105 °C to constant weight		Association of Official Analytical Chemists [21]
Crude lipid	Soxhlet extractor method (Petroleum ether)		
Crude ash	Combustion to a constant weight at 550 °C		
Crude protein	Dumas's combustion method		Jean Baptiste Dumas (2018) [22]
Serum biochemical indexes			
Total cholesterol (TC, mmol/L)	Single reagent GPO-PAP method	A111-1-1	Assay kits (Nanjing Jiancheng Bioengineering Institute, Nanjing, China)
Triglyceride (TG, mmol/L)	Single reagent GPO-PAP method	A110-1-1	
Low-density lipoprotein cholesterol (LDL-C, mmol/L)	Microplate method	A113-1-1	
High-density lipoprotein cholesterol (HDL-C, mmol/L)	Microplate method	A112-+1-1	
Glucose (GLU, mmol/L)	GOD-PAP method	A154-1-1	
Malondialdehyde (MDA, nmol/mgprot)	TBA method	A003-1	
Superoxide dismutase (SOD, U/mL)	Hydroxylamine method	A001-2-2	
Catalase (CAT, U/mL)	Ammonium molybdate method	A007-1-1	
Glutathione peroxidase (GSH-PX, U/L)	Colorimetric method	A005-1-2	
Hepatic enzyme activity			
Total protein (TP, mgprot/mL)	Microplate method	A045-4-1	
Liver glycogen (LG, mg/g)	Colorimetric method	A043-1-1	
Total lipase U/mgprot)	Colorimetric method	A067-1	
Hepatic lipase (HL, U/mgprot)	Colorimetric method	A067-1-1	
Lipoprotein lipase (LPL, U/mgprot)	Colorimetric method	A067-1-2	
Glucokinase (GK, ng/mL)	Competition method, ELISA kit	H439-1	
Hexokinase (HK, nmol/min/mgprot)	Spectrophotometric method	A077-3	
Pyruvate kinase (PK, U/mgprot)	Ultraviolet colorimetric method	A076-1-1	
Histochemistry observation			
Liver Section	Periodic Acid Schiff (PAS)		Made by Wuhan Service Biotechnology Co, China.

2.4. *Quantitative RT-PCR Analysis of Gene Expression*

Total RNA in the liver was extracted with Trizol reagent (Invitrogen, Carlsbad, CA, USA). The cDNA was synthesized by Prime Script RT kit (Takara, Osaka, Japan), and qRT-PCR was performed using SYBR Premix Ex Taq kit (Takara, Osaka, Japan) and carried out using a quantitative thermal cycler (Light Cycler480II, Roche Diagnostics, Basel, Switzerland). The reaction volume was 10 μL, containing 3.2 μL sterilized double-distilled water, 1 μL cDNA, 0.4 μL each primer and 5 μL SYBR Premix Ex Taq (Takara, Osaka, Japan). The cycle conditions were 30 s at 95 °C, then 35 cycles of 5 s at 95 °C, 25 s at 60 °C and 30 s at 72 °C. *β-actin* was as the reference gene to correct the results of different batches. The results were calculated using the $2^{-\Delta\Delta Ct}$ method in relative expression analysis. The primers used for qRT-PCR analysis are listed in Table 3.

Table 3. Primers pair sequences for qRT-PCR.

Target Gene	Nucleotide Sequence (5′-3′)	Accession No.
Lipoprotein lipase (*lpl*)	F: CCACCTGTTCATCGACTCCC R: TCGGACGGACCTTGTTGAT	EU683732.1
Adipose triglyceride lipase (*atgl*)	F: GAGGACAATAAAGGCGGTGAG R: AGCTTTGTGCAGGGTGGGT	KY649281.1
Peroxisome alpha (*ppar α*)	F: TGCTCGCCTCCAGTATGAA R: GTCCAGCTCCAGCGTGTTA	FJ196234.1
Glucose transporter protein 2 (*glut 2*)	F: TGTTCTGCTTTTCGGCTTC R: CAGTTCCGCATTGTCTATG	KY656467
Glucokinase (*gk*)	F: TGGGTTTTACCTTCTCCTT R: AGTCCCCTCGTCTCTTGAT	MH213270
Phosphofructokinase type b (*pfk b*)	F: AAACGCCCATGCAAACTAC R: CAACCTCTCTGACAGCCAC	MH213271
Catalase (*cat*)	F: GCGTTTGGTTACTTTGAGGTGA R: GAGAAGCGGACAGCAATAGGT	XM_033635388.1
Superoxide dismutase (*sod*)	F: GGAGACAATACAAACGGGTGC R: CCAGCGTTGCCAGTCTTTA	NM001303360.1
Glutathione reductase (*gr*)	F: CTTTCACTCCGATGTATCACGC R: GCTTTGGTAGCACCCATTTTG	XM_033633504.1
MHC class II molecule (*mhc ii*)	F: CAGGTTCAGCAGCAGTTTGG R: AGCAGCCTGGTAGTCAATCCC	JF796053.1
Transforming growth factor-β (*tgf-β*)	F: CGATGTCACTGACGCCCTGC R: AGCCGCGGTCATCACTTATC	GQ205390.1
Interleukin-10 (*il-10*)	F: ACACAGCGCTGCTAGACGAG R: GGGCAGCACCGTGTTCAGAT	KJ741852.1
Interferon-gamma (*ifn-γ*)	F: CCACCAAGATGGAGGCTAAG R: CTGCCACCTCACCATTGCT	JX013936.1
Interleukin-6 (*il-6*)	F: CCGACAGCCCGACAGG R: CTGCTTTTCGTGGCGTTT	JN806222.1
Tumor necrosis factor-α (*tnf-α*)	F: CTGGTGATGTGGAGATGGGTC R: CGTCGTGATGTCTGGCTTTC	HQ011925.1
β-actin	F: GGCTACTCCTTCACCACCACA R: TCTGGGCAACGGAACCTCT	AY510710.2

2.5. Calculation and Statistical Analysis

Survival rate (SR, %) = 100 × final number of fish/initial number of fish.

Percent weight gain (PWG, %) = 100 × (final mean weight − initial mean weight)/initial mean weight.

Specific growth rate (SGR, %/d) = 100 × (final mean weight − initial mean weight)/number of rearing days.

Protein efficiency ratio (PER) = (final mean weight − initial mean weight)/(total feed intake × content of dietary protein).

Protein deposition rate (PDR, %) = 100 × (final mean weight × content of final body protein- initial mean weight × content of initial body protein)/(total feed intake × content of dietary protein).

Feed conversion ratio (FCR) = consumed feed weight/(final weight − initial weight + dead weight).

Feeding rate (FR, %BW/d) = 100 × consumed feed weight/[feeding days × (final fish weight + initial fish weight)/2].

Condition factor (CF, g/cm^3) = 100 × body weight/body $length^3$.

Hepatopancreas index (HSI, %) = 100 × liver weight/body weight.

Viscerosomatic index (VSI, %) = 100 × viscera weight/body weight.

The units of fish quantity, weight, length and experimental time were tail, g, cm and d, in that order.

All data were analyzed with a one-way analysis of variance (ANOVA) and significant differences among dietary groups were estimated by Tukey's multiple comparison test. The results were expressed as means ± standard deviation (SD). Significant differences were chosen at $p < 0.05$. The images of the experimental results were drawn with Graph Pad Prism 8.0 software (8.0.2.263).

3. Results

3.1. Growth Performance

As indicated in Table 4, the FBW, PWG, PER, PDR and SGR of the fish in the $S_{0.1}$ group were not significantly different from the S_0 group and were significantly higher than those in $S_{0.2}$ group ($p < 0.05$). The number of fish that died per replicate in group S_0 was 0, 3 and 1, respectively; in group $S_{0.1}$, it was 1, 1 and 1, respectively; and in group $S_{0.2}$, it was 3, 3 and 2, respectively. The HSI and VSI of fish in the S_0 group were not significantly different from the $S_{0.2}$ group, but were significantly lower than those in the $S_{0.1}$ group ($p < 0.05$).

Table 4. Effects of dietary steroid saponins on the growth performance of hybrid groupers.

Items	S_0	$S_{0.1}$	$S_{0.2}$
IBW/g	22.73 ± 0.18	22.62 ± 0.04	22.77 ± 0.10
FBW/g	102.63 ± 2.20 ^{b}	102.67 ± 2.51 ^{b}	96.52 ± 2.24 ^{a}
SR/%	94.67 ± 6.11	96.00 ± 0.00	89.33 ± 2.31
PWG/%	351.50 ± 12.75 ^{b}	353.96 ± 11.80 ^{b}	323.72 ± 7.37 ^{a}
SGR(%/d)	2.69 ± 0.06 ^{b}	2.70 ± 0.05 ^{b}	2.58 ± 0.03 ^{a}
PER	1.77 ± 0.10 ^{b}	1.81 ± 0.06 ^{b}	1.61 ± 0.02 ^{a}
PDR/%	31.57 ± 0.41 ^{b}	31.83 ± 1.05 ^{b}	27.9 ± 1.36 ^{a}
FCR	1.05 ± 0.01	1.02 ± 0.02	1.03 ± 0.01
FR(%BW/d)	2.45 ± 0.15	2.40 ± 0.04	2.58 ± 0.06
CF(g/cm^3)	2.96 ± 0.07	3.04 ± 0.11	2.82 ± 0.00
HSI/%	3.40 ± 0.03 ^{a}	4.68 ± 0.08 ^{b}	3.86 ± 0.54 ^{ab}
VSI/%	10.31 ± 0.00 ^{a}	13.38 ± 0.38 ^{b}	10.35 ± 0.28 ^{a}

Note: Significant differences ($p < 0.05$) were indicated by different letters. The same as the following tables.

3.2. Whole-Body Proximate Chemical Analysis

Body composition in the initial fish in each group was similar. After 8 weeks of a feeding trial, no significant difference was found in the moisture, crude protein and crude lipid levels of the whole fish among the experimental groups ($p > 0.05$) (Table 5).

Table 5. Effects of dietary steroid saponins on the whole fish composition of hybrid groupers.

Items		S_0	$S_{0.1}$	$S_{0.2}$
Initial	Moisture	72.70 ± 0.15	72.53 ± 0.02	72.59 ± 0.06
	Crude protein (% DM)	61.15 ± 0.99	60.74 ± 0.13	59.47 ± 0.64
	Crude lipid (% DM)	21.43 ± 0.20	21.44 ± 0.09	21.14 ± 0.16
Final	Moisture	72.43 ± 0.09	73.06 ± 0.43	72.74 ± 1.00
	Crude protein (% DM)	62.66 ± 0.61	62.79 ± 0.33	63.14 ± 0.74
	Crude lipid (% DM)	21.69 ± 0.61	21.56 ± 0.59	21.11 ± 0.46

3.3. Serum Biochemical Indexes

The content of TG in the S_0 group was significantly higher than that in the $S_{0.1}$ group, but significantly lower than that in the $S_{0.2}$ group ($p < 0.05$). The content of serum TC and LDL-C was significantly lower in the $S_{0.1}$ group than in the $S_{0.2}$ group ($p < 0.05$). The level of HDL-C and GLU were significantly higher in the $S_{0.1}$ and $S_{0.2}$ groups than in the S_0 group ($p < 0.05$) (Table 6).

Table 6. Effects of dietary steroid saponins on serum biochemical indexes of hybrid groupers.

Indexes	S_0	$S_{0.1}$	$S_{0.2}$
TG/(mmol/L)	0.54 ± 0.01^{b}	0.43 ± 0.02^{a}	0.66 ± 0.02^{c}
TC/(mmol/L)	2.00 ± 0.14^{ab}	1.67 ± 0.14^{a}	2.13 ± 0.05^{b}
LDL-C/(mmol/L)	0.25 ± 0.04^{ab}	0.12 ± 0.13^{a}	0.36 ± 0.03^{b}
HDL-C/(mmol/L)	1.11 ± 0.15^{a}	2.49 ± 0.50^{b}	1.86 ± 0.20^{b}
GLU/(mmol/L)	6.31 ± 0.13^{a}	8.96 ± 0.26^{b}	8.45 ± 0.55^{b}

3.4. Serum Antioxidative Index

The MDA contents in the S_0 and $S_{0.2}$ groups were significantly higher than that in the $S_{0.1}$ group ($p < 0.05$) (Table 7). SOD activity of the S_0 group was significantly lower than in the $S_{0.1}$ group, but significantly higher than in the $S_{0.2}$ group ($p < 0.05$). The CAT activity in the $S_{0.1}$ and $S_{0.2}$ groups were significantly higher than that in the S_0 group ($p < 0.05$). Compared to the $S_{0.1}$ group, GSH-PX activity in S_0 and $S_{0.2}$ groups was significantly decreased ($p < 0.05$).

Table 7. Effects of dietary steroid saponins on serum antioxidative indexes of hybrid groupers.

Indexes	S_0	$S_{0.1}$	$S_{0.2}$
MDA/(nmol/mL)	4.44 ± 0.22^{b}	3.24 ± 0.11^{a}	4.24 ± 0.21^{b}
SOD/(U/mL)	309.52 ± 8.19^{b}	427.55 ± 18.60^{c}	251.81 ± 33.39^{a}
CAT/(U/mL)	5.42 ± 0.31^{a}	7.70 ± 0.56^{b}	7.15 ± 0.89^{b}
GSH-PX/(U/L)	266.52 ± 3.77^{a}	285.00 ± 6.46^{b}	274.57 ± 6.46^{a}

3.5. Liver Histochemistry by PAS Stain

The results of the liver PAS stain are shown in Figure 1. The blue substance is the nucleus, and the purple substance is glycogen (Figure 1A). The percentage of nucleus in all groups was not significantly different ($p > 0.05$). The percentage of glycogen in the control group was significantly higher than in the $S_{0.1}$ group, but was significantly lower than in the $S_{0.2}$ group ($p < 0.05$) (Figure 1B).

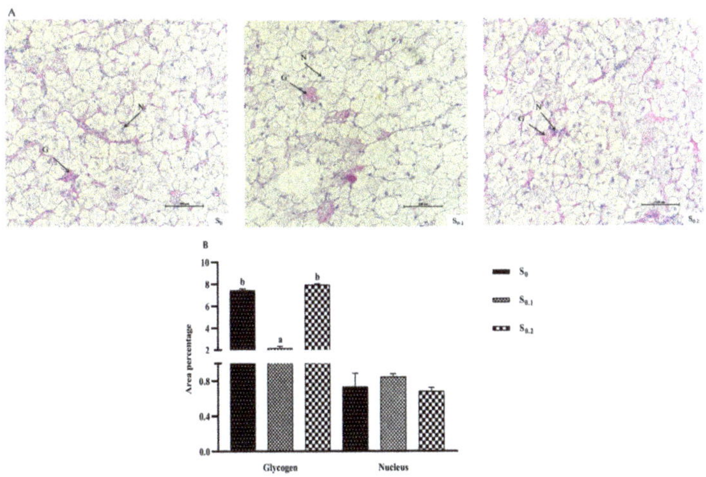

Figure 1. Effects of steroid saponin diets on the liver histochemistry of hybrid groupers after 8 weeks. Note: (**A**) Observation of liver sections (PAS staining, 200×). G, glycogen; N, nucleus. (**B**) Quantitative data on glycogen and nucleus. Different letters on the bars indicated significant difference ($p < 0.05$).

3.6. Liver Biochemical Indexes

There was no significant difference in hepatic TP among the three groups ($p > 0.05$) (Figure 2). Both the TG and TC contents of the $S_{0.1}$ group were significantly lower than those of the S_0 group ($p < 0.05$). The LG of the S_0 group was significantly higher than that of the $S_{0.1}$ group ($p < 0.05$) and was not significantly different from the $S_{0.2}$ group ($p > 0.05$).

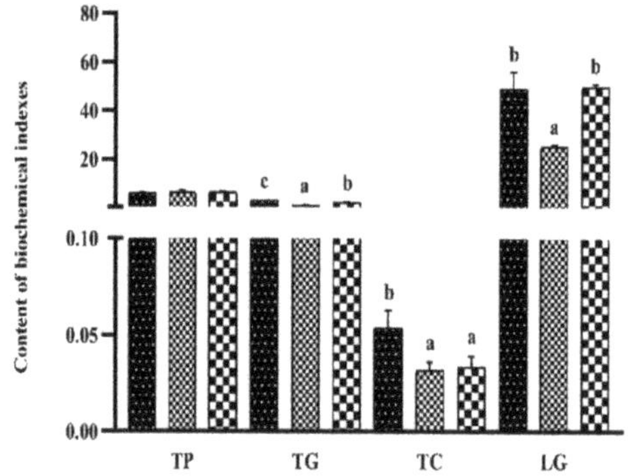

Figure 2. Effect of dietary steroid saponins on the biochemical indexes of fish livers. Note: (a–c): Values from smallest to largest. 3.7. Enzyme Activities and Genes Expression of Glucose and Lipid Metabolism in Liver.

From Figure 3A, enzyme activities of LPL and TL in the $S_{0.1}$ and $S_{0.2}$ groups were significantly higher than those in the S_0 group ($p < 0.05$). Higher HL activity was found in the $S_{0.2}$ group ($p < 0.05$), while it was not significantly different between the S_0 and $S_{0.1}$ groups ($p > 0.05$). The activities of GK and PK were significantly higher in the $S_{0.1}$ group than in the other two groups ($p < 0.05$). No significant difference was found in the HK activity in all groups ($p > 0.05$).

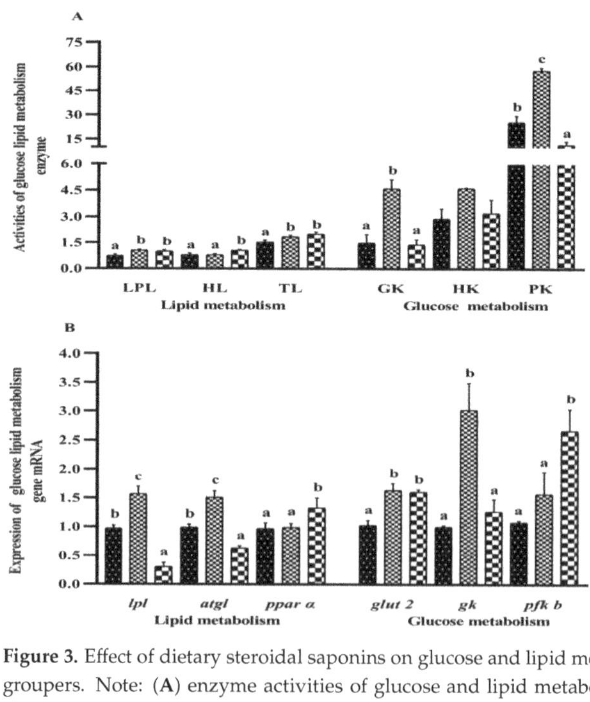

Figure 3. Effect of dietary steroidal saponins on glucose and lipid metabolism in the liver in hybrid groupers. Note: (**A**) enzyme activities of glucose and lipid metabolism; (**B**) gene expressions of glucose and lipid metabolism. (a–c): Values from smallest to largest.

The *lpl* and *atgl* mRNA expressions were significantly higher in the $S_{0.1}$ group than in the other two groups ($p < 0.05$) (Figure 3B), while gene expression levels of *lpl* and *atgl* in the $S_{0.2}$ group were significantly lower than those in the S_0 group ($p < 0.05$). The *ppar α* mRNA expression did not show significant differences between the S_0 and $S_{0.1}$ groups ($p > 0.05$), while it was significantly higher in the $S_{0.2}$ group ($p < 0.05$). The expression of *glut2* mRNA was significantly higher in the $S_{0.1}$ and $S_{0.2}$ groups than in the S_0 group ($p < 0.05$). The *gk* mRNA expression was significantly lower in the S_0 and $S_{0.2}$ groups than in the $S_{0.1}$ group ($p < 0.05$). The expression of *pfk b* mRNA was not significantly different between the $S_{0.1}$ and S_0 groups and was significantly lower than in the $S_{0.2}$ group ($p < 0.05$).

3.7. Liver Immune Molecules

As Figure 4A shows, the mRNA expressions of the genes *cat* and *sod* were significantly higher in the $S_{0.1}$ group than in the other two groups ($p < 0.05$). The mRNA expression of the *gr* was not significantly different between the S_0 and $S_{0.1}$ groups, but it was higher than in the $S_{0.2}$ group ($p < 0.05$). The *mhc* II and *il*-10 mRNA expressions in the $S_{0.1}$ group were significantly upregulated when compared with other groups ($p < 0.05$). The *tgf*-β mRNA expression in the S_0 and $S_{0.1}$ groups was lower than in the $S_{0.2}$ group ($p < 0.05$). The expressions of *ifn*-γ and *il*-6 mRNA in $S_{0.2}$ group were significantly higher than in other groups ($p < 0.05$). The *tnf*-α mRNA expression in the $S_{0.1}$ and $S_{0.2}$ groups were significantly lower than in the S_0 group ($p < 0.05$).

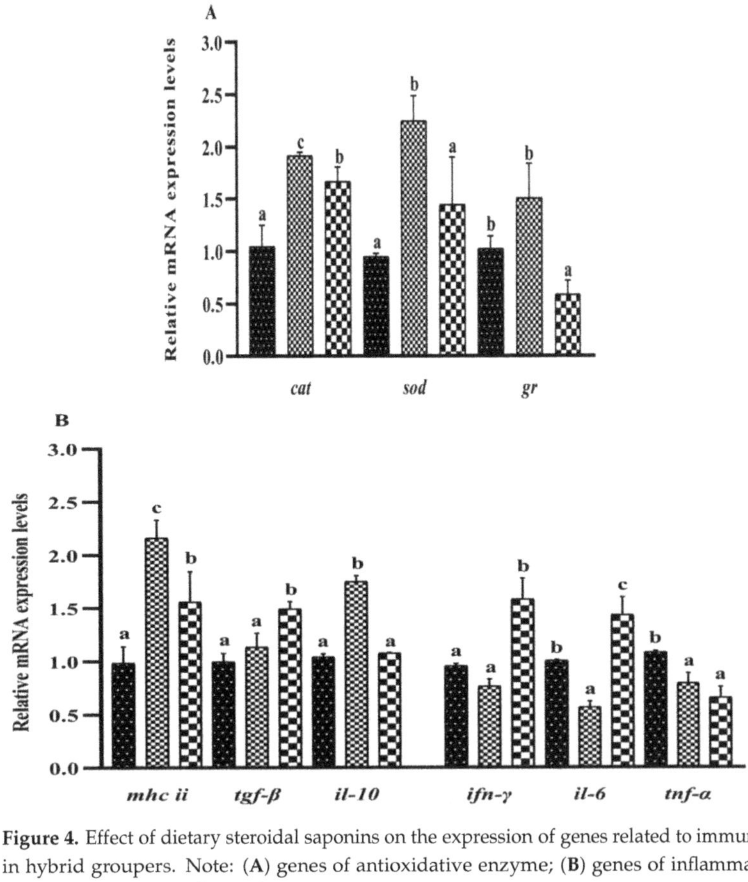

Figure 4. Effect of dietary steroidal saponins on the expression of genes related to immune molecules in hybrid groupers. Note: (**A**) genes of antioxidative enzyme; (**B**) genes of inflammatory factors. (a–c): Values from smallest to largest.

4. Discussion

In most fish, dietary lipids are able to promote protein deposition efficiently in vivo, which is known as the protein-saving effect. Although the appropriate dietary lipid level for hybrid groupers was 10% [17,23,24], a higher lipid level in the diet was used in the practical feed in order to realize the protein-saving effect. However, excessive dietary lipids would cause lipid accumulation, inflammation and decreased PER and PDR in hybrid groupers, hybrid yellow catfish (*Pelteobagrus fulvidraco* × *P. vachelli*) and yellow croakers (*Larimichthys crocea*), inhibiting fish growth [25–27]. In this experiment, when 0.1% steroidal saponins was added to the diet with 14% crude lipids, more PER and PDR were obtained, but when steroidal saponins was up to 0.2%, fish growth was significantly inhibited. Studies in Atlantic salmon (*Salmo salar* L) [28] found that dietary addition of 0.2% soya-saponins increased fish PER and PDR, but supplementation of 0.2% or 0.32% soya-saponins decreased the weight-gain rate of carnivorous field eels (*Monopterus albus*) [29] and Japanese flounder [12]. Research in omnivorous carp [30] found that growth performance was significantly inhibited when dietary Momordica charantia saponins were above 0.64%. Thus, high doses of saponins were toxic to fish and decreased growth. Dietary soya-saponins over 0.25% significantly caused a decrease in the specific growth ratio and feed efficiency ratio in juvenile turbot (*Scophthalmus maximus*) [20]. However, these differences in the results might be related to saponin dose, species habits, feeding habits and the saponin type. The addition of 0.05–1% *Panax notoginseng* extract (with 80% saponins) to high-lipid diets (15%) promoted the growth of hybrid groupers. Although the growth was increased above 0.5% [31], in which the initial grouper size was larger than in the present study, we thought that the larger fish would have greater tolerance. However, the groupers of a similar size to those in the present study were fed a diet containing 0.4% saikosaponin d, which gave the best growth, and fish fed a diet containing 0.8% saikosaponin d had similar weight gain compared to the control group. Saikosaponin is a triterpenoid saponin. The saponin used in this experiment is a steroidal saponin extracted from mulberry leaves. The two types of saponin have different functions. Perhaps the type of the saponin made a difference in the dietary dose and effect on fish. The above research was combined to show that the addition of appropriate levels or type of saponins to high-lipid diets was helpful for playing a role in the conservation of protein and to increase PER and PDR, thus resulting in promoting fish growth.

Compared to terrestrial animals, aquaculture animals are characterized by a shorter digestive tract that limits the ability to digest and utilize food. High-lipid diets would increase the burden on fish and cause problems such as disorders of lipid metabolism in the serum and liver, thus affecting body health [32–34]. In a non-high-lipid diet, saponin reduced TC, TG and LDL-C and increase HDL-C in the serum of pacific white shrimp (*Litopenaeus vannamei*) [35] and juvenile turbot [36]. In the present experiment, the addition of steroidal saponins to the high-lipid diet significantly increased the enzyme activities of LPL and HL and upregulated the expression of *lpl*, *atgl* and *ppar α* genes in livers of hybrid groupers. Meanwhile, a decrease in TG and TC levels in serum and the liver of hybrid groupers was also observed, which is consistent with the trend of lipid-metabolizing enzymes in the liver. The saponin could activate the PPAR signaling pathway to upregulate the gene expression of *lpl*, *atgl* and *ppar α* to stabilize lipid metabolism disorders of hybrid groupers fed with a high-lipid diet [19]. It was indicated that steroidal saponins probably enhanced the hydrolysis of triglycerides to fatty acids through activating the PPAR signaling pathway, which subsequently activated the LDL and HDL receptor families in the cholesterol metabolic pathway to mediate cholesterol transport [18,37,38].

In addition, the intake of a high-lipid diet is often accompanied by disturbances in the glucose metabolism in fish, mainly in terms of wave blood glucose and hepatic glycogen content [33,34,39]. Saponins were found to significantly reduce hepatic glycogen and blood glucose levels and elevated the enzyme activities of glycolysis in carp that were fed with a high-starch diet [30]. In this experiment, the addition of steroidal saponins to the high-lipid diets significantly upregulated hepatic *GLUT2*, *GK* and *PFK b* gene expression, increased

GK and PK enzyme activities, and decreased the hepatic glycogen content. When the organism was stimulated, *GK*, *HK* and *PFK b* expression in the AMPK signaling pathway was able to be activated, which subsequently regulated gene expression and glycolysis and provided energy to the organism [30,40]. The steroidal saponins might promote grouper growth by activating the expression of the *PFK b* gene in the AMPK signaling pathway and subsequently upregulating gene expression and the activity of enzymes in the glycolysis pathway. Dietary carbohydrates could be better energized by steroidal saponin stimulation, which was more conducive to the protein-sparing effect.

Saponins have been confirmed to reduce MDA content and enhance the antioxidant capacity of carp [6], white shrimp [41] and swimming crab (*Portunus trituberculatus*) [42], by increasing SOD, CAT, GSH-PX and GR activities and genes expression. In this study, enzyme activities and gene expression of SOD, CAT, GSH -PX and GR in groupers fed diets containing 0.1% steroidal saponins were significantly increased, and MDA levels in serum also decreased. At the same time, the expression of cell factors *mhc II*, *il-10* and *tgf-β* genes were upregulated, and *inf-γ*, *il-6* and *tnf-α* gene expressions were downregulated in groupers fed diets containing 0.1% steroidal saponins. In a similar trend, higher expressions of *mhc II*, *il-10* and *tgf-β* genes and lower expressions of *inf-γ*, *il-6* and *tnf-α* genes in the liver were found in turbot [43], sea bream [44] and carp [45,46] fed the diet with saponins. The expression of *mhc* II in the antigen presentation pathway might be upregulated by the saponins in groupers, which enables T cells to recognize antigens [47], together with the higher expression of *il-10* and *tgf-β* genes to exhibit the anti-inflammation in the body's immune system. Saponins could act on the NF-κB signaling pathway to play a role in inflammation in the enterocytes [48] and also enhance the antioxidant capacity of carp [45] and mice [49] by acting on the Nrf2 signaling pathway and upregulating *sod*, *cat* and *gr* gene expressions, which then alleviate the liver damage. As a result, the steroidal saponins effectively improved the antioxidant ability and removed peroxide products, etc., in the body through the nonspecial immune factors maintaining the oxidative homeostasis, and thus enabled the orderly metabolism of nutrients in the body.

In conclusion, compared to the fish that were fed a diet without steroidal saponins, groupers that ingested a high-lipid diet with 0.1% steroidal saponins could achieve a better protein efficiency ratio, lower blood lipids and a healthier liver, which were all derived from raising the efficiency of the glycolipid metabolism for energy supply. Meanwhile, 0.1% steroidal saponins helped increase the nonspecial immune-defense mechanism of the fish by regulating immune molecules.

Author Contributions: H.D.: writing—original draft preparation, investigation and formal analysis; J.Z.: writing—review and editing, methodology and conceptualization; Q.Y.: conceptualization and purchasing of experimental materials; X.D.: methodology and assistant in culture management; S.Z.: methodology and instruction in sample analysis; W.L.: methodology and conceptualization; B.T.: funding acquisition; S.C.: funding acquisition, investigation, project administration and writing—review and editing. All authors have read and agreed to the published version of the manuscript.

Funding: This study was supported by the China Agriculture Research System of MOF and MARA (CARS-47); the Guangdong Provincial Higher Education Key Field Special Project (2020ZDZX1034); the Zhanjiang Science and technology plan project (2020A02001); and the Key laboratory construction project of Zhanjiang (2020A05003).

Institutional Review Board Statement: Hybrid groupers used in this experiment were provided by the Fish Species Technology Extension Station (Zhanjiang, China) and approved by the Animal Research and Ethics Committee of Guangdong Ocean University (Approval ID: GDOUIACUC-2021-A0207; Approval date: 8 May 2021).

Informed Consent Statement: Not applicable.

Data Availability Statement: The data presented in this study are available in the main article.

Conflicts of Interest: The authors declared that they had no known competing financial interest or personal relationships that could have influenced the work reported in this paper.

References

1. Food and Agriculture Organization of the United Nations. *The State of World Fisheries and Aquaculture: Towards Blue Transition*; Food and Agriculture Organization of the United Nations: Rome, Italy, 2022.
2. Gasco, L.; Gai, F.; Maricchiolo, G.; Genovese, L.; Ragonese, S.; Bottari, T.; Caruso, G. Fishmeal Alternative Protein Sources for Aquaculture Feeds. In *Feeds for the Aquaculture Sector*; Springer Briefs in Molecular Science; Springer: Cham, Switzerland, 2018; pp. 1–28.
3. Oliveira, G.L.T.; Schneider, M. The politics of flexing soybeans: China, Brazil, and global agro-industrial restructuring. *J. Peasant. Stud.* **2016**, *43*, 167–194. [CrossRef]
4. Tan, X.; Sun, Z.; Liu, Q.; Ye, H.; Zou, C.; Ye, C.; Wang, A.; Lin, H. Effects of dietary ginkgo biloba leaf extract on growth performance, plasma biochemical parameters, fish composition, immune responses, liver histology, and immune and apoptosis-related genes expression of hybrid grouper (*Epinephelus lanceolatus*♂ × *Epinephelus fuscoguttatus*♀) fed high lipid diets. *Fish Shellfish. Immunol.* **2018**, *72*, 399–409. [PubMed]
5. Francis, G.; Makkar, H.P.S.; Becker, K. Effects of Quillaja saponins on growth, metabolism, egg production and muscle cholesterol in individually reared Nile tilapia (*Oreochromis niloticus*). *Comp. Biochem. Physiol. Part C Toxicol. Pharmacol.* **2001**, *129*, 105–114. [CrossRef] [PubMed]
6. Dawood, M.A.; Gewaily, M.S.; Monier, M.N.; Younis, E.M.; Van Doan, H.; Sewilam, H. The regulatory roles of yucca extract on the growth rate, hepato-renal function, histopathological alterations, and immune-related genes in common carp exposed with acute ammonia stress. *Aquaculture* **2021**, *534*, 736287. [CrossRef]
7. Lei, R.J.; Li, J.; Jin, S.X.; Xu, S.Y.; Yan, G.M.; He, Q.J. Hyperlipidemic effect of total steroidal saponins extracted from Allium chinense G. Don in high-fat diet-induced hyperlipidemia rats. *Chin. Tradit. Pat. Med.* **2013**, *35*, 1615–1619.
8. Lu, B.; Yin, L.; Xu, L.; Peng, J. Application of proteomic and bioinformatic techniques for studying the hepatoprotective effect of dioscin against CCl4-induced liver damage in mice. *Planta Med.* **2010**, *77*, 407–415. [CrossRef] [PubMed]
9. Moses, T.; Papadopoulou, K.K.; Osbourn, A. Metabolic and functional diversity of saponins, biosynthetic intermediates and semi-synthetic derivatives. *Crit. Rev. Biochem. Mol. Biol.* **2014**, *49*, 439–462. [CrossRef]
10. Upadhyay, S.; Jeena, G.S.; Shukla, R.K. Recent advances in steroidal saponins biosynthesis and in vitro production. *Planta* **2018**, *248*, 519–544. [CrossRef]
11. Yu, C.-H.; Xie, G.; He, R.-R.; Zhai, Y.-J.; Li, Y.-F.; Tsoi, B.; Kurihara, H.; Yang, D.-P. Effects of a purified saponin mixture from alfalfa on plasma lipid metabolism in hyperlipidemic mice. *J. Health Sci.* **2011**, *57*, 401–405. [CrossRef]
12. Chen, W.; Ai, Q.H.; Mai, K.M.; Xu, W.; Liufu, Z.G.; Zhang, W.B.; Cai, Y.H. Effects of dietary soybean saponins on feed intake, growth performance, digestibility, and intestinal structure in juvenile Japanese flounder (*Paralichthys olivaceus*). *Aquaculture* **2011**, *318*, 95–100. [CrossRef]
13. Couto, A.; Kortner, T.M.; Penn, M.; Bakke, A.M.; Krogdahl, A.; Oliva-Teles, A. Effects of dietary soy saponins and phytosterols on gilthead sea bream (*Sparus aurata*) during the on-growing period. *Anim. Feed. Sci. Technol.* **2014**, *198*, 203–214. [CrossRef]
14. Serrano, A.E., Jr. Effects of Quillaja saponins on growth, feed efficiency, digestive enzyme activities and metabolism of common carp (*Cyprinus carpio* L). *Aquac. Nutr.* **2013**, *19*, 468–474. [CrossRef]
15. Francis, G.; Makkar, H.P.S.; Becker, K. Dietary supplementation with a Quillaja saponin mixture improves growth performance and metabolic efficiency in common carp (*Cyprinus carpio* L.). *Aquaculture* **2002**, *203*, 311–320. [CrossRef]
16. Fisheries Administration of the Ministry of Agriculture and Rural Areas; National Fisheries Technology Extension Center; China Society of Fisheries. *Fishery Statistical Yearbook*; Chinese Agricultural Press: Beijing, China, 2022. (In Chinese)
17. Xie, R.; Amenyogbe, E.; Chen, G.; Huang, J. Effects of feed fat level on growth performance, body composition and serum biochemical indices of hybrid grouper (*Epinephelus fuscoguttatus* × *Epinephelus polyphekadion*). *Aquaculture* **2021**, *530*, 735813. [CrossRef]
18. Li, T.; Yan, X.; Dong, X.; Pan, S.; Tan, B.; Zhang, S.; Zhang, H. Choline Alleviates Disorders of Lipid Metabolism in Hybrid Grouper (♀*Epinephelus fuscoguttatus* × ♂*E. lanceolatus*) Caused by High-Lipid. *Diet. Aquac. Nutr.* **2022**, *2022*, 998849.
19. Zou, C.; Du, L.; Wu, J.; Gan, S.; Li, Q.; Babu, V.S.; Wu, Y.; Lin, L. Saikosaponin d alleviates high-fat-diet induced hepatic steatosis in hybrid grouper (*Epinephelus lanceolatus*♂ × *Epinephelus fuscoguttatus*♀) by targeting AMPK/PPARα pathway. *Aquaculture* **2022**, *553*, 738088. [CrossRef]
20. Gu, M.; Jia, Q.; Zhang, Z.; Bai, N.; Xu, X.; Xu, B. Soya-saponins induce intestinal inflammation and barrier dysfunction in juvenile turbot (*Scophthalmus maximus*). *Fish Shellfish. Immunol.* **2018**, *77*, 264–272. [CrossRef]
21. Mol, S.; Turan, S. Comparison of proximate, fatty acid and amino acid compositions of various types of fish roes. *Int. J. Food Prop.* **2008**, *11*, 669–677. [CrossRef]
22. Wang, H. Determination of crude protein content in grain and oil seeds by dumas combustion method. *Sci. Technol. Cereals Oils Foods* **2018**, *26*, 49–53.
23. Wang, J.; Han, T.; Li, X.; Yang, Y.; Yang, M.; Hu, S.; Jiang, Y.; Harpaz, S. Effects of dietary protein and lipid levels with different protein-to-energy ratios on growth performance, feed utilization and body composition of juvenile, red-spotted grouper (*Epinephelus akaara*). *Aquac. Nutr.* **2017**, *23*, 994–1002. [CrossRef]
24. Li, S.; Mai, K.; Xu, W.; Yuan, Y.; Zhang, Y.; Zhou, H.; Ai, Q. Effects of dietary lipid level on growth, fatty acid composition, digestive enzymes, and expression of some lipid metabolism related genes of orange-spotted grouper larvae (*Epinephelus coioides* H.). *Aquac. Res.* **2016**, *47*, 2481–2495. [CrossRef]

25. Pan, S.; Yan, X.; Dong, X.; Li, T.; Suo, X.; Tan, B.; Zhang, S.; Li, Z.; Yang, Y.; Zhang, H. The positive effects of dietary inositol on juvenile hybrid grouper (♀*Epinephelus fuscoguttatus* × ♂E. *lanceolatu*) fed high-lipid diets: Growth performance, antioxidant capacity and immunity. *Fish Shellfish. Immunol.* **2022**, *26*, 84–95.
26. Fei, S.; Xia, Y.; Chen, Z.; Liu, C.; Liu, H.; Han, D.; Jin, J.; Yang, Y.; Zhu, X.; Xie, S. A high-fat diet alters lipid accumulation and oxidative stress and reduces the disease resistance of overwintering hybrid yellow catfish (*Pelteobagrus fulvidraco*♀ × P. *vachelli*♂). *Aquac. Rep.* **2022**, *23*, 101043. [CrossRef]
27. Ding, T.; Xu, N.; Liu, Y.; Du, J.; Xiang, X.; Xu, D.; Liu, Q.; Yin, Z.; Li, J.; Mai, K.; et al. Effect of dietary bile acid (BA) on the growth performance, body composition, antioxidant responses and expression of lipid metabolism-related genes of juvenile large yellow croaker (*Larimichthys crocea*) fed high lipid diets. *Aquaculture* **2020**, *518*, 734768. [CrossRef]
28. Gu, M.; Gu, J.N.; Penn, M.; Bakke, A.M.; Lein, I.; Krogdahl, A. Effects of diet supplementation of soya-saponins, isoflavones and phytosterols on Atlantic salmon (*Salmo salar* L) fry fed from start -feeding. *Aquac. Nutr.* **2015**, *21*, 604–613. [CrossRef]
29. Hu, Y.; Zhang, J.; Xue, J.; Chu, W.; Hu, Y. Effects of dietary soy isoflavone and soy saponin on growth performance, intestinal structure, intestinal immunity, and gut microbiota community on rice field eel (*Monopterus albus*). *Aquaculture* **2021**, *537*, 736506. [CrossRef]
30. Fan, Z.; Li, J.; Wu, D.; Li, C.; Cao, D.; Miao, L.; Ge, X.; Wang, L. Preliminarily curative effectiveness of long-term bitter melon Momordica charantia saponins administration for the glucose homeostasis of juvenile common carp (*Cyprinus carpio*) fed a high-starch diet. *Aquac. Rep.* **2022**, *25*, 101232. [CrossRef]
31. Sun, Z.; Tan, X.; Ye, H.; Zou, C.; Ye, C.; Wang, A. Effects of dietary *Panax notoginseng* extract on growth performance, fish composition, immune responses, intestinal histology, and immune related genes expression of hybrid grouper (*Epinephelus lanceolatus*♂ × *Epinephelus fuscoguttatus*♀) fed high lipid diets. *Fish Shellfish. Immunol.* **2018**, *73*, 234–244. [CrossRef]
32. Li, A.; Yuan, X.; Liang, X.-F.; Liu, L.; Li, J.; Li, B.; Fang, J.; Li, J.; He, S.; Xue, M.; et al. Adaptations of lipid metabolism and food intake in response to low and high fat diets in juvenile grass carp (*Ctenopharyngodon idellus*). *Aquaculture* **2016**, *457*, 43–49. [CrossRef]
33. Tang, T.; Hu, Y.; Peng, M.; Chu, W.; Hu, Y.; Zhong, L. Effects of high-fat diet on growth performance, lipid accumulation and lipid metabolism related MicroRNA/gene expression in the liver of grass carp (*Ctenopharyngodon idella*). *Comp. Biochem. Physiol. Part B Biochem. Mol. Biol.* **2019**, *234*, 34–40. [CrossRef]
34. Lv, H.-B.; Ma, Y.-Y.; Hu, C.-T.; Lin, Q.-Y.; Yue, J.-J.; Chen, L.-Q.; Zhang, M.-L.; Du, Z.-Y.; Qiao, F. The individual and combined effects of hypoxia and high-fat diet feeding on nutrient composition and flesh quality in Nile tilapia (*Oreochromis niloticus*). *Food Chem.* **2021**, *343*, 128479. [CrossRef] [PubMed]
35. Yang, Q.H.; Tan, B.P.; Dong, X.H.; Chi, S.Y.; Liu, H.Y. Effects of different levels of Yucca schidigera extract on the growth and nonspecific immunity of Pacific white shrimp (*Litopenaeus vannamei*) and on culture water quality. *Aquaculture* **2015**, *439*, 39–44. [CrossRef]
36. Yun, B.; Ai, Q.; Qian, X.; Mai, K. Effects of soya saponins on feed intake, growth performance, and cholesterol metabolism in juvenile turbot (*Scophthalmus maximus* L). *Isr. J. Aquac. Bamidgeh* **2015**, *67*, 20734. [CrossRef]
37. Zhang, M.; Shao, Y.; Gao, B.; Chen, J.; Zhang, P.; Hu, Y.; Ding, S. Erchen decoction mitigates lipid metabolism disorder by the regulation of PPAR γ and LPL gene in a high-fat diet C57BL/6 mice model. *Evid. Based Complement. Altern. Med.* **2020**, *2020*, 1–8.
38. Fougerat, A.; Schoiswohl, G.; Polizzi, A.; Régnier, M.; Wagner, C.; Smati, S.; Fougeray, T.; Lippi, Y.; Lasserre, F.; Raho, I.; et al. ATGL-dependent white adipose tissue lipolysis controls hepatocyte PPARα activity. *Cell Rep.* **2022**, *39*, 10910. [CrossRef]
39. Li, R.-X.; Qian, Y.-F.; Zhou, W.-H.; Wang, J.-X.; Zhang, Y.-Y.; Luo, Y.; Qiao, F.; Chen, L.-Q.; Zhang, M.-L.; Du, Z.-Y. The Adaptive Characteristics of Cholesterol and Bile Acid Metabolism in Nile Tilapia Fed a High-Fat Diet. *Aquac. Nutr.* **2022**, *2022*, 1–17. [CrossRef]
40. Feige, J.N.; Gelman, L.; Michalik, L.; Desvergne, B.; Wahli, W. From molecular action to physiological outputs: Peroxisome proliferator-activated receptors are nuclear receptors at the crossroads of key cellular functions. *Prog. Lipid Res.* **2006**, *45*, 120–159. [CrossRef]
41. Su, B.K.; Chen, J.C. Effect of saponin immersion on enhancement of the immune response of white shrimp *Litopenaeus vannamei* and its resistance against *Vibrio alginolyticus*. *Fish Shellfish. Immunol.* **2008**, *24*, 74–81. [CrossRef]
42. Ng'ambi, J.W.; Li, R.; Mu, C.; Song, W.; Liu, L.; Wang, C. Dietary administration of saponin stimulates growth of the swimming crab *Portunus trituberculatus* and enhances its resistance against Vibrio alginolyticus infection. *Fish Shellfish. Immunol.* **2016**, *59*, 305–311. [CrossRef]
43. Gu, M.; Pan, S.; Li, Q.; Qi, Z.; Deng, W.; Bai, N. Protective effects of glutamine against soy saponins-induced enteritis, tight junction disruption, oxidative damage, and autophagy in the intestine of *Scophthalmus maximus* L. *Fish Shellfish. Immunol.* **2021**, *114*, 49–57. [CrossRef]
44. Couto, A.; Kortner, T.M.; Penn, M.; Bakke, A.M.; Krogdahl, Å.; Oliva-Teles, A. Effects of dietary phytosterols and soy saponins on growth, feed utilization efficiency and intestinal integrity of gilthead sea bream (*Sparus aurata*) juveniles. *Aquaculture* **2014**, *432*, 295–303. [CrossRef]
45. Wang, L.; Wu, D.; Fan, Z.; Li, H.; Li, J.; Zhang, Y.; Xu, Q.; Wang, G.; Zhu, Z. Effect of Yucca schidigera extract on the growth performance, intestinal antioxidant status, immune response, and tight junctions of mirror carp (*Cyprinus carpio*). *Fish Shellfish. Immunol.* **2020**, *103*, 211–219. [CrossRef] [PubMed]

46. Urán, P.A.; Gonçalves, A.A.; Taverne-Thiele, J.J.; Schrama, J.W.; Verreth, J.A.J.; Rombout, J.H.W.M. Soybean meal induces intestinal inflammation in common carp (*Cyprinus carpio* L.). *Fish Shellfish. Immunol.* **2008**, *25*, 751–760. [CrossRef] [PubMed]
47. Evans, A.M.; Salnikov, M.; Tessier, T.M.; Mymryk, J.S. Reduced MHC Class I and II Expression in HPV-Negative vs. HPV-Positive Cervical Cancers. *Cells* **2022**, *11*, 3911. [CrossRef] [PubMed]
48. Gu, M.; Pan, S.; Deng, W.; Li, Q.; Qi, Z.; Chen, C.; Bai, N. Effects of glutamine on the IKK/IκB/NF-κB system in the enterocytes of turbot *Scophthalmus maximus* L. stimulated with soya-saponins. *Fish Shellfish. Immunol.* **2021**, *119*, 373–378. [CrossRef] [PubMed]
49. Dong, H.Y.; Zhang, H.; Shuai, Z.F.; Rong, H.; Wang, J.P.; Li, C.H.X. Effects of β-asarone and tenuigenin on expression of SOD, GSH-PX, CAT, MDA and HO-1 in APP/PS1 double transgenic mice. *Nat. Prod. Res. Dev.* **2017**, *29*, 1551–1557.

metabolites

MDPI

Article

4-PBA Attenuates Fat Accumulation in Cultured Spotted Seabass Fed High-Fat-Diet via Regulating Endoplasmic Reticulum Stress

Tian Xia [1], Yan-Qin Liao [2,3], Lei Li [1], Lu-Yu Sun [2,3], Neng-Shui Ding [2,3], You-Lin Wu [2,3,*] and Kang-Le Lu [1,*]

[1] Key Laboratory of Healthy Mariculture for the East China Sea, Ministry of Agriculture and Rural Affairs, Fisheries College, Jimei University, Xiamen 361021, China
[2] Fujian Aonong Biological Science and Technology Group Co., Ltd., Zhangzhou 363000, China
[3] Fujian Yixinbao Biopharmaceutical Co., Ltd., Zhangzhou 363000, China
* Correspondence: wuyoulin@aonong.com.cn (Y.-L.W.); lukangle@jmu.edu.cn (K.-L.L.)

Abstract: Excessive fat accumulation is a common phenomenon in cultured fish, which can cause metabolic disease such as fatty liver. However, the relative regulatory approach remains to be explored. Based on this, two feeding trials were conducted. Firstly, fish were fed either a normal-fat diet (NFD) or a high-fat diet (HFD) for eight weeks and sampled at the 2nd, 4th, 6th, and 8th week after feeding (*Experiment I*). In the first four weeks, fish fed an HFD grew faster than those fed an NFD. Conversely, the body weight and weight gain were higher in the NFD group at the 6th and 8th weeks. Under light and transmission electron microscopes, fat accumulation of the liver was accompanied by an obvious endoplasmic reticulum (ER) swell. Accordingly, the expressions of *atf-6*, *ire-1*, *perk*, *eif-2α*, *atf-4*, *grp78*, and *chop* showed that ER stress was activated at the 6th and 8th weeks. In *Experiment II*, 50 mg/kg 4-PBA (an ERs inhibitor) was supplemented to an HFD; this was named the 4-PBA group. Then, fish was fed with an NFD, an HFD, and a 4-PBA diet for eight weeks. As the result, the excessive fat deposition caused by an HFD was reversed by 4-PBA. The expression of ER stress-related proteins CHOP and GRP78 was down-regulated by 4-PBA, and the transmission electron microscope images also showed that 4-PBA alleviated ER stress induced by the feeding of an HFD. Furthermore, 4-PBA administration down-regulated SREBP-1C/ACC/FAS, the critical pathways of fat synthesis. In conclusion, the results confirmed that ER stress plays a contributor role in the fat deposition by activating the SREBP-1C/ACC/FAS pathway. 4-PBA as an ER stress inhibitor could reduce fat deposition caused by an HFD via regulating ER stress.

Keywords: endoplasmic reticulum stress; high-fat diet; fat deposition; spotted seabass; 4-PBA

Citation: Xia, T.; Liao, Y.-Q.; Li, L.; Sun, L.-Y.; Ding, N.-S.; Wu, Y.-L.; Lu, K.-L. 4-PBA Attenuates Fat Accumulation in Cultured Spotted Seabass Fed High-Fat-Diet via Regulating Endoplasmic Reticulum Stress. *Metabolites* **2022**, *12*, 1197. https://doi.org/10.3390/metabo12121197

Academic Editor: Cholsoon Jang

Received: 3 October 2022
Accepted: 24 November 2022
Published: 30 November 2022

1. Introduction

Fat plays a key role in the energy supply and storage of animals due to its high calorie level [1]. In fish farming, the use of high-fat diets has become a common practice to achieve a sparing protein effect [2,3]. However, recent studies have indicated that the long-time intake of a high-fat diet often induces excess fat deposition referred to as fatty liver. Then, it can cause metabolic disorders, immune dysfunction, and higher mortality [4–6]. Moreover, the substitution of fish oil with plant oil also caused the development of fatty liver and increased the inflammation of fish [7]. Accordingly, it is vital to recognize the mechanisms involved in the development of fatty liver.

The endoplasmic reticulum (ER) is an organelle in eukaryotes vital for protein maturation, and ER stress (ERs) in the liver plays a causative role in the fatty liver of mammals [8]. Additionally, fish fed with a high-fat diet exhibit the disturbance of ER [9]. ER disturbance often leads to misfolded protein accumulation, which induces ERs [10]. In the mice study, ER stress is linked to the development of obesity, and the protein CHOP may be crucial in the interaction between fat deposition and inflammation [11]. It is well-known that fat deposition in the liver represents complex processes. Till now, how ER stress affects the development of fatty liver in fish is still not very clear. Moreover, some therapies acting on ER are suggested as novel strategies for the prevention of fatty liver [12].

Spotted seabass (*Lateolabrax maculatus*) is a carnivorous species with rapid growth speed and a high economic value and thus it has become the second-largest cultivated marine fish in China [13]. For its artificial rearing, the high-fat diet (crude lipid > 15%) is widely used and often leads to the development of fatty liver [3]. Hence, spotted seabass is a good fish model to study the relationship between ER stress and fatty liver. Moreover, we want to explore the role of 4-PBA (an inhibitor of ER stress) in protection against excess fat deposition in fish.

4-PBA is an FDA (The United States Food and Drug Administration)-approved drug currently used for the treatment of urea cycle disorders (UCDs). 4-PBA is an endoplasmic reticulum stress inhibitor that is helpful for protein folding in the endoplasmic reticulum and alleviates the negative effects of endoplasmic reticulum stress [14]. 4-PBA alleviates weight gain and fat production in mice exposed to a high-fat diet by inhibiting endoplasmic reticulum stress [15]. Other studies have shown that 4-PBA can reverse the apoptosis of mouse embryos exposed to high glucose, which is related to endoplasmic reticulum stress [16]. There are many studies on 4-PBA in mammals, which is closely related to endoplasmic reticulum stress. However, there are few studies on the relationship between 4-PBA and endoplasmic reticulum stress and fat deposition in fish. 4-PBA has great potential in green and healthy fish culture.

2. Materials and Methods

2.1. Fish and Experimental Diets

The animal study protocol was approved by The Committee on the Ethics of Animal Experiments of Jimei University, China (protocol code 2011-58 and approved by 20 December 2011).

A local fish breeding farm (Zhangzhou, Fujian, China) provided juvenile *L. maculatus*. Fish were transported to a recirculating aquaculture system in the laboratory located at Jimei University. These juveniles were maintained in a 1200 L tank for two weeks to acclimatize to the experimental condition. During this period, the fish were fed a commercial diet twice daily (8:00 and 17:00).

Two experimental diets were formulated with a different crude fat level and named normal-fat diet (NFD, 11% fat) and high-fat diet (HFD, 17% fat). A 4-PBA supplementation diet was formulated by adding 4-PBA to the HFD (deemed 4-PBA), at a dose of 50 mg/kg. The feed formulation and nutritional composition can be observed in Supplementary Materials Table S1. The method of feed preparation and nutrient level determination was described in our recent study [13].

2.2. Experimental Design

2.2.1. Experiment I: High-Fat-Diet Feeding Trial

A total of 180 healthy fish of a similar size (13.05 $\pm$ 0.10 g) were randomly distributed into six tanks (30 fish per tank). Fish were fed the NFD or HFD diets for 8 weeks twice daily (8:00 and 17:00). The light duration was 12 h/d. During the experiment, the fish were fed to satiation. Each treatment had three replicates. Fish were sampled at the 2nd, 4th, 6th, and 8th week of the feeding trial. For each sample time, all fish from each tank were weighed and counted to calculate the average body weight; after that, three fish from each tank were captured randomly and their liver and blood were sampled. The liver samples were put into liquid nitrogen and transferred to −80 °C until analysis. The blood samples were centrifuged (850$\times$ g, 10 min) at 4 °C, and then the serum was separated and transferred to −80 °C.

2.2.2. Experiment II: Effects of 4-PBA on Fat Deposition

To further determine the effect of ERs in the pathological process of fatty liver, an ERs inhibitor, 4-PBA, was used. Fish from the same batch used in *Expt I* were randomly allocated to three groups (30 fish per tank, three tanks per group, size 41.00 ± 0.01 g). After being fed an NFD, HFD, and 4-PBA diet, respectively, for eight weeks, fish were sampled with the same method used in *Expt I*.

During the feeding trials, water conditions were maintained as below: dissolved oxygen (>6 mg/L), pH (6.9–7.2) kept stable, and the optimum water temperature (25–27 °C). The light duration was 12 h/d. For sampling, fish were fasting for 24 h and euthanized using 100 mg/L MS-222, then measured the body weight for the calculation of growth indices. Then, five fish were randomly selected and captured from each tank for sampling liver and serum, as per our earlier work [17].

2.3. *Liver Histology*

For oil red O staining, liver samples were fixed in 4% paraformaldehyde for 24 h and then dehydrated in a 15% and 30% sugar solution at 4 °C. The dehydrated samples were sliced into 8 μm-thick sections using a cryostat (Cryostar NX50, Thermo, Waltham, MA, USA). The sections were then stained and photographed. Furthermore, the structure of the ER and mitochondria in hepatocytes was examined using transmission electron microscopy (TEM) analysis. The protocol was described in a previous study [4].

2.4. *Gene Expression*

The FastPure Cell/Tissue Total RNA Isolation Kit (Vazyme Biotech Co., Ltd. Nanjing, China) was used to extract the total RNA from the liver. The concentration and purity of isolated RNA were assayed under a spectrophotometer (NanoDrop Technologies, Waltham, MA, USA). The integrity of the RNA was examined through 1% agarose gel electrophoresis, and a NanoDrop™ One spectrophotometer (Thermo Scientific, Waltham, MA, USA) was used to determine the purity at 260/280 nm. Then, the first-strand cDNA was synthesized from 1 μg of each RNA sample using a commercial kit (Vazyme Biotech Co., Ltd. Nanjing, China), and gDNA wiper was performed to remove the remaining genomic DNA. Quantification was performed using a ABI StepOnePlus Real-Time PCR System (Life Technologies, New York, NY, USA).

According to our recent work, real-time quantitative PCR (qPCR) was carried out [18]. The primers were designed with Primer 5.0 software and synthesized by Genewiz Co. Ltd. (Suzhou, China). Table S2 displayed the primer sequences used in this study. The relative expression levels of target genes were normalized by β-actin and calculated by the $2^{-\Delta\Delta Ct}$ method. The amplification efficiency of all primers was verified among 90-110%.

2.5. *Western Blotting*

Total liver protein was extracted using RIPA buffer (Thermo Scientific, Waltham, MA, USA) for the determination of protein expression. In total, 20 μg of total protein was loaded into each well and separated by 12% sodium dodecyl sulfate-polyacrylamide gel electrophoresis, transferred to PVDF membranes, and blocked by 5% (*w/v*) skimmed milk in TBST buffer (LABLEAD Inc., Beijing, China), following primary antibody incubations with anti-CHOP (PA5-88116, Invitrogen, Carlsbad, CA, USA), anti-GRP78 (ab108613, Abcam, Cambridge, UK), or anti-GAPDH (ab181602, Abcam, UK) overnight at 4°C. Then, it was incubated for one hour with the secondary antibody Goat Anti-Rabbit IgG H&L (HRP) pre-absorbed (ab7090, Abcam, Cambridge, UK). The protein bands were detected by ECL Substrate (BIO-RAD, Hercules, California, USA). The protein concentration used for the western blot was 1.5 mg/g. ImageJ 1.44p was used to analyze the gray value of bands [19,20].

2.6. Measures of Biochemical Parameters

The contents of triacylglycerols (TG) and total cholesterol (T-CHO) were measured by commercial kits (Nanjing JianCheng Bioengineering Institute, Nanjing, China). TG was measured using the GPO-PAP enzymatic method. T-CHO was measured by the COD-PAP method. The specific assay steps are performed according to the instructions of the commercial kit.

2.7. Statistical Analysis

For *Expt I*, the two groups that sampled simultaneously were compared using a student's *t*-test. For *Expt II*, one-way ANOVA followed by Tukey's post-hoc test was used to analyze the differences among the three treatments. All of the results were presented as means ± standard errors (SE), and the significance difference level was set at $p < 0.05$ in both *Expt I* and *Expt II*.

3. Results

3.1. Experiment I: High-Fat-Diet Feeding Trial

3.1.1. Fish Growth and Fat Accumulation

The body weight of spotted seabass fed an NFD or HFD was measured at the 2nd, 4th, 6th, and 8th weeks of the feeding trial. In the 8th week, the body weight and cumulative weight gain of the NFD group were significantly higher than that of an HFD (Figure 1A,B).

Oil-red O staining was used to evaluate the fat deposition. As a result, hepatic lipid droplets increased over time both in the NFD and HFD groups. However, fish fed an HFD exhibited an enhanced number and a volume of lipid droplets compared to an NFD at the 4th, 6th, and 8th weeks (Figure 2E).

Furthermore, the expression of lipid synthesis key genes *fas* and *srebp-1c* in the liver was determined. In the NFD group, both *fas* and *srebp-1c* expressions showed no obvious difference within eight weeks. The expression of *fas* in fish fed an HFD obviously increased after six weeks of feeding and was obviously higher than the fish in the NFD group. Similarly, the HFD group up-regulated *srebp-1c* after eight weeks, and its expression was remarkably higher than in the NFD group (Figure 3A,B).

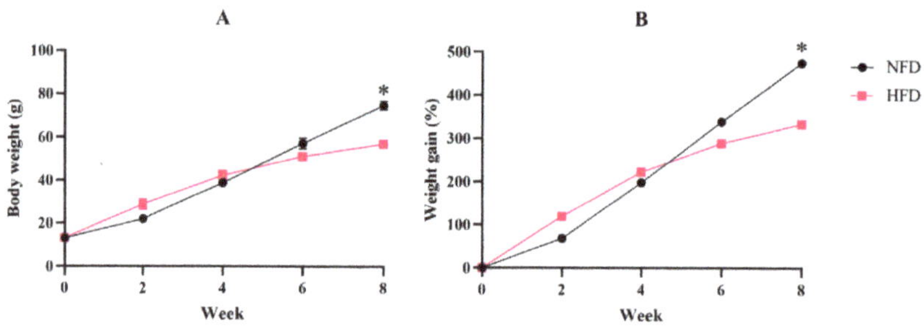

Figure 1. Experiment I. (**A**) Body weight. (**B**) Weight gain. "*" indicates a significant difference between groups ($p < 0.05$, Student's *t*-test, $n = 3$). Weight gain = final body weight/initial body weight.

In the 4th and 8th weeks, the content of serum triglyceride in the HFD group was significantly higher than that of the NFD group. The serum total cholesterol concentration of fish in the HFD group was considerably greater than the NFD group in the 4th week (Figure 2A,B). In the 8th week, the HFD group's hepatic triglyceride and total cholesterol contents were noticeably higher than those of the NFD group (Figure 2C,D).

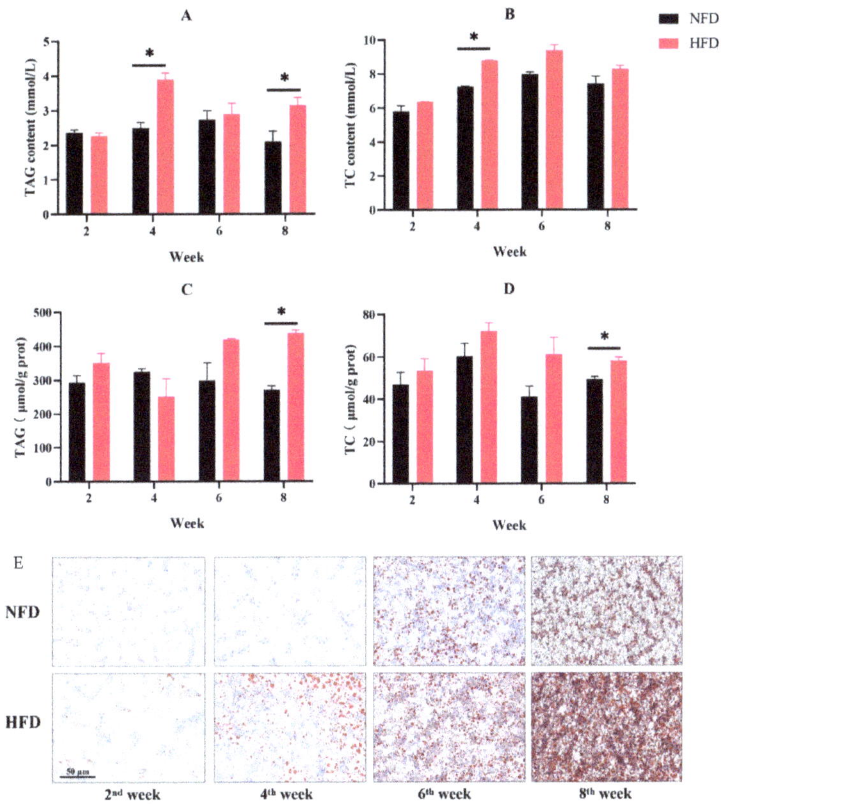

Figure 2. Experiment I. (**A**) Serum TAG content. (**B**) Serum TC content. (**C**) Hepatic TAG content. (**D**) Hepatic TC content. (**E**) Liver oil red O-staining ((**E**) scale bar = 50 μm). "*" indicates a significant difference between groups ($p < 0.05$, Student's t-test, $n = 3$).

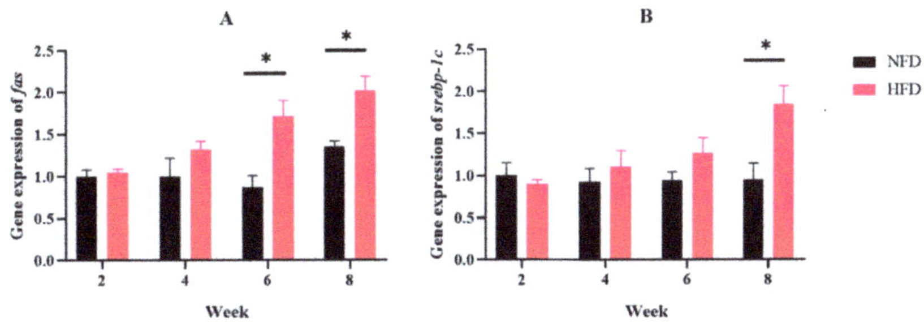

Figure 3. Experiment I. The expressions of fat metabolism-related genes in the liver. (**A**) Gene expression of *fas*. (**B**) Gene expression of *srebp-1c*. "*" indicates a significant difference between groups ($p < 0.05$, Student's t-test, $n = 3$).

3.1.2. Ultrastructure of Hepatocytes

The structure of hepatic mitochondria and ER was observed by TEM. The NFD group exhibited a normal ultrastructure during the experimental period. However, after two weeks of feeding, the mitochondrial injury was exhibited in the HFD group. Additionally, the dilated ER was observed after the 6-week feeding of an HFD. Remarkably, swell and cristae lost mitochondria were accompanied by fragmented and expanded ER in the 8th week (Figure 4).

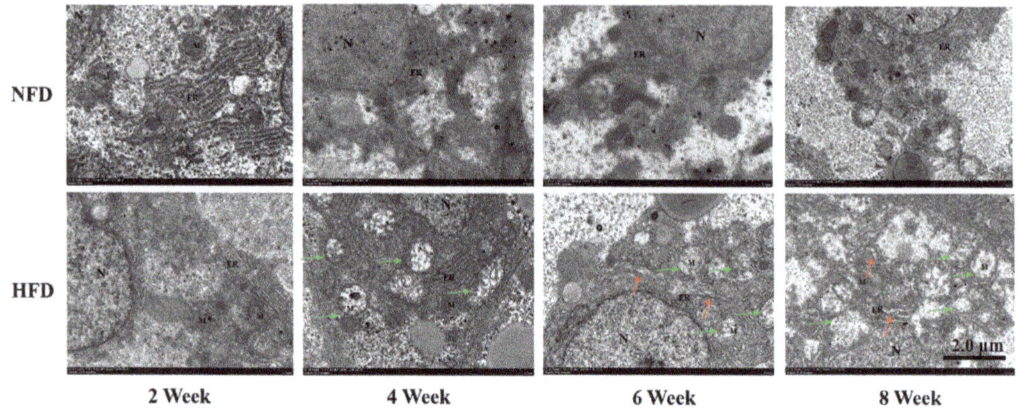

Figure 4. Experiment I. Transmission electron microscopy images of the fish liver at the 2nd, 4th, 6th, and 8th week of feeding trial ($n = 3$). scale bar = 2.0 μm; N, nucleus; ER, endoplasmic reticulum; M, mitochondria. Red arrows: damaged endoplasmic reticulum, green arrows: damaged mitochondria.

3.1.3. ERs-Related Gene Expressions

Considering that the ER ultrastructure damage was observed, we further determined the expression of ERs-related genes. NFD feeding has no appreciable impact on the expression of the ERs-related genes (*atf-6*, *ire-1*, *perk*, *eif-2α*, *atf-4*, *grp78*, and *chop*). The *atf4* gene expression was significantly up-regulated in the 8th week in the HFD group. In addition, the expression of ERs-related genes (*atf-6*, *ire-1*, *perk*, *eif-2α*, *grp78*, and *chop*) was significantly up-regulated in the HFD group at the 6th and 8th week (Figure 5).

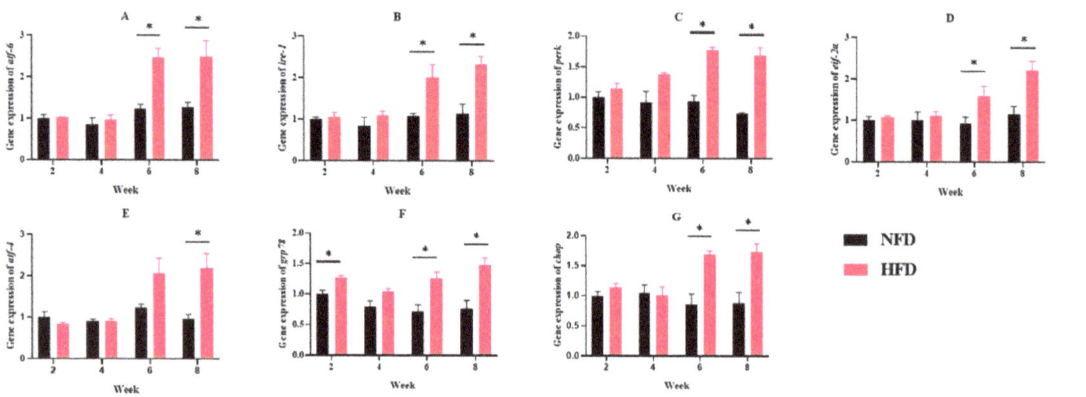

Figure 5. Experiment I. The expressions of endoplasmic reticulum stress-related genes in the liver. (**A**) Gene expression of *atf6*. (**B**) Gene expression of *ire-1*. (**C**) Gene expression of *perk*. (**D**) Gene expression of *eif-2α*. (**E**) Gene expression of *atf4*. (**F**) Gene expression of *grp78*. (**G**) Gene expression of *chop*. "*" indicates a significant difference between groups ($p < 0.05$, Student's t-test, $n = 3$).

3.2. *Experiment II: Effects of 4-PBA on Fat Deposition*

Effects of 4-PBA on Fat Accumulation Caused by an HFD

After eight weeks of feeding, the whole body and abdominal fat levels were dramatically raised by an HFD, which were significantly reversed by 4-PBA supplementation. Additionally, the increase in TAG and TC contents in serum and liver induced by an HFD was reversed by the supplementation of 4-PBA (Figure 6A–D). Additionally, fish given HFD had an overabundance of lipid droplets in the liver, shown by oil red O-stained sections. The addition of 4-PBA to an HFD markedly reduced the lipid droplet accumulation

(Figure 6E). Accordingly, the expressions of lipid synthesis-related genes *srebp-1c*, *fas*, and *acc* were significantly up-regulated by an HFD while 4-PBA down-regulated their values (Figure 7).

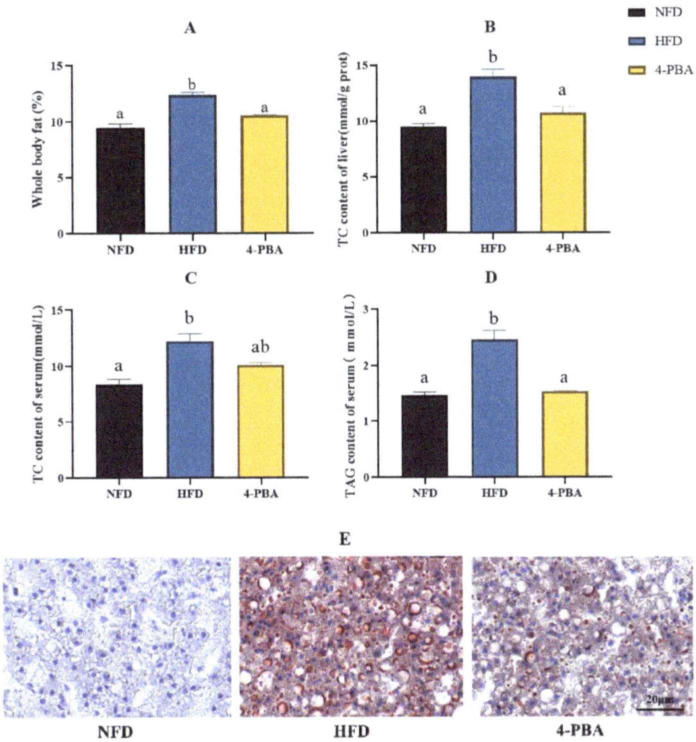

Figure 6. Experiment II. (**A**) Whole body fat. (**B**) Hepatic TC content. (**C**) Serum TC content. (**D**) Serum TAG content. (**E**) Liver oil red O-staining ((**E**) scale bar = 20 μm). All values are exhibited as mean ± SE (n = 3). Different letters indicate a significant difference among groups ($p < 0.05$, Tukey's test, n = 3). Whole body fat = abdominal fat weight/whole body weight.

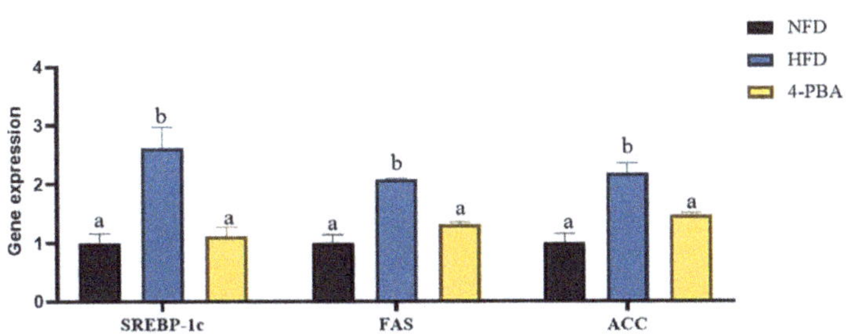

Figure 7. Experiment II. The liver's expression of genes associated with fat metabolism. Gene expression of *srebp-1c*, *fas*, and *acc*. All values are exhibited as mean ± SE. Different letters indicate a significant difference among groups ($p < 0.05$, Tukey's test, n = 3).

The ER and mitochondria in the NFD group exhibited normal morphology. The ER lumen in the HFD group expanded, and the ER network was ruptured, accompanied by swollen mitochondria. Interestingly, the supplementation of 4-PBA significantly decreased these alterations caused by an HFD (Figure 8A). Furthermore, the HFD group had a higher abundance of CHOP and GRP78, and 4-PBA significantly reduced the expression level of these two proteins (Figure 8B).

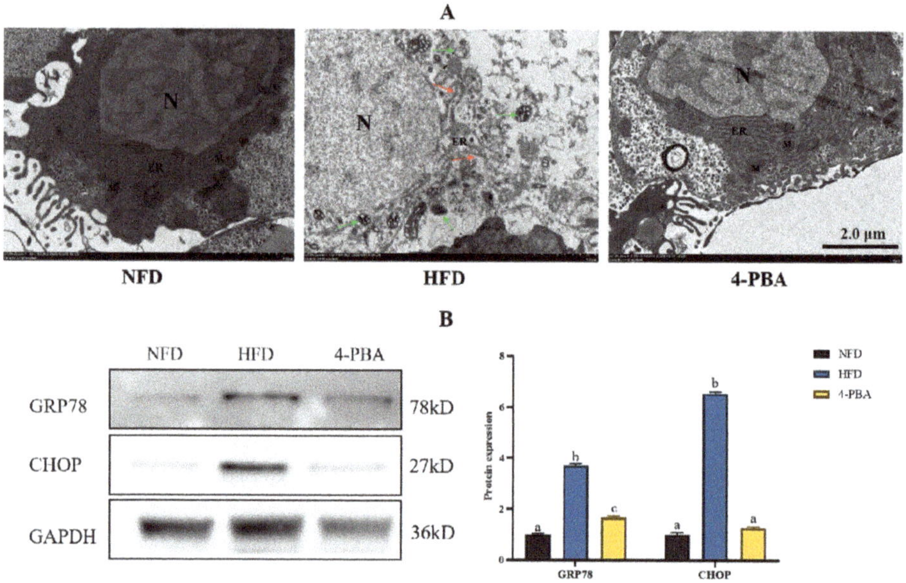

Figure 8. Experiment II. (**A**) Transmission electron microscopy images of fish liver. Scale bar = 2.0 μm. (**B**) GRP78 and CHOP were analyzed through western blotting in the liver of fish. N, Nucleus; ER, endoplasmic reticulum; M, mitochondria. Red arrows: damaged endoplasmic reticulum, green arrows: damaged mitochondria. All values are exhibited as mean ± SE. Different letters indicate a significant difference among groups ($p < 0.05$, Tukey's test, $n = 3$).

4. Discussion

In the present, we found that an HFD caused excessive fat deposition through oil red O staining. This was also evidenced by the results of liver TAG and TC contents. Similar results were also reported in some previous in *L. maculatus* [21] and other fish [20,22,23]. The excessive fat deposition was a result of the imbalance of catabolism and anabolism [24–26]. ER was found to be an important participator in cellular biosynthesis [27,28]. Hence, there is a tight correlation between fat deposition and ER health. Here, we observed the abnormal ER ultrastructure in fish fed an HFD for more than six weeks. At a similar time point, excessive hepatic fat accumulation is emerging. Similarly, ER damage was found to be accompanied by the fatty liver phenomenon in some previous studies. Collectively, we hypothesize that ER injury may play an important role in the pathological excessive fat deposition caused by long-term HFD feeding.

Recently, the increasing data have demonstrated that ERs was involved in the regulation of ER function and acted as a crucial pathogenic mechanism in metabolic disease. ER is the main location of protein folding, maturation, and secretion. ER dysfunction could be induced by ROS, inflammation-related proteins, and fat overloading, resulting in protein misfolding and unfolding [8]. During these situations, mild or moderate ERs is often regarded as a helpful protective response as the activation of UPR could promote molecular chaperone production, wiping off the unfolded/misfolded proteins [27,29]. However, the long-term energy surplus often creates severe or/and prolonged ERs, meaning that ERs turn out to be detrimental. Accordingly, the results in this study showed that long-term (≥six weeks) HFD feeding caused remarkable up-regulation of *atf6*, *perk*, *ire1*, *grp78*, and *chop*. ATF6, PERK, and IRE1 are the three promoters in the UPR pathway; these pathways could be activated by the release of molecular chaperone GRP78 and induce cell-death signaling via CHOP [30]. Furthermore, the increased GRP78 and CHOP protein levels, as well as the dilated ER lumen found by TEM, confirmed the occurrence of ERs after HFD feeding for more than six weeks.

There are a series of lipid-synthesis-related proteins bound at the ER membrane. Among them, SREBP-1C is a key transcription factor, which is responsible for activating the transcription of its downstream genes encoding rate-liming enzymes of fatty acid and triacylglycerol biosyntheses, such as FAS and ACC [31]. A recent study also indicates that the up-regulation of SREBP-1C impairs autophagic flux and thus suppresses fat degradation [32]. In the present study, we found that the expression of *srepb-1c* and *acc* were increased along with the activation of ERs. Liver fat accumulation is a complex process, which is determined by fat transportation, secretion, oxidation, and *do novo* synthesis [1]. Beyond lipid synthesis, some previous studies have demonstrated that ERs decreased hepatic fat oxidation and secretion [33]. Taken together, ERs play a role in excessive fat deposition, and targeting the inhibition of ERs may be an attractive therapeutic modality for fatty liver. 4-PBA is a short-chain fatty acid with an aromatic group, which has been proven to be a non-toxic pharmacological compound [34]. The United States Food and Drug Administration (FDA) has approved its clinical use [35]. There are many studies that have demonstrated that 4-PBA is a potent inhibitor of ERs, owing to the fact it can serve as a molecular chaperone to repair misfolding protein effectively [36–39]. Here, we added 50 mg/kg 4-PBA to an HFD to verify our conjectures. As expected, 4-PBA supplementation notably down-regulated the protein abundance of GRP78 and CHOP and eliminated the alterations of ultrastructure caused by an HFD, indicating ERs was suppressed. This is similar to the results obtained in previous mouse studies. 4-PBA can inhibit endoplasmic reticulum stress, reduce the content of endoplasmic reticulum stress markers, and alleviate the harm caused by a high-fat diet [15]. Additionally, 4-PBA administration reduced the liver, serum, and whole-body fat deposition, as well as the expression of *srebp-1c*, *fas*, and *acc*. 4-PBA can prevent the activation of UPR. Therefore, PERK, a key site of one of the three classical pathways of UPR, is also inhibited. Gene *srebp-1c* is regulated by upstream PERK [40], so its expression is down-regulated. This is strong support for our experimental results. It has been supported by the fact that the SREBP-1C/FAS/ACC pathway was a reliable regulation target of fat deposition in mammals and fish. Overall, our results confirmed that the inhibition of ERs down-regulated the SREBP-1C/FAS/ACC pathway, indicating ERs is an attractive target for the remission of excessive fat accumulation.

5. Conclusions

In conclusion, long-term ($\geq$six weeks) HFD feeding caused a hepatic fat overload of *L. maculate*, which may be largely due to the activation of ERs. Then, ERs up-regulated the gene expression of the SREBP-1C/ACC/FAS pathway. 4-PBA as an ER stress inhibitor could reduce fat deposition caused by an HFD via regulating ER stress.

Supplementary Materials: The following supporting information can be downloaded at: https://www.mdpi.com/article/10.3390/metabo12121197/s1, Table S1: Formulation and proximate composition of experimental diets; Table S2: Sequences of primers used for RT-qPCR.

Author Contributions: K.-L.L., Y.-L.W. and T.X. contributed to conception and design of the study. Y.-Q.L. and L.-Y.S. organized the database. L.L. performed the statistical analysis. T.X. wrote the first draft of the manuscript. Y.-Q.L., L.L. and N.-S.D. conducted formal analysis. All authors have read and agreed to the published version of the manuscript.

Funding: This work was funded by the National Natural Science Foundation of China (32072984), the Natural Science Foundation of Fujian Province (2020J01664), and the China Agriculture Research System (CARS-47).

Institutional Review Board Statement: The animal study protocol was conducted in accordance with the Declaration of Basel and approved by The Committee on the Ethics of Animal Experiments of Jimei University, China (protocol code 2011-58 and approved by 20 December 2011).

Informed Consent Statement: Not applicable.

Data Availability Statement: Data are contained within the article and Supplementary Materials.

Conflicts of Interest: K.L. has received research grants from Fujian Aonong Biological Science and Technology Group Co., Ltd. and Fujian Yixinbao Biopharmaceutical Co., Ltd. Moreover, Y.L., L.S., N.S. and Y.W. are employees of these companies.

References

1. Li, Y.; Ding, W.; Li, C.Y.; Liu, Y. HLH-11 modulates lipid metabolism in response to nutrient availability. *Nat. Commun.* **2020**, *11*, 5959. [CrossRef] [PubMed]
2. Kim, K.; Kim, K.; Kim, K.; Jin Kang, Y.; Lee, S. Influence of lipid level and supplemental lecithin in diet on growth, feed utilization and body composition of juvenile flounder (*Paralichthys olivaceus*) in suboptimal water temperatures. *Aquaculture* **2006**, *251*, 484–490. [CrossRef]
3. Lu, K.L.; Cai, L.S.; Wang, L.; Song, K.; Zhang, C.X.; Rahimnejad, S. Effects of dietary protein/energy ratio and water temperature on growth performance, digestive enzymes activity and non-specific immune response of spotted seabass (*Lateolabrax maculatus*). *Aquacult. Nutr.* **2020**, *26*, 2023–2031. [CrossRef]
4. Lu, K.L.; Wang, L.N.; Zhang, D.D.; Liu, W.B.; Xu, W.N. Berberine attenuates oxidative stress and hepatocytes apoptosis via protecting mitochondria in blunt snout bream Megalobrama amblycephala fed high-fat diets. *Fish Physiol. Biochem.* **2017**, *43*, 65–76. [CrossRef] [PubMed]
5. Dai, Y.J.; Jiang, G.Z.; Yuan, X.Y.; Liu, W.B. High-fat-diet-induced inflammation depresses the appetite of blunt snout bream (*Megalobrama amblycephala*) through the transcriptional regulation of leptin/mammalian target of rapamycin. *Br. J. Nutr.* **2018**, *120*, 1422–1431. [CrossRef] [PubMed]
6. Jin, Y.; Tian, L.; Zeng, S.; Xie, S.; Yang, H.; Liang, G.; Liu, Y. Dietary lipid requirement on non-specific immune responses in juvenile grass carp (*Ctenopharyngodon idella*). *Fish Shellfish Immun.* **2013**, *34*, 1202–1208. [CrossRef]
7. Li, X.; Chen, Q.; Li, Q.; Li, J.; Cui, K.; Zhang, Y.; Kong, A.; Zhang, Y.; Wan, M.; Mai, K.; et al. Effects of High Levels of Dietary Linseed Oil on the Growth Performance, Antioxidant Capacity, Hepatic Lipid Metabolism, and Expression of Inflammatory Genes in Large Yellow Croaker (*Larimichthys crocea*). *Front. Physiol.* **2021**, *12*, 631850. [CrossRef]
8. Fu, S.; Watkins, S.M.; Hotamisligil, G.S. The Role of Endoplasmic Reticulum in Hepatic Lipid Homeostasis and Stress Signaling. *Cell Metab.* **2012**, *15*, 623–634. [CrossRef]
9. Jia, R.; Cao, L.; Du, J.; He, Q.; Gu, Z.; Jeney, G.; Xu, P.; Yin, G. Effects of High-Fat Diet on Steatosis, Endoplasmic Reticulum Stress and Autophagy in Liver of Tilapia (*Oreochromis niloticus*). *Front. Mar. Sci.* **2020**, *7*, 363. [CrossRef]
10. Schr Der, M. Endoplasmic reticulum stress responses. *Cell. Mol. Life Sci.* **2008**, *65*, 862–894. [CrossRef]
11. Maris, M.; Overbergh, L.; Gysemans, C.; Waget, A.; Cardozo, A.K.; Verdrengh, E.; Cunha, J.P.M.; Gotoh, T.; Cnop, M.; Eizirik, D.L.; et al. Deletion of C/EBP homologous protein (Chop) in C57Bl/6 mice dissociates obesity from insulin resistance. *Diabetologia* **2012**, *55*, 1167–1178. [CrossRef] [PubMed]
12. Dong, Y.; Li, L.; Xia, T.; Wang, L.; Xiao, L.; Ding, N.; Wu, Y.; Lu, K. Oxidative Stress Can Be Attenuated by 4-PBA Caused by High-Fat or Ammonia Nitrogen in Cultured Spotted Seabass: The Mechanism Is Related to Endoplasmic Reticulum Stress. *Antioxidants* **2022**, *11*, 1276. [CrossRef] [PubMed]
13. Cai, L.; Wang, L.; Song, K.; Lu, K.; Zhang, C.; Rahimnejad, S. Evaluation of protein requirement of spotted seabass (*Lateolabrax maculatus*) under two temperatures, and the liver transcriptome response to thermal stress. *Aquaculture* **2020**, *516*, 734615. [CrossRef]
14. de Almeida, S.F.; Picarote, G.; Fleming, J.V.; Carmo-Fonseca, M.; Azevedo, J.E.; de Sousa, M. Chemical chaperones reduce endoplasmic reticulum stress and prevent mutant HFE aggregate formation. *J. Biol. Chem.* **2007**, *282*, 27905–27912. [CrossRef]
15. Basseri, S.; Lhotak, S.; Sharma, A.M.; Austin, R.C. The chemical chaperone 4-phenylbutyrate inhibits adipogenesis by modulating the unfolded protein response. *J. Lipid Res.* **2009**, *50*, 2486–2501. [CrossRef]
16. Li, X.; Xu, C.; Yang, P. c-Jun NH2-terminal kinase 1/2 and endoplasmic reticulum stress as interdependent and reciprocal causation in diabetic embryopathy. *Diabetes* **2013**, *62*, 599–608. [CrossRef]
17. Lu, K.; Xu, W.; Li, J.; Li, X.; Huang, G.; Liu, W. Alterations of liver histology and blood biochemistry in blunt snout bream Megalobrama amblycephala fed high-fat diets. *Fish. Sci.* **2013**, *79*, 661–671. [CrossRef]
18. Dong, Y.; Xia, T.; Lin, J.; Wang, L.; Song, K.; Zhang, C. Quercetin Attenuates High-Fat Diet-Induced Excessive Fat Deposition of Spotted Seabass (*Lateolabrax maculatus*) Through the Regulatory for Mitochondria and Endoplasmic Reticulum. *Front. Mar. Sci.* **2021**, *8*, 746811. [CrossRef]
19. Zhou, W.; Rahimnejad, S.; Tocher, D.R.; Lu, K.; Zhang, C.; Sun, Y. Metformin attenuates lipid accumulation in hepatocytes of blunt snout bream (*Megalobrama amblycephala*) via activation of AMP-activated protein kinase. *Aquaculture* **2019**, *499*, 90–100. [CrossRef]
20. Dong, Y.Z.; Li, L.; Espe, M.; Lu, K.L.; Rahimnejad, S. Hydroxytyrosol Attenuates Hepatic Fat Accumulation via Activating Mitochondrial Biogenesis and Autophagy through the AMPK Pathway. *J. Agric. Food Chem.* **2020**, *68*, 9377–9386. [CrossRef]
21. Dong, Y.; Yu, M.; Wu, Y.; Xia, T.; Wang, L.; Song, K.; Zhang, C.; Lu, K.; Rahimnejad, S. Hydroxytyrosol Promotes the Mitochondrial Function through Activating Mitophagy. *Antioxidants* **2022**, *11*, 893. [CrossRef] [PubMed]

22. Jobling, M.; Koskela, J.; Savolainen, R. Influence of dietary fat level and increased adiposity on growth and fat deposition in rainbow trout, *Oncorhynchus mykiss* (Walbaum). *Aquac. Res.* **1998**, *29*, 601–607. [CrossRef]
23. Zhou, W.; Rahimnejad, S.; Lu, K.; Wang, L.; Liu, W. Effects of berberine on growth, liver histology, and expression of lipid-related genes in blunt snout bream (*Megalobrama amblycephala*) fed high-fat diets. *Fish Physiol. Biochem.* **2018**, *45*, 83–91. [CrossRef]
24. Li, W.; Wu, K.; Liu, Y.; Yang, Y.; Tang, H. Molecular cloning of SLC35D3 and analysis of its role during porcine intramuscular preadipocyte differentiation. *BMC Genet.* **2020**, *21*, 20–28. [CrossRef] [PubMed]
25. Zhang, Y.Y.; Wang, J.X.; Ren, J.; Wang, J.; Qiao, F.; Zhang, M.L.; Du, Z.Y. Activation of peroxisome proliferator-activated receptor-α stimulated lipid catabolism in liver of largemouth bass, Micropterus salmoides. *Aquacult. Nutr.* **2021**, *27*, 2749–2758. [CrossRef]
26. Zhou, C.; Li, P.; Han, M.; Gao, X. Daidzein stimulates fatty acid-induced fat deposition in C2C12 myoblast cells via the G protein-coupled receptor 30 pathway. *Anim. Biotechnol.* **2020**, *33*, 851–863. [CrossRef]
27. Hetz, C.; Chevet, E.; Oakes, S.A. Proteostasis control by the unfolded protein response. *Nat. Cell Biol.* **2015**, *17*, 829–838. [CrossRef]
28. Wang, M.; Kaufman, R.J. Protein misfolding in the endoplasmic reticulum as a conduit to human disease. *Nature* **2016**, *529*, 326–335. [CrossRef]
29. Lerner, A.G.; Upton, J.; Praveen, P.V.K.; Ghosh, R.; Nakagawa, Y.; Igbaria, A.; Shen, S.; Nguyen, V.; Backes, B.J.; Heiman, M.; et al. IRE1α Induces Thioredoxin-Interacting Protein to Activate the NLRP3 Inflammasome and Promote Programmed Cell Death under Irremediable ER Stress. *Cell Metab.* **2012**, *16*, 250–264. [CrossRef]
30. Faitova, J.; Krekac, D.; Hrstka, R.; Vojtesek, B. Endoplasmic reticulum stress and apoptosis. *Cell. Mol. Biol. Lett.* **2006**, *11*, 488–505. [CrossRef]
31. Yan, F.; Wang, Q.; Lu, M.; Chen, W.; Song, Y.; Jing, F.; Guan, Y.; Wang, L.; Lin, Y.; Bo, T.; et al. Thyrotropin increases hepatic triglyceride content through upregulation of SREBP-1c activity. *J. Hepatol.* **2014**, *61*, 1358–1364. [CrossRef] [PubMed]
32. Nguyen, T.T.P.; Kim, D.; Lee, Y.; Lee, Y.; Truong, X.T.; Lee, J.; Song, D.; Kwon, T.K.; Park, S.; Jung, C.H.; et al. SREBP-1c impairs ULK1 sulfhydration-mediated autophagic flux to promote hepatic steatosis in high-fat-diet-fed mice. *Mol. Cell.* **2021**, *81*, 3820–3832. [CrossRef] [PubMed]
33. Cao, X.; Dai, Y.; Liu, M.; Yuan, X.; Wang, C.; Huang, Y.; Liu, W.; Jiang, G. High-fat diet induces aberrant hepatic lipid secretion in blunt snout bream by activating endoplasmic reticulum stress-associated IRE1/XBP1 pathway. *Biochim. Biophys. Acta Mol. Cell Biol. Lipids* **2019**, *1864*, 213–223. [CrossRef] [PubMed]
34. Luo, T.; Chen, B.; Wang, X. 4-PBA prevents pressure overload-induced myocardial hypertrophy and interstitial fibrosis by attenuating endoplasmic reticulum stress. *Chem. Biol. Interact.* **2015**, *242*, 99–106. [CrossRef] [PubMed]
35. Guzman, M.N.; Homma, K.; Osada, H.; Toda, E.; Ban, N.; Nagai, N.; Negishi, K.; Tsubota, K.; Ozawa, Y. Neuroprotective Effect of 4-Phenylbutyric Acid against Photo-Stress in the Retina. *Antioxidants* **2021**, *10*, 1147. [CrossRef] [PubMed]
36. Souza-Neto, F.V.; Jiménez-González, S.; Delgado-Valero, B.; Jurado-López, R.; Genty, M.; Romero-Miranda, A.; Rodríguez, C.; Nieto, M.L.; Martínez-Martínez, E.; Cachofeiro, V. The Interplay of Mitochondrial Oxidative Stress and Endoplasmic Reticulum Stress in Cardiovascular Fibrosis in Obese Rats. *Antioxidants* **2021**, *10*, 1274. [CrossRef]
37. Pereira, S.; Moore, J.; Li, J.X.; Yu, W.Q.; Ghanim, H.; Vlavcheski, F.; Joseph, Y.D.; Dandona, P.; Volchuk, A.; Cummins, C.L.; et al. 4-Phenylbutyric acid improves free fatty acid-induced hepatic insulin resistance in vivo. *Endocr. Connect.* **2021**, *10*, 861–872. [CrossRef]
38. Randhawa, R.; Bhardwaj, R.; Kaur, T. Amelioration of hyperoxaluria-induced kidney dysfunction by chemical chaperone 4-phenylbutyric acid. *Urolithiasis* **2018**, *47*, 171–179. [CrossRef]
39. Kolb, P.S.; Ayaub, E.A.; Zhou, W.; Yum, V.; Dickhout, J.G.; Ask, K. The therapeutic effects of 4-phenylbutyric acid in maintaining proteostasis. *Int. J. Biochem. Cell Biol.* **2015**, *61*, 45–52. [CrossRef]
40. Wang, T.; Wei, X.; Sun, Y.; Hu, Y.; Li, J.; Zhang, X.; Yin, S.; Shi, Y.; Zhu, Y. Copper nanoparticles induce the formation of fatty liver in Takifugu fasciatus triggered by the PERK-EIF2α-SREBP-1c pathway. *NanoImpact* **2021**, *21*, 100280. [CrossRef]

metabolites

MDPI

Article

Mechanism Analysis of Metabolic Fatty Liver on Largemouth Bass (*Micropterus salmoides*) Based on Integrated Lipidomics and Proteomics

Moyong Xue [1,2,3,†], Ting Yao [4,†], Min Xue [1], Frédéric Francis [2], Yuchang Qin [3], Ming Jia [1], Junguo Li [1] and Xu Gu [1,*]

1 Feed Research Institute, Chinese Academy of Agricultural Science, Beijing 100081, China
2 Functional & Evolutionary Entomology, Agro-Bio-Tech Gembloux, University of Liege, 5030 Gembloux, Belgium
3 Institute of Animal Science, Chinese Academy of Agriculture Sciences, Beijing 100193, China
4 Beijing Institute of Feed Control, Beijing 110108, China
* Correspondence: guxu@caas.cn
† These authors contributed equally to this work.

Abstract: Metabolic fatty liver disease caused by high-starch diet restricted the intensive and sustainable development of carnivorous fish such as largemouth bass. In this study, the combination liver proteomic and lipidomic approach was employed to investigate the key signaling pathways and identify the critical biomarkers of fatty liver in largemouth bass. Joint analysis of the correlated differential proteins and lipids revealed nine common metabolic pathways; it was determined that FABP1 were significantly up-regulated in terms of transporting more triglycerides into the liver, while ABCA1 and VDAC1 proteins were significantly down-regulated in terms of preventing the transport of lipids and cholesterol out of the liver, leading to triglyceride accumulation in hepatocyte, eventually resulting in metabolic fatty liver disease. The results indicate that FABP1, ABCA1 and VDAC1 could be potential biomarkers for treating metabolic fatty liver disease of largemouth bass.

Keywords: metabolic liver disease; integrated analysis; proteomics; lipidomics; largemouth bass

Citation: Xue, M.; Yao, T.; Xue, M.; Francis, F.; Qin, Y.; Jia, M.; Li, J.; Gu, X. Mechanism Analysis of Metabolic Fatty Liver on Largemouth Bass (*Micropterus salmoides*) Based on Integrated Lipidomics and Proteomics. *Metabolites* **2022**, *12*, 759. https://doi.org/10.3390/metabo12080759

Academic Editors: Qingchao Wang, Yan He and Qingsong Tan

Received: 19 July 2022
Accepted: 14 August 2022
Published: 17 August 2022

1. Introduction

Largemouth bass (*Micropterus salmoides*), originally a popular sport fish in North America, is now reared in many other countries worldwide and also is an important freshwater carnivorous species in China for its high market value as a food [1–4]. At present, more than 80% of cultured largemouth bass are fed on chilled fish [5]. However, the high costs of chilled fish as well as the potential role as disease vectors hinder the healthy development of largemouth bass aquaculture worldwide [2].

Previous studies discovered that fishes (especially carnivorous ones such as largemouth bass) generally have a low ability to use carbohydrate in compound feed [6–8]. Starch is the cheapest energy source in practical diet ingredients, and it contributes to good binders for aquatic feeds [9]. Traditional puffed floating feeds usually require at least 20% starch to obtain adequate swelling and floating properties. However, researchers demonstrated that excess dietary carbohydrate could cause fat deposition, glucose and lipid metabolism disorder, hepatopathy, apoptosis and finally result in liver damage [8,10–13]. Previous studies suggested that the starch content in the diet should be lower than 10% to ensure the liver health of largemouth bass [14,15]. Therefore, high carbohydrate is recognized as the primary factor that induced metabolic liver disease (MLD) in largemouth bass [8,9]. MLD of largemouth bass was the first limiting factor for its sustainable development worldwide [7]. Thus, it is crucial to understand the processes and progression of MLD, as well as the key factors in largemouth bass.

Proteomic technologies could be used for identifying and quantifying target proteins to comprehensively explore the etiology of liver disease, due to their high-resolution and high-throughput advantages [16]. Currently, it has become a powerful tool for understanding pathology of liver disease [17]. By identifying groups of changed proteins, we could gain insight into potential pathways and regulatory networks that might contribute to the development of MLD. Many studies have found that proteomics could be used for biomarker identification for early diagnosis in MLD [18–20]. To date, no study has ever comprehensively investigated the protein expression changes in fishes of MLD. Lipids are essential components for maintaining various homeostasis, physiological and cellular processes in animals [21]. Lipid metabolism disorders would lead to many major health problems, such as obesity and nonalcoholic fatty liver disease (NAFLD) [22]. Lipidomics, based on UPLC-MS/MS technology, is a novel omics strategy for investigating lipid metabolism and identifying lipid biomarkers [23,24], which enables large-scale and comprehensive studies of lipids [25]. However, there is a lack of studies that systematically assess the protein and lipid changes in the diseased liver of fish for the discovery of the biomarkers of metabolic liver disease.

In this study, we constructed a fatty liver model of largemouth bass by feeding high levels of dietary starch (16.2%) to investigate the complex biological processes and pathogenesis of MLD with proteomic and lipidomic techniques. The mass spectrometry technique was used for proteome-level quantification analyses. Some characteristic changes to different proteins and lipid metabolites are documented for the first time in this study and provide a comprehensive multi-omics framework for MLD biomarker discovery in largemouth bass.

2. Materials and Methods

During the experiment period, all fishes were maintained in compliance with the Laboratory Animal Welfare Guidelines of China (Decree No. 2 of Ministry of Science and Technology, issued in 2021).

2.1. Experimental Diets

Two experimental diets were formulated to be isonitrogenous and isoenergetic. A basal diet was used as the control containing 10.8% starch (named the Normal group), whereas another diet was prepared with 16.2% starch (named the MLD group). Each diet was extruded into 2 mm diameter pellets using a twin-screwed extruder (EXT50A, Yang gong Machine, Beijing, China). The diet formulation and analyzed chemical composition are shown in Table 1.

Table 1. Formulation and composition of experimental diets (%).

Ingredients (in Dry Matter Basis, %)	Normal Group	MLD Group
[a] Fish meal	30.0	30.0
Tapioca starch	5	5
Wheat flour	9.0	18
Microbial protein	4	4
Cottonseed concentrate protein	23.5	22.6
Wheat gluten meal	4	4
Soybean meal	2	-
Spay-dried blood cell powder	4	4
α-cellulose	4.6	-
$Ca(H_2PO4)_2$	1.7	1.7
Lecithin oil	2.0	2.0
Fish oil	3.5	3.5
Soybean oil	3.5	3.5
Kelp powder	1.5	0
L-Threonine	0.1	0.1
DL-Methionine	0.2	0.2

Table 1. *Cont.*

Ingredients (in Dry Matter Basis, %)	Normal Group	MLD Group
[b] Vitamin and mineral premix	1.4	1.4
Total	100	100
Nutrients compositions		
Crude protein	50.83	51.15
Ether extract	12.36	12.33
Crude ash	10.08	10.04
Moisture	6.10	7.43
Gross energy (MJ/Kg)	20.45	20.15

[a] Fish meal: crude protein content was 68.8%; soybean protein concentrate: crude protein content was 65.2%; cottonseed protein concentrate: crude protein content was 61.5%. [b] Vitamin and mineral premix (mg/kg diets): VA 20 mg, VK_3 20 mg, VD_3 10 mg, VE 400 mg, VB_1 10 mg, VB_2 15 mg, VB_6 15 mg, VB_{12} (1%) 8 mg, VC (35%) 1000 mg, Calcium pantothenate 40 mg, Inositol 200 mg, Niacinamide 100 mg, Biotin (2%) 2 mg, Folic acid 10 mg, Corn gluten meal 150 mg; Choline chloride (50%) 4000 mg; $FeSO_4{\cdot}H_2O$ 300 mg, $ZnSO_4{\cdot}H_2O$ 200 mg, $MnSO_4{\cdot}H_2O$ 100 mg, $CoCl_2{\cdot}6H_2O$ (10%Co) 5 mg, KI (10%) 80 mg, Na_2SeO_3 (10% Se) 10 mg, $MgSO_4{\cdot}5H_2O$ 2000 mg, Zeolite 4995 mg, NaCl 100 mg, Antioxidant 200 mg, $CuSO_4{\cdot}5H_2O$ 10 mg.

2.2. Experimental Fish, Feeding and Sampling

Largemouth bass were obtained from the Tangshan Aquafarm (Tangshan, Hebei, China). The experiment was conducted in the indoor circulating water system at the National Aquatic Feed Safety Evaluation Base (Nan Kou, Beijing, China). Prior to the formal experiment, the fish were acclimatized for 4 weeks by being fed the Normal diet. The water temperature was maintained at 21–25 °C, pH = 7.2–8.0, dissolved oxygen (DO) >6.0 mg/L, ammonia nitrogen content <0.3 mg/L and NO_2^- < 0.1 mg/L. The fish were fed until apparent satiation twice a day (8:00 am and 4:00 pm) for 10 weeks. Food intake was measured daily.

At the end of the growth trial, fishes from each group were randomly selected and anaesthetized with chlorobutanol (300 mg/mL) after 24 h starvation. The specific growth rate (SGR), final body weight (FBW) and feed conversion ratio (FCR) were detected by weighing the fish at the end of the 10 weeks. The body weight, body length, liver and viscera weight of the fish were recorded individually to calculate condition factor (CF), hepatosomatic index (HSI), viscerosomatic index (VSI) and hepatic liquid (HL), respectively. Blood samples were drawn from the caudal part of the sedated fish using anticoagulant syringes with 2% NaF and 4% potassium oxalate and centrifuged at 4000× *g* rpm for 10 min at 4 °C to obtain serum. Two parts of liver samples near to the bile duct were collected for histology examination (fixed in 4% paraformaldehyde solution) and biochemical criterion analysis (frozen in liquid nitrogen). All samples (except for histological samples) were stored at −80 °C until analysis.

2.3. Hematological and Liver Homogenate Parameters

Hematological parameters included total protein (TP), total cholesterol (TC), triglyceride (TG), high-density lipoprotein cholesterol (HDL-C), low-density lipoprotein cholesterol (LDL-C), alkaline phosphatase (AKP), aspartate aminotransferase (AST), alanine aminotransferase (ALT), total bile acid (TBA) and glucose (GLU). Hepatic total antioxidative capability (T-AOC), superoxide dismutase (SOD), glutathione peroxidase (GSH-Px), catalase (CAT) and malondialdehyde (MDA) were determined by assay kits (Nanjing jiancheng Co., Nanjing, China) following the protocols given by the supplier. The reactive oxygen species (ROS) was determined by assay kit (Beyotime Biotechnology, Shanghai, China) following the protocols given by the supplier.

2.4. Histopathological Examination of the Liver Tissue

After 24 h of fixation, all liver samples were dehydrated by the standard procedures, and the samples were embedded in paraffin and cut to 6 μm sections. Liver sections were stained following the protocols of hematoxylin and eosin (H&E) staining and observed with light microscopy (Leica DM2500, Leica, Solms, Germany).

2.5. Proteomics Analysis

2.5.1. Protein Extraction

The sample (3 mm × 3 mm) was grinded with liquid nitrogen into cell powder and then transferred to a 5 mL centrifuge tube. After that, four volumes of lysis buffer (8 M urea, 1% Triton-100, 10 mM dithiothreitol and 1% Protease Inhibitor Cocktail) were added to the cell powder, followed by sonication three times on ice using a high-intensity ultrasonic processor. The remaining debris was removed by centrifugation at 20,000× *g* at 4 °C for 10 min. Finally, the protein was precipitated with cold 20% TCA for 2 h at −20 °C. After centrifugation at 12,000× *g* 4 °C for 10 min, the supernatant was discarded. The remaining precipitate was washed with cold acetone three times. The protein was redissolved in 8 M urea and the protein concentration was determined with BCA kit according to the manufacturer's instructions.

2.5.2. Trypsin Digestion

For digestion, the protein solution was reduced with 5 mM dithiothreitol for 30 min at 56 °C and alkylated with 11 mM iodoacetamide for 15 min at room temperature in darkness. The protein sample was then diluted to urea concentration less than 2 M. Finally, trypsin was added at 1:50 trypsin-to-protein mass ratio for the first digestion overnight and 1:100 trypsin-to-protein mass ratio for a second 4 h digestion.

2.5.3. TMT Labeling

After trypsin digestion, peptide was desalted with Strata X C18 SPE column (Phenomenex) and vacuum dried. Peptide was reconstituted in 0.5 M TEAB and processed according to the manufacturer's protocol for TMT kit. Briefly, one unit of TMT reagent was thawed and reconstituted in acetonitrile. The peptide mixtures were then incubated for 2 h at room temperature and pooled, desalted and dried by vacuum centrifugation.

2.5.4. LC-MS/MS Analysis

The hydrolysated peptides were separated by HPLC connected with a reverse capillary column. The specific steps and related conditions were as follows [19,26,27]: (1) sample desalination: 0.1% TFA cleaning for 5 min (20 μL/min); (2) gradient elution: 2–35% ACN gradient elution for 45 min (350 nL/min), 80% ACN elution for 15 min; (3) the eluent composition of HPLC: A, 0.1% formic acid solution (*v*/*v*). B, 0.1% formic acid acetonitrile solution (*v*/*v*); (4) the peptides were subjected to NSI source followed by tandem mass spectrometry (MS/MS) in Q Exactive^{TM} Plus (Thermo Fisher Scientific Co., LTD, Shanghai, China) coupled online with the UPLC. The electrospray voltage applied was 2.0 kV. The m/z scan range was 350 to 1000 for full scan, and intact peptides were detected in the Orbitrap at a resolution of 35,000. Peptides were then selected for MS/MS using NCE setting as 27 and the fragments were detected in the Orbitrap at a resolution of 17,500. A data-independent procedure alternated between one MS scan followed by 20 MS/MS scans. Automatic gain control (AGC) was set at 3E6 for full MS and 1E5 for MS/MS. The maximum IT was set at 20 ms for full MS and auto for MS/MS. The isolation window for MS/MS was set at 2.0 m/z.

2.5.5. Protein Identification

Protein identifications were performed by using the ProteinPliot software (AB Sciex). For protein quantitation, proteins were required to contain at least two unique peptides.

To demonstrate the reproducibility of the replicates, protein differential expressions between various biological replicates were compared. Then *p*-value was calculated by using the two-sample and two-tail *t*-test. When p-value < 0.05, the change of differential expression exceeding 1.3 was regarded as the threshold for significant up-regulation, and that less than 0.77 was regarded as the threshold for significant down-regulation.

2.6. Lipidomics Analysis

2.6.1. Lipid Extraction

For the lipidomics analysis, the total lipids were extracted from the liver in the Normal and MLD groups. Liver tissues (20 mg) were homogenized in a 2 mL centrifuge tube with 1 mL extracting solution (methyl tertiary butyl ether: methyl alcohol = 3:1, v/v, including internal standard mixture). After homogenization, 200 μL deionized water was added and then centrifuged for 10 min at 12,000× g r/min, 4 °C. The supernatant solution was dried under nitrogen and redissolved in 200 μL mobile phase B for further analysis.

2.6.2. UPLC-MS/MS Analysis

Lipid profiles were determined using UPLC-MS/MS (AB Sciex, Framingham, MA, USA). Samples (2 μL) were separated on Thermo Accucore^{TM}C30 column (2.6 μm, 2.1 mm × 100 nm i.d.) with a flow rate of 0.35 mL/min and column temperature of 45 °C. The mobile phase consisted of a mixture of acetonitrile/water (60:40, v/v) (A) and a mixture of acetonitrile/isopropanol (10:90, v/v) (B), both containing 0.1% acetic acid and 10 mmol/L ammonium formate. The elution gradient was set stepwise as follows: 0–2 min, 20–30% B; 2–4 min, 30–60% B; 4–9 min, 60–85% B; 9–14 min, 85–90% B; 14–17.3 min, 90–95% B; 17.3–20 min, 95–20% B.

2.6.3. Lipid Identification and Quantitation

The data were analyzed by multi-reaction detection mode (MRM). Conditions of MS were set as follows: ion spray voltage: +5500 V (positive) and −4500 V (negative); collision energy (CE): 35 V (positive) and −35 V (negative); temperature of electrospray ionization (ESI): 500 °C; gas 1 (GS1): 45 psi; gas 2 (GS2): 50 psi; curtain gas: 35 psi.

Fifteen standard solutions with different concentrations were prepared for quantitative analysis (concentration: 0.0002, 0.0005, 0.001, 0.002, 0.005, 0.01, 0.02, 0.05, 0.1, 0.2, 0.5, 1, 2, 5, 10 nmol/mL). Table 2 lists the lipid types and information of internal and external standards.

Table 2. Lipid types and information of part of internal and external standards.

	Lipids	Abbreviation	CAS	Standards Type
1	Lysophosphatidyl choline	LPC (13:0)	20559-17-5	Internal
2	Cholesterol heptadecanoate	CE (17:0)	24365-37-5	Internal
3	Ceramide C4	Cer (d18:1/4:0)	74713-58-9	Internal
4	Diester of glycerol dodecanoate	DG (12:0/12:0)	60562-15-4	Internal
5	Lysophosphatidyl ethanolamine	LPE (14:0)	123060-40-2	Internal
6	Phosphatidylcholine	PC (13:0/13:0)	71242-28-9	Internal
7	Phosphatidyl ethanolamine	PE (12:0/12:0)	59752-57-7	Internal
8	Diphosphatidyl glycerol	PG (12:0/12:0)	322647-27-8	Internal
9	Phosphatidylserin	PS (14:0/14:0)	105405-50-3	Internal
10	Triglyceride dodecyl	TG (12:0/12:0/12:0)	555-44-2	Internal
11	Phosphatidyl inositol	PI (16:0/16:0)	34290-57-8	Internal
12	Palmitic acid -d31	FFA (16:0)-d31	39756-30-4	Internal
13	Cholesterol linoleate	CE (18:2)	604-33-1	External
14	Ceramide C17	Cer (d18:1/17:0)	67492-16-4	External
15	Diester of glycerol hexadecanoate	DG (16:0/16:0)	30334-71-5	External
16	Lysophosphatidyl choline	LPC (17:0)	50930-23-9	External
17	Lysophosphatidyl ethanolamine	LPE (16:0)	53862-35-4	External
18	Phosphatidyl choline	PC (17:0/17:0)	70897-27-7	External
19	Phosphatidyl ethanolamine	PE (17:0/17:0)	140219-78-9	External
20	Phosphatidyl glycerol	PG (17:0/17:0)	799268-52-3	External
21	Phosphatidylserine	PS (17:0/17:0)	799268-51-2	External
22	Sphingomyelin	SM (d18:1/17:0)	121999-64-2	External
23	Triglyceride heptadecanoate	TG (17:0/17:0/17:0)	2438-40-6	External
24	Phosphatidyl inositol	PI (16:0/18:1)	50730-13-7	External
25	Palmitic acid	FFA (16:0)	57-10-3	External

2.7. Statistical Analysis

Independent *t*-test was performed by data of growth performance. All data are presented as the mean value ± standard error of the mean (S.E.M). $p < 0.05$ was considered significantly different. The correlation and linear regression analysis of biological repetitions were carried out using GraphPad Prism 8.0 (Graph- Pad Software Inc., San Diego, CA, USA).

Functional annotations of the proteins and genetic ontology were conducted using the Blast2GO program. The related pathways of each protein and bioinformation mining were analyzed by searching the KEGG pathway database.

R software was used for Principal Component Analysis (PCA), cluster analysis, grouping Principal Component Analysis, differential metabolite screening and KEGG functional annotation of different samples.

3. Results

3.1. High-Starch Diet (HSD) Induced Fatty Liver of Largemouth

Fish liver sections were examined after H&E staining and Sirius red staining for collagen. Typical phenotypes of normal liver and fatty liver are shown in Figure 1. The results showed that HSD led to enlarged lipid droplets and vacuolated cells.

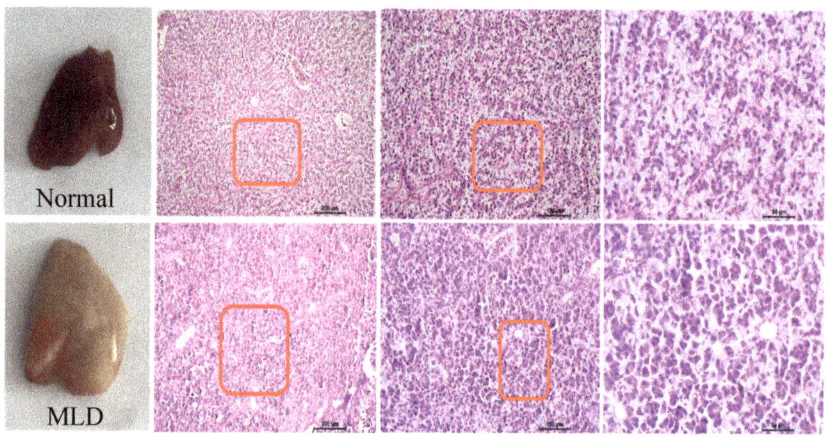

Figure 1. Liver histopathology, enlarged lipid droplets (marked with red boxes) were clearly observed in MLD group. (MLD: metabolic liver disease).

Both Normal and MLD groups showed high survival (99%) and there was no significant difference between the two groups. Compared with the Normal group, fish in the MLD group showed significantly higher HSI and HL ($p < 0.05$) but significantly lower SGR and FBW ($p < 0.05$, Table 3).

LDL-C indicators (in plasma and liver, Tables 4 and 5) and the hepatic function indicators (AKP, AST and ALT) increased significantly in the MLD group ($p < 0.05$, Table 4). Higher ROS level and lower CAT content were recorded in the MLD group (Table 6).

Table 3. Changes of MLD on morphometric parameters of largemouth bass (means ± SEM).

Items	Normal	MLD
CF (g/cm^3)	2.03 ± 0.09	2.01 ± 0.08
VSI (%)	7.17 ± 1.86	7.36 ± 0.20
HSI (%)	1.66 ± 0.08 ^{b}	2.36 ± 1.17 ^{a}
FBW (g)	105.83 ± 1.68 ^{b}	95.54 ± 1.30 ^{a}
SGR	2.00 ± 0.31 ^{b}	1.80 ± 0.03 ^{a}
FCR	0.98 ± 0.01	1.00 ± 0.18
HL	6.54 ± 0.29 ^{b}	7.39 ± 0.10 ^{a}

$^{a, b}$ Within the same column, values with different superscripts are significantly different ($p < 0.05$). The same as below, $n = 8$; CF (condition factor, g/cm^3) = 100 × average body weight/average body length^3; VSI (viscerosomatic index, %) = 100 × visceral weight/whole body weight; HSI (hepatosomatic index, %) = 100 × liver weight/whole body weight; FBW: final body weight, $n = 8$; SGR (specific growth rate, %) = 100 × [ln (FBW/initial body weight)]/days, $n = 8$; FCR (feed conversion rate) = FI_{abs}/ [(final total weight − initial total weight)/days]; Where, FI_{abs} is the daily absolute feed intake; HL (hepatic liquid, %) = fat weight of liver/liver weight.

Table 4. Changes of MLD on plasma liver function parameters of largemouth bass (means ± SEM).

Items	Normal	MLD
TP (g/L)	16.49 ± 0.47	15.69 ± 0.50
GLU (mmol/L)	5.71 ± 0.35 ^{b}	4.25 ± 0.45 ^{a}
TG (mmol/L)	5.64 ± 1.18	5.48 ± 0.45
TC (mmol/L)	7.70 ± 0.79	8.16 ± 0.62
HDL-C (mmol/L)	1.75 ± 0.38	1.64 ± 0.32
LDL-C (mmol/L)	2.00 ± 0.20 ^{b}	2.24 ± 0.13 ^{a}
AKP (U/L)	49.10 ± 5.56 ^{b}	77.34 ± 5.13 ^{a}
AST (U/L)	5.87 ± 1.05 ^{b}	11.15 ± 1.88 ^{a}
ALT (U/L)	5.87 ± 1.04 ^{b}	12.65 ± 1.97 ^{a}
TBA (umol/L)	74.41 ± 1.12 ^{b}	78.11 ± 4.49 ^{a}

$^{a, b}$ Within the same column, values with different superscripts are significantly different ($p < 0.05$). The same as below; TP, total protein; GLU, glucose; TG, triglyceride; TC, total cholesterol; HDL-C, high-density lipoprotein cholesterol; LDL-C, low-density lipoprotein cholesterol; AKP, alkaline phosphatase; AST, aspartate aminotransferase; ALT, aminotransferase; TBA, total bile acid. Within the same row, values with different superscripts are significantly different.

Table 5. Effects of MLD on hepatic lipid metabolism of largemouth bass (means ± SEM).

Items	Normal	MLD
TG (mmol/g·prot)	0.17 ± 0.01 ^{b}	0.21 ± 0.05 ^{a}
TC (mmol/g·prot)	0.15 ± 0.02	0.15 ± 0.01
TBA (umol/mg·prot)	2.33 ± 0.28	2.33 ± 0.65
LDL-C (umol/g·prot)	30.03 ± 2.92 ^{b}	43.92 ± 4.89 ^{a}
LDL-C/TC	0.22 ± 0.03	0.29 ± 0.03

$^{a, b}$ Within the same column, values with different superscripts are significantly different ($p < 0.05$).

Table 6. Effects of MLD on hepatic antioxidant responses of largemouth bass (means ± SEM).

Items	Normal	MLD
ROS (U/mg·prot)	66.78 ± 4.92 ^{b}	88.14 ± 4.85 ^{a}
T-AOC (umol/g·prot)	76.78 ± 5.96	89.14 ± 9.02
CAT (U/mg·prot)	46.62 ± 2.07 ^{a}	31.13 ± 3.42 ^{b}
GSH-Px (U/ug·prot)	4.07 ± 0.51	4.38 ± 0.57
SOD (U/mg·prot)	182.40 ± 9.45	168.75 ± 19.81
MDA (nmol/mg·prot)	2.69 ± 0.71	3.00 ± 0.38

$^{a, b}$ Within the same column, values with different superscripts are significantly different ($p < 0.05$).

3.2. Proteomics Analysis

Liver proteome profiles of largemouth bass with fatty liver ($n = 8$) and normal liver were compared ($n = 8$) (Table S1). This revealed an obvious shift in the proteome composition, reflected by 99 significantly differentially abundant proteins (Fold change >1.3 or <0.77), of which 38 proteins were up-regulated and 61 down-regulated (Figure 2A).

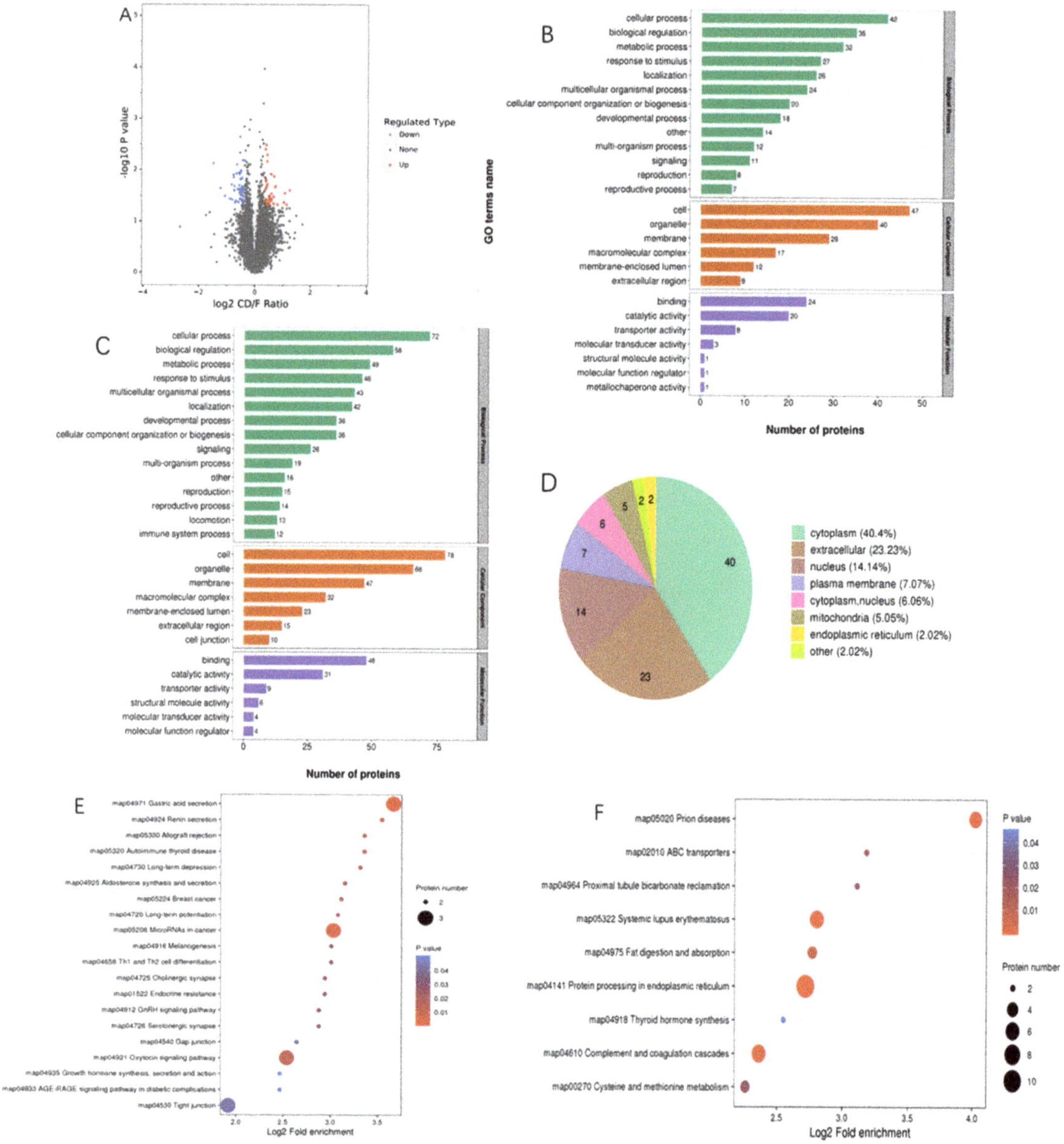

Figure 2. Significantly different expressions between MLD and Normal groups. (**A**) Protein significantly different expression between MLD and Normal groups (red means up-regulated, blue means down-regulated). (**B**,**C**) Differentially expressed proteins in GO functional classification (**B**: up; **C**: down). (**D**) The subcellular localization of differentially expressed proteins. (**E**,**F**) KEGG enrichment pathways of different expressed proteins (**E**: up-regulated; **F**: down-regulated).

To investigate the functions of the identified and quantified proteins, we annotated their functions and features, gene ontology (GO), subcellular localization, GO-based enrichment, Kyoto Encyclopedia of Genes and Genomes (KEGG) pathway analysis, protein domain and involvement in a protein complex of the characteristics we annotated [26].

For the biological process category, the differential proteins between the Normal and MLD groups were related to cellular processes, biological regulation processes, metabolic processes, response to stimulus processes, multicellular organismal processes, localization processes and development processes. For the cellular component category, these differential proteins were mainly enriched in the organelle, membrane, macromolecular complex and membrane-enclosed lumen. For the molecular function category, fatty liver induced by high-starch diet was primarily enriched in functional clusters, such as catalytic activity, transporter activity and structural molecule activity. (Figure 2B,C).

Subcellular structure localizations of differentially expressed proteins were predicted and classified, the results showed that most of these differential proteins were located in the nucleus, cell membrane, cytoplasmic cells and external matrix, and a few were located in mitochondria and endoplasmic reticulum. Forty were located in the cytoplasm of the 99 differential expressed proteins, with 23 in the extracellular matrix, 14 in the nucleus, 7 in the cell membrane, 6 in the cytoplasm and nucleus, 5 in the mitochondria, 2 in the endoplasmic reticulum and 2 in other cellular structures (Figure 2D).

For the KEGG pathway analysis, differential proteins were primarily associated with enrichment in pathways related to the fat digestion and absorption, gastric acid secretion, ABC transport, cholesterol metabolism, glycolysis/gluconeogenesis, PPAR and MAPK (Figure 2E,F). The differential pathways that may be related to fatty liver metabolism and the important differentially expressed proteins in these pathways are listed in Table 7.

Table 7. Selected KEGG pathway analysis of differentially expressed proteins related to metabolic fatty liver.

KEGG Pathway	Related Proteins (Up-Regulated)	Related Proteins (Down-Regulated)
Protein processing in endoplasmic reticulum	RRBP1	TRAPα, PDIA4
Fat digestion and absorption	FABP1	ABCA1, MTTP
ABC transporters		ABCA1
Cholesterol metabolism		ABCA1, VDAC1, AK1R1D1
PPAR signal pathway	HRAs, FABP1	PEPCK
FoxO signaling pathway;	HRAS	PEPCK
mTOR signaling pathway	HRAs	
Glycolysis/Gluconeogenesis		PEPCK
Phosphatidylinositol signalingsystem		PI4Kβ
Metabolic pathways	FA-CoA, UGT, lipocalin	PEPCK, CBS, L2HGDH, B3GNT3, PI4Kβ, CYP2U1, AK1R1D1
Insulin signaling pathway	HRAs, GNAQ	PEPCK
Thyroid hormone synthesis	GNAQ	PDIA4
Primary bile acids synthesis		AK1R1D1
Themogenesis	HRAs	NPR-A

3.3. Lipidomics Analysis

For liver lipid analysis, 15 lipid species including phosphatidylcholine (PC), triacylglycerol (TG), phosphatidylethanolamine (DPPE), lysophosphatidyl choline (LPC), ceramide (Cer), sphingomyelin (SM), lysophosphatidylethanolamine (LPE), diacylglycerol (DG), free fatty acid (FFA), phosphatidylserine (PS), acyl carnitine (CAR), cholesteryl ester (CE), phosphatidylinositol (PI), phosphatidyl glycerol (PG) and lipid peroxidn (LPO) were detected with lipidomics analysis (74 PCs, 69 TGs, 55 DPPEs, 37 LPCs, 36 Cers, 32 SMs, 21 LPEs, 17 DGs, 16 FFAs, 15 PSs, 14 CARs, 14 CEs, 13 PIs, 10 PGs, 10 LPOs). The major lipid compositions were analyzed (Table 8 and Figure 3).

PCA analysis showed discrimination between the MLD group and the Normal group (Figure 4A), indicating that there was difference in the lipid composition between the two groups. OPLS-DA analysis was applied to screen out the differential metabolites between the two groups. Using the criteria of VIP $\geq$ 1 and FC $\geq$ 1.5 or $\leq$ 0.67, 164 significantly differential lipid species were identified in the Normal versus the MLD group (Figure 4B). As shown in the heatmap (Figure 4C), the differential lipid species were mainly enriched in TC, PC, DPPE, Cer, SM, PI and DG classes, and all of them were upregulated in the MLD group. Notably, the top 20 differential upregulated lipids were mainly TGs, including TG (51:0), TG (51:1), TG (54:1), TG (50:0), TG (52:0), TG (49:1) and TG (54:0). However, the three downregulated lipids were mainly LPC (20:2), Cer (d34:1) and PE (P-34:2) (Table 9).

Table 8. Major lipid composition in the liver of largemouth bass.

Classification	Subclass	Composition
Fatty acyl	Free fatty acid	FFA(18:1), FFA(16:0), FFA(18:2), FFA(18:0), FFA(16:1), FFA(20:1), FFA(14:0), FFA(22:1), FFA(20:2), FFA(24:1)
	Acyl carnitine	CAR18:1, CAR16:1, CAR16, CAR18:2, CAR18, CAR14, CAR12-OH, CAR14:1, CAR18:1-OH, CAR12
Glyveride	Diacylglycerol	DG(34:1), DG(36:2), DG(36:3), DG(34:4), DG(34:0), DG(40:7), DG(32:1), DG(34:2), DG(32:0), DG(36:0)
	Triacylglycerol	TG(52:2), TG(54:3), TG(52:3), TG(54:5), TG(50:2), TG(54:2), TG(52:4), TG(50:3), TG(58:8), TG(50:1)
Glyceryl phosphatide	Lysophosphatidyl choline	LPC(22:6), LPC(18:2), LPC(16:0), LPC(18:1), LPC(18:0), LPC(20:5), LPC(16:1), LPC(20:3), LPC(22:5), LPC(18:3)
	Lysophosphatidyl ethanolamine	PE(40:6), PE(40:7), PE(38:6), PE(36:2), PE(38:4), PE(36:3), PE(42:7), PE(34:2), PE(P-38:5), PE(40:8), LPE(22:6), LPE(18:0), LPE(18:1), LPE(16:0), LPE(18:2), LPE(20:5), LPE(20:4), LPE(22:5), LPE(20:1), LPE(20:0)
	Phosphatidyl choline	PC(34:1), PC(34:2), PC(38:6), PC(36:3), PC(36:1), PC(32:1), PC(40:6), PC(O-40:7), PC(36:2), PC(O-38:6)
	Phosphatidyl glycerol	PG(40:8), PG(36:2), PG(34:1), PG(36:3), PG(40:7), PG(36:1), PG(36:4), PG(38:3), PG(34:2), PG(34:3)
	Phosphatidylinositol	PI(38:4), PI(38:5), PI(40:6), PI(38:6), PI(36:2), PI(40:7), PI(38:3), PI(36:1), PI(36:3), PI(40:5)
	Phosphatidylserine	PS(40:6), PS(38:6), PS(40:7), PS(36:2), PS(36:1), PS(40:5), PS(38:4), PS(38:5), PS(38:3), PS(34:1)
Sphingolipid	Sphingomyelin	SM(d42:2), SM(d34:1), SM(d42:1), SM(d40:1), SM(d36:1), SM(d32:1), SM(d41:1), SM(d42:3), SM(d42:5), SM(d35:0)
	Ceramide	Cer(d42:2), Cer(d42:1), Cer(d34:1), Cer(d40:1), HerCer(d34:1), Cer(d32:1), Cer(d41:1), Cer(d36:1), CerP(d36:1), Cer(d44:2)
Cholesterol	Cholesteryl ester	CE(22:6), CE(18:0), CE(18:1), CE(18:2), CE(20:5), CE(22:5), CE(20:4), CE(16:0), CE(22:3), CE(20:2), CE(20:1)

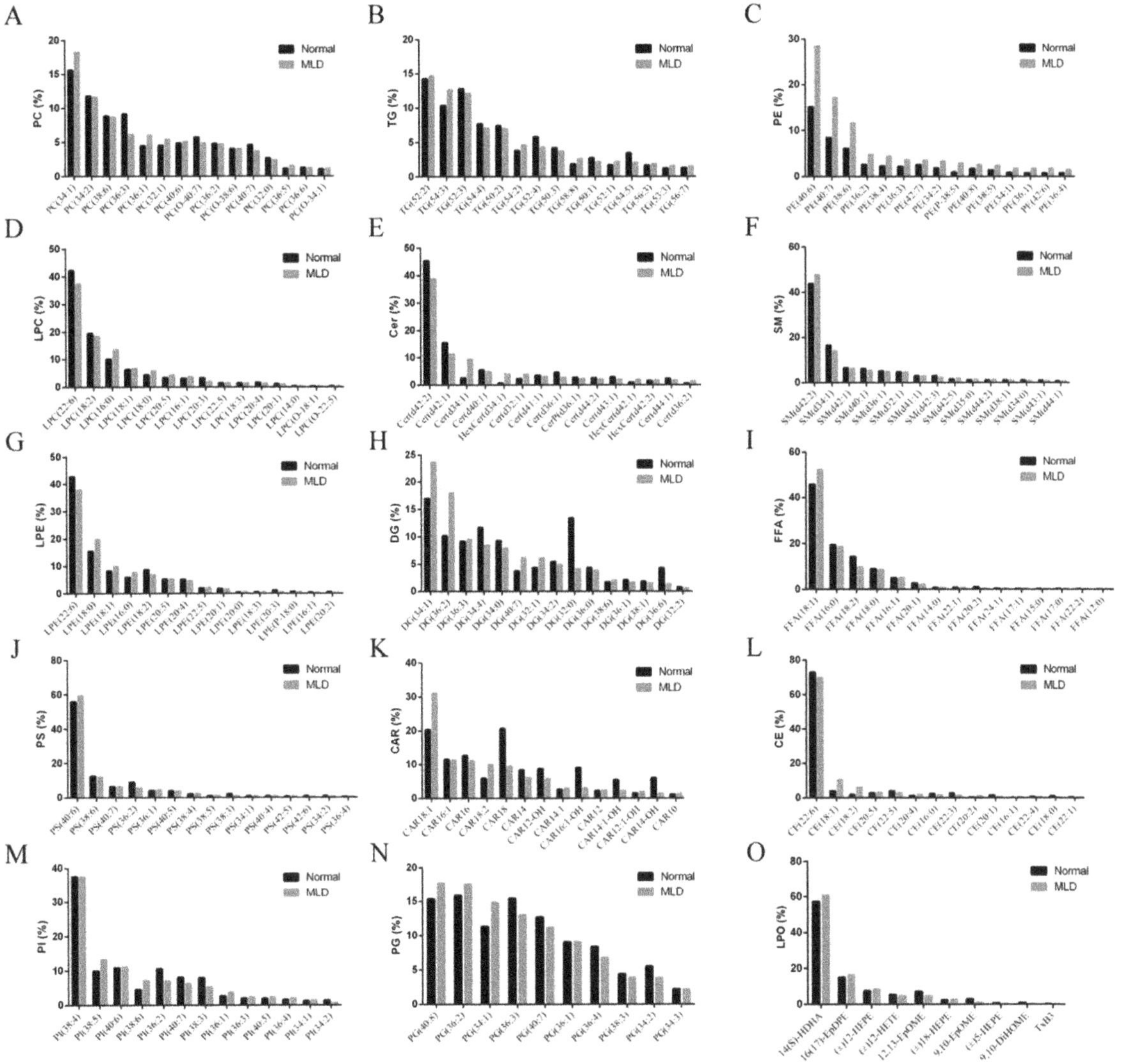

Figure 3. Major lipid in the liver of largemouth bass (Mean ± SEM, $n = 8$). (**A**) PC: phosphatidylcholine. (**B**) TG: triacylglycerol. (**C**) PE: phosphatidylethanolamine. (**D**) LPC: lysophosphatidyl choline. (**E**) Cer: ceramide. (**F**) SM: sphingomyelin. (**G**) LPE: lysophosphatidylethanolamine. (**H**) DG: diacylglycerol. (**I**) FFA: free fatty acid. (**J**) PS: phosphatidylserine. (**K**) CAR: acyl carnitine. (**L**) CE: cholesteryl ester. (**M**) PI: phosphatidylinositol. (**N**) PG: phosphatidyl glycerol. (**O**) LPO: lipid peroxidation.

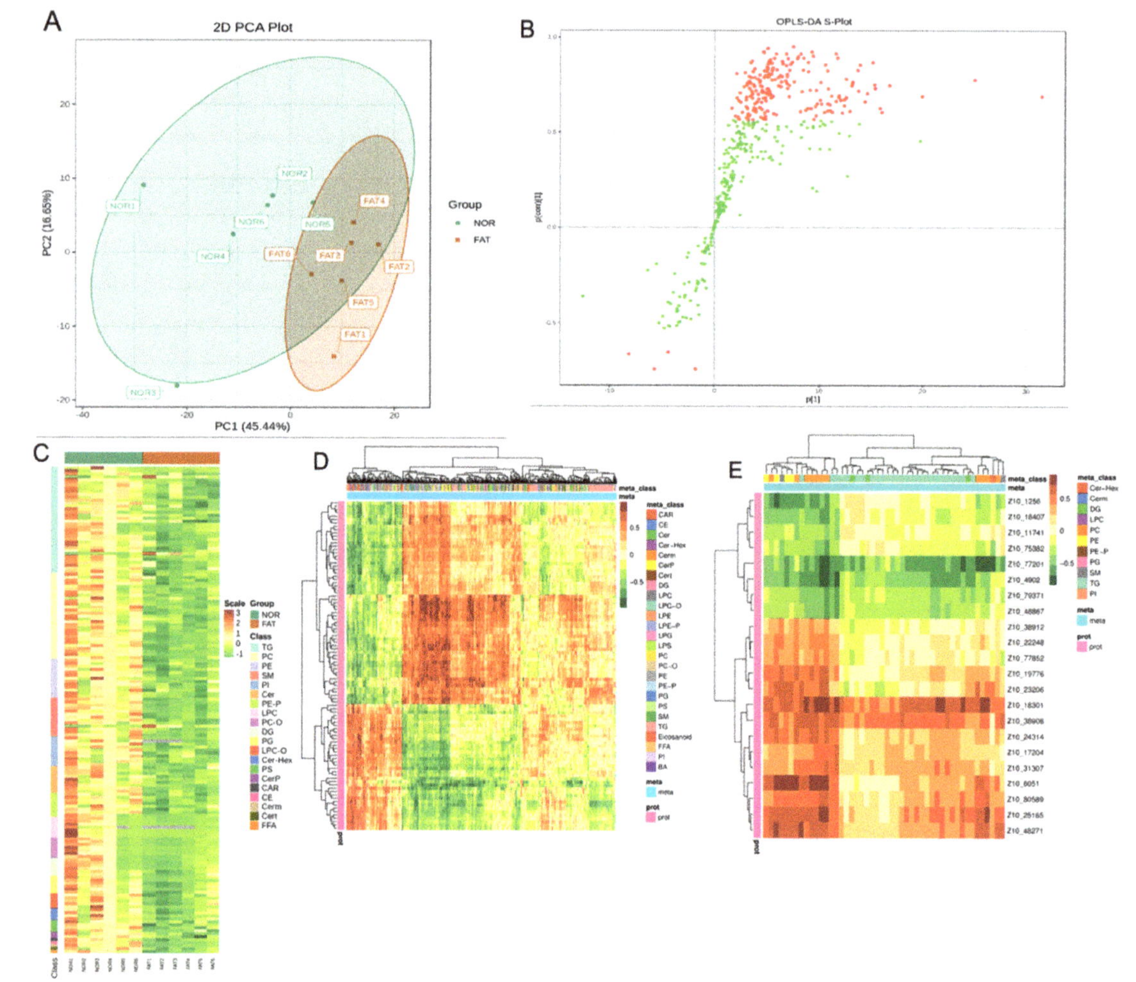

Figure 4. The lipidomics analysis of largemouth bass liver. (**A**) Principal component analysis (PCA) score plots. (**B**) OPLS-DA analysis (red points and green points represent VIP $\geq$ 1 and VIP $\leq$ 1, respectively) and (**C**) heatmap. (Mean $\pm$ SEM, n = 8). (**D**) Heatmap of all lipids and proteins (red represents high correlation coefficient; green represents low correlation coefficient). (**E**) Heatmap of correlation coefficient of differential lipids and differential proteins (red represents a positive correlation between proteins and lipids and green represents a negative correlation). Each row and column represents a protein and a metabolite, respectively.

A total of 29 differential metabolic pathways were found with KEGG pathway enrichment analysis. These metabolic pathways mainly involved fat digestion and absorption, cholesterol metabolism, vitamin digestion and absorption, sphingolipid metabolism and insulin resistance, all of which are related to lipid metabolism (Table 10).

Table 9. Top 20 differential up-regulated and down-regulated metabolites.

Metabolites	Class	VIP	*p*-Value	Log2FC	Type
TG (51:0)	TGs	1.22	N/A	4.50	up
TG (51:1)	TGs	1.22	0.09	3.44	up
TG (54:1)	TGs	1.11	0.09	2.93	up
TG (50:0)	TGs	1.25	0.07	2.79	up
TG (52:0)	TGs	1.19	0.13	2.64	up
TG (49:1)	TGs	1.29	0.03	2.62	up
TG (54:0)	TGs	1.09	0.16	2.53	up
TG (52:1)	TGs	1.17	0.06	2.49	up
TG (48:0)	TGs	1.32	0.03	2.43	up
TG (46:0)	TGs	1.34	0.02	2.39	up
TG (56:0)	TGs	1.09	0.12	2.37	up
TG (52:7)	TGs	1.52	0.02	2.33	up
TG (53:2)	TGs	1.07	0.06	2.26	up
TG (58:7)	TGs	1.32	0.04	2.21	up
TG (44:0)	TGs	1.18	0.03	2.14	up
TG (56:9)	TGs	1.21	0.07	2.02	up
TG (58:10)	TGs	1.14	0.05	1.92	up
DG (38:6)	DGs	1.33	0.00	2.03	up
PI (38:5)	PIs	1.54	0.00	2.25	up
PE (P-40:5)	PEs	1.47	0.02	2.37	up
LPC (20:2)	LPCs	1.32	0.03	0.96	down
Cer (d34:1)	Cers	1.16	0.05	0.72	down
PE (P-34:2)	PEs	1.17	0.05	1.38	down

Table 10. Differential metabolic pathways.

Pathways	ko_ID	Unique Compound
Metabolic pathways	ko01100	99
Insulin resistance	ko04931	36
Sphingolipid signaling pathway	ko04071	21
Fat digestion and absorption	ko04975	29
Cholesterol metabolism	ko04979	29
Glycerophospholipid metabolism	ko00564	50
Vitamin digestion and absorption	ko04977	29
Regulation of lipolysis in adipocytes	ko04923	28
Necroptosis	ko04217	21
Neurotrophin signaling pathway	ko04722	8
Adipocytokine signaling pathway	ko04920	8
Sphingolipid metabolism	ko00600	25
AGE-RAGE signaling pathway in diabetic complications	ko04933	8
Leishmaniasis	ko05140	8
Glycerolipid metabolism	ko00561	33
Inositol phosphate metabolism	ko00562	5
Phosphatidylinositol signaling system	ko04070	5
Long-term depression	ko04730	5
Choline metabolism in cancer	ko05231	32
Arachidonic acid metabolism	ko00590	27
Linoleic acid metabolism	ko00591	27
alpha-Linolenic acid metabolism	ko00592	27
Retrograde endocannabinoid signaling	ko04723	40
Glycosylphosphatidylinositol (GPI)-anchor biosynthesis	ko00563	13
Autophagy—other	ko04136	13
Autophagy	ko04140	13
Pathogenic Escherichia coli infection	ko05130	13
Kaposi sarcoma-associated herpesvirus infection	ko05167	13
Thermogenesis	ko04714	28

3.4. Integrative Proteomic and Lipidomic Analysis

To best observe the correlation between lipid metabolites and proteins, all the lipids and proteins were analyzed by clustering analysis as illustrated in a heatmap (Figure 4D). The result indicated that there are correlations between many lipids and proteins. All the differential proteins and differential lipids were selected, and the heatmap clustering was drawn. As shown in Figure 4E, these differential proteins were negatively correlated with differential lipids, including aldo-keto reductase family 1 member D1 (AK1R1D1), phosphoenolpyruvate carboxykinase (PEPCK), ATP binding cassette transporter A1 (ABCA1), beta-1,3-N-acetylglucosaminyltransferase 3 (B3GNT3), L-2-hydroxyglutarate dehydrogenase (L2HGDH), cytochrome P450 2U1 (CYP2U1), mercaptopyruvate transferase, phosphatidylinositol 4-kinase beta (PI4K2β), hexose transferase, cystathionine beta-synthase (CBS) and voltage-dependent anion channels1 (VDAC1). These differential proteins were positively correlated with differential lipids, namely hexose tetraphosphate kinase, natriuretic peptide receptor-A (NPR-A), fatty acid binding protein 1 (FABP1), UDP-glucuronosyltransferase (UGT), fatty acid acetyl-CoA synthase (FA-CoA) and Guanine nucleotide-binding protein G(q) subunit alpha (GNAQ). Among these differential lipid metabolites, we observed that the strongest relationships with differential proteins were triglyceride (TG) and phosphatidylcholine (PC).

The above correlated differential lipids and differential proteins were simultaneously mapped to the KEGG pathway, and were enriched into nine common metabolic pathways, as shown in Table 11. Four KEGG pathways with the highest concentration were selected for subsequent analysis, including thermogenesis (ko04714), fat digestion and absorption (ko04975), cholesterol metabolism (ko04979) and metabolic pathways (ko01100).

Table 11. Nine KEGG enrichment pathways of lipidomics and proteomics.

Pathway	Number of Lipids	Number of Proteins	Differential Lipids	Differential Proteins (Up-Regulated)	Differential Proteins (Down-Regulated)
Themogenesis	27	2	TG	NPR-A, HRAs	
Fat digestion and absorption	27	2	TG	FABP1	ABCA1
Cholesterol metabolism	27	2	TG		ABCA1, VDCA1
Metabolic pathways	48	17	TG	NPR-A, FA-CoA, UGT	PI4Kβ, AK1R1D1, PEPCK, L2HGDH, CBS
Arachidonic acid metabolism	11	2	TG		CYP2U1
Inositol phosphate metabolism	3	2	TG	ITPK1	PI4Kβ
Phosphatidylinositol signaling system	3	2	TG	ITPK1	PI4Kβ
Long-term depression	3	2	TG	GNAQ, HRAs	
Sphingolipid signaling pathway	2	2	SM	GNAQ, HRAs	

4. Discussion

Carnivorous fish have a low starch utilization rate, and high digestible starch intake leads to accumulation of liver glycogen and persistent high blood sugar, finally causing fish liver disease [28,29]. Generally, the level of digestible carbohydrate above 10% reduced the growth performance of largemouth [9,15,30,31], induced glycogen and lipid accumulation and dysfunction of antioxidant capabilities, leading to MLD in largemouth [9,32]. In this study, compared with the Normal group (starch content: 10.8%), the MLD group (starch content: 16.2%) decreased the growth performance of largemouth bass and induced liver lipid accumulation, liver function injury, oxidative stress and higher hepatosomatic index, which is similar to the non-alcoholic fatty liver symptom. Therefore, we used HSD to construct the fatty liver phenotype in largemouth bass successfully.

4.1. Proteomic Analysis

The differentially expressed proteins induced by a high-starch diet of largemouth bass were mainly involved in endoplasmic reticulum protein processing, fat digestion and absorption, cholesterol metabolism, phosphatidylinositol signal system and insulin signal transduction, which affected the occurrence and development of fatty liver through the comprehensive action of multiple metabolic pathways.

Endoplasmic reticulum (ER) is an important site involved in the regulation of substance transport, metabolism and protein synthesis in eukaryotic cells [33]; ER stress caused by various reasons is closely related to the occurrence and development of diseases [34]. The evidence implied that ER played a role in the development of steatosis and nonalcoholic steatohepatitis. ER stress occurred in liver and adipose tissue in patients with non-alcoholic fatty liver disease [35,36]. RRBP1 (reticulum ribosome-binding protein 1) is one of the important proteins involved in the ER unfolded protein reaction, and abnormal up-regulated expression of RRBP1 has been found in the study of human cancer-related diseases [37–39]. Xiong found that the expression of RRBP1 in liver cancer tissues was significantly higher than that in normal liver tissues, and the expression of RRBP1 in liver cancer tissues with metastasis was significantly higher than that in liver cancer tissues without metastasis, indicating that RRBP1 is closely related to liver cancer metastasis in the state of liver injury [40]. In this study, RRBP1 was found to be highly expressed in liver in the MLD group, indicating that it may have an important relationship with liver injury and the development of MLD. In addition, TRAPα (translocon-associated protein subunit alpha) is a typical glycosylated membrane protein which plays an important role in the process of signal modification recognition and the transport of new peptide chains. It synergizes with AMPK (adenosine monophosphate activated protein kinase) and ROS pathway to cause resistance to oxidative stress-induced apoptosis [41]. TRAPα also plays an important role in maintaining cell homeostasis and may become a target drug for intervention in related diseases [42]. The low expression of TRAPα may be due to the abnormality in the modification and transport of the peptide chain, resulting in liver damage, which affects the normal progress of subsequent life activities and leads to the occurrence and development of fatty liver. At present, the TRAPα sequence of fish is relatively conserved, but the research on its related functions is still in the blank stage and needs to be further studied.

The pathway of fat digestion and absorption has the important function of maintaining the homeostasis of lipid metabolism. In this study, we found the differential expression of three important proteins FABP1, MTTP (microsomal triglyceride transfer protein) and ABCA1 in its pathway. FABP1 is expressed mainly in the liver and could combine with long-chain fatty acid to regulate lipid absorption and fatty acid metabolism in the cytoplasm [43]. FABP1 activity changed significantly in the occurrence and development of fatty liver, liver cirrhosis, liver cancer and other liver diseases [44–46]. In this study, HSD induced fatty liver of largemouth bass, leading to significant upregulation of FABP1. Similar results also showed that FABP1 levels were higher in patients with non-alcoholic fatty liver disease [43,47]. Therefore, FABP1 is expected to become a diagnostic marker of liver injury [48–50]. The MTTP is a key protein for lipid excretion in liver and affects the metabolism of lipids and lipoproteins, leading to the occurrence and development of many diseases [51–54]. Specific knockout of the MTTP gene in hepatocytes leads to the accumulation of a large amount lipids in mouse hepatocytes, and eventually results in hepatic steatosis and fatty liver [55]. Some evidence showed that the variability of MTTP is associated with the development of NAFLD, raised lipid and risk of atherosclerotic cardiovascular disease [56,57]. A previous study indicated that in the NAFLD of rat induced by a high-fat diet, the expression level of MTTP is markedly decreased [58]. Similarly, HSD-induced fatty liver of largemouth bass decreased MTTP level in liver in our results. ABCA1 played an essential role in the regulation of high-density lipoproteins (HDLs) and reverse cholesterol transport [59]. ABCA1 overexpression increased the concentration of cholesterol and high-density lipoprotein and decreased hepatic lipid contents [60,61]. The deposition of cholesterol in tissues induced the absence of ABCA1 [62,63]. A previous study reported that decreased hepatic ABCA1 expression can cause steatohepatitis in obese adult patients [61]. It also presented the evidence that the expression level of ABCA1 was significantly lower in the NAFLD patients than in the healthy controls [64]. Consistent with these observations,

ABCA1 was down-regulated in the MLD group induced by HSD. These findings suggest that the expression level of ABCA1 was associated with the development of fatty liver disease. In this study, the differential expression of FABP1, ABCA1 and MTTP in the liver of largemouth bass in the MLD group showed a high-starch diet increased hepatic lipid contents and affected fat digestion and absorption pathways, and eventually promoted the development of metabolic fatty liver.

4.2. Lipidomic Analysis

Significant changes in hepatic lipids are important pathophysiological markers of fatty liver [65]. Lipidomic studies of the mouse liver have shown that NAFLD is associated with an increase in TG [66]. Similar results were observed in our HSD treatment largemouth bass. In this study, most TG species were increased by HSD. During the occurrence of fatty liver, the up-regulation of TGs is attributed to the rate of TG synthesis in the liver exceeding the catabolic rate of TGs [67,68]. Therefore, preventing TG accumulation in liver may contribute to the attenuating effect of fatty liver disease [66,69].

KEGG pathway enrichment analysis indicated that the insulin-resistance pathway was significantly enriched. Many studies believe that NAFLD is closely related to insulin resistance [70–72]. The relationship between NAFLD and insulin resistance is bidirectional. Insulin resistance may promote the development of NAFLD, while NAFLD may promote the development of insulin resistance [71,73]. The related research results of our team also found that the persistent hyperglycemia of largemouth bass after a meal is caused by insufficient insulin secretion (unpublished), which is also the reason for the starch intolerance of largemouth bass. Therefore, insulin resistance is a key factor in the development and progression of NAFLD [74] and may be the direction of targeted intervention of MLD in largemouth bass in the future.

Interestingly, in this study we also found that differential lipids were mainly enriched in the sphingolipid metabolism signaling pathway. Liver is the center of ceramide production and usually contains much higher sphingolipid than all other tissues, especially ceramide and sphingomyelin [75–77]. Therefore, liver is prone to sphingolipid toxicity [78]. For example, sphingolipid content was significantly increased in the liver of rats fed with a high-fat diet [79]. Mice exposure to exogenous microbiome sphingolipids no longer had lipid accumulation and rescued diet-induced hepatic steatosis [80]. Evidence suggested sphingolipid levels contribute to the development of NAFLD in a variety of ways, including inflammation, insulin resistance and oxidative stress [78].

4.3. Integrative Proteomic and Lipidomic Analysis

In this study, using integrative proteomic and lipidomic analysis, we demonstrated the hepatic TG accumulation was important for the development progress of fatty liver induced by HSD. Theoretically, the primary cause of lipid accumulation is the increased uptake of free fatty acids (FFAs) by the liver [56,81]. Undoubtedly, regulation of TG accumulation is important for improving fatty liver disease [82]. Many studies have shown that overexpression of FABP1 enhances fatty acid uptake [83–85]. Researchers also found FABP1-deficient mice were protected from TG accumulation in liver induced by a high-fat diet [86,87]. It is expected that reducing the expression or function of FABP1 could inhibit TG accumulation in liver and ameliorate NAFLD [85]. In this study, we similarly observed HSD increased the expression level of FABP1, resulting in more TG accumulation in liver, ultimately leading to fatty liver in largemouth bass.

Meanwhile, accumulated evidence suggested cholesterol homeostasis and hepatic cholesterol accumulation associated with the risk of liver cirrhosis and other liver diseases [88–90]. ABCA1 could prevent cholesterol accumulation by transporting cholesterol and phospholipids from cells to apolipoproteins [91–93]. Recently, a report showed that ABCA1 were tightly associated with the lipid variables and lipid metabolism [94]. Human studies also suggest the elevation of TG levels usually develop fatty liver disease, and there is an inverse association between TG concentrations and dysfunctional ABCA1 [95].

In addition, a study also reported when the function of ABCA1 is diminished, the TG secretion from liver increased, suggesting there is a strong association between ABCA1 and fatty liver disease [59]. Consistent with this finding, this study also showed the content of ABCA1 was significantly reduced, which exacerbated the accumulation of cholesterol in the liver and worsened metabolic fatty liver disease.

Furthermore, VDAC1 was involved in cholesterol transport and was thought to be part of a complex process that mediates fatty acid transport [96–98]. A lack of VDAC1 caused mitochondria to stop oxidizing fatty acids [99]. It is also indicated that the absence of ABCA1 lead to steatosis in metabolic liver disease [100,101]. Recent study had found hepatocyte fat droplet and glycogen accumulation in high-fat diet-treated mice. However, the VDAC1-based peptide decreased serum TG content and hepatocyte fat droplets and increased hepatocyte glycogen accumulation [98]. Moreover, VDAC1 could interact with many proteins, such as metabolism, apoptosis and anti-oxidation-related proteins [102,103], so it could be regarded as a potential therapeutic for metabolic liver disease, such as NAFLD [98]. Here, we demonstrated that the reduction in VDAC1 expression led to cholesterol and lipid accumulation in liver of largemouth bass induced by HSD. Importantly, in addition to hepatic TG accumulation, abnormal transport of cholesterol also plays a critical role in the occurrence and development of fatty liver disease.

5. Conclusions

We successfully induced metabolic fatty liver phenotype of largemouth bass with high-starch diet. We found that three differential proteins (FABP1, ABCA1 and VDAC1) in fat digestion and absorption and cholesterol metabolism pathways played important roles in the occurrence and development of MLD in largemouth bass after the combination of targeted lipidomics and proteomics analyses. The three differential proteins may be promising targets for therapeutic intervention in fatty liver disease. Our findings propose the possible key regulatory factors in the occurrence and development of metabolic fatty liver, which provide a theoretical basis for the efficient use of compound feed and targeted nutritional regulation in commercial aquaculture.

Supplementary Materials: The following supporting information can be downloaded at: https://www.mdpi.com/article/10.3390/metabo12080759/s1, Table S1: Protein accession.

Author Contributions: Conceptualization, M.X. (Min Xue); Data curation, T.Y. and M.J.; Project administration, Y.Q.; Supervision, X.G.; Visualization, F.F. and J.L.; Writing—review & editing, M.X. (Moyong Xue). All authors have read and agreed to the published version of the manuscript.

Funding: This study was supported by the National Key Research and Development Program of China (2018YFD0900400 and 2019YFD0900200); The Agricultural Science and Technology Innovation Program of CAAS, China (CAAS-ASTIP-2017-FRI-08).

Institutional Review Board Statement: During our whole research process, all handling of largemouth bass was conducted in strict accordance with the Guidelines for the Experimental Animal Ethics Committee, Institute of Feed Research, Chinese Academy of Agricultural Sciences. Approval Code: FRI-CAAS-20200506; Approval Date: 7 May 2020.

Informed Consent Statement: Not applicable.

Data Availability Statement: Data available on request.

Conflicts of Interest: The authors declare no competing financial interest.

References

1. Sun, C.F.; Li, J.; Dong, J.J.; Niu, Y.C.; Hu, J.; Lian, J.M.; Li, W.H.; Li, J.; Tian, Y.Y.; Shi, Q.; et al. Chromosome-level genome assembly for the largemouth bass *Micropterus salmoides* provides insights into adaptation to fresh and brackish water. *Mol. Ecol. Resour.* **2020**, *21*, 301–315. [CrossRef] [PubMed]
2. Li, X.Y.; Zheng, S.X.; Ma, X.K.; Cheng, K.M.; Wu, G.Y. Effects of dietary protein and lipid levels on the growth performance, feed utilization, and liver histology of largemouth bass (*Micropterus salmoides*). *Amino Acids* **2020**, *52*, 1043–1061. [CrossRef] [PubMed]

3. Zhou, Y.; Liu, C. Largemouth bass Micropterus salmoides production in China. In *Largemouth Bass Aquaculture*; Tidwell, J.H., Coyle, S.D., Bright, L.A., Eds.; 5M Publishing Ltd., Benchmark House: Sheffield, UK, 2019.
4. Bai, J.; Li, S. *Development of Largemouth Bass (Micropterus salmoides) Culture*; John Wiley & Sons, Ltd.: Hoboken, NJ, USA, 2018.
5. Yu, H.H.; Liang, X.F.; Chen, P.; Wu, X.F.; Zheng, Y.H.; Luo, L.; Qin, Y.C.; Long, X.C.; Xue, M. Dietary supplementation of Grobiotic®-A increases short-term inflammatory responses and improves long-term growth performance and liver health in largemouth bass (*Micropterus salmoides*). *Aquaculture* **2018**, *500*, 327–337. [CrossRef]
6. Chen, Y.F.; Sun, Z.Z.; Liang, Z.M.; Xie, Y.D.; Tan, X.H.; Su, J.L.; Luo, Q.L.; Zhu, J.Y.; Liu, Q.Y.; Wang, A.L. Addition of L-carnitine to formulated feed improved growth performance, antioxidant status and lipid metabolism of juvenile largemouth bass, *Micropterus salmoides*. *Aquaculture* **2019**, *518*, 734434. [CrossRef]
7. Yu, L.L.; Yu, H.H.; Liang, X.F.; Li, N.; Wang, X.; Li, F.H.; Wu, X.F.; Zheng, Y.H.; Xue, M.; Liang, X.F. Dietary butylated hydroxytoluene improves lipid metabolism, antioxidant and anti-apoptotic response of largemouth bass (*Micropterus salmoides*). *Fish Shellfish. Immunol.* **2018**, *72*, 220–229. [CrossRef]
8. Amoah, A.; Coyle, S.D.; Webster, C.D.; Durborow, R.M.; Bright, L.A.; Tidwell, J.H. Effects of graded levels of carbohydrate on growth and survival of largemouth bass, *Micropterus salmoides*. *J. World Aquac. Soc.* **2008**, *39*, 397–405. [CrossRef]
9. Lin, S.M.; Shi, C.M.; Mu, M.M.; Chen, Y.J.; Luo, L. Effect of high dietary starch levels on growth, hepatic glucose metabolism, oxidative status and immune response of juvenile largemouth bass, *Micropterus salmoides*. *Fish Shellfish. Immunol.* **2018**, *78*, 121–126. [CrossRef]
10. Asaoka, Y.; Terai, S.; Sakaida, I.; Nishina, H. The expanding role of fish models in understanding non-alcoholic fatty liver disease. *Dis. Models Mech.* **2013**, *6*, 905–914. [CrossRef]
11. Han, T.; Li, X.Y.; Wang, J.T.; Hu, S.X.; Jiang, Y.D.; Zhong, X.D. Effect of dietary lipid level on growth, feed utilization and body composition of juvenile giant croaker *Nibea japonica*. *Aquaculture* **2014**, *434*, 145–150. [CrossRef]
12. Shi, C.M.; Zhao, H.; Zhai, X.L.; Chen, Y.J.; Lin, S.M. Linseed oil can decrease liver fat deposition and improve antioxidant ability of juvenile largemouth bass, *Micropterus salmoides*. *Fish Physiol. Biochem.* **2019**, *45*, 1513–1521. [CrossRef]
13. Goodwin, A.E.; Lochmann, R.T.; Tieman, D.M.; Mitchell, A.J. Massive hepatic necrosis and nodular regeneration in largemouth bass fed diets high in available carbohydrate. *J. World Aquac. Soc.* **2010**, *33*, 466–477. [CrossRef]
14. Xu, X.T.; Chen, N.S.; Liu, Z.K.; Gou, S.P.; Jia, Y. Effects of dietary starch sources and levels on liver histology in largemouth bass, *Micropterus salmoide*. *J. Shanghai Ocean. Univ.* **2016**, *25*, 61–70.
15. Ma, H.J.; Mou, M.M.; Pu, D.C.; Lin, S.M.; Chen, Y.J.; Luo, L. Effect of dietary starch level on growth, metabolism enzyme and oxidative status of juvenile largemouth bass, *Micropterus salmoides*. *Aquaculture* **2019**, *498*, 482–487. [CrossRef]
16. Wang, Z.; Kavdia, K.; Dey, K.K.; Pagala, V.R.; Kodali, K.; Liu, D.T.; Lee, D.G.; Sun, H.; Chepyala, S.R.; Cho, J.H. High-throughput and deep-proteome profiling by 16-plex tandem mass tag labeling coupled with two-dimensional chromatography and mass spectrometry. *J. Vis. Exp.* **2020**, *162*, e61684. [CrossRef] [PubMed]
17. Liu, M.W.; Ge, R.; Liu, W.L.; Liu, Q.M.; Xia, X.; Lai, M.; Liang, L.Z.; Li, C.; Song, L.; Zhen, B.; et al. Differential proteomics profiling identifies LDPs and biological functions in high-fat diet-induced fatty livers. *J. Lipid Res.* **2017**, *58*, 681–694. [CrossRef] [PubMed]
18. Willians, E.G.; Wu, Y.B.; Jha, P.; Dubuis, S.; Blattmann, P.; Argmann, C.A.; Houten, S.M.; Amariuta, T.; Wolski, W.; Zamboni, N.; et al. Systems proteomics of liver mitochondria function. *Science* **2016**, *352*, aad0189. [CrossRef] [PubMed]
19. Niu, L.; Geyer, P.E.; Albrechtsen, N.J.W.; Gluud, L.L.; Santos, A.; Doll, S.; Treit, P.V.; Holst, J.J.; Knop, F.K.; Vilsbol, T.; et al. Plasma proteome profiling discovers novel proteins associated with non-alcoholic fatty liver disease. *Mol. Syst. Biol.* **2019**, *15*, e8793. [CrossRef]
20. Zhang, Y.; Zhan, C.; Chen, G.W.; Sun, J.Y. Labelfree quantitative proteomics and bioinformatics analyses of alcoholic liver disease in a chronic and binge mouse model. *Mol. Med. Rep.* **2018**, *18*, 2079–2087.
21. Assini, J.M.; Mulvihill, E.E.; Huff, M.W. Citrus flavonoids and lipid metabolism. *Curr. Opin. Lipidol.* **2013**, *24*, 34–40. [CrossRef]
22. Feng, K.L.; Lan, Y.Q.; Zhu, X.A.; Li, J.; Chen, T.; Huang, Q.R.; Ho, C.T.; Chen, Y.J.; Cao, Y. Hepatic Lipidomics analysis reveals the antiobesity and cholesterol-lowering effects of tangeretin in high-fat diet-fed rats. *J. Agric. Food Chem.* **2020**, *68*, 6142–6153. [CrossRef]
23. Yang, K.; Han, X. Lipidomics: Techniques, applications, and outcomes related to biomedical sciences. *Trends Biochem. Sci.* **2016**, *41*, 954–969. [CrossRef] [PubMed]
24. Han, X.L.; Gross, R.W. Global analyses of cellular lipidomes directly from crude extracts of biological samples by ESI mass spectrometry: A bridge to lipidomics. *J. Lipid Res.* **2003**, *44*, 1071–1079. [CrossRef] [PubMed]
25. Chen, H.; Wei, F.; Dong, X.Y.; Xiang, J.Q.; Quek, S.Y.; Wang, X.M. Lipidomics in food science. *Curr. Opin. Food Sci.* **2017**, *16*, 80–87. [CrossRef]
26. Li, Y.X.; Luo, Z.P.; Wu, X.L.; Zhu, J.; Yu, K.; Jin, Y.; Zhang, Z.W.; Zhao, S.H.; Zhou, L. Proteomic analyses of cysteine redox in high-fat-fed and fasted mouse livers: Implications for liver metabolic homeostasis. *J. Proteome Res.* **2017**, *14*, 129–140. [CrossRef]
27. Kwon, O.K.; Kim, S.J.; Lee, Y.M.; Lee, Y.H.; Bae, Y.S.; Kim, J.Y.; Peng, X.J.; Cheng, Z.Y.; Zhao, Y.M.; Lee, S. Global analysis of phosphoproteome dynamics in embryonic development of zebrafish (*Danio rerio*). *Proteomics* **2016**, *16*, 136–149. [CrossRef]
28. Enes, P.; Panserat, S.; Kaushik, S.; Oliva-Teles, A. Nutritional regulation of hepatic glucose metabolism in fish. *Fish Physiol. Biochem.* **2008**, *35*, 519–539. [CrossRef]
29. Hemre, G.I.; Mommsen, T.P.; Krogdahl, A. Carbohydrates in fish nutrition: Effects on growth, glucose metabolism and hepatic enzymes. *Aquac. Nutr.* **2002**, *8*, 175–194. [CrossRef]

30. Liang, X.F.; Chen, P.; Wu, X.L.; Xing, S.J.; Morais, S.; He, M.L.; Gu, X.; Xue, M. Effects of High Starch and Supplementation of an Olive Extract on the Growth Performance, Hepatic Antioxidant Capacity and Lipid Metabolism of Largemouth Bass. *Antioxidants* **2022**, *11*, 577. [CrossRef]
31. Zhang, Y.M.; Xie, S.W.; Wei, H.L.; Zheng, L.; Liu, Z.L.; Fang, H.H.; Xie, J.J.; Liao, S.Y.; Tian, L.X.; Liu, Y.J. High dietary starch impaired growth performance, liver histology and hepatic glucose metabolism of Juvenile largemouth bass, *Micropterus salmoides*. *Aquac. Nutr.* **2020**, *26*, 1083–1095. [CrossRef]
32. Yu, H.H.; Zhang, L.L.; Chen, P.; Liang, X.F.; Cao, A.Z.; Han, J.; Wu, X.F.; Zheng, Y.H.; Qin, Y.C.; Xue, M. Dietary Bile Acids Enhance Growth, and Alleviate Hepatic Fibrosis Induced by a High Starch Diet via AKT/FOXO1 and cAMP/AMPK/SREBP1 Pathway in *Micropterus salmoides*. *Front. Physiol.* **2019**, *10*, 1430. [CrossRef]
33. English, A.R.; Voeltz, G.K. Endoplasmic reticulum structure and interconnections with other organelles. *Cold Spring Harb. Perspect. Biol.* **2013**, *5*, a013227. [CrossRef] [PubMed]
34. Juan, R.C.; Sarah, E.B.; Laurie, H.G. Tumorigenic and Immunosuppressive Effects of Endoplasmic Reticulum Stress in Cancer. *Cell* **2017**, *168*, 692–706.
35. Michael, J. Endoplasmic reticulum stress in nonalcoholic fatty liver disease. *Annu. Rev. Nutr.* **2012**, *32*, 17–33.
36. Christopher, L.; Melinda, A.; Michael, J. Fatty acids and the endoplasmic reticulum in nonalcoholic fatty liver disease. *BioFactors* **2011**, *37*, 8–16.
37. Pan, Y.; Cao, F.; Guo, A.; Chang, W.; Ma, W.; Gao, X.; Guo, S.; Fu, C.; Zhu, J. Endoplasmic reticulum ribosome-binding protein 1, RRBP1, promotes progression of colorectal cancer and predicts an unfavourable prognosis. *Br. J. Cancer* **2015**, *113*, 763–772. [CrossRef] [PubMed]
38. Tsai, H.Y.; Yang, Y.F.; Wu, A.T.; Yang, C.J.; Liu, Y.P.; Jan, Y.H.; Lee, C.H.; Hsiao, Y.W.; Yeh, C.T.; Shen, C.N. Endoplasmic reticulum ribosome-binding protein1 (RRBP1) overexpression is frequently found in lung cancer patients and alleviates intracellular stress-induced apoptosis through the enhancement of GRP78. *Oncogene* **2013**, *32*, 4921–4931. [CrossRef]
39. Liang, X.S.; Sun, S.S.; Zhang, X.Y.; Wu, H.; Tao, W.Y.; Liu, T.; Wei, W.; Geng, J.S.; Pang, D. Expression of ribosome-binding protein 1 correlates with shorter survival in Her-2 positive breast cancer. *Cancer Sci.* **2015**, *106*, 740–746. [CrossRef]
40. Xiong, L.; Zhou, X.R.; Liu, D.; Liu, G.C. Effect of overexpression of RRBP1 on metastasis and invasion of liver cancer cells. *J. Chengdu Med. Coll.* **2021**, *16*, 5.
41. Li, H.Y.; Wang, Y.Z.; Yang, H.G.; Zhang, Y.D.; Xing, L.; Wang, J.Q.; Zheng, N. Furosine, a maillard reaction product, triggers necroptosis in hepatocytes by regulating the RIPK1/RIPK3/MLKL pathway. *Int. J. Mol. Sci.* **2019**, *20*, 2388. [CrossRef]
42. Wu, T.; Zhang, Y.Z. Transport Transporter on the Endoplasmic Reticulum-TRAP. *Chin. J. Biochem. Mol. Biol.* **2008**, *24*, 95–100.
43. Lu, Y.C.; Chang, C.C.; Wang, C.P.; Hung, W.C.; Tsai, I.T.; Tang, W.H.; Wu, C.C.; Wei, C.T.; Chung, F.M.; Lee, Y.J. Circulating fatty acid-binding protein 1 (FABP1) and nonalcoholic fatty liver disease in patients with type 2 diabetes mellitus. *Int. J. Med. Sci.* **2020**, *17*, 182–190. [CrossRef] [PubMed]
44. Wang, G.Q.; Shen, H.; Ganesh, R.; Roberts, M.S.; Gong, Y.W.; Jiang, P.; Burczynski, F. Expression and antioxidant function of liver fatty acid binding protein in normal and bile-duct ligated rats. *Eur. J. Pharmacol.* **2007**, *560*, 61–68. [CrossRef] [PubMed]
45. Waikar, S.S.; Bonventre, J.V. Biomarkers for the diagnosis of acute kidney injury. *Nephron Clin. Pract.* **2008**, *109*, c192–c197. [CrossRef] [PubMed]
46. Carla, G.; Marta, B.; Sandra, P.V.; Marta, M.M.; Victoria, G.M.; M.Luz, M.C.; Javer, G.G.; Jose, V.C.; Sonia, S.C.; Ramiro, J. The human liver fatty acid binding protein (FABP1) gene is activated by FOXA1 and PPARα; and repressed by C/EBPα: Implications in FABP1 down-regulation in nonalcoholic fatty liver disease. *Biochim. Biophys. Acta* **2013**, *1831*, 803–818.
47. Zhang, Y. Anti-Hepatic Fibrosis Effects of Bifendate in Rat and Its Relationship with CBR1 and FABP1. Master's Thesis, Hebei Medical University, Shijiazhuang, China, 2015.
48. Akbal, E.; Köklü, S.; Koçak, E.; Cakal, B.; Gunes, F.; Basar, O.; Tuna, Y.; Senes, M. Liver fatty acid-binding protein is a diagnostic marker to detect liver injury due to chronic hepatitis C infection. *Arch. Med. Res.* **2013**, *44*, 34–38. [CrossRef]
49. Ozenirler, S.; Degertekin, C.K.; Erkan, G.; Elbeg, S.; Tuncer, C.; Kandilci, U.; Akyol, G. Serum liver fatty acid binding protein shows good correlation with liver histology in NASH. *Hepatogastroenterology* **2013**, *60*, 1095–1100.
50. Akbal, E.; Koçak, E.; Akyürek, Ö.; Koklu, S.; Batgi, H.; Senes, M. Liver fatty acid-binding protein as a diagnostic marker for non-alcoholic fatty liver disease. *Wien. Klin. Wochenschr.* **2016**, *128*, 48–52. [CrossRef]
51. Hussain, M.M.; Shi, J.; Dreizen, P. Microsomal triglyceride transfer protein and its role in apoB-lipoprotein assembly. *J. Lipid Res.* **2003**, *44*, 22–32. [CrossRef]
52. Khatun, I.; Walsh, M.T.; Hussain, M.M. Loss of both phospholipid and triglyceride transfer activities of microsomal triglyceride transfer protein in abetalipoproteinemia. *J. Lipid Res.* **2013**, *54*, 1541–1549. [CrossRef]
53. Love, J.D.; Suzuki, T.; Robinson, D.B.; Harris, C.M.; Johnson, J.E.; Mohler, P.J.; Jerome, W.G.; Swift, L.L. Microsomal triglyceride transfer protein (MTP) associates with cytosolic lipid droplets in 3T3-L1 adipocytes. *PLoS ONE* **2015**, *10*, e0135598. [CrossRef]
54. Iqbal, J.; Walsh, M.T.; Hammad, S.M.; Cuchel, M.; Tarugi, P.; Hegele, R.A.; Davidson, N.O.; Rader, D.J.; Klein, R.L.; Hussain, M.M. Microsomal triglyceride transfer protein transfers and determines plasma concentrations of ceramide and sphingomyelin but not glycosylceramide. *J. Biol. Chem.* **2015**, *290*, 25863–25875. [CrossRef] [PubMed]
55. Ueshima, K.; Akihisa-Umeno, H.; Nagayoshi, A.; Takakura, S.; Matsuo, M.; Mutoh, S. Implitapide, a microsomal triglyceride transfer protein inhibitor, reduces progression of atherosclerosis in apolipoprotein E knockout mice fed a Western-type diet: Involvement of the inhibition of postprandial triglyceride elevation. *Biol. Pharm. Bull.* **2005**, *28*, 247–252. [CrossRef] [PubMed]

56. Koo, S.H. Nonalcoholic fatty liver disease: Molecular mechanisms for the hepatic steatosis. *Clin. Mol. Hepatol.* **2013**, *19*, 210–215. [CrossRef] [PubMed]
57. Kawano, Y.; Cohen, D.E. Mechanisms of hepatic triglyceride accumulation in non-alcoholic fatty liver disease. *J. Gastroenterol.* **2013**, *48*, 434–441. [CrossRef]
58. Chang, X.X.; Yan, H.M.; Fei, J.; Jiang, M.H.; Zhu, H.G.; Lu, D.R.; Gao, X. Berberine reduces methylation of the MTTP promoter and alleviates fatty liver induced by a high-fat diet in rats. *J. Lipid Res.* **2010**, *51*, 2504–2515. [CrossRef]
59. Liu, M.; Chung, S.; Shelness, G.S.; Parks, J.S. Hepatic ABCA1 and VLDL triglyceride production. *Biochim. Biophys. Acta* **2012**, *1821*, 770–777. [CrossRef]
60. Vaisman, B.L.; Lambert, G.; Amar, M.; Joyce, C.; Ito, T.; Shamburek, R.D.; Cain, W.J.; Fruchart, N.J.; Neufeld, E.D.; Remaley, A.T. ABCA1 overexpression leads to hyperalphalipoproteinemia and increased biliary cholesterol excretion in transgenic mice. *J. Clin. Investig.* **2001**, *108*, 303–309. [CrossRef]
61. Vega-Badillo, J.; Gutiérrez-Vidal, R.; Hernández-Pérez, H.A.; Villanil, R.Z.; Leon, M.P.; Sanchez, M.F.; Moran, R.S.; Larriieta, C.E.; Fernandez, S.I.; Mendez, S.N.; et al. Hepatic miR-33a/miR-144 and their target gene ABCA1 are associated with steatohepatitis in morbidly obese subjects. *Liver Int.* **2016**, *36*, 1383–1391. [CrossRef]
62. Ma, K.L.; Ruan, X.Z.; Powis, S.H.; Chen, Y.; Moorhead, J.F.; Varghese, Z. Inflammatory stress exacerbates lipid accumulation in hepatic cells and fatty livers of apolipoprotein E knockout mice. *Hepatology* **2008**, *48*, 770–778. [CrossRef]
63. Fukunaga, K.; Imachi, H.; Lyu, J.; Dong, T.; Sato, S.; Ibata, T.; Kobayashi, T.; Yoshimoto, T.; Yonezaki, K.; Matsunaga, T. IGF1 suppresses cholesterol accumulation in the liver of growth hormone-deficient mice via the activation of ABCA1. *Am. J. Physiol. Endocrinol. Metab.* **2018**, *315*, 1232–1241. [CrossRef]
64. Wang, C.; Liu, S.S.; Lu, L.L.; Liao, S.L.; Yue, H.Y.; Dong, Q.J.; Xin, Y.N.; Xuan, S.Y. Association Between Four ABCA1 gene Polymorphisms and Risk of Non-Alcoholic Fatty Liver Disease in a Chinese Han Population. *Hepat. Mon.* **2018**, *6*, e66149. [CrossRef]
65. Farrell, G.C.; Larter, C.Z. Nonalcoholic fatty liver disease: From steatosis to cirrhosis. *Hepatology* **2006**, *43*, S99–S112. [CrossRef] [PubMed]
66. Feng, S.M.; Gan, L.; Yang, C.S.; Liu, A.B.; Lu, W.Y.; Shao, P.; Dai, Z.Q.; Sun, P.L.; Luo, Z.S. Effects of Stigmasterol and β-Sitosterol on Nonalcoholic Fatty Liver Disease in a Mouse Model: A Lipidomic Analysis. *J. Agric. Food Chem.* **2018**, *66*, 3417–3425. [CrossRef] [PubMed]
67. Perla, F.M.; Prelati, M.; Lavorato, M.; Visicchio, D.; Anania, C. The role of lipid and lipoprotein metabolism in non-alcoholic fatty liver disease. *Children* **2017**, *4*, 46. [CrossRef]
68. Zhai, R.N.; Feng, L.; Zhang, Y.; Liu, W.; Li, S.L.; Hu, Z.Y. Combined Transcriptomic and Lipidomic Analysis Reveals Dysregulated Genes Expression and Lipid Metabolism Profiles in the Early Stage of Fatty Liver Disease in Rats. *Front. Nutr.* **2021**, *8*, 733197. [CrossRef]
69. Goldberg, I.J.; Ginsberg, H.N. Ins and outs modulating hepatic triglyceride and development of nonalcoholic fatty liver disease. *Gastroenterology* **2006**, *130*, 1343–1346. [CrossRef]
70. Ma, L.N.; Feng, M.; Zhou, Y.Z. The relationship between nonalcoholic fatty liver disease and insulin resistance. *J. Clin. Hepatol.* **2010**, *26*, 173–175.
71. Zhang, C.H.; Zhou, B.G.; Sheng, J.Q.; Chen, Y.; Cao, Y.Q.; Chen, C. Molecular mechanisms of hepatic insulin resistance in nonalcoholic fatty liver disease and potential treatment strategies. *Pharmacol. Res.* **2020**, *159*, 104984. [CrossRef]
72. Marieke, D.V.; Westerink, J.; EL-Morabit, F.; Kaasjager, H.A.H.; De Valk, H.W. Prevalence of non-alcoholic fatty liver disease (NAFLD) and its association with surrogate markers of insulin resistance in patients with type 1 diabetes. *Diabetes Res. Clin. Pract.* **2022**, *186*, 109827.
73. Byrne, C.D. Dorothy Hodgkin Lecture: Non-alcoholic fatty liver disease, insulin resistance and ectopic fat: A new problem in diabetes management. *Diabet. Med.* **2019**, *29*, 1098–1107. [CrossRef]
74. Marjot, T.; Moolla, A.; Cobbold, J.F.; Hodson, L.; Tomlinson, J.W. Nonalcoholic Fatty Liver Disease in Adults: Current Concepts in Etiology, Outcomes, and Management. *Endocr. Rev.* **2020**, *41*, 66–117. [CrossRef] [PubMed]
75. Holland, W.L.; Summers, S.A. Sphingolipids, insulin resistance, and metabolic disease: New insights from in vivo manipulation of sphingolipid metabolism. *Endocr. Rev.* **2008**, *29*, 381–402. [CrossRef] [PubMed]
76. Kotronen, A.; Seppänen-Laakso, T.; Westerbacka, J.; Kiviluoto, T.; Arola, J.; Ruskeepaa, A.L.; Yki-Jarvinen, H.; Oresic, M. Comparison of lipid and fatty acid composition of the liver, subcutaneous and intra-abdominal adipose tissue, and serum. *Obesity* **2010**, *18*, 937–944. [CrossRef] [PubMed]
77. Llacuna, L.; Marí, M.; Garcia-Ruiz, C.; Fernandez-Checa, J.C.; Morales, A. Critical role of acidic sphingomyelinase in murine hepatic ischemia-reperfusion injury. *Hepatology* **2006**, *44*, 561–572. [CrossRef]
78. Regnier, M.; Polizzi, A.; Guillou, H.; Loiseau, N. Sphingolipid metabolism in non-alcoholic fatty liver diseases. *Biochimie* **2019**, *1590*, 9–22. [CrossRef]
79. Chocian, G.; Chabowski, A.; Żendzian-Piotrowska, M.E.; Harasim, E.; Lukaszuk, B.; Gorski, J. High fat diet induces ceramide and sphingomyelin formation in rat's liver nuclei. *Mol. Cell. Biochem.* **2010**, *340*, 125–131. [CrossRef]
80. Brown, E.M. Fatty liver? Microbiome sphingolipids to the rescue. *Cell Host Microbe* **2022**, *30*, 755–757. [CrossRef]
81. Dowman, J.K.; Tomlinson, J.W.; Newsome, P.N. Pathogenesis of non-alcoholic fatty liver disease. *Int. J. Med.* **2010**, *103*, 71–83. [CrossRef]

82. Amrutkar, M.; Cansby, E.; Chursa, U.; Nunez-Duran, E.; Chanclon, B.; Stahlman, M.; Friden, V.; Manneras-Holm, L.; Wickman, A.; Smith, U.; et al. Genetic disruption of protein kinase STK25 ameliorates metabolic defects in a diet-induced type 2 diabetes model. *Diabetes* **2015**, *64*, 2791–2804. [CrossRef]
83. Martin, G.G.; Danneberg, H.; Kumar, L.S.; Atshaves, B.P.; Erol, E.; Bader, M.; Schroeder, F.; Binas, B. Decreased liver fatty acid binding capacity and altered liver lipid distribution in mice lacking the liver fatty acid-binding protein gene. *J. Biol. Chem.* **2003**, *278*, 21429–21438. [CrossRef]
84. Atshaves, B.P.; McIntosh, A.M.; Lyuksyutova, O.I.; Zipfel, W.; Webb, W.W.; Schroeder, F. Liver fatty acid-binding protein gene ablation inhibits branched-chain fatty acid metabolism in cultured primary hepatocytes. *J. Biol. Chem.* **2004**, *279*, 30954–30965. [CrossRef] [PubMed]
85. Mukai, T.; Egawa, M.; Takeuchi, T.; Yamashita, H.; Kusudo, T. Silencing of FABP1 ameliorates hepatic steatosis, inflammation, and oxidative stress in mice with nonalcoholic fatty liver disease. *FEBS Open Bio* **2017**, *7*, 1009–1016. [CrossRef] [PubMed]
86. Newberry, E.P.; Xie, Y.; Kennedy, S.M.; Luo, J.; Davidson, N.O. Protection against Western diet-induced obesity and hepatic steatosis in liver fatty acid-binding protein knockout mice. *Hepatology* **2006**, *44*, 1191–1205. [CrossRef] [PubMed]
87. Martin, G.G.; Atshaves, B.P.; Huang, H.; Mclntosh, A.L.; Williams, B.J.; Pai, P.J.; Russell, D.H.; Kier, A.B.; Schroeder, F. Hepatic phenotype of liver fatty acid binding protein gene-ablated mice. *Am. J. Physiol. Gastrointest. Liver Physiol.* **2009**, *297*, G1053–G1065. [CrossRef] [PubMed]
88. Puri, P.; Baillie, R.A.; Wiest, M.M.; Mirshahi, F.; Sanyal, A.J. A lipidomic analysis of nonalcoholic fatty liver disease. *Hepatology* **2007**, *46*, 1081–1090. [CrossRef]
89. Ioannou, G.N. The Role of Cholesterol in the Pathogenesis of NASH. *Trends Endocrinol. Metab.* **2016**, *27*, 84–95. [CrossRef]
90. Sozen, E.; Ozer, N.K. Impact of high cholesterol and endoplasmic reticulum stress on metabolic diseases: An updated mini-review. *Redox Biol.* **2017**, *12*, 456–461. [CrossRef] [PubMed]
91. Oram, J.F.; Heinecke, J.W. ATP-binding cassette transporter A1: A cell cholesterol exporter that protects against cardiovascular disease. *Physiol. Rev.* **2005**, *85*, 1343–1372. [CrossRef]
92. Oram, J.F. The ins and outs of ABCA. *J. Lipid Res.* **2008**, *49*, 1150–1151. [CrossRef]
93. Tang, C.R.; Liu, Y.H.; Yang, W.; Storey, C.; McMillen, T.S.; Houston, B.A.; Heinecke, J.W.; LeBoeuf, R.C. Hematopoietic ABCA1 deletion promotes monocytosis and worsens diet-induced insulin resistance in mice. *J. Lipid Res.* **2016**, *57*, 100–108. [CrossRef]
94. Kolovou, V.; Marvaki, A.; Boutsikou, M.; Vasilopoulos, G.; Degiannis, D.; Marvaki, C.; Kolovou, G. Effect of ATP-binding Cassette Transporter A1 (ABCA1) Gene Polymorphisms on Plasma Lipid Variables and Common Demographic Parameters in Greek Nurses. *Open Cardiovasc. Med. J.* **2016**, *10*, 233–239. [CrossRef] [PubMed]
95. Clee, S.M.; Kastelein, J.J.; van Dam, M.; Marcil, M.; Roomp, K.; Zwarts, K.Y.; Collins, J.A.; Roelants, R.; Tamasawa, N.; Stulc, T.; et al. Age and residual cholesterol efflux affect HDL cholesterol levels and coronary artery disease in ABCA1 heterozygotes. *J. Clin. Investig.* **2000**, *106*, 1263–1270. [CrossRef]
96. Lee, K.; Kerner, J.; Hoppel, C.L. Mitochondrial carnitine palmitoyltransferase 1a (CPT1a) is part of an outer membrane fatty acid transfer complex. *J. Biol. Chem.* **2011**, *286*, 25655–25662. [CrossRef] [PubMed]
97. Tonazzi, A.; Giangregorio, N.; Console, L.; Indiveri, C. Mitochondrial carnitine/acylcarnitine translocase: Insights in structure/function relationships. Basis for drug therapy and side effects prediction. *Mini Rev. Med. Chem.* **2015**, *15*, 396–405. [CrossRef] [PubMed]
98. Pittala, S.; Krelin, Y.; Kuperman, Y.; Shoshan-Barmatz, V. A Mitochondrial VDAC1-Based Peptide Greatly Suppresses Steatosis and NASH-Associated Pathologies in a Mouse Model. *Mol. Ther.* **2019**, *27*, 1848–1862. [CrossRef] [PubMed]
99. Turkaly, P.; Kerner, J.; Hoppel, C. A 22 kDa polyanion inhibits carnitine-dependent fatty acid oxidation in rat liver mitochondria. *FEBS Lett.* **1999**, *460*, 241–245. [CrossRef]
100. Zhong, Z.; Lemasters, J.J. A Unifying Hypothesis Linking Hepatic Adaptations for Ethanol Metabolism to the Proinflammatory and Profibrotic Events of Alcoholic Liver Disease. *Alcohol. Clin. Exp. Res.* **2018**, *42*, 2072–2089. [CrossRef]
101. Holmuhamedov, E.L.; Czerny, C.; Beeson, C.C.; Lemasters, J.J. Ethanol suppresses ureagenesis in rat hepatocytes: Role of acetaldehyde. *J. Biol. Chem.* **2012**, *287*, 7692–7700. [CrossRef]
102. Shoshan-Barmatz, V.; Krelin, A.Y.; Shteinfer-Kuzmine, T. Voltage-Dependent Anion Channel 1 As an Emerging Drug Target for Novel Anti-Cancer Therapeutics. *Front. Oncol.* **2017**, *7*, 154. [CrossRef]
103. Maldonado, E.N.; Shoshan-Barmatz, V.; Krelin, Y. VDAC1 at the crossroads of cell metabolism, apoptosis and cell stress. *Cell Stress* **2017**, *1*, 11–36.

metabolites

MDPI

Article

Dietary Soybean Lecithin Improves Growth, Immunity, Antioxidant Capability and Intestinal Barrier Functions in Largemouth Bass *Micropterus salmoides* Juveniles

Jiaojiao Wu [1,2,†], Wenxue Yang [1,2,†], Rui Song [1,2], Zhe Li [1,2], Xiaowei Jia [1,2], Hao Zhang [1,2], Penghui Zhang [1,2], Xinyu Xue [1,2], Shenghui Li [1,2], Yuanyuan Xie [1,2], Rongfei Zhang [1,2], Jinyun Ye [1,2], Zhijin Zhou [3] and Chenglong Wu [1,2,*]

1 National-Local Joint Engineering Laboratory of Aquatic Animal Genetic Breeding and Nutrition (Zhejiang), Huzhou University, 759 East 2nd Road, Huzhou 313000, China
2 Zhejiang Provincial Key Laboratory of Aquatic Resources Conservation and Development, College of Life Science, Huzhou University, 759 East 2nd Road, Huzhou 313000, China
3 Huzhou Agricultural Science and Technology Development Center, 768 Luwang Road, Huzhou 313000, China
* Correspondence: 01998@zjhu.edu.cn; Tel./Fax: +86-572-2389817
† These authors contributed equally to this work.

Abstract: This study evaluated the effects of dietary soybean lecithin (SBL) on the growth, haematological indices, immunities, antioxidant capabilities, and inflammatory and intestinal barrier functions because little information of dietary SBL could be obtained in juvenile largemouth bass (*Micropterus salmoides*). The fish were fed identical diets except for SBL added at 0, 2, 4 and 8%. It was found that 4 and 8% SBL significantly increased fish weight gain and daily growth rate ($p < 0.05$), while 4% SBL was optimal for enhancing RBC, HGB, PLT, MCV, MCH, WBC and MON in blood, and ALB and ALP in serum ($p < 0.05$). SBL (4%) also significantly elevated the antioxidant enzymes activities of T-SOD, CAT, GR, GPx, GST and T-AOC and GSH contents; increased mRNA transcription levels of Nrf2, Cu/Zn-SOD, CAT, GR, GST3 and GPx3; and decreased MDA contents. Keap1a and Keap1b levels were markedly down-regulated ($p < 0.05$). SBL (4%) significantly enhanced levels of the immune factors (ACP, LZM and C3) and the mRNA expression levels of innate immune-related genes (C3, C4, CFD, HEPC and MHC-I) compared with the control groups (0%) ($p < 0.05$). SBL (4%) significantly increased IgM and T-NOS in the intestine ($p < 0.05$) and significantly decreased levels of TNF-α, IL-8, IL-1β and IFN-γ and increased TGF-β1 at both transcription and protein levels in the liver and intestine ($p < 0.05$). The mRNA expression levels of MAPK13, MAPK14 and NF-κB P65 were significantly decreased in the intestine in the 4% SBL groups ($p < 0.05$). Histological sections also demonstrated that 4% SBL protected intestinal morphological structures compared with controls. This included increased intestinal villus height and muscular thickness ($p < 0.05$). Furthermore, the mRNA expression levels of the intestinal epithelial cell tight junction proteins (TJs) (ZO-1, claudin-3, claudin-4, claudin-5, claudin-23 and claudin-34) and mucin-5AC were significantly up-regulated in the 4% SBL groups compared with the controls ($p < 0.05$). In conclusion, these results suggested that 4% dietary SBL could not only improve growth, haematological indices, antioxidant capabilities, immune responses and intestinal functions, but also alleviate inflammatory responses, thereby providing reference information for the feed formulations in cultured largemouth bass.

Keywords: *Micropterus salmoides*; soybean lecithin; haematology; antioxidant capabilities; immune responses

Citation: Wu, J.; Yang, W.; Song, R.; Li, Z.; Jia, X.; Zhang, H.; Zhang, P.; Xue, X.; Li, S.; Xie, Y.; et al. Dietary Soybean Lecithin Improves Growth, Immunity, Antioxidant Capability and Intestinal Barrier Functions in Largemouth Bass *Micropterus salmoides* Juveniles. *Metabolites* **2023**, *13*, 512. https://doi.org/10.3390/metabo13040512

Academic Editors: Qingchao Wang, Yan He and Qingsong Tan

Received: 9 February 2023
Revised: 28 March 2023
Accepted: 31 March 2023
Published: 2 April 2023

1. Introduction

As by-product of soybean oil production, soybean lecithin (SBL) plays important nutritional roles for its high content of phosphatidyl choline, phosphatidyl ethanolamine, phosphatidyl inositol, phospholipids (PL) contents and trace amounts of glycolipids and triglycerides [1]. SBL could function as a food additive and improve growth, bolster immunity and increase antistress abilities in humans [2] and in terrestrial animals (cattle, chickens and turkeys) [1,3,4]. However, the limited biosynthetic capacity of PL could not meet the requirements for the normal growth and development in larval and juvenile aquatic animals [5]. Therefore, exogenous SBL was added into aquatic animal feeds as a primary non-protein raw material owing its ready availability [6–8]. Moreover, it has been widely used in many cultured fish species including golden mahseer (*Tor putitora*) [6], gilthead seabream (*Sparus aurata*) [8], Caspian brown trout (*Salmo trutta caspius*) [9], common carp (*Cyprinus carpio*) [10], stellate sturgeon (*Acipenser stellatus*) [11], rock bream (*Oplegnathus fasciatus*) [12] and channel catfish (*Ictalurus punctatus*) [13].

Many studies have found dietary SBL could not only increase growth, but also enhance health status, including innate immune and antioxidant responses in animals [9,11]. It is well-known that animal immune status could be monitored using haematological and antioxidant parameters [9,14]. Higher values of haemoglobin (HGB), red blood cells (RBC) and white blood cells (WBC) induced by adequate nutrients and probiotics could reflect better immunity and oxygen-carrying capacity via the haematopoietic system in many fish species [15,16]. Meanwhile, antioxidant capabilities can reflect the animal responses to reactive oxygen species (ROS) produced in normal metabolic and immune defence processes, while excessive ROS is deleterious, resulting in oxidative damage and impairment of physiological functions [17]. These antioxidant enzymes activities could be regulated by the Nrf2/Keap1 signalling pathway to maintain redox homeostasis [18], which is also closely linked to fish health status and innate immunity [19]. However, little information could be obtained about the relationships among the antioxidant capabilities, Nrf2/Keap1 pathway and dietary SBL in fish.

Encounters with nutrient-deficient, pathogens or environmental toxins could induce inflammatory processes and ROS products, which are the primary immune defence lines for protecting animals [20]. In particular, the p38 MAPK/NF-κB signalling pathway is responsible for pro- and anti-inflammatory cytokines production [21], while excessive immune stimulation could cause typical 'cytokine storm' mediated by pro-inflammatory mediators, including interleukin-1β (IL-1β), interferon-γ (IFN-γ) and tumour necrosis factor-α (TNF-α), etc. [22,23]. Many studies have demonstrated that overproduction of inflammatory cytokines could result in tissue damage and inappropriate physiological responses [24]. However, these processes are normally counteracted by increasing contents of anti-inflammatory cytokines, such as interleukin 10 (IL-10) and transforming growth factor β 1 (TGF-β1), that participate in tissue repair processes [5,24,25]. Therefore, it is essential to alleviate inflammation and improve immunity through regulating the balance between pro- and anti-inflammatory cytokine productions mediated by adequate nutrients and/or additives in animals [25].

In addition, the intestinal barrier function of fish was closely related to its health status. Intestinal barrier functions and integrity were associated with elevated levels of the epithelial cell tight junction proteins (TJs) occludin, zonula occludens and claudin [5]. These cells also maintain the mucus barrier using mucin-linked carbohydrates that form a protective gel that directly interacts with the intestinal microbiota [26]. These barrier protective functions could be altered by balanced nutrition in largemouth bass (*Micropterus salmoides*) [27], grass carp (*Ctenopharyngodon idella*) [28], jian carp (*Cyprinus carpio* var. Jian) [29] and Cobia (*Rachycentron canadum*) [30]. In contrast, the effects of dietary SBL on intestinal barrier functions were little described in aquatic animals.

Largemouth bass (*Micropterus salmoides*) is a typical carnivorous freshwater fish that is widely cultivated due to its rapid growth, strong adaptability and better meat quality. This species is one of the most important commercial freshwater species cultivated in China and its production reached 0.7 million tons in 2021 [31]. Preliminary studies using dietary phospholipids and intestinal health indices for largemouth bass larvae have been conducted but adults or juveniles were not included [5]. In the current study, we evaluated how SBL affects *M. salmoides* both at the transcriptional and protein levels, and further explored the mechanism of dietary SBL on the intestinal barrier functions and immune health using haematological indices and measures of inflammation and redox stress as well as intestinal tight junction (TJ) and mucin protein expression. These data provide a reference to optimize compound feed formulations for largemouth bass.

2. Materials and Methods

2.1. Experimental Diets

Four diet formulations were produced that included defatted fish meal (1.8% lipid), casein and gelatine as protein sources, and soybean oil and SBL as lipid sources. The ingredients were mechanically forced through 60-mesh screens for mixing. The 4 formulations provided identical levels of nitrogen, energy sources and lipids but contained graded levels of SBL and 0, 2, 4 and 8% substituted *w/w* for soybean oil, respectively (Table 1). SBL contained 196, 132 and 146 mg/g phosphatidylcholine, phosphatidylethanolamine and phosphatidylinositol, respectively, and was analysed using high-performance liquid chromatography with evaporative light scattering detector (HPLCELSD). The components were processed into 2.5 mm diameter pellets using a commercial F-26 twin-screw extruder (South China University of Technology Machinery Factory, Guangzhou, China). The pellets were air-dried at 35 °C and stored at −20 °C. The proximate composition of these four experimental diets were determined according to the methods described by Jia et al. [17].

Table 1. Ingredient and proximate composition of the experimental diets (on dry weight basis).

Ingredients (%)	Diets (%)			
	0	2	4	8
Defatted fish meal [a]	20.00	20.00	20.00	20.00
Casein [b]	35.00	35.00	35.00	35.00
Gelatine [c]	5.00	5.00	5.00	5.00
Soybean oil [d]	10.00	8.00	6.00	2.00
Soybean lecithin [e]	0.00	2.00	4.00	8.00
Dextrin [c]	10.00	10.00	10.00	10.00
Mineral premix [f]	2.40	2.40	2.40	2.40
Vitamin premix [g]	1.20	1.20	1.20	1.20
Choline chloride [h]	0.40	0.40	0.40	0.40
Microcrystalline cellulose [c]	16.00	16.00	16.00	16.00
Proximate analysis (%)				
Moisture	6.49	6.64	6.61	6.76
Crude protein	50.67	50.59	50.65	50.53
Crude lipid	9.58	9.53	9.57	9.49
Crude ash	5.61	5.57	5.67	5.78
Gross energy (MJ kg^{-1})	17.43	17.34	17.38	17.41

[a] Defatted fish meal, Pesquera Diamante S.A. crude protein 76.7%; crude lipid 1.8%. [b] Casein obtained from Gansu Hualing Dairy Co., Ltd., Lanzhou, China; crude protein 80.56%. [c] Gelatine, Dextrin and Microcrystalline cellulose, obtained from Sinopharm Chemical Reagent Co., Ltd., Shanghai, China, crude protein 99.76%. [d] Soybean oil, produced using oil press. [e] Soybean lecithin, Jiangsu Yuanshengyuan Biological Engineering Co., Ltd., Nanjing, China. [f] Mineral Premix (mg kg^{-1} diet): Na_2SeO_3, 1.2 mg; KI, 2 mg; $CoCl_2·6H_2O$, 13 mg; $CuSO_4·5H_2O$, 5 mg; $FeSO_4·7H_2O$, 70 mg; $ZnSO_4·7H_2O$, 100 mg; $MnSO_4·H_2O$, 45 mg; $MgSO_4·7H_2O$, 130 mg; NaCl, 50 mg. [g] Vitamin Premix (mg kg^{-1} diet): vitamin A, 35 mg; vitamin D, 6 mg; vitamin C, 1000 mg; vitamin E, 300 mg; thiamine, 30 mg; riboflavin, 50 mg; pyridoxine HCl, 20 mg; vitamin B12, 0.1 mg; vitamin K3, 10 mg; inositol, 800 mg; pantothenic acid, 60 mg; folic acid, 20 mg; niacin acid, 200 mg; biotin, 60 mg. [h] Zhejiang Yixing Feed Group Co. Ltd., Jiaxing, China.

2.2. Experimental Fish and Feeding Trial

M. salmoides juveniles were obtained from Deqing Longshenli Biotech Co., Ltd., (Huzhou, China) and cultured in the recirculating aquaculture system of Huzhou University. Before the start of the normal feeding experiments, the fish was disinfected and placed in 500 L tanks for one week and domesticated using commercial feed (Deqing Longshenli). Disease- and injury-free juveniles (360) weighing 9.33 ± 0.06 g were selected and distributed randomly into 12 tanks (90 fish per experimental diet). The fish were fed twice a day at 08:00 and 17:00 for 8 weeks. Uneaten feed was removed by siphoning every 2 days, and the same amount of fresh water was returned to the tanks. Water temperature was maintained at 26–29 °C, and lighting was synchronized with natural light. Over the experimental course, water pH was ~7.2, dissolved oxygen > 5.8 mg/L and total ammonia nitrogen was kept < 0.04 mg/L.

2.3. Sample Preparation

At the end of the experiment, all fish were fasted for 24 h, then anesthetized with 100 ppm tricane methane sulfonate (MS-222, Sigma, St. Louis, MO, USA) and weighed to determine growth performance. Fifteen fish were randomly selected from each tank and used for blood sampling from the caudal tail vessels; 0.2–0.3 mL was added to heparinized Eppendorf tubes for haematological assays. The remainder of the blood samples were placed at 4 °C for 24 h and then centrifuged at 3500 rpm for 15 min; the serum was removed and stored in liquid nitrogen. The liver and intestine samples were collected from the fish and quick-frozen in liquid nitrogen and then stored at −80 °C for further use.

2.4. Haematological and Biochemical Analyses

Blood sample total counts and differential cell types were performed using a TEK 8500 VET automatic blood analyzer (Jiangxi Tekang Technology, Nanchang, China) for white blood cells (WBC), red blood cells (RBC), Hb (HGB), platelets (PLT), mean corpuscular Hb (MCH), mean corpuscular volume (MCV), mean corpuscular Hb concentration (MCHC), monocytes (MON), lymphocytes (LYM) and neutrophils (NEU). Blood sample analyses consisted of 15 replicates in each group. Serum concentrations of albumin (ALB) and activities of alkaline phosphatase (ALP) were measured using a TC6010L autoanalyzer (Tekang Technology. Nanchang, China). Serum analyses consisted of at least six replicates.

2.5. Antioxidant Analysis

Frozen tissue specimens (liver and intestine) were pulverized in liquid nitrogen and then suspended in a 9-fold volume (*w/v*) of 0.9% NaCl at 4 °C and then centrifuged at 2500 rpm for 20 min at 4 °C. The supernatants were used for the determination of superoxide dismutase (SOD), catalase (CAT), glutathione reductase (GR), glutathione peroxidase (GPx) and glutathione s-transferase (GST) activities; total antioxidant capacity (T-AOC); and glutathione (GSH) and malondialdehyde (MDA) contents using commercial kits (Jiancheng Bioengineering, Nanjing, China).

Kelch-like ECH-associated protein 1a (Keap1a) amounts were measured using an enzyme-linked immunosorbent assay (ELISA) according to the manufacturer's instructions (Jiancheng, Nanjing, China). Protein contents were determined using the Coomassie brilliant blue method. All analyses consisted of at least three replicates.

2.6. Immunological Analyses

Commercial kits were used for the following enzyme and immunological (ELISA) analyses using methods provided by the manufacturer (Jiancheng, Nanjing, China) These tests included Lysozyme (LZM), acid phosphatase (ACP), ALP and total nitric oxide synthase (T-NOS) activities. ELISA kits were used for complement component 3 (C3), immunoglobulin M (IgM), IL-1β, IL-8, TNF-α, IFN-γ and TGF-β1 measurements. All analyses consisted of at least three replicates.

2.7. Gene Expression Measurements

Total RNA was extracted from the intestine and liver using a BioFast Simply P Total RNA Extraction kit (BioFlux, Hangzhou, China), and RNA integrity was examined using 1% agarose gel electrophoresis. RNA was quantified using UV spectroscopy (Nano Drop 2000, Thermo Scientific, Pittsburgh, PA, USA). RNA was reverse transcribed using a PrimeScript RT Kit with gDNA Eraser (Novoprotein, Shanghai, China) following the manufacturer's instructions. The obtained cDNA templates were then kept frozen at −80 °C until analysis. Primers for quantitative real-time PCR were designed from the largemouth bass genome [32] (Table 2). Real time qPCR was conducted using a NovoStart SYBR qPCR Super Mix Plus (E096-01B, Novoprotein,) using a CFX96 instrument (Bio-Rad,

Hercules, CA, USA) following the manufacturer's protocol. Expression variations were assessed using the $2^{-\Delta\Delta CT}$ method and β-actin was used as an internal control [17]. All analyses consisted of at least three replicates.

2.8. Histomorphometry

Posterior intestine tissue samples were washed with 0.6% saline and fixed in 4% paraformaldehyde for 48 h. After dehydration in a graded ethanol series, the samples were cleared in xylene and embedded in paraffin wax. Histological sections were stained with haematoxylin and eosin (H & E). Micrographs of the intestine were taken at a final magnification of 100× and documented photographically with a digital camera. The images were analysed using K-Viewer (https://kv.kintoneapp.com/en/user/, accessed on 28 August 2022) 1.0 software (1.0.4) for villus height, villus width, muscular thickness and crypt depth measures [33]. All analyses consisted of at least four replicates.

2.9. Data Analysis

All results were reported as mean ± SD (standard deviation). One-way analysis of variance (ANOVA) was performed for comparing groups using SPSS 25.0 (IBM, Chicago, IL, USA). Tukey's multiple interval test was used as a multiple comparison test between different dietary treatments. Moreover, a follow-up trend analysis was conducted using orthogonal polynomial contrasts to determine the significant effects (linear and/or quadratic). $p < 0.05$ indicated a significant difference.

Table 2. Primer sequences for real-time PCR analysis.

Gene	Primers	Primer Sequence (5′–3′)	Reference
Cu/Zn-SOD	F	AGGTAGTGAGGCTTGTCTGC	XM_038708943.1
	R	TCGCCCTCCTGCTCAAAATA	
Mn-SOD	F	AAACCTGGTAGCTGCATCCA	XM_038727054.1
	R	TGTGGTTAATGTGGCCTCCT	
CAT	F	CTGGCGAATGTGTCTACTGC	XM_038704976.1
	R	CTTGGTCAGATCGAAGGGGT	
GPx3	F	GCCTTTCCCACCAACCAATT	XM_038699914.1
	R	TCGTTGATCTTCAGAGGCCC	
GR	F	TGTAAAACGGACAGAGGGCT	XM_038700350.1
	R	CTCGCCGATGTTCAATCCAG	
GST3	F	CGTGGAGAGATGGACGTTCT	XM_038729946.1
	R	CTTGTCGCTGTACATGGTGG	
Keap1a	F	TTCAGACGGCAGGAGATGTT	XM_038728593.1
	R	ATGTGGGAGGTGTAGGCAAA	
Keap1b	F	AGCCGGGTACAATGCTTAGT	XM_038713665.1
	R	GGAGTTTGGTTCGTTGGCTT	
Nrf2	F	AGCGAACTGGACTCACTGAA	XM_038720536.1
	R	CTGCTGGAGGGAGTAGTCTG	
C3	F	ATGACCAACTGCAACGTGAC	XM_038714039.1
	R	CTCGCTGGCTTTCAATCTCC	
C4	F	CTCTAAGACAGGCACGTGGA	XM_038711699.1
	R	CAAACTGGCAGTGGAAAGCT	
CFD	F	GGTGCCCATTCTCTGAGTGA	XM_038729044.1
	R	CCTCTGCATTTGTGTCAGGG	
HEPC	F	TTGTGGTGCTCTTTTGGTGG	XM_038710826.1
	R	CCGCAACTGGAGTGTCATTG	
MHC-I	F	TCATTGCTGGGGTGTTTGTG	XM_038725863.1
	R	ACTACATCTGGACCTTGGGC	
TGF-β1	F	GCTCAAAGAGAGCGAGGATG	XM_038693206.1
	R	TCCTCTACCATTCGCAATCC	
IL-10	F	CGGCACAGAAATCCCAGAGC	XM_038696252.1
	R	CAGCAGGCTCACAAAATAAACATCT	
IL-1β	F	TTCAAACTGCAGGCCATCAC	XM_038733429.1
	R	CAGCTTTACCGCCTTCCATG	
IL-8	F	GCTGCAGAGGTTGAATGACTAC	XM_038713529.1
	R	TCCAATGGGCCTTTTCTCCT	
TNF-α	F	TGTGACGGAGACCAGATGAC	XM_038699437.1
	R	CAGACCGGCTTCAACAAACA	
IFN-γ	F	AGATCAGAGGCTTTCAAATCCC	XM_026298863.1
	R	CAACATGTGGCTAATCAGCTT	
NF-κB P65	F	GCGAGCCCCTTCATAGAGAT	XM_038730802.1
	R	ACCAATGAGATGCGAACACG	
MAPK13	F	CTGCTTGAGAAGATGCTGGTT	XM_038723459.1
	R	AGGCTGTCGAAATATGGGTG	

Table 2. *Cont.*

Gene	Primers	Primer Sequence (5′–3′)	Reference
MAPK14	F	TTCGATGGAGACGAGATGG	XM_038696748.1
	R	GAGATGAATGACCGCAGGC	
ZO-1	F	ATCTCAGCAGGGATTCGACG	XM_038701018.1
	R	CTTTTGCGGTGGCGTTGG	
Occludin	F	ATGTGCCAGAACCTGTACCA	XM_038715419.1
	R	GTTTCCCGTAACGCCAGATC	
Claudin-1	F	CCAGGGAAGGGGAGCAATG	XM_038713307.1
	R	GCTCTTTGAACCAGTGCGAC	
Claudin-3	F	GGCTTTGCTCATCTCCATCG	XM_038693400.1
	R	GATTATGGTGTTGGCCGACC	
Claudin-4	F	TAATCGCTATGGTGGGAGCC	XM_038708626.1
	R	GCCCCGATCTCCATCTTCTG	
Claudin-5	F	TGGGACTATCGATGTCGATA	XM_038704228.1
	R	CCACAAGCCGTCCCAGATAG	
Claudin-23	F	GTCCTGACCTCATTGCTCCT	XM_038729173.1
	R	TACACCCCAAGAGCAGTCAG	
Claudin-34	F	TGCATTTCCTGTGTTACCGC	XM_038715311.1
	R	TAGGATTGTCTCTGCCGTGG	
Mucin-2	F	TTGTAATGAGGCCGGTGTCT	XM_038706114.1
	R	CCAGGTGGGGTATAGGGTTG	
Mucin-5AC	F	TGGGCCTTCTAGAGTCTCCT	XM_038696526.1
	R	CTGATTGGTTTAGCGCCCTC	
Mucin-17	F	AAGTCCACATCCAGTCCCAG	XM_038725272.1
	R	TCCTTGTCCTGTCCTTGTCC	
β-actin	F	TTCACCACCACAGCCGAAAG	XM_038695351.1
	R	TCTGGGCAACGGAACCTCT	

F: forward, R: reverse. Cu/Zn-Superoxide dismutase, Cu/Zn-SOD; Mn-Superoxide dismutase, Mn-SOD; catalase, CAT; glutathione peroxidase 3, GPx3; glutathione reductase, GR; glutathione S-transferase 3, GST3; kelch-like ECH-associated protein 1a, Keap1a; kelch-like ECH-associated protein 1b, Keap1b; NF-E2-related factor 2, Nrf2; complement component 3, C3; complement component 4, C4; complement factor D, CFD; hepcidin, HEPC; major histocompatibility complex class I, MHC-I; interleukin 10, IL-10; transforming growth factor beta 1, TGF-β1; interleukin -1β, IL-1β; interleukin 8, IL-8; tumour necrosis factor-α, TNF-α; interferon-γ, IFN-γ; nuclear factor-kappa B p65, NF-κB P65; mitogen-activated protein kinase 13, MAPK13; mitogen-activated protein kinase 14, MAPK14; zonula occludens-1, ZO-1.

3. Results

3.1. Growth Performance and Haematological Indices

The growth performance of the juvenile largemouth bass fed our experimental diets for 8 weeks were positively affected by the inclusion of SBL to the feed. The 4 and 8% SBL significantly increased weight gain (WG) and daily growth rate (DGR) of the fish compared with 0 and 2% SBL ($p < 0.05$) (Figure 1). WBC and MON values were also significantly ($p < 0.05$) increased in the 2 and 4% groups compared with the controls, and interestingly, decreased in the 8% group. Similarly, RBC, HGB and PLT levels were the maximum with 4% SBL. MCV and MCH levels displayed an increasing trend positively associated with SBL doses ($p < 0.05$), while MCHC, NEU and LYM values were not altered among these 4 experimental groups (Table 3). However, serum activities of ALP were significantly ($p < 0.05$) greater for the 2 to 8% treatment compared with the controls (0%) that possessed the lowest activities. Serum ALB levels in 2 and 4% SBL were significantly elevated ($p < 0.05$) compared with the controls (Figure 2).

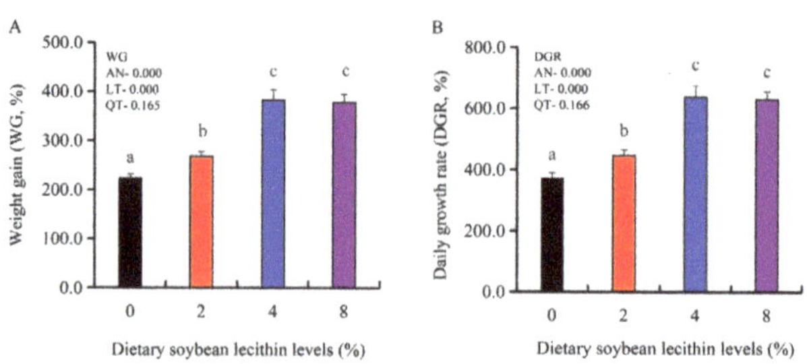

Figure 1. Effect of dietary soybean lecithin on weight gain (%) (**A**) and daily growth rate (%) (**B**) of juvenile largemouth bass. Bars with different letters indicate significant differences ($p < 0.05$), while that with the same letter or no letter superscripts indicate no significant differences ($p > 0.05$), AN: ANOVA, LT: linear trend, QT: quadratic trend.

Table 3. Haematological parameters of juvenile largemouth bass fed diet trials.

	Lecithin (%) in Diets						
Parameters	0	2	4	8	AN	LT	QT
WBC (10^9/L)	171.30 ± 7.79 [a]	187.70 ± 6.26 [b]	190.80 ± 3.87 [b]	175.21 ± 3.99 [a]	0.001	0.929	0.000
RBC (10^{12}/L)	2.73 ± 0.09 [ab]	2.84 ± 0.17 [ab]	3.00 ± 0.14 [b]	2.66 ± 0.16 [a]	0.027	0.533	0.006
HGB (g/L)	86.75 ± 1.94 [a]	89.00 ± 3.54 [ab]	93.00 ± 0.91 [b]	90.88 ± 3.20 [ab]	0.031	0.061	0.036
PLT (10^9/L)	123.50 ± 9.57 [ab]	132.38 ± 11.88 [ab]	140.88 ± 7.38 [b]	116.00 ± 7.35 [a]	0.013	0.324	0.002
MCV (fL)	125.68 ± 3.05 [a]	134.65 ± 6.73 [ab]	146.73 ± 7.26 [b]	161.73 ± 5.21 [c]	0.000	0.000	0.505
MCH (pg)	30.40 ± 0.86 [a]	32.03 ± 0.93 [ab]	32.58 ± 0.53 [b]	33.53 ± 0.89 [b]	0.001	0.000	0.120
MCHC (g/L)	224.50 ± 8.34	221.50 ± 7.77	220.00 ± 11.07	217.50 ± 7.59	0.727	-	-
NEU (10^9/L)	28.95 ± 2.75	27.73 ± 1.67	27.59 ± 1.49	24.89 ± 1.95	0.081	-	-
LYM (10^9/L)	134.90 ± 3.36	137.21 ± 4.00	140.32 ± 1.41	136.97 ± 2.49	0.052	-	-
MON (10^9/L)	27.72 ± 2.50 [ab]	28.84 ± 1.61 [b]	29.64 ± 1.40 [b]	25.21 ± 1.08 [a]	0.018	0.078	0.008

Note: [a–c] Values in the same row with different letters indicate significant differences ($p < 0.05$), while that with the same letter or no letter superscripts indicate no significant differences ($p > 0.05$). AN: ANOVA, LT: linear trend, QT: quadratic trend.

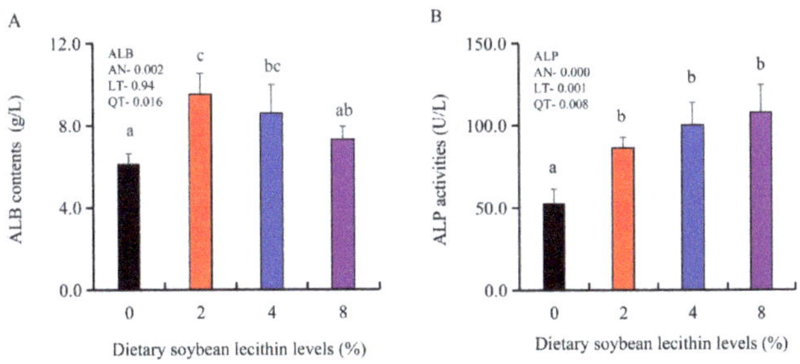

Figure 2. Effect of dietary soybean lecithin on serum albumin (ALB) contents (**A**) and alkaline phosphatase (ALP) activities (**B**) of juvenile largemouth bass. Bars with different letters indicate significant differences ($p < 0.05$), while that with the same letter or no letter superscripts indicate no significant differences ($p > 0.05$), AN: ANOVA, LT: linear trend, QT: quadratic trend.

3.2. Antioxidant and Oxidant Analyses

3.2.1. Antioxidant and Oxidant Parameters in the Liver

Antioxidative indices in the fish also significantly differed among these test groups. Activities of CAT and GPx were significantly higher in the 4% SBL groups compared with these indices in other three groups ($p < 0.05$). In contrast, MDA contents notably declined with SBL supplementation from 2% to 8% and was the lowest in the 4 and 8% groups ($p < 0.05$). T-AOC, GST activities and GSH levels were significantly higher in liver samples of the 4 and 8% groups ($p < 0.05$). T-SOD and GR activities were enhanced in the 2% group ($p < 0.05$) and then plateaued. However, Keap1a levels were significantly reduced and significantly negatively correlated with SBL doses ($p < 0.05$) (Table 4).

Table 4. Effects of dietary soybean lecithin on the antioxidative and oxidative indices in the liver of largemouth bass.

	Lecithin (%) in Diets						
Parameters	0	2	4	8	AN	LT	QT
T-SOD (U/mgprot)	507.67 ± 10.41 [a]	575.54 ± 15.07 [b]	630.14 ± 34.80 [b]	603.26 ± 28.86 [b]	0.002	0.019	0.001
CAT (U/mgprot)	15.37 ± 0.28 [a]	23.97 ± 0.72 [c]	26.47 ± 0.98 [d]	20.77 ± 0.75 [b]	0.000	0.277	0.000
GPX (U/mgprot)	51.96 ± 1.52 [a]	63.75 ± 3.32 [b]	88.23 ± 3.58 [d]	80.22 ± 0.70 [c]	0.000	0.004	0.001
GR (U/gprot)	7.55 ± 0.33 [a]	9.18 ± 0.50 [b]	9.46 ± 0.79 [b]	9.38 ± 0.70 [b]	0.015	0.033	0.017
GST (U/mgprot)	39.69 ± 1.29 [a]	41.97 ± 1.52 [a]	48.69 ± 1.34 [b]	49.08 ± 1.07 [b]	0.000	0.000	0.034
GSH (μmol/gprot)	58.09 ± 1.13 [a]	59.16 ± 3.07 [a]	76.88 ± 1.55 [b]	111.12 ± 8.32 [b]	0.000	0.000	0.015
Keap1a (ng/mgprot)	151.58 ± 3.52 [b]	147.46 ± 7.02 [ab]	137.52 ± 6.53 [a]	136.44 ± 2.85 [a]	0.000	0.002	0.061
T-AOC (μmol/gprot)	30.83 ± 1.78 [a]	30.66 ± 2.16 [a]	90.99 ± 7.00 [b]	117.31 ± 4.70 [c]	0.000	0.000	0.679
MDA (nmol/mgprot)	1.46 ± 0.12 [c]	0.92 ± 0.10 [b]	0.52 ± 0.04 [a]	0.68 ± 0.03 [a]	0.000	0.006	0.000

Note: [a–c] Values in the same row with different letters indicate significant differences ($p < 0.05$), while that with the same letter or no letter superscripts indicate no significant differences ($p > 0.05$). AN: ANOVA, LT: linear trend, QT: quadratic trend.

3.2.2. Antioxidant and Oxidant Parameters in the Intestine

In the intestine of our experimental fish, CAT, T-SOD, GPx and GR activities were gradually elevated at 4% SBL ($p < 0.05$) and then declined. GSH contents were the highest in the 8% group compared with the controls ($p < 0.05$). T-AOC activities displayed a significant upward trend to 4 and 8% SBL ($p < 0.05$). GST activities were also significantly increased to the maximum at 4% and then plateaued. In contrast, MDA contents and Keap1a levels were significantly ($p < 0.05$) negatively correlated with SBL doses (Table 5).

Table 5. Effects of dietary soybean lecithin on the antioxidative and oxidative indices in the intestine of largemouth bass.

	Lecithin (%) in Diets						
Parameters	**0**	**2**	**4**	**8**	**AN**	**LT**	**QT**
T-SOD (U/mgprot)	128.31 ± 7.01 ^{a}	146.30 ± 2.59 ^{b}	196.84 ± 11.31 ^{d}	165.17 ± 2.06 ^{c}	0.000	0.056	0.002
CAT (U/mgprot)	14.31 ± 1.05 ^{a}	20.82 ± 1.94 ^{b}	27.71 ± 1.02 ^{c}	22.39 ± 0.65 ^{b}	0.000	0.050	0.000
GPx (U/mgprot)	67.89 ± 3.07 ^{a}	95.10 ± 3.49 ^{b}	114.24 ± 4.61 ^{c}	97.81 ± 0.74 ^{b}	0.000	0.219	0.000
GR (U/gprot)	59.14 ± 2.46 ^{a}	84.51 ± 5.32 ^{b}	105.38 ± 8.67 ^{c}	56.80 ± 3.99 ^{a}	0.000	0.745	0.000
GST (U/mgprot)	191.24 ± 14.27 ^{a}	261.73 ± 12.84 ^{b}	352.95 ± 10.54 ^{c}	347.04 ± 32.31 ^{c}	0.000	0.001	0.001
GSH (μmol/gprot)	49.86 ± 4.02 ^{a}	72.25 ± 4.75 ^{b}	116.63 ± 8.96 ^{c}	182.59 ± 9.71 ^{d}	0.000	0.000	0.526
Keap1a (ng/mgprot)	167.90 ± 9.84 ^{b}	146.01 ± 7.93 ^{a}	141.10 ± 6.07 ^{a}	147.43 ± 5.10 ^{a}	0.000	0.087	0.004
T-AOC (μmol/gprot)	18.00 ± 1.393 ^{a}	18.67 ± 2.052 ^{a}	47.05 ± 2.506 ^{b}	67.69 ± 3.015 ^{c}	0.000	0.000	0.794
MDA (nmol/mgprot)	8.79 ± 0.30 ^{c}	7.83 ± 0.10 ^{b}	5.25 ± 0.40 ^{a}	5.47 ± 0.42 ^{a}	0.000	0.001	0.008

Note: $^{a–c}$ Values in the same row with different letters indicate significant differences ($p < 0.05$), while that with the same letter or no letter superscripts indicate no significant differences ($p > 0.05$). AN: ANOVA, LT: linear trend, QT: quadratic trend.

3.3. *Immunological Analysis*

3.3.1. Liver Immunological Analysis

Immunological parameters in the liver of our experimental fish were also measured. ACP activities were maximum in the 2 and 4% dietary SBL compared with 0 and 8% diets ($p < 0.05$). Similarly, the fish fed 4% SBL possessed significantly higher levels of C3 and LZM than those in the other three groups ($p < 0.05$). The 4% SBL group also displayed significant decreases for the pro-inflammatory cytokines (IL-8, IL-1β and IFN-γ) ($p < 0.05$), and TNF-α values were also significantly decreased in the 8% SBL groups ($p < 0.05$). Interestingly, the anti-inflammatory cytokine, TGF-β1, displayed a trend that was the opposite as that seen for TNF-α ($p < 0.05$). However, there were no significant differences on T-NOS and ALP activities in the liver among these experimental groups (Tables 6 and 7).

Table 6. Effects of dietary soybean lecithin on the immune parameters in the liver of largemouth bass.

	Lecithin (%) in Diets						
Parameters	**0**	**2**	**4**	**8**	**AN**	**LT**	**QT**
LZM (U/mgprot)	29.49 ± 2.08 ^{a}	33.27 ± 2.28 ^{ab}	40.32 ± 1.89 ^{c}	37.61 ± 2.92 ^{bc}	0.002	0.016	0.010
ACP (U/gprot)	143.61 ± 13.68 ^{a}	249.80 ± 14.57 ^{c}	254.52 ± 16.34 ^{c}	190.50 ± 7.86 ^{b}	0.000	0.584	0.000
ALP (U/gprot)	49.61 ± 1.11	49.10 ± 3.84	49.42 ± 2.30	48.42 ± 1.79	0.937	-	-
T-NOS (U/mgprot)	3.60 ± 0.31	3.84 ± 0.22	3.82 ± 0.24	3.91 ± 0.36	0.600	-	-
C3 (μg/mgprot)	87.89 ± 1.58 ^{a}	91.69 ± 3.07 ^{a}	113.96 ± 3.16 ^{c}	102.62 ± 4.57 ^{b}	0.000	0.043	0.014

Note: $^{a–c}$ Values in the same row with different letters indicate significant differences ($p < 0.05$), while that with the same letter or no letter superscripts indicate no significant differences ($p > 0.05$). AN: ANOVA, LT: linear trend, QT: quadratic trend.

3.3.2. Intestinal Immunological Analysis

Intestinal immunological and anti-inflammatory indices were significantly up-regulated for LZM, ACP, ALP, T-NOS activities as well as for C3, IgM and TGF-β1. All these indices reached the maximum at 4% dietary SBL ($p < 0.05$). According to these results, the amounts of pro-inflammatory cytokines (IL-8, IL-1β, TNF-α and IFN-γ) were significantly decreased ($p < 0.05$) and negatively correlated with dietary SBL contents (Tables 8 and 9).

Table 7. Effects of dietary soybean lecithin on the anti-inflammatory and pro-inflammatory cytokines in the liver of largemouth bass.

	Lecithin (%) in Diets						
Parameters	0	2	4	8	AN	LT	QT
TGF-β1 (ng/mgprot)	0.20 ± 0.01 [a]	0.20 ± 0.01 [a]	0.27 ± 0.03 [b]	0.31 ± 0.02 [b]	0.002	0.004	0.007
IL-8 (ng/gprot)	42.38 ± 1.33 [b]	38.73 ± 2.71 [ab]	34.31 ± 3.19 [a]	36.22 ± 1.92 [ab]	0.016	0.033	0.019
IL-1β (ng/gprot)	15.56 ± 0.98 [b]	14.70 ± 0.82 [ab]	12.31 ± 0.81 [a]	13.28 ± 1.10 [ab]	0.012	0.036	0.040
TNF-α (ng/gprot)	38.73 ± 0.94 [c]	37.37 ± 0.97 [bc]	35.06 ± 1.11 [ab]	34.29 ± 1.32 [a]	0.004	0.001	0.144
IFN-γ (ng/gprot)	37.69 ± 0.59 [c]	36.37 ± 0.46 [c]	29.70 ± 1.52 [a]	33.10 ± 0.19 [b]	0.000	0.038	0.010

Note: [a–c] Values in the same row with different letters indicate significant differences ($p < 0.05$), while that with the same letter or no letter superscripts indicate no significant differences ($p > 0.05$). AN: ANOVA, LT: linear trend, QT: quadratic trend.

Table 8. Effects of dietary soybean lecithin on the immune parameters in the intestine of largemouth bass.

	Lecithin (%) in Diets						
Parameters	0	2	4	8	AN	LT	QT
LZM (U/mgprot)	203.42 ± 12.77 [a]	217.28 ± 7.97 [ab]	278.79 ± 13.72 [c]	241.49 ± 15.16 [b]	0.000	0.081	0.011
ACP (U/gprot)	141.01 ± 2.37 [a]	177.22 ± 8.89 [b]	303.27 ± 11.91 [c]	150.17 ± 1.85 [a]	0.000	0.786	0.001
ALP (U/gprot)	296.56 ± 5.42 [a]	322.56 ± 8.05 [b]	693.55 ± 6.86 [d]	434.08 ± 5.80 [c]	0.000	0.176	0.010
T-NOS (U/mgprot)	5.53 ± 0.35 [a]	7.24 ± 0.45 [b]	9.10 ± 0.77 [c]	7.94 ± 0.51 [bc]	0.000	0.031	0.000
C3 (μg/mgprot)	93.96 ± 3.33 [a]	105.79 ± 4.97 [b]	137.75 ± 3.56 [c]	105.41 ± 5.61 [ab]	0.000	0.376	0.001
IgM (μg/mgprot)	167.73 ± 7.63 [a]	178.06 ± 6.54 [a]	241.60 ± 10.34 [b]	176.95 ± 11.76 [a]	0.000	0.568	0.004

Note: [a–c] Values in the same row with different letters indicate significant differences ($p < 0.05$), while that with the same letter or no letter superscripts indicate no significant differences ($p > 0.05$). AN: ANOVA, LT: linear trend, QT: quadratic trend.

Table 9. Effects of dietary soybean lecithin on the anti-inflammatory and pro-inflammatory cytokines in the intestine of largemouth bass.

	Lecithin (%) in Diets						
Parameters	0	2	4	8	AN	LT	QT
TGF-β1 (ng/mgprot)	1.85 ± 0.05 [a]	2.08 ± 0.18 [ab]	2.35 ± 0.19 [b]	2.03 ± 0.11 [a]	0.019	0.337	0.004
IL-8 (ng/gprot)	68.10 ± 1.50 [c]	63.42 ± 4.91 [bc]	50.57 ± 1.93 [a]	59.75 ± 2.85 [b]	0.001	0.114	0.005
IL-1β (ng/gprot)	17.34 ± 0.68 [c]	16.29 ± 0.99 [bc]	14.27 ± 0.65 [ab]	12.36 ± 1.34 [a]	0.001	0.000	0.584
TNF-α (ng/gprot)	37.64 ± 1.01 [b]	35.01 ± 2.06 [b]	28.50 ± 1.80 [a]	26.29 ± 0.95 [a]	0.000	0.000	0.079
IFN-γ (ng/gprot)	28.59 ± 0.82 [b]	23.84 ± 1.62 [a]	22.68 ± 1.32 [a]	22.44 ± 1.05 [a]	0.001	0.006	0.003

Note: [a–c] Values in the same row with different letters indicate significant differences ($p < 0.05$), while that with the same letter or no letter superscripts indicate no significant differences ($p > 0.05$). AN: ANOVA, LT: linear trend, QT: quadratic trend.

3.4. Gene Expression Measurements

3.4.1. Gene Expression in the Liver

The expression of antioxidant-related genes in the liver tissue for Cu/Zn-SOD, CAT, GR and GST3 were significantly enhanced and positively correlated with dietary SBL level and the maximum at 4% SBL ($p < 0.05$) (Figure 3). The mRNA levels of Mn-SOD and GPx3 were increased at 4% dietary SBL and peaked at 8% SBL ($p < 0.05$). Moreover, the 4 to 8% SBL diets significantly up-regulated the expression of Nrf2 compared with controls ($p < 0.05$). In contrast, Keap1a and Keap1b mRNA levels were negatively correlated with dietary SBL doses ($p < 0.05$). As MAPK signalling pathway key genes, NF-κB P65 and MAPK13 were significantly lower at 4% and 8% SBL ($p < 0.05$) (Figure 4). With the SBL dose increasing, MAPK14 mRNA expression levels were only significantly decreased at 8% SBL ($p < 0.05$). The expression amounts of C3, C4, CFD, HEPC, MHC-I, IL-10 and TGF-β1 were all significantly enhanced compared with the controls and reached the maximum at 4% SBL ($p < 0.05$) (Figures 5 and 6). In contrast, the pro-inflammatory cytokines (IL-8, IL-1β, TNF-α and IFN-γ) were significantly lower in the 4% and 8% groups compared with the controls ($p < 0.05$) (Figure 5).

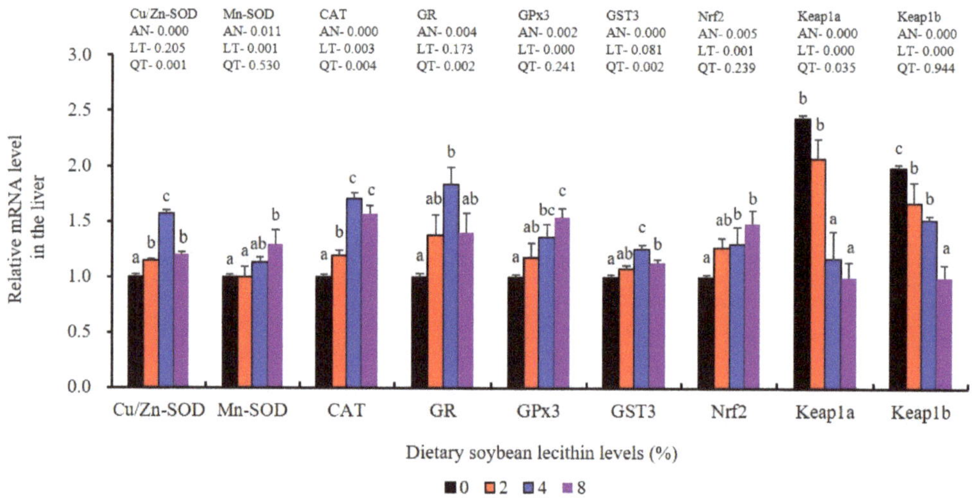

Figure 3. Effect of dietary soybean lecithin on expression levels of antioxidant-related genes in the liver of juvenile largemouth bass. Bars with different letters indicate significant differences ($p < 0.05$), while that with the same letter or no letter superscripts indicate no significant differences ($p > 0.05$), AN: ANOVA, LT: linear trend, QT: quadratic trend.

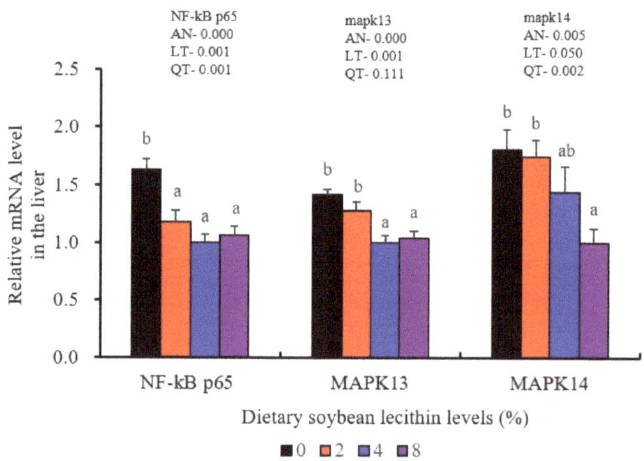

Figure 4. Effects of dietary soybean lecithin on relative gene expression levels of mitogen-activated protein kinase (MAPK) signal pathway in the liver of juvenile largemouth bass. Bars with different letters indicate significant differences ($p < 0.05$), while that with the same letter or no letter superscripts indicate no significant differences ($p > 0.05$), AN: ANOVA, LT: linear trend, QT: quadratic trend.

3.4.2. Expression of Immune-Related Genes in the Intestine

Antioxidant-related gene expression assays in the intestine of our experimental fish indicated that GST3 and GPx3 were significantly increased at 4% dietary SBL ($p < 0.05$). CAT mRNA levels were constantly increased and reached the maximum level in the 8% SBL groups ($p < 0.05$). However, Mn-SOD mRNA expression in the intestine was not significant among these four groups. The 2–8% groups displayed significant up-regulation of Cu/Zn-SOD, Nrf2 and GR ($p < 0.05$) versus the control groups but did not differ between the experimental groups. Keap1a and Keap1b levels were significantly decreased and negatively correlated with SBL doses ($p < 0.05$) (Figure 7).

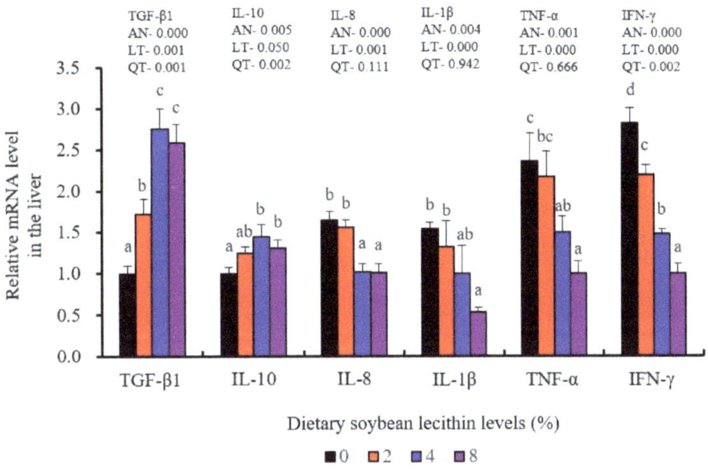

Figure 5. Effect of dietary soybean lecithin on relative gene expression levels of anti-inflammatory and pro-inflammatory cytokines in the liver of juvenile largemouth bass. Bars with different letters indicate significant differences ($p < 0.05$), while that with the same letter or no letter superscripts indicate no significant differences ($p > 0.05$), AN: ANOVA, LT: linear trend, QT: quadratic trend.

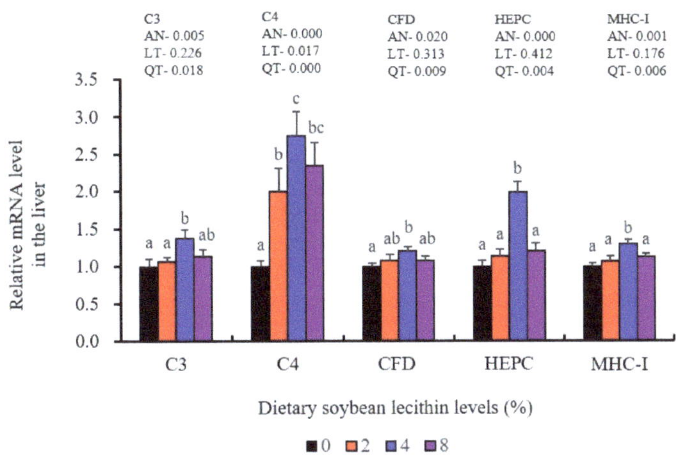

Figure 6. Effect of dietary soybean lecithin on relative gene expression levels of immune-related factors in the liver of juvenile largemouth bass. Bars with different letters indicate significant differences ($p < 0.05$), while that with the same letter or no letter superscripts indicate no significant differences ($p > 0.05$), AN: ANOVA, LT: linear trend, QT: quadratic trend.

The expression levels of NF-κB P65, MAPK13 and MAPK14 were all significantly decreased to the lowest at 4% SBL ($p < 0.05$) (Figure 8). In addition, the 4 and 8% SBL groups displayed significantly decreased levels of IL-1β and TNF-α ($p < 0.05$). Although IL-8 and IFN-γ mRNA expression levels decreased with increasing SBL doses, their levels were only significantly lower at 8% SBL ($p < 0.05$) (Figure 9). The mRNA levels for C4 were significantly enhanced compared with controls and especially at the 2 to 4% levels ($p < 0.05$). The 4% SBL group displayed significantly ($p < 0.05$) elevated levels of C3, HEPC, MHC-I, IL-10 and TGF-β1 while CFD level was significantly higher in the 8% group versus the control groups ($p < 0.05$) (Figures 9 and 10). The measurements of mRNA levels for TJs genes in the intestine indicated significant increases for ZO-1, claudin-3, claudin-4, claudin-5, claudin-23 and claudin-34 in the SBL addition groups ($p < 0.05$). In contrast, claudin-1 expression was negatively related with the dose ($p < 0.05$) while the expression of occludin genes was unaffected (Figure 11). Although mRNA levels were constantly improved with increasing SBL doses, mucin-2 and mucin-17 mRNA levels were significantly up-regulated only in the 8% SBL groups compared with the control groups ($p < 0.05$). The mucin-5AC gene in the intestine were significantly up-regulated and reached the highest level in 4% SBL ($p < 0.05$) compared with the control groups ($p < 0.05$) (Figure 12).

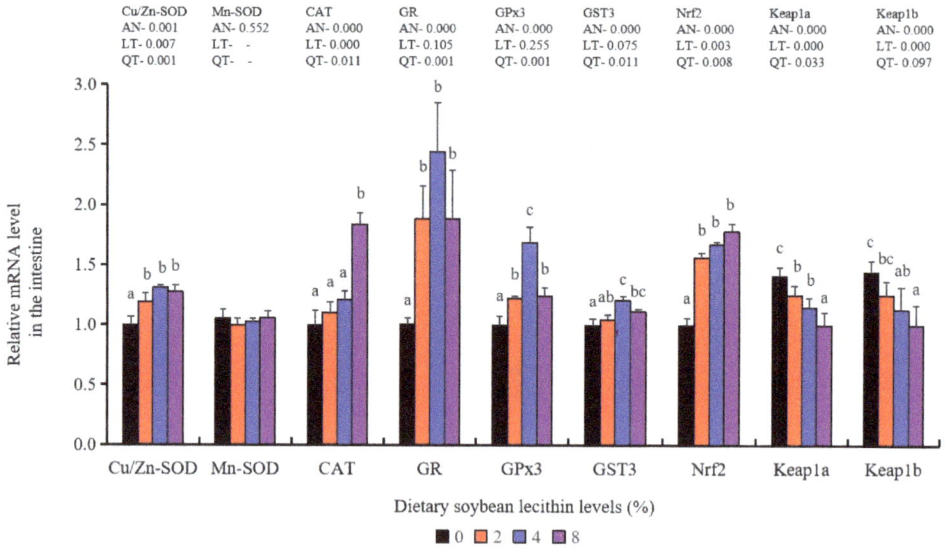

Figure 7. Effect of dietary soybean lecithin on expression levels of antioxidant-related genes in the intestine of juvenile largemouth bass. Bars with different letters indicate significant differences ($p < 0.05$), while that with the same letter or no letter superscripts indicate no significant differences ($p > 0.05$), AN: ANOVA, LT: linear trend, QT: quadratic trend.

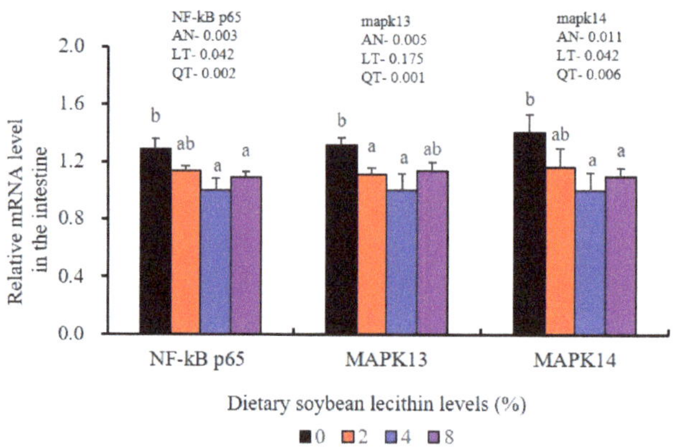

Figure 8. Effects of dietary soybean lecithin levels on relative gene expression levels of mitogen-activated protein kinase (MAPK) signal pathway in the intestine of juvenile largemouth bass. Bars with different letters indicate significant differences ($p < 0.05$), while that with the same letter or no letter superscripts indicate no significant differences ($p > 0.05$), AN: ANOVA, LT: linear trend, QT: quadratic trend.

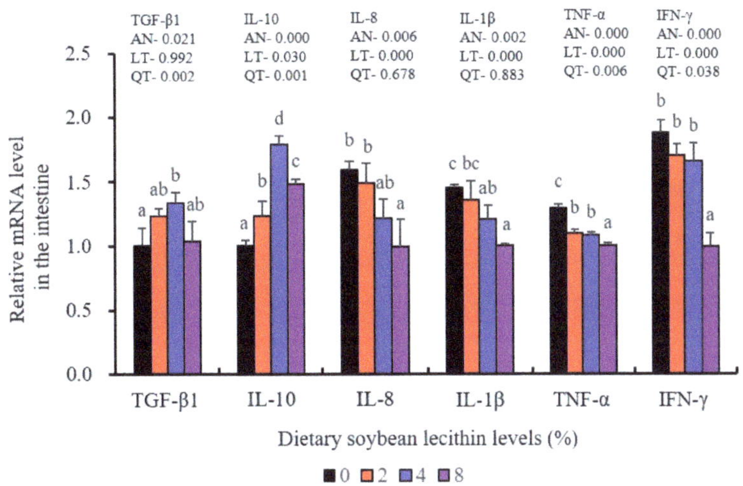

Figure 9. Effect of dietary soybean lecithin on relative gene expression levels of anti-inflammatory and pro-inflammatory cytokines in the intestine of juvenile largemouth bass. Bars with different letters indicate significant differences ($p < 0.05$), while that with the same letter or no letter superscripts indicate no significant differences ($p > 0.05$), AN: ANOVA, LT: linear trend, QT: quadratic trend.

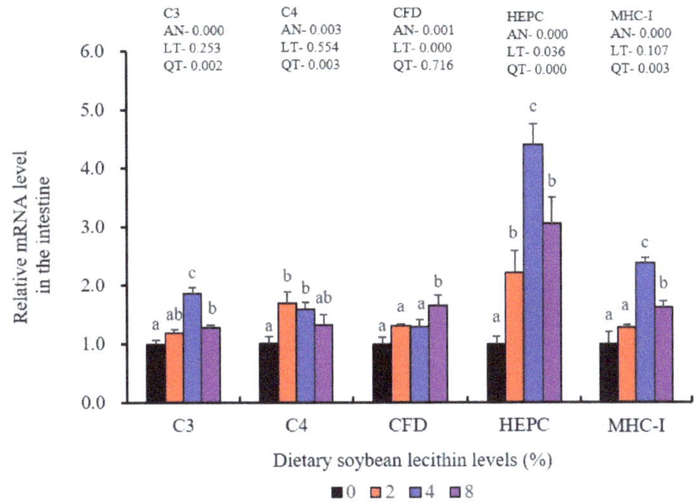

Figure 10. Effect of dietary soybean lecithin on relative gene expression levels of immune-related factors in the intestine of juvenile largemouth bass. Bars with different letters indicate significant differences ($p < 0.05$), while that with the same letter or no letter superscripts indicate no significant differences ($p > 0.05$), AN: ANOVA, LT: linear trend, QT: quadratic trend.

3.5. Histomorphometry

Our analysis of experimental fish also included examinations of the microstructure of intestinal sections (Figure 13). Villi height and muscle thickness were all decreased in the 0 and 2% SBL groups compared with the 4 and 8% groups ($p < 0.05$). Crypt depths and villus widths were not significant (Table 10). The morphology was characterized by detached villi and villous atrophy in the 0 and 2% SBL groups. In contrast, the morphology of the intestine in the 4 and 8% groups was well-developed with increased villi heights and muscular thickness that was the highest in the 4% SBL group (Figure 13).

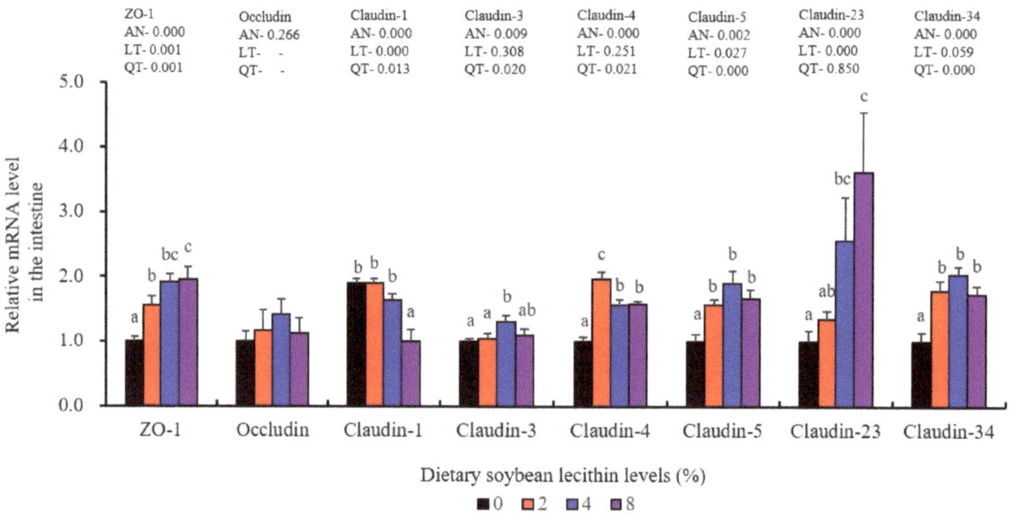

Figure 11. Effects of dietary soybean lecithin on intestinal tight junction-related gene expression levels of juvenile largemouth bass. Bars with different letters indicate significant differences ($p < 0.05$), while that with the same letter or no letter superscripts indicate no significant differences ($p > 0.05$), AN: ANOVA, LT: linear trend, QT: quadratic trend.

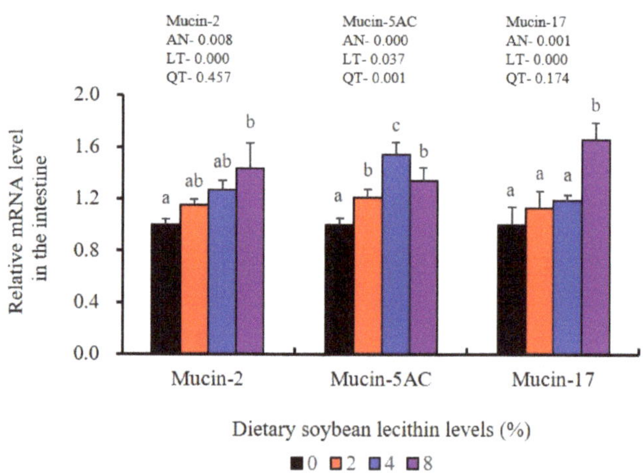

Figure 12. Effects of dietary soybean lecithin levels on intestinal mucins gene expression levels of juvenile largemouth bass. Bars with different letters indicate significant differences ($p < 0.05$), while that with the same letter or no letter superscripts indicate no significant differences ($p > 0.05$), AN: ANOVA, LT: linear trend, QT: quadratic trend.

Table 10. The effects of soybean lecithin on the intestinal morphology of juvenile largemouth bass.

	Lecithin (%) in Diets						
Parameters	**0**	**2**	**4**	**8**	**AN**	**LT**	**QT**
Villi height (μm)	456.96 ± 17.21 [a]	459.69 ± 22.44 [a]	508.00 ± 15.11 [b]	463.15 ± 9.68 [a]	0.043	0.704	0.088
Villi width (μm)	128.45 ± 15.54	131.00 ± 13.11	145.00 ± 10.52	133.11 ± 11.22	0.313	-	-
Muscular thickness (μm)	99.17 ± 5.00 [a]	103.53 ± 4.27 [ab]	112.87 ± 7.80 [b]	99.25 ± 6.83 [a]	0.025	0.584	0.026
Crypt depth (μm)	60.23 ± 5.46	52.92 ± 6.09	47.28 ± 5.92	51.54 ± 4.11	0.112	-	-

Note: [a–c] Values in the same row with different letters indicate significant differences ($p < 0.05$), while that with the same letter or no letter superscripts indicate no significant differences ($p > 0.05$). AN: ANOVA, LT: linear trend, QT: quadratic trend.

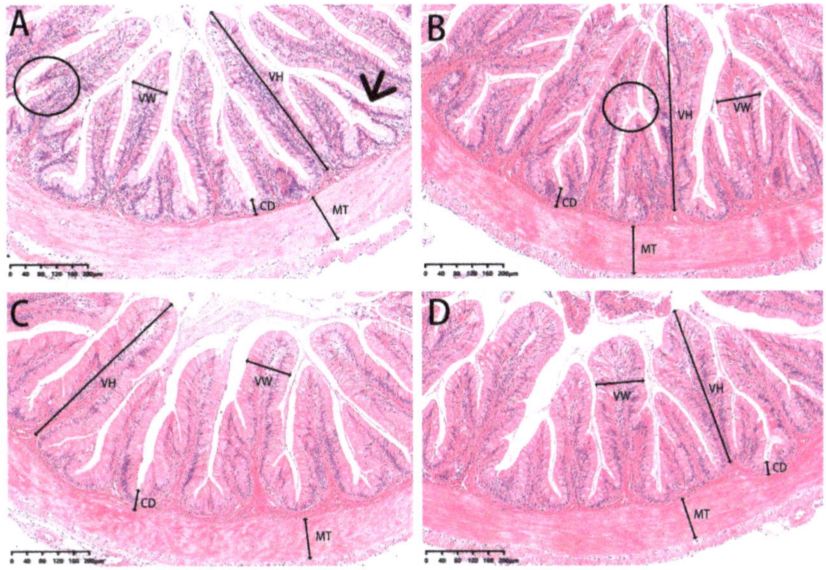

Figure 13. HE staining of the intestine sections of juvenile largemouth bass fed with diets containing 0% (**A**), 2% (**B**), 4% (**C**) and 8% (**D**) SBL (magnification × 100). VH: villus height, VW: villus width, MT: muscular thickness, CD: crypt depth. Circles represent intestinal villi fall off. Black arrow represents villous atrophy.

4. Discussion

As a functional lipid source, SBL has been widely used in the compound feeds of terrestrial and aquatic animals. Many studies have found that dietary SBL could influence the growth performances in different animals. For example, adequate dietary SBL could increase the growth in Caspian brown trout (*S. trutta caspius*) [9], common carp (*C. carpio*) [10], stellate sturgeon (*A. stellatus*) [11], rock bream (*O. fasciatus*) [12], sea urchin (*Strongylocentrotus intermedius*) [34], pacific white shrimp (*Litopenaeus vannamei*) [35], cattle [1] and chickens [3]. Consistent with these studies, it was found that SBL added at 4 and 8% significantly increased WG and DGR for the largemouth bass, which indicated that SBL dietary supplementation have had positive effects on the growth of this species. However, other studies have found no significant growth effects of SBL addition in golden mahseer (*T. putitora*) [6] and channel catfish (*I. punctatus*) [13]. These differences may be related to the different fish species, developmental periods and rearing conditions [6,8–13].

Haematological parameters are important indicators of the health status and nutritional conditions of aquatic animals. Consistent with previous results in trout [9] and sturgeon [11], it was found that 4% SBL elevated WBC and MON levels in blood, indicating enhanced immunity [36]. It was also found that 4% SBL increased RBC, HGB, PLT, MCV and MCH levels suggesting that SBL could reduce inflammation and enhance the oxygen-carrying capacity in largemouth bass, which was consistent with the results in Nile tilapia (*O. niloticus*) [15,36]. ALB content is also an effective indicator of liver and immune dysfunction and was also significantly increased with dietary SBL supplementation, which is similar to the results in rainbow trout (*Oncorhynchus mykiss*) [37]. ALP activities were significantly boosted in the serum and intestine, similar to the results of previous studies in largemouth bass (*M. salmoides*) [38]. Therefore, these results suggest that the addition of adequate SBL could enhance the immune status by improving the related haematology and serum biochemistry indices in juvenile largemouth bass.

It is well-known that ROS could be constantly produced during the metabolism of all nutrients and influence the activities of antioxidant enzymes in aerobic organisms. In the current study, SBL addition enhanced antioxidant capabilities by increasing T-SOD, T-AOC,

CAT, GPx and GSH. These results were similar to the results reported in golden mahseer (*T. putitora*) [6], Caspian brown trout (*S. trutta caspius*) [9], common carp (*C. carpio*) [10] and sea urchin (*S. intermedius*) [34]. These data suggest that adequate SBL alleviates ROS toxicity via increasing antioxidant capabilities [39]. In both the liver and the intestine, GST activities were positively correlated with SBL dose while MDA content presented the opposite trend as GST. These data were consistent with SBL reducing the damage to the body caused by oxidative stress. Interestingly, Nrf2 and its downstream targets Cu/Zn-SOD, CAT, GR, GST and GPx3 were all up-regulated and positively associated with SBL doses, while Keap1a and Keap1b expression displayed the opposite effect. These results were similar to the results of previous studies in grass carp (*C. idella*) [40] and *Sillago sihama* [19]. Nrf2/Keap1 signalling is a primary regulator of the antioxidant response where Nrf2 binds to Keap1 to maintain an inhibitory state under normal physiological conditions [41]. Under oxidative stress, this association is broken and Nrf2 is translocated to the nucleus to activate antioxidant enzyme genes [21]. Therefore, dietary SBL could promote the expression of relevant antioxidant genes via Nrf2/Keap1 signalling that serve to minimize oxidative damage in juvenile largemouth bass.

LZM and ACP are additional diagnostic indexes used to evaluate the health status in fish [25]. Complement C3 release enhances phagocytosis as well as the macrophage respiratory burst that also includes NO production that serves a key immune function in fish [42]. The elevated LZM, ACP and complement components in fish fed SBL have been previously reported in stellate sturgeon (*A. stellatus*) [11], Caspian brown trout (*S. trutta caspius*) [9] and common carp (*C. carpio*) [10]. T-NOS activities were also significantly increased in the intestine for the 4% SBL group, and similar findings were reported in largemouth bass (*M. salmoides*) fed adequate fibre [43]. Similarly, IgM levels were also significantly enhanced in the intestine of our experimental fish fed 4% SBL and were consistent with the results for Caspian brown trout (*S. trutta caspius*) [9]. IgM is a primary component of fish humoral response and is essential for immune clearance of pathogens [44]. HEPC also participates in this process as an antimicrobial peptide and enhances phagocytic endocytosis of pathogens [45], and MHC-I plays a pivotal role in antigen presentation [46]. The elevated expression levels of HEPC, MHC-I, C3, C4 and CFD in our experimental fish fed adequate dietary SBL were consistent with results in black carp (*Mylopharyngodon piceus*) [47]. Together, these results indicated that adequate dietary SBL improved systemic and mucosal immunity via enhancing direct defence effectors in the intestine of juvenile largemouth bass.

Furthermore, chronic inflammation is associated with impaired health status [25]. Inflammatory responses are primarily regulated by pro-inflammatory cytokines (IL-1β, TNF-α, IFN-γ and IL-8) and anti-inflammatory cytokines (IL-10 and TGF-β1) [22]. We found that adequate dietary SBL inhibited the expression levels of these four inflammatory cytokines (IL-1β, TNF-α, IFN-γ and IL-8) and promoted expression of TGF-β1, which is in agreement with previous results in largemouth bass (*M. salmoides*) [27] and rainbow trout (*O. mykiss*) [37]. These cytokines levels were also mirrored at the gene expression levels, indicating that SBL could modulate the critical p38 MAPK/NF-κB inflammatory signalling pathway [23]. In our study, compared with the controls, 4 and 8% dietary SBL significantly down-regulated the expression of MAPK13, MAPK14 and NF-κB P65 in the liver and intestine of largemouth bass, which is consistent with previous studies in this fish [5] and grass carp (*C. idella*) [40]. The overall effect of adequate SBL was to enhance innate immunity while inhibiting a hyperactive p38 MAPK/ NF-κB responses.

The intestinal tract is the primary barrier that limits the entry of anti-nutritional factors and pathogenic substances. Largemouth bass fed with the trace mineral supplement azomite displayed enhanced barrier health [48], which is similar with our results in 4 and 8% SBL. Well-developed intestinal structures with increased villus height and muscular thickness in our results indicated better intestinal health for largemouth bass [33]. However, there were no effects associated with soybean oil replacement by SBL in the intestinal tract of broiler chickens [3], which might be due to differences between terrestrial and aquatic animals. We also found that the integrity and selective permeability of the intestinal cell barrier is maintained in epithelial cells linked via transmembrane TJs including occludin, zonula occludens-1 (ZO-1) and claudin [29]. Similar with previous results in largemouth bass (*M. salmoides*) fed phospholipid and starch [5,27], our present study found that 4 and 8% dietary SBL elevated expression of ZO-1, claudin-3, claudin-4, claudin-5, claudin-23 and claudin-34, while suppressed expression of claudin-1, which indicated that higher levels of TJs could improve intestinal barrier functions and integrity [27–30]. Meanwhile, the expression of mucin-2, mucin-5AC and mucin-17 was also enhanced with the higher doses of SBL, suggesting SBL could enhance the mucus barrier [26]. Taken together, these results indicate that adequate SBL supplement resulted in a positive and protective effect in maintaining intestinal barrier function and increased mucin contents [43].

In conclusion, 4% dietary SBL could improve growth as well as haematological and serum biochemical indicators of health. Adequate dietary SBL could up-regulate antioxidant capabilities by increasing antioxidant enzyme activities through the Nrf2/Keap1 pathway, boost immunity by raising direct defensive effectors and alleviate the inflammatory responses through p38 MAPK/NF-κB regulation. By modulating the expression of TJs and mucin-5AC, adequate SBL also improved intestinal barrier properties in largemouth bass. These findings supplied the benefits of adequate SBL used in functional artificial feed of largemouth bass.

Author Contributions: Conceptualization, C.W., J.W. and W.Y.; methodology, J.W., W.Y., Z.L., X.J. and P.Z.; software, J.W., W.Y., R.S. and P.Z.; validation, J.W., W.Y., Z.L., H.Z., P.Z. and Y.X.; formal analysis, J.W., W.Y. and R.Z.; investigation, J.W., W.Y., P.Z. and X.X.; resources, Z.Z. and J.Y.; data curation, J.W. and W.Y.; writing-original draft preparation, J.W., W.Y. and P.Z.; writing-review and editing, J.W., W.Y., Z.L., R.S. and C.W.; visualization, J.W., W.Y., Z.L. and S.L.; supervision, J.W., W.Y. and R.Z.; project administration, C.W.; funding acquisition, C.W. and Z.Z. All authors have read and agreed to the published version of the manuscript.

Funding: This research was funded by the Key Research and Development Program of Zhejiang Province (2019C02046) and Aquatic Industry Technology project for regional test station of Zhejiang Province (2022-2024-003).

Institutional Review Board Statement: Animal procedures were strictly performed in accordance with the Regulations of the Experimental Animal Ethics Committee of Huzhou University and approved by the Institutional Animal Care and Use Committee (approval ID: HUZJ-DW-2021-019; approval date: 26 February 2021).

Informed Consent Statement: Not applicable.

Data Availability Statement: The data presented in this study are available in the main article.

Conflicts of Interest: The authors declare no conflict of interest.

References

1. Chen, G.J.; Zhang, R.; Wu, J.H.; Shang, Y.S.; Li, X.D.; Qiong, M.; Wang, P.C.; Li, S.G.; Gao, Y.H.; Xiong, X.Q. Effects of soybean lecithin supplementation on growth performance, serum metabolites, ruminal fermentation and microbial flora of beef steers. *Livest. Sci.* **2020**, *240*, 104121.
2. Shamshiri, M.K.; Momtazi-Borojeni, A.A.; Shahraky, M.K.; Rahimi, F. Lecithin soybean phospholipid nano-transfersomes as potential carriers for transdermal delivery of the human growth hormone. *J. Cell. Biochem.* **2019**, *120*, 9023–9033. [CrossRef]
3. Viñado, A.; Castillejos, L.; Barroeta, A.C. Soybean lecithin as an alternative energy source for grower and finisher broiler chickens: Impact on performance, fatty acid digestibility, gut health, and abdominal fat saturation degree. *Poult. Sci.* **2020**, *99*, 5653–5662. [PubMed]
4. Nemati, M.; Ghasemi, H.A.; Hajkhodadadi, I.; Moradi, M.H. De-oiled soy lecithin positively influenced growth performance, nutrient digestibility, histological intestinal alteration, and antioxidant status in turkeys fed with low energy diets. *Br. Poult. Sci.* **2021**, *62*, 858–867. [CrossRef] [PubMed]
5. Wang, S.L.; Han, Z.H.; Turchini, G.M.; Wang, X.Y.; Fang, Z.S.; Chen, N.S.; Xie, R.T.; Zhang, H.T.; Li, S.L. Effects of dietary phospholipids on growth performance, digestive enzymes activity and intestinal health of largemouth bass (*Micropterus salmoides*) larvae. *Front. Immunol.* **2022**, *12*, 827946. [PubMed]

6. Ciji, A.; Akhtar, M.S.; Tripathi, P.H.; Pandey, A.; Rajesh, M.; Samkamalam, B. Dietary soy lecithin augments antioxidative defense and thermal tolerance but fails to modulate non-specific immune genes in endangered golden mahseer (*Tor putitora*) fry. *Fish Shellfish Immunol.* **2021**, *109*, 34–40. [PubMed]
7. Siyal, F.A.; El-Hack, M.E.A.; Alagawany, M.; Wang, C.; Wan, X.; He, J.; Wang, M.; Zhang, L. Effect of soy lecithin on growth performance, nutrient digestibility and hepatic antioxidant parameters of broiler chickens. *Int. J. Pharmacol.* **2017**, *13*, 396–402.
8. Saleh, N.E.; Wassef, E.A.; Kamel, M.A.; El-Haroun, E.R.; El-Tahan, R.A. Beneficial effects of soybean lecithin and vitamin C combination in fingerlings gilthead seabream (*Sparus aurata*) diets on; Fish performance, oxidation status and genes expression responses. *Aquaculture* **2022**, *546*, 737345.
9. Haghparast, R.J.; Moghanlou, K.S.; Mohseni, M.; Imani, A. Effect of dietary soybean lecithin on fish performance, hemato-immunological parameters, lipid biochemistry, antioxidant status, digestive enzymes activity and intestinal histomorphometry of pre-spawning Caspian brown trout (*Salmo trutta caspius*). *Fish Shellfish Immunol.* **2019**, *91*, 50–57.
10. Adel, M.; Gholaghaie, M.; Khanjany, P.; Citarasu, T. Effect of dietary soybean lecithin on growth parameters, digestive enzyme activity, antioxidative status and mucosal immune responses of common carp (*Cyprinus carpio*). *Aquac. Nutr.* **2017**, *23*, 1145–1152.
11. Jafari, F.; Agh, N.; Noori, F.; Tokmachi, A.; Gisbert, E. Effects of dietary soybean lecithin on growth performance, blood chemistry and immunity in juvenile stellate sturgeon (*Acipenser stellatus*). *Fish Shellfish Immunol.* **2018**, *80*, 487–496. [PubMed]
12. Tan, P.; Zhang, P.; Zhang, L.; Zhu, W.L.; Wang, L.G.; Chen, R.Y.; Zhu, Q.H.; Xu, D.D. Effects of soybean lecithin on growth performance, intestine morphology, and liver tissue metabolism in rock bream (*Oplegnathus fasciatus*) larvae. *Front. Mar. Sci.* **2022**, *9*, 942259.
13. Sink, T.D.; Lochmann, R.T. The effects of soybean lecithin supplementation to a practical diet formulation on juvenile channel catfish, *Ictalurus punctatus*: Growth, survival, hematology, innate immune activity, and lipid biochemistry. *J. World Aquacult. Soc.* **2014**, *45*, 163–172. [CrossRef]
14. Ahmed, I.; Ahmad, I. Dietary lysine modulates growth performance, haemato-biochemical indices, non-specific immune response, intestinal enzymatic activities and antioxidant properties of rainbow trout, *Oncorhynchus mykiss* fingerlings. *Aquacult. Nutr.* **2021**, *27*, 124–139. [CrossRef]
15. Hassaan, M.S.; Soltan, M.A.; Jarmołowicz, S.; Abdo, H.S. Combined effects of dietary malic acid and Bacillus subtilis on growth, gut microbiota and blood parameters of Nile tilapia (*Oreochromis niloticus*). *Aquac. Nutr.* **2018**, *24*, 83–93. [CrossRef]
16. Harikrishnan, R.; Naafar, A.; Musthafa, M.S.; Ahamed, A.; Arif, I.A.; Balasundaram, C. Effect of *Agaricus bisporus* enriched diet on growth, hematology, and immune protection in *Clarias gariepinus* against *Flavobacterium columnare*. *Fish Shellfish Immunol.* **2018**, *73*, 245–251.
17. Jia, X.W.; Qian, P.C.; Wu, C.L.; Xie, Y.Y.; Yang, W.X.; Song, R.; Wu, J.J.; Ye, J.Y. Effects of dietary pantothenic acid on growth, antioxidant ability and innate immune response in juvenile black carp. *Aquac. Rep.* **2022**, *24*, 101131.
18. Chen, S.; Lu, Y.; Chen, Y.; Xu, J.D.; Chen, L.L.; Zhao, W.B.; Wang, T.T.; Wang, H.; Wang, P.J. The effect of Bu Shen Huo Xue Tang on autoimmune premature ovarian insufficiency via Modulation of the Nrf2/Keap1 signaling pathway in mice. *J. Ethnopharmacol.* **2021**, *273*, 113996.
19. Huang, Q.C.; Zhang, S.; Du, T.; Yang, Q.H.; Chi, S.Y.; Liu, H.Y.; Yang, Y.Z.; Dong, X.H.; Tan, B.P. Effects of dietary vitamin E on growth, immunity and oxidation resistance related to the Nrf2/Keap1 signalling pathway in juvenile *Sillago sihama*. *Anim. Feed Sci. Technol.* **2020**, *262*, 114403. [CrossRef]
20. Chen, X.; Yang, T.; Wang, W.; Xi, W.; Zhang, T.; Li, Q.; Yang, A.; Wang, T. Circular RNAs in immune responses and immune diseases. *Theranostics* **2019**, *9*, 588–607. [CrossRef]
21. Habte-Tsion, H. A review on fish immuno-nutritional response to indispensable amino acids in relation to TOR, NF-κB and Nrf2 signaling pathways: Trends and prospects. *Comp. Biochem. Physiol. Part B Biochem. Mol. Biol.* **2020**, *241*, 110389.
22. Xu, Q.; Deng, D.; Guo, H.; Yuan, H.; Zhang, W.; Wang, B.; Lu, Y.; Chen, D.; Zhang, S. Comprehensive comparison of thirteen kinds of cytokine receptors from the endangered fish Chinese sturgeon (*Acipenser sinensis*). *Dev. Comp. Immunol.* **2021**, *123*, 104132. [CrossRef]
23. Li, H.; Gao, C.; Liu, C.; Liu, L.; Zhuang, J.; Yang, J.; Zhou, C.; Feng, F.; Sun, C.; Wu, J. A review of the biological activity and pharmacology of cryptotanshinone, an important active constituent in Danshen. *Biomed. Pharmacother.* **2021**, *137*, 111332. [PubMed]
24. Nordgreen, J.; Munsterhjelm, C.; Aae, F.; Popova, A.; Boysen, P.; Ranheim, B.; Heinonen, M.; Raszplewicz, J.; Piepponen, P.; Lervik, A.; et al. The effect of lipopolysaccharide (LPS) on inflammatory markers in blood and brain and on behavior in individually-housed pigs. *Physiol. Behav.* **2018**, *195*, 98–111. [PubMed]
25. Zhao, H.F.; Jiang, W.D.; Liu, Y.; Jiang, J.; Wu, P.; Kuang, S.Y.; Tang, L.; Tang, W.N.; Zhang, Y.A.; Zhou, X.Q.; et al. Dietary choline regulates antibacterial activity, inflammatory response and barrier function in the gills of grass carp (*Ctenopharyngodon idella*). *Fish Shellfish Immunol.* **2016**, *52*, 139–150. [CrossRef]
26. Malachowicz, M.; Wenne, R.; Burzynski, A. De novo assembly of the sea trout (*Salmo trutta* m. *trutta*) skin transcriptome to identify putative genes involved in the immune response and epidermal mucus secretion. *PLoS ONE* **2017**, *12*, e172282.
27. Zhou, Y.L.; He, G.L.; Jin, T.; Chen, Y.J.; Dai, F.Y.; Luo, L.; Lin, S.M. High dietary starch impairs intestinal health and microbiota of largemouth bass, *Micropterus salmoides*. *Aquaculture* **2021**, *534*, 736261. [CrossRef]

28. Chen, L.; Feng, L.; Jiang, W.D.; Jiang, J.; Wu, P.; Zhao, J.; Kuang, S.Y.; Tang, L.; Tang, W.N.; Zhang, Y.A.; et al. Intestinal immune function, antioxidant status and tight junction proteins mRNA expression in young grass carp (*Ctenopharyngodon idella*) fed riboflavin deficient diet. *Fish Shellfish Immunol.* **2015**, *47*, 470–484.
29. Zhao, J.; Feng, L.; Liu, Y.; Jiang, W.D.; Wu, P.; Jiang, J.; Zhang, Y.G.; Zhou, X.Q. Effect of dietary isoleucine on the immunity, antioxidant status, tight junctions and microflora in the intestine of juvenile Jian carp (*Cyprinus carpio* var. Jian). *Fish Shellfish Immunol.* **2014**, *41*, 663–673. [CrossRef]
30. Yang, E.J.; Zhang, J.D.; Yang, L.T.; Amenyogbe, E.; Wang, W.Z.; Huang, J.S.; Chen, G. Effects of hypoxia stress on digestive enzyme activities, intestinal structure and the expression of tight junction proteins coding genes in juvenile cobia (*Rachycentron canadum*). *Aquac. Res.* **2021**, *52*, 5630–5641.
31. Wang, D.; Wu, F.X.; Song, D.D.; Gao, H.Q. *China Fishery Statistical Yearbook*; Article in China; China Agriculture Press: Beijing, China, 2022; p. 25.
32. Sun, C.; Li, J.; Dong, J.; Niu, Y.; Hu, J.; Lian, J.; Li, W.; Li, J.; Tian, Y.; Shi, Q.; et al. Chromosome-level genome assembly for the largemouth bass *Micropterus salmoides* provides insights into adaptation to fresh and brackish water. *Mol. Ecol. Resour.* **2021**, *21*, 301–315. [PubMed]
33. Yang, H.; Bian, Y.H.; Huang, L.L.; Lan, Q.; Ma, L.Z.; Li, X.Q.; Leng, X.J. Effects of replacing fish meal with fermented soybean meal on the growth performance, intestinal microbiota, morphology and disease resistance of largemouth bass (*Micropterus salmoides*). *Aquacult. Rep.* **2022**, *22*, 100954. [CrossRef]
34. Zhang, F.; Ning, Y.C.; Yuan, R.; Ding, J.; Chang, Y.Q.; Zuo, R.T. Effects of soya lecithin addition on the growth, antioxidant capacity, gonad development and nutritional quality of adult sea urchin (*Strongylocentrotus intermedius*). *Aquacult. Rep.* **2022**, *22*, 100990.
35. Liang, X.L.; Luo, X.L.; Lin, H.X.; Han, F.G.; Qin, J.G.; Chen, L.Q.; Xu, C.; Li, E. Growth, health, and gut microbiota of female pacific white shrimp, *Litopenaeus vannamei* broodstock fed different phospholipid sources. *Antioxidants* **2022**, *11*, 1143. [CrossRef] [PubMed]
36. El Sayed, A.F.M.; Tammam, M.S.; Makled, S.O. Lecithin-containing bioemulsifier boosts growth performance, feed digestion and absorption and immune response of adult Nile tilapia (*Oreochromis niloticus*). *Aquac. Nutr.* **2021**, *27*, 757–770. [CrossRef]
37. Taghavizadeh, M.; Hosseini Shekarabi, S.P.; Mehrgan, M.S.; Islami, H.R. Efficacy of dietary lysophospholipids (Lipidol™) on growth performance, serum immuno-biochemical parameters, and the expression of immune and antioxidant-related genes in rainbow trout (*Oncorhynchus mykiss*). *Aquaculture* **2020**, *525*, 735315.
38. Zhong, Y.; Shi, C.; Zhou, Y.; Chen, Y.; Lin, S.; Tang, R. Optimum dietary fiber level could improve growth, plasma biochemical indexes and liver function of largemouth bass, *Micropterus salmoides*. *Aquaculture* **2020**, *518*, 734661. [CrossRef]
39. Kumar, S.; Saxena, J.; Srivastava, V.K.; Kaushik, S.; Singh, H.; Abo-El-Sooud, K.; Abdel-Daim, M.M.; Jyoti, A.; Saluja, R. The interplay of oxidative stress and ROS scavenging: Antioxidants as a therapeutic potential in sepsis. *Vaccines* **2022**, *10*, 1575. [CrossRef]
40. Huang, D.Y.; Maulu, S.Y.; Ren, M.C.; Liang, H.L.; Ge, X.P.; Ji, K.; Yu, H. Dietary lysine levels improved antioxidant capacity and immunity via the TOR and p38 MAPK signaling pathways in grass carp, *Ctenopharyngodon idellus* fry. *Front. Immunol.* **2021**, *12*, 635015. [CrossRef]
41. Zhang, D.M.; Guo, Z.X.; Zhao, Y.L.; Wang, Q.J.; Gao, Y.S.; Yu, T.; Chen, Y.K.; Chen, X.M.; Wang, G.Q. L-carnitine regulated Nrf2/Keap1 activation in vitro and in vivo and protected oxidized fish oil-induced inflammation response by inhibiting the NF-κB signaling pathway in *Rhynchocypris lagowski* Dybowski. *Fish Shellfish Immunol.* **2019**, *93*, 1100–1110. [CrossRef]
42. Liang, H.; Ji, K.; Ge, X.; Ren, M.; Liu, B.; Xi, B.; Pan, L. Effects of dietary arginine on antioxidant status and immunity involved in AMPK-NO signaling pathway in juvenile blunt snout bream. *Fish Shellfish Immunol.* **2018**, *78*, 69–78.
43. Lin, S.M.; Zhou, X.M.; Zhou, Y.L.; Kuang, W.M.; Chen, Y.J.; Luo, L.; Dai, F.Y. Intestinal morphology, immunity and microbiota response to dietary fibers in largemouth bass, *Micropterus salmoide*. *Fish Shellfish Immunol.* **2020**, *103*, 135–142.
44. Wang, J.; Wu, C.; Hu, Y.; Yang, L.; Zhang, X.; Zhang, Y. Plasmablasts induced by chitosan oligosaccharide secrete natural IgM to enhance the humoral immunity in grass carp. *Carbohyd. Polym.* **2022**, *281*, 119073.
45. Alvarez, C.A.; Acosta, F.; Montero, D.; Guzman, F.; Torres, E.; Vega, B.; Mercado, L. Synthetic hepcidin from fish: Uptake and protection against *Vibrio anguillarum* in sea bass (*Dicentrarchus labrax*). *Fish Shellfish Immunol.* **2016**, *55*, 662–670. [CrossRef]
46. Rock, K.L.; Reits, E.; Neefjes, J. Present yourself! By MCH class I and MCH class II molecules. *Trends Immunol.* **2016**, *37*, 724–737. [PubMed]
47. Wu, C.L.; Chen, L.; Lu, Z.B.; Gao, J.E.; Chu, Y.D.; Li, L.; Wang, M.; Zhang, G.Y.; Zhang, M.T.; Ye, J.Y. The effects of dietary leucine on the growth performances, body composition, metabolic abilities and innate immune responses in black carp *Mylopharyngodon piceus*. *Fish Shellfish Immunol.* **2017**, *67*, 419–428. [PubMed]
48. Xu, X.Y.; Li, X.Q.; Xu, Z.; Yao, W.X.; Leng, X.J. Dietary Azomite, a natural trace mineral complex, improved the growth, immunity response, intestine health and resistance against bacterial infection in largemouth bass (*Micropterus salmoides*). *Fish Shellfish Immunol.* **2021**, *108*, 53–62. [PubMed]

metabolites

MDPI

Article

Dietary Lactoferrin Supplementation Improves Growth Performance and Intestinal Health of Juvenile Orange-Spotted Groupers (*Epinephelus coioides*)

Tao Song [1], Yingmei Qin [1], Liner Ke [1], Xuexi Wang [2], Kun Wang [1], Yunzhang Sun [1] and Jidan Ye [1,*]

1 Xiamen Key Laboratory for Feed Quality Testing and Safety Evaluation, Fisheries College of Jimei University, Xiamen 361021, China

2 Key Laboratory of Marine Biotechnology of Fujian Province, College of Marine Sciences, Fujian Agriculture and Forestry University, Fuzhou 350002, China

* Correspondence: yjyjd@jmu.edu.cn; Tel.: +86-59-2618-1420

Abstract: A 56-day feeding trial was conducted to investigate the effects of dietary lactoferrin (LF) supplementation on the growth performance and intestinal health of juvenile orange-spotted groupers fed high-soybean-meal (SBM) diets. The control diet (FM) and high-soybean-meal diet (SBM60) were prepared to contain 480 g/kg protein and 110 g/kg fat. Three inclusion levels of 2, 6, and 10 g/kg LF were added into the SBM60 to prepare three diets (recorded as LF2, LF6, and LF10, respectively). The results showed that the supplementation of LF in SBM60 increased the growth rate in a dose-dependent manner. However, the feed utilization, hepatosomatic index, whole-body proximate composition, and the abundance and diversity of intestinal microbiota did not vary across the dietary treatments ($p > 0.05$). After the dietary intervention with LF, the contents of the intestinal malondialdehyde, endotoxin, and d-lactic acid, as well as the plasma low-density lipoprotein cholesterol, high-density lipoprotein cholesterol, and total cholesterol were lower, and the intestinal activities of the glutathione peroxidase, lipase, trypsin, and protease were higher in the LF2-LF10 groups than that in the SBM60 group ($p < 0.05$). The supplementation of LF in SBM60 increased the muscle layer thickness of the middle and distal intestine and the mucosal fold length of the middle intestine vs. the SBM60 diet ($p < 0.05$). Furthermore, the supplementation of LF in SBM60 resulted in an up-regulation of the mRNA levels for the *IL-10* and *TGF-β1* genes and a down-regulation of the mRNA levels of the *IL-1β*, *IL-12*, *IL-8*, and *TNF-α* genes vs. the SBM60 diet ($p < 0.05$). The above results showed that a dietary LF intervention improves the growth and alleviates soybean meal-induced enteritis in juvenile orange-spotted groupers. The dietary appropriate level of LF was at 5.8 g/kg, through the regression analysis of the percent weight gain against the dietary LF inclusion levels.

Keywords: lactoferrin; *Epinephelus coioides*; growth performance; intestinal damage

Citation: Song, T.; Qin, Y.; Ke, L.; Wang, X.; Wang, K.; Sun, Y.; Ye, J. Dietary Lactoferrin Supplementation Improves Growth Performance and Intestinal Health of Juvenile Orange-Spotted Groupers (*Epinephelus coioides*). *Metabolites* **2022**, *12*, 915. https://doi.org/10.3390/metabo12100915

Academic Editors: Qingchao Wang, Yan He and Qingsong Tan

Received: 20 August 2022
Accepted: 25 September 2022
Published: 28 September 2022

1. Introduction

Fish meal (FM) has always been the main important protein source in the feeds of farmed marine fish due to its high-quality protein, balanced amino acids profile, and less anti-nutritional factors (ANFs) in comparison with terrestrial animals and plant protein sources [1]. However, over the past three decades, the stagnant global marine fishery catches have also led to stagnant FM production due to the shift in climate change [2]. At the same time, the rapid expansion of the aquaculture industry has stimulated a huge demand for FM, resulting in a sustained shortage of FM supply [3]. Therefore, research into FM replacement with other protein sources has always been one of the most priority issues in aquafeeds. Soybean meal (SBM), as the major plant protein source, is widely used in aquafeeds due to its relatively balanced amino acids profile, huge output and availability, and reasonable price [1,4,5]. However, when SBM is used in aquafeeds in a large proportion, its side effects are also considerable. It is well known that SBM can cause the reduction in the ability of fish to digest and absorb nutrients, due to many ANFs that induce enteritis [6,7], the so-called soybean meal-induced enteritis (SBMIE) [8]. Therefore, how to prevent and control the widespread SBMIE is the key to maintain a normal daily fish culture and reduce disease risk.

In the last two decades, a great deal of research effort has been devoted to alleviating or/and preventing SBMIE. An effective strategy to counteract fish intestinal inflammation is the dietary use of functional substances that exert metabolic regulation, antioxidant properties, and immune promotion [4,9–12]. Lactoferrin (LF), a glycoprotein with a molecular weight of 78–80 kDa, is rich in milk [13,14] and has many biological functions [14–16]. A previous study showed that dietary LF supplementation could enhance the growth performance of early-weaned piglets by promoting the proliferation of intestinal beneficial bacteria, such as lactic acid bacteria; inhibiting the proliferation of *E. coli*; and improving the intestinal mucosal morphological structure and intestinal function [17]. Dietary LF supplementation could improve the growth performance of early-weaned piglets and zebrafish by reducing pathogenic bacteria and diarrhea, enriching beneficial bacteria, and affecting intestinal morphology [15,16]. LF was found to inhibit the activation of the TOR signaling pathway induced by lipopolysaccharide (LPS), thereby inhibiting the production of pro-inflammatory factors such as *IL-1β* and *IL-8* or/and indirectly promoting the production of anti-inflammatory factors such as *IL-10* in terrestrial animals [18]. However, little is known about the dietary LF intervention effect on the improvement of SBMIE in farmed fish.

An orange-spotted grouper (*Epinephelus coioides*) is a marine economic carnivorous fish, which is widely cultured in Southeast Asian countries, including China [19,20]. According to the China Fishery Statistics Yearbook (2021), this fish has become the third largest mariculture fish in China, with an annual output of 192,045 tons in 2020 [21]. Although there are many reports about its nutrition, feed research, and development [20,22,23], there is still a lack of nutritional regulation research on the prevention of SBMIE in the fish species. The recent studies also showed that a high-SBM diet caused juvenile orange-spotted grouper enteritis [1,5]. Therefore, in this study, the growth performance, intestinal health, and prevention or/and control of SBMIE were investigated to evaluate the effects of dietary LF supplementation in juvenile *E. coioides* fed high-SBM diets. For this purpose, five experiment diets were prepared, including the FM diet, SBM60 diet without LF, and three SBM60 diets with LF at inclusion levels of 2, 6, and 10 g/kg. This study provided a new technical reference for the prevention of SBMIE in the fish species.

2. Materials and Methods

2.1. Experimental Diets

A control diet (FM) was formulated using FM, casein, and gelatin as the protein sources and fish oil, soybean oil, and soybean lecithin as the lipid sources to contain 480 g/kg crude protein and 110 g/kg crude lipid (Table 1). On the basis of the FM diet, SBM was used to replace 600 g/kg FM protein to prepare a high-SBM diet (SBM60). LF was then added to SBM60 diets at 2, 6, and 10 g/kg to prepare another three experimental diets (LF2, LF6, and LF10, respectively), according to a previous study [17]. The coarse dry feed ingredients were pulverized using a grinder (ZFJ-300, Jiangyin Ruizong Machinery Manufacturing Co., Ltd., Jiangyin, Jiangsu, China) and sifted through a 60-mesh sieve (250 μm particle size), weighed and homogenized. The liquid ingredients (water, soybean oil, fish oil, and soy lecithin) were then added to the dry feed ingredients and a mash was prepared. This dough was extruded into strands and pelletized through 2.5 and 4 mm die using cold press extrusion (F-76, Guangzhou Huagong Optical Mechanical and Electrical Technology Co., Ltd., Guangzhou, Guangdong, China) and a feed pellet shaping machine (GY-500, Changzhou Beicheng Drying Equipment Engineering Co., Ltd., Changzhou, Jiangsu, China). The pellets were dried in a ventilated oven at 55 °C for 24 h to reduce the moisture of the feed to less than

100 g/kg and then stored at room temperature for 24 h, before being sealed in plastic bags and stored in a refrigerator at −20 °C.

Table 1. Formulations and nutrient levels of the experimental diets (on an as-fed basis, g/kg).

Items	Diets [1]				
	FM	SBM60	LF2	LF6	LF10
Ingredients					
Fish meal [2]	520	220	220	220	220
Casein	119.8	112.7	112.7	112.7	112.7
Gelatin	30	28.2	28.2	28.2	28.2
Soybean meal [3]	—	470	470	470	470
Soybean oil	35	35	35	35	35
Fish oil	8.2	35.2	35.2	35.2	35.2
Soybean lecithin	20	20	20	20	20
Lactoferrin 4	—	—	2	6	10
Corn starch	177.2	32.6	30.6	26.6	22.6
Sodium alginate	10	10	10	10	10
$Ca(H_2PO_4)_2$	15	15	15	15	15
Choline chloride	4	4	4	4	4
Stay-C (350 g/kg)	0.3	0.3	0.3	0.3	0.3
Vitamin premix [5]	4	4	4	4	4
Mineral premix [5]	5	5	5	5	5
Taurine	5	8	8	8	8
Microcrystalline cellulose	46.5	—	—	—	—
Total	1000	1000	1000	1000	1000
Nutrient level (analyzed values)					
Dry matter	950.6	957.8	955.6	952.7	954.7
Crude protein	480.5	503.4	517.2	513.3	515
Crude lipid	120.6	114.2	116.3	116.4	116.3
Ash	91.7	84.2	84.8	84	82.9

[1] FM, fish meal diet (control diet); SBM60, high-soybean-meal diet with 600 g/kg fish meal protein replacement and without LF supplementation; LF2, LF6, and LF10 were added 2, 6, and 10 g/kg LF in SBM60 diets, respectively. [2] Fish meal was obtained from Austral Group S.A.A.Peru. (crude protein 703.4 g/kg, crude lipid 90.6 g/kg). [3] Soybean meal was obtained from Jiakang Feed Co., Ltd., Xiamen, China. (crude protein 466.0 g/kg, crude lipid 7.2 g/kg). [4] Lactoferrin (LF, food grade, the purity of lactoferrin was 970 g/kg) was obtained from Fujian Furun Pharmaceutical Co., Ltd. (Fuzhou, China). [5] The vitamin and mineral premixes were obtained from Guangzhou Feixite Aquatic Technology Co., Ltd (Guangzhou, China). Vitamin premix (per kg diet): VA, 10 mg; VD, 10 mg; VE, 100 mg; VB1, 10 mg; VB2, 20 mg; VB6, 20 mg; VB12, 0.05 mg; nicotinic acid, 50 mg; calcium-D-pantothenate, 100 mg; D-biotin, 1 mg; meso-inositol, 500 mg; folic acid, 4 mg. Mineral premix (per kg diet): ferric citrate, 497 mg; $CuSO_4 \cdot 5H_2O$, 24 mg; $ZnSO_4 \cdot 7H_2O$, 176 mg; $MnSO_4 \cdot 4H_2O$, 122 mg; $CoCl_2 \cdot 6H_2O$, 0.18 mg; KIO_3, 0.51 mg; Na_2SeO_3, 0.33 mg.

2.2. Feeding Trial

This experiment was conducted at Fujian Dabeinong Fisheries Technology Company (Zhaoan County, Zhangzhou City, Fujian, China). Prior to the trial, orange-spotted grouper juveniles were kept in a concrete pond and fed with diet FM for a 2-week acclimatization. At the beginning of the experiment, 450 orange-spotted grouper juveniles (initial average weight of 33.82 ± 0.03 g) were randomly distributed into 15 blue polypropylene tanks (500 L/tank), at a density of 30 fish per tank in a water temperature-controlled recirculating culture system. Groups of triplicate tanks were hand fed one of the diets to apparent satiation twice daily (8:00 and 17:00) under a natural photoperiod across a feeding period 56 days. Excess feed was collected 30 min after each meal and dried for 24 h at 65 °C and weighed for the calculation of feed intake. Because of the water loss of the aquaculture system caused by daily sewage discharge, fresh sea water was refilled until the original water level of tanks was reached. During the experimental period, water salinity ranged from 32 to 36, temperature was about 28.5 °C, and dissolved oxygen level was >5.7 mg/L.

2.3. Sample Collection and Chemical Analysis

At the end of the growth trial, fish in each tank were caught and anaesthetized with a dose of 100 μl/L solution of eugenol (Nanjing wensenbao International Trade Co., Ltd., Nanjing, Jiangsu, China). Fish weight and number were then recorded for each tank to measure weight gain (WG), feed efficiency (FE), specific growth rate (SGR), and survival. Three fish from each tank were randomly sampled and pooled in plastic bags and stored at −20 °C for whole-body proximate composition determination. A total of 9 fish per tank (27 fish each group) were weighed individually after anesthesia with eugenol (100 μl/L) to calculate the hepatosomatic index (HSI) and condition factor (CF). Blood was drawn from the caudal vein, using 1 mL heparinized syringe, and centrifuged at 1027× g, 4 °C, 10 min. Plasma was then collected, pooled by tank, and stored in 1.5 mL Eppendorf tubes at −80 °C for the subsequent biochemical analysis. The intestines of nine fish per tank were aseptically removed and pooled into one tube by tank, stored at −80 °C for the analysis of biochemical components, microbiota analysis, and gene expression.

Prior to component analysis, whole-fish samples were prepared according to the method described by Ye et al. [24]. The proximate composition of diet and whole-body fish samples were determined according to standard methods [25]. Dry matter was determined by drying the samples in an oven at 105 °C to a constant weight. Crude protein was determined by the Kjeldahl method ($N \times 6.25$), using Kjeltec TM 8400 Auto Sample Systems (Foss Teacher AB). The crude lipid content was determined by the Soxtec extraction method by using Soxtec Avanti 2050 (Foss Teacher AB). Ash was measured in the residues of samples burned in a muffle furnace at 550 °C for 8 h.

The activities of intestinal protease, trypsin, and amylase were determined according to the method described by Hu et al. [26]. Casein was used as the reaction substrate, and the reaction product was determined by Folin's reagent. At 37 °C, intestinal protease degraded casein for 1 min to produce 1 μg tyrosine as an enzyme activity unit, expressed as U/mg protein. *Na*-benzoyl-arginine-*p*-nitroanilide (BAPNA) was used as the reaction substrate. Degradation of BAPNA by the trypsin for 1 min at 37 °C resulted in 1 μmol *p*-nitroanilide as an enzyme activity unit, expressed as U/g protein. Starch was used as the reaction substrate, and the degradation of 10 mg of starch by amylase at 37 °C at 30 min was taken as an enzyme activity unit, expressed as U/mg protein. The total cholesterol (TC), triglycerides (TG), high-density lipoprotein cholesterol (HDL-C), and low-density lipoprotein cholesterol (LDL-C) contents in plasma samples; the contents of diamine oxidase (DAO), d-lactic acid (D-Lac), endotoxin (ET), endothelin-1 (ET-1), glutathione peroxidase (GSH-Px), superoxide dismutase (SOD), total antioxidant capacity (T-AOC), and malondialdehyde (MDA) in the intestinal samples; and the activity of lipase in the intestinal samples were determined using commercial kits (Nanjing jiancheng Bioengineering Institute, Nanjing, Jiangsu, China), according to the manufacturer's instructions.

2.4. Intestinal Histology Observation

One fish was caught from each tank and dissected to obtain the whole gut, then divided into proximal, middle, and distal intestine (i.e., PI, MI, and DI, respectively), according to the method described by Anguiano et al. [27]. All the segments were washed with normal saline, fixed in Bouin's solution for 24 h, rinsed with 70% (v/v) ethanol solution, and finally immersed in 70% (v/v) ethanol until histological processing was performed [19]. The fixed gut segments were embedded in paraffin and 5 μm sections were cut by using a rotary microtome (KD-2258S, China). The serial histological sections were then mounted on glass slides and stained with hematoxylin and eosin for morphometric analysis. Pictures were examined under a light microscope (Leica DM5500B, Germany), and digital images were taken and processed with a digital camera (Leica DFC450) equipped with the image program LAS AF (Version 4.3.0 Leica). Five slides were prepared for each gut segment sample and 30 measurements were made to determine the number of mucosal folds, muscle layer thickness, and length of the complete mucosal fold.

2.5. Intestinal Microbiota Analysis

The total DNA in the distal intestine (DI) of juvenile orange-spotted groupers were extracted using a DNA extraction kit (Omega Bio-teK, Norcross, GA, USA) according to the manufacturer's instructions. The integrity and quality, purity, and quantity of DNA samples were assessed by electrophoresis on a 1% (*w*/*v*) agarose gel and spectrophotometer method (NanoDrop 2000, Wilmington, DE, USA 260 nm/280 nm optical density ratio), respectively. The V3-V4 region of the 16S rDNA gene of DI bacterial was amplified by polymerase chain reaction (PCR) using the forward primer 338F (5′-ACTCCTACGGGAGGCAGCAG-3′) and the reverse primer 806R (5′-GGACTACNNGGGTATCTAAT-3′). The PCR reaction system included pre-denaturation at 95 °C for 5 min; denaturation at 95 °C for 45 s, annealing at 55 °C for 50 s, and extension at 72 °C for 45 s, 32 cycles; extension at 72 °C for 10 min. Subsequently, high-throughput sequencing was performed using Illumina Miseq PE300 at Beijing Allwegene technology Co., Ltd. (Beijing, China). A library of small fragments was constructed using paired-end for sequencing, and the data were passed through QIIME (v1.8.0) for removal low-quality sequences and chimeras. Based on 97% sequence similarity, similar sequences were assigned to the same operational taxonomic units (OTU). Species classification information corresponding to each OTU was obtained by comparing with the sliva database, and alpha diversity analysis (Shannon, ACE, and Chao1) was performed using Mothur software (version 1.31.2). Based on the weighted unifrace distance, the pheatmap of R (v3.1.1) software package was used for clustering analysis. After the UniFrac algorithm, the information of system evolution was used to compare the difference in species communities among samples and Beta diversity analysis was performed.

2.6. RNA Extraction and Gene Expression

The total RNA was extracted from the intestinal samples using TRIzol® reagent (Takara Co., Ltd., Tokyo, Japan) according to the manufacturer's instructions. Isolated RNA was quantified using the NanoDrop ND-2000 Spectrophotometer, and its integrity was confirmed by agarose gel electrophoresis. The cDNA was generated from 1 μg DNase-treated RNA and synthesized by a PrimeScript™ RT Reagent Kit with gDNA Eraser (Perfect Real Time) (Takara Co., Ltd., Tokyo, Japan). Real-time PCR was employed to determine mRNA levels based on the TB Green™ Premix Ex Taq™ II (Tli RNaseH Plus) (Takara Co., Ltd., Tokyo, Japan) using a QuantStudio™ Real-Time PCR System (ABI) quantitative thermal cycler. The fluorescent quantitative PCR solution consisted of 10 μL TB Green Premix Ex Taq™ II (Tli RNaseH Plus) (2×), 0.8 μL PCR forward primer (10 μM), 0.8 μL PCR reverse primer (10 μM), 2.0 μL RT reaction (cDNA solution), and 6 μL dH_2O. The thermal program included 30 s at 95 °C, 40 cycles at 95 °C for 5 s, and 60 °C for 30 s. The sequences of primers are showed in Table 2. All amplicons were initially separated by agarose gel electrophoresis to ensure that they were of the correct size. β-actin served as the internal reference gene to normalize cDNA loading. The gene expression levels of the target genes were analyzed by the $2^{-\Delta\Delta Ct}$ method [28] after verifying that the primers were amplified with an efficiency of approximately 100% [29], and the data for all treatment groups were compared with the data for the control group.

2.7. Statistical Analysis

All data were presented as mean and standard error of the mean (SEM). The data were analyzed using a one-way analysis of variance (ANOVA) to test for differences between treatments and then the Student–Neuman–Keuls multiple comparison test was performed after confirming the normality and homogeneity of variance using the Kolmogorov–Smirnov test and Levene's test in SPSS Statistics 25.0 (SPSS, Michigan Avenue, Chicago, IL, USA). The data expressed as percentages or ratios were subjected to data conversion prior to statistical analysis. *p*-value < 0.05 was deemed as significant difference.

Table 2. Primers sequences for lipid-related genes and reference genes used for real-time PCR of juvenile orange-spotted groupers.

Genes [1]	Primer Sequence (5′ to 3′) [2]	E-Value (%)	Accession Number
IL-8	F: AAGTTTGCCTTGACCCCGAA R: TGAAGCAGATCTCTCCCGGT	94.0	FJ913064.1
IL-1β	F: GCAACTCCACCGACTGATGA R: ACCAGGCTGTTATTGACCCG	116.0	EF582837.1
IL-10	F: GTCCACCAGCATGACTCCTC R: AGGGAAACCCTCCACGAATC	99.0	KJ741852.1
TGF-β1	F: GCTTACGTGGGTGCAAACAG R: ACCATCTCTAGGTCCAGCGT	102.0	GQ503351.1
IL-12	F: CCAGATTGCACAGCTCAGGA R: CCGGACACAGATGGCCTTAG	115.0	KC662465.1
TNF-α	F: GGATCTGGCGCTACTCAGAC R: CGCCCAGATAAATGGCGTTG	91.0	FJ009449.1
β-actin	F: TGCTGTCCCTGTATGCCTCT R: CCTTGATGTCACGCACGAT	104.0	AY510710.2

[1] IL-8, interleukin-8; IL-1β, interleukin-1β; TNF-α, tumor necrosis factor-α; IL-12, interleukin-12; TGF-β1, transforming growth factor-β1; IL-10, interleukin-10. [2] F, forward; R, reverse.

3. Results

3.1. Growth Performance and Whole-Body Proximate Composition

The results of the growth performance and whole-body proximate composition are presented in Table 3. The diets SBM60 had a lower WG and SGR compared with the diets FM ($p < 0.05$), but the diets LF2-LF10 showed an improved WG and SGR vs. the diets SBM60 ($p < 0.05$) and returned to the level of the diets FM ($p > 0.05$). The WG and FE were in a dose-dependent relationship with the dietary LF inclusion levels (Figure 1). The maximum WG and FE were observed for the diet LF6 and diet LF10, respectively. However, there were no differences in the FE, HSI, CF, survival, and whole-body proximate composition among the dietary treatments ($p > 0.05$).

Table 3. Effects of lactoferrin (LF) supplementation in high-SBM diets on growth performance and whole-body proximate composition of juvenile orange-spotted groupers in a 56-day feeding period [1].

Items	Diets [2]				
	FM	SBM60	LF2	LF6	LF10
			Growth performance		
IBW (g/fish) [3]	33.82 ± 0.10	33.76 ± 0.06	33.74 ± 0.05	33.80 ± 0.07	33.98 ± 0.01
FBW (g/fish) [3]	113.24 ± 0.66 [b]	95.47 ± 2.59 [a]	107.49 ± 4.28 [b]	109.85 ± 3.46 [b]	106.65 ± 2.96 [b]
WG (%) [3]	234.81 ± 2.54 [b]	182.81 ± 7.24 [a]	218.56 ± 12.96 [b]	224.98 ± 9.89 [b]	208.68 ± 5.73 [ab]
SGR (%/d) [3]	2.16 ± 0.01 [b]	1.86 ± 0.05 [a]	2.07 ± 0.07 [b]	2.10 ± 0.05 [b]	2.04 ± 0.05 [b]
FE [3]	0.98 ± 0.00	0.95 ± 0.00	0.96 ± 0.12	0.99 ± 0.07	1.00 ± 0.12
Survival (%) [3]	100.00 ± 0.00	97.78 ± 1.11	100.00 ± 0.00	100.00 ± 0.00	98.89 ± 1.11
HSI (%) [4]	1.31 ± 0.09	1.24 ± 0.04	1.28 ± 0.07	1.24 ± 0.05	1.17 ± 0.02
CF (g/cm^3) [4]	3.16 ± 0.07	3.05 ± 0.11	3.19 ± 0.02	2.93 ± 0.12	2.94 ± 0.03
			Proximate composition (%)		
Moisture	67.05 ± 0.21	67.27 ± 0.22	67.56 ± 0.37	67.42 ± 0.34	68.26 ± 0.39
Crude protein	18.01 ± 0.49	17.95 ± 0.27	17.93 ± 0.90	19.20 ± 0.33	17.75 ± 0.42
Crude lipid	8.25 ± 0.17	7.90 ± 0.29	8.18 ± 0.40	7.85 ± 0.32	7.49 ± 0.11
Ash	5.00 ± 0.15	4.96 ± 0.07	4.90 ± 0.06	4.87 ± 0.22	4.93 ± 0.09

Abbreviations: IBW, initial body weight (g/fish); FBW, final body weight (g/fish); FI, feed intake (g/fish); FN, final number; IN, initial number; LW, liver weight (g/fish); BW, body weight (g/fish); BL, body length (cm/fish). [1] Statistical analysis was performed by one-way ANOVA, followed by S–N–K test. [2] FM, fish meal diet (control diet); SBM60, high-soybean-meal diet with 600 g/kg fish meal protein replacement and without LF supplementation; LF2, LF6, and LF10 were added 2, 6, and 10 g/kg LF in SBM60 diets, respectively. [3] Values are presented as the means ± SEM (n = 3 tanks). [4] Values are presented as the means ± SEM (n = 27 fish). [a,b] Values in the same row with different superscripts indicate significant differences ($p < 0.05$), while that with the same letter or no letter superscripts indicate no significant differences ($p > 0.05$). WG, weight gain (%) = 100 × (FBW − IBW)/IBW. SGR, specific growth rate (%/d) = 100 × (lnFBW − lnIBW)/days. FE, feed efficiency = 100 × (FBW − IBW)/FI (as fed basis, g/fish). Survival (%) = 100 × FN/IN. HSI, hepatosomatic index (%) = 100 × LW/BW. CF, condition factor (g/cm^3) = 100 × BW/(BL)^3.

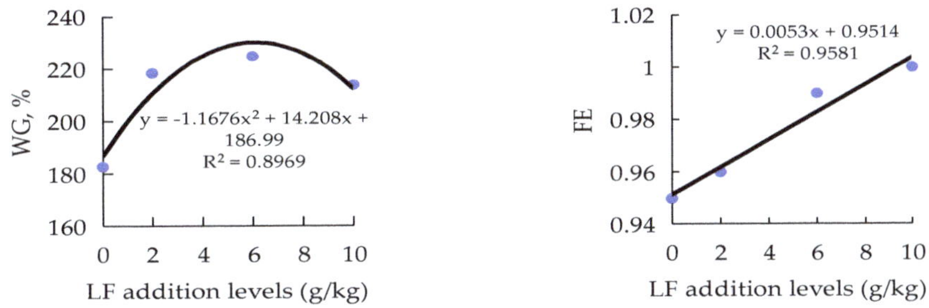

Figure 1. The relationship between weight gain (WG) and feed efficiency (FE) of juvenile orange-spotted groupers and the lactoferrin (LF) inclusion levels in SBM60 diets in a 56-day feeding period. Values are means of 3 triplicates per dietary treatment. SBM60, high-soybean-meal diet with 600 g/kg fish meal protein replacement and without LF supplementation; LF2, LF6, and LF10 were added 2, 6, and 10 g/kg LF in SBM60 diets, respectively.

3.2. Intestinal Antioxidant Capacity

As shown in Table 4, the FM group had higher intestinal GSH-Px activity and lower MDA content compared with the SBM60 group ($p < 0.05$), but the values of the two parameters were not different between the LF2-LF10 groups and the FM group ($p > 0.05$). The MDA content showed a negative quadratic response to the increasing dietary LF inclusion levels; a minimum value was observed for the diet LF6. However, the dietary treatments did not affect the intestinal SOD and T-AOC activities ($p > 0.05$).

3.3. Plasma Components

Table 5 shows that plasma HDL-C content had an irregular change with increasing dietary LF inclusion levels, and the value in the LF6 group was similar ($p > 0.05$) to that of the SBM60 and FM groups but was higher than that of the LF2 and LF10 groups ($p < 0.05$). The FM group had higher plasma LDL-C content ($p < 0.05$) and comparable plasma TC content compared with the SBM60 group ($p > 0.05$), but the plasma TC content was not different between the LF2-LF10 groups and SBM60 group ($p > 0.05$), while the SBM60 and LF2 groups had higher plasma LDL-C contents ($p < 0.05$) vs. the LF6 and LF10 groups. Both the plasma LDL-C and TC contents showed a linear decreasing trend with the dietary increase in the LF inclusion levels and reached the minimum values at LF6 and LF10, respectively. However, the plasma TG content did not differ across all the dietary treatments ($p > 0.05$).

Table 4. Effect of lactoferrin (LF) supplementation in high-SBM diets on intestinal antioxidant indices of juvenile orange-spotted groupers in a 56-day feeding period [1].

Items [3]	Diets [2]				
	FM	SBM60	LF2	LF6	LF10
SOD (U/mg protein)	71.67 ± 5.50	68.92 ± 1.76	62.16 ± 5.85	64.26 ± 3.23	60.70 ± 2.15
GSH-Px (U/mg protein)	79.58 ± 3.31 ^{bc}	65.29 ± 2.97 ^{a}	82.52 ± 1.76 ^{bc}	72.99 ± 1.35 ^{b}	86.76 ± 4.00 ^{c}
T-AOC (U/mg protein)	0.19 ± 0.01	0.19 ± 0.01	0.21 ± 0.01	0.19 ± 0.02	0.18 ± 0.01
MDA (nmol/mg protein)	3.00 ± 0.28 ^{a}	4.56 ± 0.88 ^{b}	1.97 ± 0.21 ^{a}	1.86 ± 0.12 ^{a}	2.55 ± 0.05 ^{a}
$Y_{MDA} = 0.0783X^2 - 0.9351X + 4.2006$, $R^2 = 0.8297$, X = LF supplementation levels (g/kg)					

[1] Data were presented as means ± SEM ($n = 3$ tanks). Statistical analysis was performed by one-way ANOVA, followed by S–N–K test. [2] FM, fish meal diet (control diet); SBM60, high-soybean-meal diet with 600 g/kg fish meal protein replacement and without LF supplementation; LF2, LF6, and LF10 were added 2, 6, and 10 g/kg LF in SBM60 diets, respectively. [3] Abbreviations: T-AOC, total antioxidant capacity; SOD, superoxide dismutase; MDA, malondialdehyde; GSH-Px, glutathione peroxidase. $^{a-c}$ Values in the same row with different superscripts indicate significant differences ($p < 0.05$), while that with the same letter or no letter superscripts indicate no significant differences ($p > 0.05$).

Table 5. Effect of lactoferrin (LF) supplementation in high-SBM diets on plasma components of juvenile orange-spotted groupers in a 56-day feeding period [1].

Items [3]	Diets [2]				
	FM	SBM60	LF2	LF6	LF10
HDL-C (mmol/L)	1.06 ± 0.05 [b]	1.00 ± 0.03 [b]	0.81 ± 0.05 [a]	1.03 ± 0.03 [b]	0.83 ± 0.09 [a]
LDL-C (mmol/L)	0.28 ± 0.01 [c]	0.19 ± 0.01 [b]	0.18 ± 0.01 [b]	0.12 ± 0.01 [a]	0.12 ± 0.01 [a]
TC (mmol/L)	3.77 ± 0.21 [b]	3.49 ± 0.23 [ab]	3.35 ± 0.19 [ab]	3.09 ± 0.09 [ab]	2.90 ± 0.09 [a]
TG (mmol/L)	1.61 ± 0.17	1.36 ± 0.08	1.25 ± 0.06	1.57 ± 0.10	1.55 ± 0.07
$Y_{LDL\text{-}C} = -0.0079X + 0.188$, $R^2 = 0.8573$, X = LF supplementation levels (g/kg)					
$Y_{TC} = -0.0592x + 3.4741$, $R^2 = 0.9931$, X = LF supplementation levels (g/kg)					

[1] Data were presented as means ± SEM (n = 3 tanks). Statistical analysis was performed by one-way ANOVA, followed by S–N–K test. [2] FM, fish meal diet (control diet); SBM60, high-soybean-meal diet with 600 g/kg fish meal protein replacement and without LF supplementation; LF2, LF6, and LF10 were added 2, 6, and 10 g/kg LF in SBM60 diets, respectively. [3] Abbreviations: TG, triglyceride; TC, total cholesterol; HDL-C, high-density lipoprotein cholesterol; LDL-C, low-density lipoprotein cholesterol. [a–c] Values in the same row with different superscripts indicate significant differences ($p < 0.05$), while that with the same letter or no letter superscripts indicate no significant differences ($p > 0.05$).

3.4. Intestinal Digestive Enzyme Activity

As shown in Table 6, the intestinal lipase and protease activities were comparable, and the intestinal trypsin activity was lower in the SBM60 group than that in the FM group ($p < 0.05$). The intestinal trypsin and lipase activities were higher in the LF2-LF10 groups than that in the SBM60 group ($p < 0.05$) and returned to the level of the FM group and even higher than that of the FM group. The intestinal activities of the lipase, trypsin, and protease showed a linear increasing trend with increasing dietary LF inclusion levels. However, the intestinal amylase activity was not affected by the dietary treatments ($p > 0.05$).

Table 6. Effect of lactoferrin (LF) supplementation in high-SBM diets on activities of intestinal digestive enzymes of juvenile orange-spotted groupers in a 56-day feeding period [1].

Items	Diets [2]				
	FM	SBM60	LF2	LF6	LF10
Lipase (U/mg protein)	0.68 ± 0.00 [a]	0.61 ± 0.00 [a]	0.84 ± 0.05 [b]	0.86 ± 0.02 [b]	0.95 ± 0.05 [bc]
Amylase (U/mg protein)	0.76 ± 0.06	0.73 ± 0.11	0.90 ± 0.04	0.73 ± 0.07	0.82 ± 0.06
Trypsin (U/g protein)	256.07 ± 17.23 [b]	175.55 ± 17.55 [a]	238.95 ± 17.46 [b]	235.03 ± 9.36 [b]	283.57 ± 8.83 [bc]
Protease (U/mg protein)	20.54 ± 0.87	15.91 ± 2.04	17.37 ± 2.91	23.54 ± 2.56	26.54 ± 2.53
$Y_{Lipase} = 0.0283X + 0.6876$, $R^2 = 0.7515$, X = LF supplementation levels (g/kg)					
$Y_{Trypsin} = 8.8954X + 193.25$, $R^2 = 0.7917$, X = LF supplementation levels (g/kg)					
$Y_{Protease} = 1.1231X + 15.786$, $R^2 = 0.9775$, X = LF supplementation levels (g/kg)					

[1] Data were presented as means ± SEM (n = 3 tanks). Statistical analysis was performed by one-way ANOVA, followed by S–N–K test. [2] FM, fish meal diet (control diet); SBM60, high-soybean-meal diet with 600 g/kg fish meal protein replacement and without LF supplementation; LF2, LF6, and LF10 were added 2, 6, and 10 g/kg LF in SBM60 diets, respectively. [a–c] Values in the same row with different superscripts indicate significant differences ($p < 0.05$), while that with the same letter or no letter superscripts indicate no significant differences ($p > 0.05$).

3.5. Intestinal Permeability

Table 7 shows that the intestinal D-Lac and ET contents were higher in the SBM60 group than that in the FM group ($p < 0.05$). However, the intestinal D-Lac and ET contents were reduced when given the SBM60 diet with the LF supplementation ($p < 0.05$), and the values in the fish receiving the SBM60 diets with the LF supplementation returned to the level and even lower than that of the FM group. The intestinal D-Lac and ET contents did not differ with the dietary LF levels from 2 to 10 g/kg ($p > 0.05$). Both the intestinal ET-1 and DAO contents were not affected by the dietary treatments ($p > 0.05$).

Table 7. Effect of lactoferrin (LF) supplementation in high-SBM diets on the biochemical indices of intestinal mucosal permeability of orange-spotted groupers in a 56-day feeding period [1].

Items [3]	Diets [2]				
	FM	SBM60	LF2	LF6	LF10
DAO (U/L)	19.75 ± 1.39	20.59 ± 1.05	16.09 ± 1.22	17.77 ± 2.40	17.72 ± 1.47
D-Lac (nmol/mL)	2.03 ± 0.20 [a]	4.05 ± 0.23 [b]	1.41 ± 0.18 [a]	1.59 ± 0.21 [a]	1.41 ± 0.07 [a]
ET-1 (ng/L)	1.91 ± 0.07	2.12 ± 0.09	2.27 ± 0.10	1.92 ± 0.13	2.36 ± 0.20
ET (EU/L)	1.51 ± 0.03 [b]	1.70 ± 0.10 [c]	1.23 ± 0.01 [a]	1.25 ± 0.01 [a]	1.25 ± 0.04 [a]

[1] Data were presented as means ± SEM (n = 3 tanks). Statistical analysis was performed by one-way ANOVA, followed by S–N–K test. [2] FM, fish meal diet (control diet); SBM60, high-soybean-meal diet with 600 g/kg fish meal protein replacement and without LF supplementation; LF2, LF6, and LF10 were added 2, 6, and 10 g/kg LF in SBM60 diets, respectively. [3] Abbreviations: DAO, diamine oxidase; D-Lac, d-lactic acid; ET, endotoxin; ET-1, endothelin-1. [a–c] Values in the same row with different superscripts indicate significant differences ($p < 0.05$), while that with the same letter or no letter superscripts indicate no significant differences ($p > 0.05$).

3.6. Intestinal Histomorphology

Table 8 shows the effects of the dietary treatments on the mucosal fold number (nMF), muscle layer thickness (tML), and mucosal fold length (lMF) in the three intestinal segments (PI, MI, and DI). The nMF of the PI, MI, and DI; the lMF of the PI and DI; and the tMF of the PI remained unaffected by the dietary treatments ($p > 0.05$), but the tML of the DI showed a positive quadratic response to increasing dietary LF inclusion levels, with a maximum value observed for diet LF6. Diets LF2 to LF10 displayed a higher ($p < 0.05$) tML of the DI vs. the diet SBM60, and the value returned to the level of the diet FM ($p > 0.05$). The lMF and tML of the MI had an irregular change in response to the dietary LF inclusion levels, but the maximum values observed all for the diet LF2.

Table 8. Effect of lactoferrin (LF) supplementation in high-SBM diets on the intestinal morphology of juvenile orange-spotted groupers in a 56-day feeding period [1].

Items [3]	Diets [2]				
	FM	SBM60	LF2	LF6	LF10
			PI		
lMF (μm)	577.30 ± 87.68	489.10 ± 54.31	574.92 ± 35.62	513.26 ± 50.67	737.53 ± 95.20
tML (μm)	63.24 ± 6.74	64.56 ± 8.11	79.05 ± 2.27	86.60 ± 9.31	86.00 ± 2.51
nMF (unit)	42.50 ± 4.25	45.83 ± 3.09	51.67 ± 1.36	50.33 ± 8.62	48.00 ± 3.55
			MI		
lMF (μm)	465.12 ± 50.20 [ab]	381.90 ± 42.42 [a]	580.47 ± 9.06 [b]	356.66 ± 9.37 [a]	540.48 ± 47.06 [b]
tML (μm)	53.53 ± 2.44 [ab]	44.96 ± 4.06 [a]	76.61 ± 7.02 [b]	63.13 ± 4.61 [ab]	69.62 ± 7.81 [b]
nMF (unit)	34.33 ± 2.20	31.67 ± 1.01	43.00 ± 4.36	34.83 ± 3.49	39.17 ± 0.33
			DI		
lMF (μm)	417.87 ± 63.72	337.13 ± 44.48	437.82 ± 22.32	397.03 ± 4.38	466.67 ± 53.64
tML (μm)	87.58 ± 7.61 [b]	51.53 ± 1.48 [a]	74.80 ± 3.34 [b]	86.49 ± 1.35 [b]	69.78 ± 7.20 [b]
nMF (unit)	32.00 ± 5.20	37.00 ± 4.00	40.83 ± 5.33	35.00 ± 1.04	34.00 ± 3.21
DI: $Y_{tML} = -1.0492X^2 + 12.167X + 52.619$, $R^2 = 0.9884$, X = LF supplementation levels (g/kg)					

[1] Data were presented as means ± SEM (n = 3 tanks). Statistical analysis was performed by one-way ANOVA, followed by S–N–K test. [2] FM, fish meal diet (control diet); SBM60, high-soybean-meal diet with 600 g/kg fish meal protein replacement and without LF supplementation; LF2, LF6, and LF10 were added 2, 6, and 10 g/kg LF in SBM60 diets, respectively. [3] Abbreviations: PI, proximal intestine; MI, middle intestine; DI, distal intestine; nMF, mucosal fold number; tML, muscle layer thickness; lMF, mucosal fold length. [a,b] Values in the same row with different superscripts indicate significant differences ($p < 0.05$), while that with the same or no letter superscripts indicate no significant differences ($p > 0.05$).

3.7. Abundance and Difference in Intestinal Microbiota

The Firmicutes, Bacteroidetes, and Proteobacteria in the DI were the dominant phyla of all the dietary treatments (Figure 2A). Compared with the diets FM, the abundance of Firmicutes was increased and the abundances of the Bacteroidetes and Proteobacteria were decreased by the diets SBM60. The abundances of the Firmicutes and Bacteroidetes generally increased, but the abundance of the Proteobacteria decreased with increasing dietary LF-levels supplementation in the SBM60 diets (Figure 2A and Supplementary Table S1). However, no significant differences in the abundances of the dominant phyla were observed between the dietary treatments at the phylum level ($p > 0.05$).

At the genus level, the DI bacteria of all the dietary treatments mainly contained the genera *Photobacterium*, *Selenomonas-1*, *Prevotella-1*, *Vibrio*, and *Rikenellaceae-RC9-gut-group* (Figure 2B and Supplementary Table S1). The decreased abundances of the genera *Photobacterium*, *Selenomonas-1*, and *Prevotella-1* and the increased abundances of the genera *Vibrio* and *Rikenellaceae-RC9-gut-group* were observed in the diets SBM60 vs. the diets FM. The abundances of the genera *Selenomonas-1*, *Prevotella-1*, and *Rikenellaceae-RC9-gut-group* showed an open downward parabola response, but the abundances of the genera *Photobacterium* and *Vibrio* showed an open upward parabola in response to the increasing dietary LF inclusion levels. However, there was no significant difference in the abundances of the dominant genera between the dietary treatments at the genus level ($p > 0.05$).

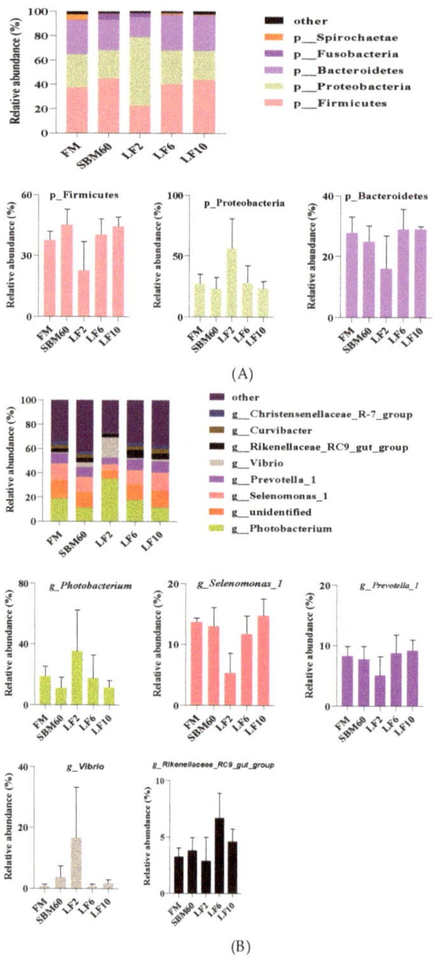

Figure 2. Relative abundances of the dominant bacterial at phylum (**A**) and genus (**B**) in distal intestine (DI) of juvenile orange-spotted grouper fed different diets in a 56-day feeding period. Bars bearing the different letters are significantly different ($p < 0.05$). FM, fish meal diet (control diet); SBM60, high-soybean-meal diet with 600 g/kg fish meal protein replacement and without LF supplementation; LF2, LF6, and LF10 were added 2, 6, and 10 g/kg LF in SBM60 diets, respectively.

3.8. Expression of Intestinal Inflammatory Factor Genes

As shown in Figure 3, the FM group had lower mRNA levels for the *IL-1β*, *IL-12*, *IL-8*, and *TNF-α* in comparison with the SBM60 group ($p < 0.05$), and no differences in the value for the *TGF-β1* and *IL-10* were observed between them ($p > 0.05$). Although the intestinal mRNA levels for the *IL-1β*, *IL-12*, *IL-8*, *TGF-β1*, and *TNF-α* did not differ across the diets from LF2 to LF10 ($p > 0.05$), the values for the *IL-1β*, *IL-12*, *IL-8*, and *TNF-α* were lower, but the value for the *TGF-β1* was higher in the diets LF2–LF10 than in the diet SBM60 ($p < 0.05$), and returned to the level of the FM group ($p > 0.05$) and even lower than that of the FM group. The mRNA level for the *IL-10* was not different in the LF2 and LF10 groups ($p > 0.05$) but higher than the FM, SBM60, and LF6 groups ($p < 0.05$).

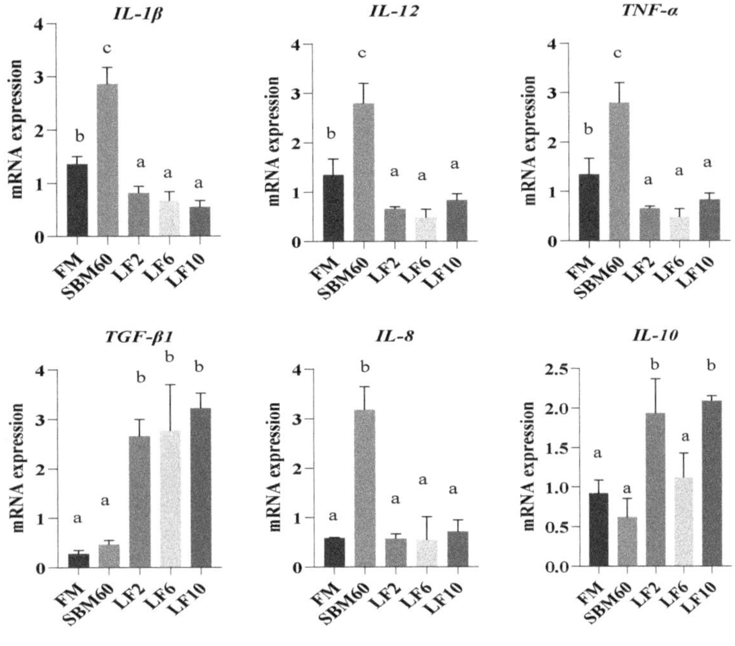

Figure 3. Effect of lactoferrin (LF) inclusion levels in SBM60 diets on mRNA levels of intestinal inflammatory factor genes of juvenile orange-spotted groupers in a 56-day feeding period. Data are presented as means ± SEM ($n = 3$ tanks). Statistical analysis was performed by one-way ANOVA, followed by S–N–K test. Bars bearing the different letters are significantly different ($p < 0.05$). FM, fish meal diet (control diet); SBM60, high-soybean-meal diet with 600 g/kg fish meal protein replacement and without LF supplementation; LF2, LF6, and LF10 were added 2, 6, and 10 g/kg LF in SBM60 diets, respectively. Abbreviations: IL-8, interleukin-8; IL-1β, interleukin-1β; TNF-α, tumor necrosis factor-α; IL-12, interleukin-12; TGF-β1, transforming growth factor-β1; IL-10, interleukin-10.

4. Discussion

Our previous study and other studies have shown that high-SBM diets resulted in poor fish growth performance [1,5,30,31], which supported our current results that the growth rate was reduced when juvenile orange-spotted groupers were fed high-SBM diets vs. FM diets, as evidenced by a significant decrease in the WG and SGR. Consistent with our previous results [1], no variations between the high-SBM diets and FM diets in the HIS [31–33], CF [30,33,34], and FE [5] were observed. In contrast, feeding a high-SBM diet led to a reduction in the HSI, CF, and FE. The restricted growth of fish caused by high-SBM diets might be due to the imbalanced amino acids profile in high-SBM diets [5,31] and SBMIE, a resultant low feed utilization [1,35]. It is clear that dietary LF administration shows a strong effect on improving growth for terrestrial animals. For example, LF administration could enhance the growth performance of early-weaned piglets and neonatal calves [36,37].

Recently, Olyayee reported a growth promotion and carcass yield of broiler chicken when fed a diet with 0.8 g/kg LF [38]. In fish, LF was used as a feed additive to promote the growth of farmed fish [39–41]. However, in other studies, this promoting effect was not observed in Atlantic salmon [42], Nile tilapia [43], gilthead sea bream [44], African cichlid [45], grouper [46], and Siberian sturgeon [47]. This discrepancy may be ascribed to the differences in fish species and physiological status, growth stage, LF dosage, water temperature, etc.

There were no differences in the whole-body proximate composition between fish fed high-SBM diets and fish fed FM diets in the present study and previous studies with largemouth bass [31], rockfish [48], and pompano [49]. In contrast, previous studies showed that feeding high SBM resulted in an increase in moisture content [33,50] and a decrease in crude protein content [33,50] and crude lipid content [1,33,50]. After dietary LF intervention, no notable changes were observed in the whole-body proximate composition in the present study and a previous study on silvery-black porgy [51], which was inconsistent with El-Sayed and Al-Kenawy [40], who reported that dietary LF increased the whole-body protein content and decreased the whole-body lipid content of Nile tilapia.

The activities of digestive enzymes reflect the ability to digest feed [52]. In the present study, the intestinal activities of trypsin and protease were decreased in fish fed with high-SBM diets vs. fish fed FM diets, which supported the results in previous studies with drum fish [53], Japanese seabass [54], hybrid tilapia [55], as well as our previous study [1]. The reduced intestinal trypsin and the protease activities were the result of the presence of trypsin inhibitors or other ANFs in the SBM, resulting in a poor growth performance and feed utilization in fish [55–58]. After the dietary intervention of LF, the intestinal activities of the lipase, protease, and trypsin showed a linear response to the dietary LF inclusion levels and reached or even exceeded the original levels of the FM diet, indicating that LF could promote the digestion of feed. Dietary LF administration alone or in combination with other functional feed additives was also reported to have an enhancement effect on the intestinal trypsin and protease activities of silvery-black porgy [59,60]. Dietary LF administration could promote the proliferation of intestinal epithelial cells and protect the intestinal crypt and villous structure of piglets [61,62], reflecting the integrity of the intestinal mucosa and the stabilization of the intestinal brush border. Therefore, it is natural to improve the digestive capacity through the intervention of dietary LF on the enteritis response of fish.

The intestinal tract is not only the place where nutrients are digested and absorbed [63] but also a site of immune defense, an important barrier against exogenous pathogens [64]. D-Lac and ET are the products secreted by the inherent bacteria of the gastrointestinal tract, and increases in their concentrations in the blood reflect the dysregulation of the intestinal flora and impaired permeability [65–68]. In this study, the intestinal D-Lac and ET contents increased in the diets SBM60 vs. that in the diets FM. This finding was consistent with previous results [1,54,69], indicating the intestinal mucosal injury of fish induced by dietary high SBM. So far, there is no report on the intervention of LF in the SBMIE of fish. In the present study, we observed reduced intestinal ET and D-Lac contents in juvenile orange-spotted groupers administrated with dietary LF vs. those of fish fed SBM60 diets, reduced to the level of the control or even lower. This indicated that dietary LF supplementation in a high-SBM diet could reduce the intestinal mucosal permeability of the fish species.

An intact intestinal histomorphology is a prerequisite for maintaining a good digestion and absorption state and intestinal immune disease resistance [70]. The histomorphological indicators, such as the lMF, tML, and nMF, were used to evaluate the intestinal digestion and absorption capacity in previous studies [1,19]. In the present study, a significant reduction in the tML in the DI was observed in the SBM diets vs. the FM diets, as evidenced by a previous study [1]. Many studies showed that fish fed high-SBM diets had shortened villus height and thinned muscular thickness [5,54,71], which indicated a potential abnormal intestinal histomorphology caused by dietary high SBM. The intestinal histopathological changes were improved when fish were intervened by the dietary LF administration in our current study. We observed an increase in the lMF in the MI, and the tML in the MI and DI in a dose-dependent manner with increasing LF levels in high-SBM diets. Consistent with the results of our current study, dietary LF administration increased the height, width, and surface area of intestinal villi, the depth of the crypts, and the thickness of the muscular layer of broiler chickens [38]. The positive effect of dietary LF administration on the intestinal histomorphology of terrestrial animals has been widely recognized [72–74]. Therefore, the intervention effect of dietary LF administration on the SBMIE of juvenile orange-spotted groupers can also be achieved by improving the histomorphology.

The intestine is also the so-called "second brain" that links between the intestinal microbiota and diseases. Dysbacteriosis of the intestinal flora increases the susceptibility to intestinal pathogens, and in severe cases, it will further develop into intestinal infection and reduce immune function [75,76]. In the present study, fish fed high-SBM diets showed an increased abundance of the phylum Firmicutes and a decreased abundance of the phylum Bacteroidetes and Proteobacteria vs. the fish fed the FM diets. At the genus level, the fish fed high-SBM diets exhibited a decreased abundance of *Photobacterium*, *Selenomonas_1*, and *Prevotella_1* and an increased abundance of *Vibrio* and *Rikenellaceae_RC9_gut_group* vs. the fish fed FM diets. Nevertheless, a high-SBM substitution for FM did not alter the intestinal microbial abundance and diversity of groupers at either the phylum level or the genus level, which was inconsistent with previous studies in largemouth bass [77], gilthead sea bream [78], and large yellow croaker [79]. The difference in the relative abundance of intestinal microbiota at the phylum and genus levels may be directly or indirectly influenced by external environmental conditions, such as the living environment, fish species, growth stage, and feed composition. In the present study, the phylum Firmicutes, Bacteroidetes, and Proteobacteria were identified as the dominant phyla of the intestine in fish fed high-SBM diets with LF administration. The relative abundances of the Firmicutes and Bacteroidetes were generally promoted, while the abundance of the Proteobacteria was reduced by the intervention of the dietary LF administration. At the genus level, the intestinal microbiota diversity did not vary, but the relative abundance of the genus bacteria varied with the dietary LF inclusion levels. As mentioned above, thus far, there are few reports regarding the intervention of LF on fish enteritis.

The presence of an inflammatory response is a complex pathophysiological process, which is mediated by the activation of a variety of cytokines and complement factors secreted by macrophages and leukocytes [80]. LF can specifically bind B-lymphocytes and macrophages, thus inhibiting the production of pro-inflammatory factors, such as *IL-1β*, *IL-8*, and *TNF-α*, or/and indirectly promoting the production of anti-inflammatory factors such as *IL-10* through preventing the activation of the TOR signaling pathway induced by lipopolysaccharide [18,81,82]. LF was used to modulate the inflammatory response, with an emphasis on protection against intestinal infections and inflammatory bowel diseases of mammals [82–84]. Like mammals, fish have pro-inflammatory cytokines, including *IL-1β*, *TNF-α*, *IL-8*, and *IL-12*, and the anti-inflammatory ones are *IL-10* and *TGF-β1* in fish immune responses [1,85,86]. It is clear that the up-regulation of pro-inflammatory gene expression and the down-regulation of anti-inflammatory gene expression are caused by SBMIE [87,88]. In the present study, the fish fed with the high-SBM diets promoted the expression of such pro-inflammatory genes as *IL-1β*, *TNF-α*, *IL-8*, and *IL-12* vs. the fish fed with the FM diets. The findings agreed with what has been reported in previous studies with an SBM substitution for FM in different fish species [1,5,77]. In the present study, the mRNA levels of the *IL-1β*, *TNF-α*, *IL-8*, and *IL-12* genes were down-regulated, while the mRNA levels of the *IL-10* and *TGF-β1* genes up-regulated after the fish were fed high-SBM diets administrated with LF. This finding indicated that a dietary LF intervention could alleviate the SBMIE of fish through promoting the production of anti-inflammatory factors and preventing the production of pro-inflammatory factors in this study.

The antioxidant capacity reflects the physiological status of aquatic animals [89]. It is well known that the antioxidant enzyme SOD catalyzes the dismutation of the superoxide anion into O_2 and H_2O_2. H_2O_2 is subsequently degraded into H_2O by the antioxidant enzyme GSH-Px in the cytosol. These enzymes are easily induced by oxidative stress and their activity is usually used to reflect the ability of cleaning free radicals in cells [90]. T-AOC reflects the protective capacity of non-enzymatic antioxidant defense system [91]; and MDA as one of the end-products of lipid peroxidation reflects the degree of lipid peroxidation caused by free radicals and indirectly reflects the degree of cell damage [92]. In the present study, the intestinal GSH-Px activity was decreased, and the MDA content was increased in fish fed the high-SBM diets vs. those fed the FM diets. Similarly, fish showed decreased activities of SOD [69,77,93], GSH-Px [48], and T-AOC [77] as well as increased MDA content [69,77] when fed the diets with an FM replacement of high SBM. This indicated that fish suffered from oxidative stress. The decline in the immunity and antioxidant capacity of animals is associated with the presence of ANFs in the SBM [94,95]. After a dietary intervention with LF, fish had higher intestinal GSH-Px activity and lower MDA content vs. with high-SBM diets. Furthermore, the intestinal GSH-Px activity showed an increasing trend with increasing dietary LF inclusion levels and even returned to the level of the diet FM. The findings suggest a dietary LF intervention could promote the intestinal antioxidant capacity and prevent the intestinal lipid peroxidation in juvenile orange-spotted groupers with SBMIE, through influencing both enzymatic and non-enzymatic antioxidants. Consistent with our results, a reduction in the liver MDA content on day 15 and an increase in the liver T-AOC on day 30 were observed after *Aeromonas veronii*-induced Nile tilapias were administrated with LF at 0.8 g/kg diet [96]; increased SOD activity and decreased MDA content in the hepatopancreas were observed in shrimp receiving the diets with 1.5–2.5 g/kg LF [97]; and similar effects occurred for weaned piglets when they were fed diets with LF [37]. The above results validated that a dietary LF intervention has a protective effect against oxidative stress resulting from different sources of stress. However, higher dietary LF inclusion levels did not improve the oxidative stress of the fish species. This was also reflected in the growth, that is, higher dietary LF inclusion levels did not further enhance the growth performance.

Besides the antioxidant capacity, another important indicator to measure the health status of fish is the plasma biochemical components [24,98]. HDL-C participates in the transportation of lipids from peripheral tissues to the liver for catabolism, whereas LDL-C transports cholesterol from the liver to peripheral tissues [22]. High-SBM diets could decrease the plasma TC, TG, LDL-C, and/or HDL-C contents vs. the control (FM diet) in many previous studies [1,69,99–101], as observed in our current study. This indicates an inferior nutritional status of fish caused by high-SBM inclusion. Not as expected, plasma contents of TC, HDL-C and LDL-C were still lower in high-SBM diets with LF intervention than that in FM diets. As a result, the dietary intervention of LF on the malnutrition of grouper caused by high-SBM diets is limited, though it improved the overall antioxidant capacity and growth.

5. Conclusions

The supplementation of LF in high-SBM diets not only improves the growth performance and intestinal morphology but also reduces the permeability of intestinal mucosal cells and attenuates the intestinal inflammatory response in juvenile orange-spotted groupers, an improvement of the nutritional status under the intervention of LF. The optimal appropriate supplementation level of LF was 5.8 g/kg based on the quadratic regression analysis of the percent weight gain against dietary LF inclusion levels. This is the first report on the intervention effect of dietary LF on grouper enteritis induced by dietary SBM. Our current study will provide a basis for LF use as a functional feed additive to alleviate fish SBMIE.

Supplementary Materials: The following supporting information can be download at: https://www.mdpi.com/article/10.3390/metabo12100915/s1, Table S1: MetaStat analysis of the abundance of DI bacterial phyla and genera ($\times 10^{-4}$) of juvenile orange-spotted grouper in a 56-day feeding period.

Author Contributions: T.S. and Y.Q. conducted the experiments and analyzed the data and wrote the manuscript; L.K. and K.W. contributed to the data analysis and curation; X.W. and Y.S. contributed to the formal analysis and the writing—review; and J.Y. supervised the experiments and reviewed the manuscript. All authors have read and agreed to the published version of the manuscript.

Funding: This study was supported by the funding from the National Natural Science Foundation of China (Grant Nos. 31772861 and 31372546) and the Science and Technology Project of Fujian Province, China (Grant No. 2020N0012).

Institutional Review Board Statement: Experimental design and procedures in this study were reviewed and approved by the Animal Ethics Committee of Jimei University, Xiamen, China (Approval number: 2011-58).

Informed Consent Statement: Not applicable.

Data Availability Statement: The data that support the findings of this study are available from the corresponding author upon reasonable request.

Conflicts of Interest: The authors declare on conflict of interest.

References

1. Zhao, X.; Wang, Y.; Wang, X.; Ye, J. Growth performance, plasma components, and intestinal barrier in grouper (*Epinephelus coioides*) are altered by dietary fish meal replacement with extruded soybean meal. *Aquac. Rep.* **2021**, *21*, 100863. [CrossRef]
2. Ghosh, K.; Ray, A.K.; Ringø, E. Applications of plant ingredients for tropical and subtropical freshwater finfish: Possibilities and challenges. *Rev. Aquac.* **2019**, *11*, 793–815. [CrossRef]
3. Shao, J.; Zhao, W.; Liu, X.; Wang, L. Growth performance, digestive enzymes, and TOR signaling pathway of *Litopenaeus vannamei* are not significantly affected by dietary protein hydrolysates in practical conditions. *Front. Physiol.* **2018**, *9*, 998. [CrossRef]
4. Bravo-Tello, K.; Ehrenfeld, N.; Solís, C.J.; Ulloa, P.E.; Hedrera, M.; Pizarro-Guajardo, M.; Paredes-Sabja, D.; Feijóo, C.G. Effect of microalgae on intestinal inflammation triggered by soybean meal and bacterial infection in zebrafish. *PLoS ONE* **2017**, *12*, e0187696. [CrossRef] [PubMed]
5. Wang, Y.; Wang, L.; Zhang, C.; Song, K. Effects of substituting fishmeal with soybean meal on growth performance and intestinal morphology in orange-spotted grouper (*Epinephelus coioides*). *Aquac. Rep.* **2017**, *5*, 52–57. [CrossRef]
6. Chen, Y.; Ma, J.; Huang, H.; Zhong, H. Effects of the replacement of fishmeal by soy protein concentrate on growth performance, apparent digestibility, and retention of protein and amino acid in juvenile pearl gentian grouper. *PLoS ONE* **2019**, *14*, e0222780. [CrossRef]
7. Dam, C.T.M.; Ventura, T.; Booth, M.; Pirozzi, I.; Salini, M.; Smullen, R.; Elizur, A. Intestinal transcriptome analysis highlights key differentially expressed genes involved in nutrient metabolism and digestion in yellowtail kingfish (*Seriola lalandi*) fed terrestrial animal and plant proteins. *Genes* **2020**, *11*, 621. [CrossRef]
8. Bai, N.; Gu, M.; Xu, X.; Xu, B.; Krogdahl, Å. Protective effects of mannan oligosaccharides on turbot *Scophthalmus maximus* suffering from soy enteropathy. *Aquaculture* **2017**, *476*, 141–151. [CrossRef]
9. Grammes, F.; Reveco, F.E.; Romarheim, O.H.; Landsverk, T.; Mydland, L.T.; Øverland, M. *Candida utilis* and *Chlorella vulgaris* counteract intestinal inflammation in Atlantic salmon (*Salmo salar* L.). *PLoS ONE* **2013**, *8*, e83213. [CrossRef]
10. Romarheim, O.H.; Hetland, D.L.; Skrede, A.; Øverland, M.; Mydland, L.T.; Landsverk, T. Prevention of soya-induced enteritis in Atlantic salmon (*Salmo salar*) by bacteria grown on natural gas is dose dependent and related to epithelial MHC II reactivity and CD8α^+ intraepithelial lymphocytes. *Br. J. Nutr.* **2013**, *109*, 1062–1070. [CrossRef]
11. Wu, N.; Xu, X.; Wang, B.; Li, X.; Cheng, Y.; Li, M.; Xia, X.; Zhang, Y. Anti-foodborne enteritis effect of galantamine potentially via acetylcholine anti-inflammatory pathway in fish. *Fish Shellfish Immunol.* **2020**, *97*, 204–215. [CrossRef]
12. Xie, J.; Li, M.; Ye, W.; Shan, J.; Zhao, X.; Duan, Y.; Liu, Y.; Unger, B.H.; Cheng, Y.; Zhang, W.; et al. Sinomenine hydrochloride ameliorates fish foodborne enteritis via α7nAchR-mediated anti-inflammatory effect whilst altering microbiota composition. *Front. Immunol.* **2021**, *12*, 766845. [CrossRef] [PubMed]
13. Rascón-Cruz, Q.; Espinoza-Sánchez, E.A.; Siqueiros-Cendón, T.S.; Nakamura-Bencomo, S.I.; Arévalo-Gallegos, S.; Iglesias-Figueroa, B.F. Lactoferrin: A glycoprotein involved in immunomodulation, anticancer, and antimicrobial processes. *Molecules* **2021**, *26*, 205. [CrossRef] [PubMed]
14. Luna Castro, S.; Ceballos Olvera, I.; Benavides González, F.; Blanco Martínez, Z.; Sánchez Martínez, G.; Vázquez Sauceda, M.D.L.L.; de la Garza, M. Bovine lactoferrin in fish culture: Current research and future directions. *Aquac. Res.* **2022**, *53*, 735–745. [CrossRef]
15. Ulloa, P.E.; Solís, C.J.; De La Paz, J.F.; Alaurent, T.G.S.; Caruffo, M.; Hernández, A.J.; Dantagnan, P.; Feijóo, C.G. Lactoferrin decreases the intestinal inflammation triggered by a soybean meal-based diet in zebrafish. *J. Immunol. Res.* **2016**, *2016*, 1639720. [CrossRef]
16. Wang, Y.; Shan, T.; Xu, Z.; Feng, J.; Wang, Z. Effects of the lactoferrin (LF) on the growth performance, intestinal microflora and morphology of weanling pigs. *Anim. Feed Sci. Technol.* **2007**, *135*, 263–272. [CrossRef]

17. Li, M.; Fang, C.; Zhang, K.; Li, Y.; Zhang, R. Effects of lactoferrin supplementation on growth performance, intestinal microflora and mucosal morphology of early weaned piglets. *Chin. J. Anim. Nutr.* **2012**, *24*, 111–116.
18. Drago-Serrano, M.E.; de la Garza-Amaya, M.; Luna, J.S.; Campos-Rodríguez, R. Lactoferrin-lipopolysaccharide (LPS) binding as key to antibacterial and antiendotoxic effects. *Int. Immunopharmacol.* **2012**, *12*, 1–9. [CrossRef]
19. Niu, X.; Qian, X.; Feng, H.; Yi, K.; Li, D.; Chen, W.; Ye, J. Growth and metabolic responses of grouper juveniles (*Epinephelus coioides*) fed diets containing varying levels of leucine. *Aquaculture* **2021**, *534*, 736281. [CrossRef]
20. Shapawi, R.; Abdullah, F.C.; Senoo, S.; Mustafa, S. Nutrition, growth and resilience of tiger grouper (*Epinephelus fuscoguttatus*) × giant grouper (*Epinephelus lanceolatus*) hybrid—A review. *Rev. Aquac.* **2019**, *11*, 1285–1296. [CrossRef]
21. Bureau of Fisheries, Ministry of Agriculture. *China Fishery Statistics Yearbook*; China Agriculture Press: Beijing, China, 2021.
22. Bai, F.; Niu, X.; Wang, X.; Ye, J. Growth performance, biochemical composition and expression of lipid metabolism related genes in groupers (*Epinephelus coioides*) are altered by dietary taurine. *Aquac. Nutr.* **2021**, *27*, 2690–2702. [CrossRef]
23. Zhuo, L.; Chen, C.; Lin, Y. Dietary supplementation of fermented lemon peel enhances lysozyme activity and susceptibility to *Photobacterium damselae* for orange-spotted grouper, *Epinephelus coioides*. *Fish Shellfish Immunol.* **2021**, *117*, 248–252. [CrossRef]
24. Ye, J.; Liu, X.; Wang, Z.; Wang, K. Effect of partial fish meal replacement by soybean meal on the growth performance and biochemical indices of juvenile Japanese flounder *Paralichthys olivaceus*. *Aquac. Int.* **2011**, *19*, 143–153. [CrossRef]
25. *Official Methods of Analysis of AOAC International*, 16th ed.; Association of Official Analytical Chemists: Arlington, VA, USA, 1995.
26. Hu, H.; Liu, J.; Li, Y.; Zhang, Y.; Mai, K.; Ai, Q.; Shao, M.; Yang, P. Effects of dietary daidzein on growth performance, activities of digestive enzymes, anti-oxidative ability and intestinal morphology in juvenile turbot (*Scophthalmus maximus* L.). *J. Fish. China* **2014**, *38*, 1503–1513.
27. Anguiano, M.; Pohlenz, C.; Buentello, A.; Gatlin, D.M. The effects of prebiotics on the digestive enzymes and gut histomorphology of red drum (*Sciaenops ocellatus*) and hybrid striped bass (*Morone chrysops* × *M. saxatilis*). *Br. J. Nutr.* **2013**, *109*, 623–629. [CrossRef] [PubMed]
28. Schmittgen, T.D.; Livak, K.J. Analyzing real-time PCR data by the comparative CT method. *Nat. Protoc.* **2008**, *3*, 1101–1108. [CrossRef]
29. Hanaki, K.; Ike, F.; Kajita, A.; Yasuno, W.; Yanagiba, M.; Goto, M.; Sakai, K.; Ami, Y.; Kyuwa, S. A broadly reactive one-step SYBR Green I real-time RT-PCR assay for rapid detection of murine norovirus. *PLoS ONE* **2014**, *9*, e98108. [CrossRef]
30. Jahan, H.; Tumpa, I.J.; Qasem, W.A.; Moniruzzaman, M.; Pervin, M.A.; Akter, R.; Omri, A.; Min, T.; Hossain, Z. Evaluation of the partial replacement of dietary fish meal with fermented or untreated soybean meal in juvenile silver barb, *Barbonymus gonionotus*. *Front. Nutr.* **2021**, *8*, 733402. [CrossRef]
31. Yang, H.; Bian, Y.; Huang, L.; Lan, Q.; Ma, L.; Li, X.; Leng, X. Effects of replacing fish meal with fermented soybean meal on the growth performance, intestinal microbiota, morphology and disease resistance of largemouth bass (*Micropterus salmoides*). *Aquac. Rep.* **2022**, *22*, 100954. [CrossRef]
32. He, M.; Li, X.; Poolsawat, L.; Guo, Z.; Yao, W.; Zhang, C.; Leng, X. Effects of fish meal replaced by fermented soybean meal on growth performance, intestinal histology and microbiota of largemouth bass (*Micropterus salmoides*). *Aquac. Nutr.* **2020**, *26*, 1058–1071. [CrossRef]
33. Liang, X.F.; Hu, L.; Dong, Y.C.; Wu, X.F.; Qin, Y.C.; Zheng, Y.H.; Shi, D.D.; Xue, M. Substitution of fish meal by fermented soybean meal affects the growth performance and flesh quality of Japanese seabass (*Lateolabrax japonicus*). *Anim. Feed Sci. Technol.* **2017**, *229*, 1–12. [CrossRef]
34. Lim, S.-J.; Kim, S.-S.; Ko, G.-Y.; Song, J.-W.; Oh, D.-H.; Kim, J.-D.; Kim, J.-U.; Lee, K.-J. Fish meal replacement by soybean meal in diets for tiger puffer, *Takifugu rubripes*. *Aquaculture* **2011**, *313*, 165–170. [CrossRef]
35. Abdel-Warith, A.A.; Younis, E.M.; Al-Asgah, N.A.; Mahboob, S. Effect of replacing fish meal by full fat soybean meal on growth performance, feed utilization and gastrointestinal enzymes in diets for African catfish *Clarias gariepinus*. *Braz. J. Biol.* **2020**, *80*, 535–543. [CrossRef] [PubMed]
36. Prenner, M.L.; Prgomet, C.; Sauerwein, H.; Pfaffl, M.W.; Brož, J.; Schwarz, F.J. Effects of lactoferrin feeding on growth, feed intake and health of calves. *Arch. Anim. Nutr.* **2007**, *61*, 20–30. [CrossRef] [PubMed]
37. Tang, Z.; Yin, Y.; Zhang, Y.; Huang, R.; Sun, Z.; Li, T.; Chu, W.; Kong, X.; Li, L.; Geng, M.; et al. Effects of dietary supplementation with an expressed fusion peptide bovine lactoferricin-lactoferrampin on performance, immune function and intestinal mucosal morphology in piglets weaned at age 21 d. *Br. J. Nutr.* **2009**, *101*, 998–1005. [CrossRef]
38. Olyayee, M.; Javanmard, A.; Janmohammadi, H.; Kianfar, R.; Alijani, S.; Mir Ghelenj, S.A. Supplementation of broiler chicken diets with bovine lactoferrin improves growth performance, histological parameters of jejunum and immune-related gene expression. *J. Anim. Physiol. Anim. Nutr.* **2022**. [CrossRef]
39. Esmaeili, A.; Sotoudeh, E.; Morshedi, V.; Bagheri, D.; Dorafshan, S. Effects of dietary supplementation of bovine lactoferrin on antioxidant status, immune response and disease resistance of yellowfin sea bream (*Acanthopagrus latus*) against *Vibrio harveyi*. *Fish Shellfish Immunol.* **2019**, *93*, 917–923. [CrossRef]
40. Badawy, T.E.; Al-Kenawy, D. Assessment of immune response supplemental immunoton and bovine lactoferrin as alternatives to antibiotics in Nile tilapia (*Oreochromis niloticus*). *Assessment* **2013**, *8*, 341–356.
41. Kakuta, I. Effects of orally administrated bovine lactoferrin on growth and blood properties of goldfish. *Aquac. Sci.* **1996**, *44*, 419–426.

42. Lygren, B.; Sveier, H.; Hjeltnes, B.; Waagbø, R. Examination of the immunomodulatory properties and the effect on disease resistance of dietary bovine lactoferrin and vitamin C fed to Atlantic salmon (*Salmo salar*) for a short-term period. *Fish Shellfish Immunol.* **1999**, *9*, 95–107. [CrossRef]
43. Welker, T.L.; Lim, C.; Yildirim-Aksoy, M.; Klesius, P.H. Growth, immune function, and disease and stress resistance of juvenile Nile tilapia (*Oreochromis niloticus*) fed graded levels of bovine lactoferrin. *Aquaculture* **2007**, *262*, 156–162. [CrossRef]
44. Esteban, M.A.; Rodríguez, A.; Cuesta, A.; Meseguer, J. Effects of lactoferrin on non-specific immune responses of gilthead seabream (*Sparus auratus* L.). *Fish Shellfish Immunol.* **2005**, *18*, 109–124. [CrossRef] [PubMed]
45. Moradian, A.M.; Dorafshan, S.; Heyrati, F.P.; Ebrahimi, E. Effects of dietary bovine lactoferrin on growth, haemato-biochemical parameters, immune functions and tolerance to air exposure stress in the African cichlid *Sciaenochromis fryeri*. *Aquac. Nutr.* **2017**, *24*, 392–399. [CrossRef]
46. Yokoyama, S.; Koshio, S.; Takakura, N.; Oshida, K.; Ishikawa, M.; Gallardo-Cigarroa, F.J.; Catacutan, M.R.; Teshima, S. Effect of dietary bovine lactoferrin on growth response, tolerance to air exposure and low salinity stress conditions in orange spotted grouper *Epinephelus coioides*. *Aquaculture* **2006**, *255*, 507–513. [CrossRef]
47. Falahatkar, B.; Eslamloo, K.; Yokoyama, S. Suppression of stress responses in siberian sturgeon, *Acipenser baeri*, juveniles by the dietary administration of bovine lactoferrin. *J. World Aquac. Soc.* **2014**, *45*, 699–708. [CrossRef]
48. Lee, S.; Mohammadi Azarm, H.; Chang, K.H. Effects of dietary inclusion of fermented soybean meal on growth, body composition, antioxidant enzyme activity and disease resistance of rockfish (*Sebastes schlegeli*). *Aquaculture* **2016**, *459*, 110–116. [CrossRef]
49. Lin, H.; Chen, X.; Chen, S.; Zhuojia, L.; Huang, Z.; Niu, J.; Wu, K.; Lu, X. Replacement of fish meal with fermented soybean meal in practical diets for pompano *Trachinotus ovatus*. *Aquac. Res.* **2012**, *44*, 151–156. [CrossRef]
50. García-Ortega, A.; Kissinger, K.R.; Trushenski, J.T. Evaluation of fish meal and fish oil replacement by soybean protein and algal meal from *Schizochytrium limacinum* in diets for giant grouper *Epinephelus lanceolatus*. *Aquaculture* **2016**, *452*, 1–8. [CrossRef]
51. Morshedi, V.; Agh, N.; Noori, F.; Jafari, F.; Ghasemi, A.; Mozanzadeh, M.T. Effects of single and combined supplementation of dietary probiotic with bovine lactoferrin and xylooligosaccharide on hemato-immunological and digestive enzymes of silvery-black porgy (*Sparidentex hasta*) fingerlings. *Ann. Anim. Sci.* **2020**, *20*, 137–155. [CrossRef]
52. Li, J.; Li, J.; Li, W.; Sun, Y.; Liu, X.; Liu, M.; Cheng, Y. Juvenile *Procambarus clarkii* farmed using biofloc technology or commercial feed in zero-water exchange indoor tanks: A comparison of growth performance, enzyme activity and proximate composition. *Aquac. Res.* **2019**, *50*, 1834–1843. [CrossRef]
53. Fuentes-Quesada, J.P.; Viana, M.T.; Rombenso, A.N.; Guerrero-Rentería, Y.; Nomura-Solís, M.; Gomez-Calle, V.; Lazo, J.P.; Mata-Sotres, J.A. Enteritis induction by soybean meal in *Totoaba macdonaldi* diets: Effects on growth performance, digestive capacity, immune response and distal intestine integrity. *Aquaculture* **2018**, *495*, 78–89. [CrossRef]
54. Zhang, C.; Rahimnejad, S.; Wang, Y.; Lu, K.; Song, K.; Wang, L.; Mai, K. Substituting fish meal with soybean meal in diets for Japanese seabass (*Lateolabrax japonicus*): Effects on growth, digestive enzymes activity, gut histology, and expression of gut inflammatory and transporter genes. *Aquaculture* **2018**, *483*, 173–182. [CrossRef]
55. Lin, S.; Luo, L. Effects of different levels of soybean meal inclusion in replacement for fish meal on growth, digestive enzymes and transaminase activities in practical diets for juvenile tilapia, *Oreochromis niloticus* × *O. aureus*. *Anim. Feed Sci. Technol.* **2011**, *168*, 80–87. [CrossRef]
56. Dias, J.; Alvarez, M.J.; Arzel, J.; Corraze, G.; Diez, A.; Bautista, J.M.; Kaushik, S.J. Dietary protein source affects lipid metabolism in the European seabass (*Dicentrarchus labrax*). *Comp. Biochem. Physiol. Part A Mol. Integr. Physiol.* **2005**, *142*, 19–31. [CrossRef]
57. Santigosa, E.; Sánchez, J.; Médale, F.; Kaushik, S.; Pérez-Sánchez, J.; Gallardo, M.A. Modifications of digestive enzymes in trout (*Oncorhynchus mykiss*) and sea bream (*Sparus aurata*) in response to dietary fish meal replacement by plant protein sources. *Aquaculture* **2008**, *282*, 68–74. [CrossRef]
58. Yaghoubi, M.; Mozanzadeh, M.T.; Marammazi, J.G.; Safari, O.; Gisbert, E. Dietary replacement of fish meal by soy products (soybean meal and isolated soy protein) in silvery-black porgy juveniles (*Sparidentex hasta*). *Aquaculture* **2016**, *464*, 50–59. [CrossRef]
59. Morshedi, V.; Agh, N.; Marammazi, J.G.; Noori, F.; Mohammadian, T.; Mozanzadeh, M.T. Combined effects of dietary bovine lactoferrin, lactobacillus plantarum, and xylooligosaccharide on hemato-immunological and digestive enzymes of silvery-black porgy (*Sparidentex hasta*) fingerlings. *Comp. Clin. Pathol.* **2019**, *28*, 731–736. [CrossRef]
60. Pagheh, E.; Marammazi, J.G.; Agh, N.; Nouri, F.; Sepahdari, A.; Gisbert, E.; Mozanzadeh, M.T. Growth performance, hemato-immunological responses, and digestive enzyme activities in silvery-black porgy (*Sparidentex hasta*) fed dietary bovine lactoferrin. *Probiotics Antimicrob. Proteins* **2017**, *10*, 399–407. [CrossRef]
61. Li, Q.; Hu, W.; Zhao, J.; Wang, J.; Dai, Y.; Zhao, Y.; Meng, Q.; Li, N. Supplementation transgenic cow's milk containing recombinant human lactoferrin enhances systematic and intestinal immune responses in piglets. *Mol. Biol. Rep.* **2014**, *41*, 2119–2128. [CrossRef]
62. Nguyen, D.N.; Li, Y.; Sangild, P.T.; Bering, S.B.; Chatterton, D.E. Effects of bovine lactoferrin on the immature porcine intestine. *Br. J. Nutr.* **2014**, *111*, 321–331. [CrossRef]
63. Ray, A.K.; Ghosh, K.; Ringø, E. Enzyme-producing bacteria isolated from fish gut: A review. *Aquac. Nutr.* **2012**, *18*, 465–492. [CrossRef]
64. Liu, C.; Chi, K.; Yang, M.; Guo, N. Staphylococcal enterotoxin a induces intestinal barrier dysfunction and activates NLRP3 inflammasome via NF-κB/MAPK signaling pathways in mice. *Toxins* **2022**, *14*, 29. [CrossRef] [PubMed]

65. Han, R.; Wang, L.; Zhao, Z.; You, L.; Pedisić, S.; Kulikouskaya, V.; Lin, Z. Polysaccharide from *Gracilaria Lemaneiformis* prevents colitis in Balb/c mice via enhancing intestinal barrier function and attenuating intestinal inflammation. *Food Hydrocoll.* **2020**, *109*, 106048. [CrossRef]
66. Peng, P.; Chen, J.; Yao, K.; Yin, Y.; Long, L.; Fang, R. The effects of dietary supplementation with porous zinc oxide on growth performance, intestinal microbiota, morphology, and permeability in weaned piglets. *Anim. Sci. J.* **2019**, *90*, 1220–1228. [CrossRef] [PubMed]
67. Ruan, P.; Gong, Z.J.; Zhang, Q.R. Changes of plasma D(-)-lactate, diamine oxidase and endotoxin in patients with liver cirrhosis. *Hepatobiliary Pancreat. Dis. Int.* **2004**, *3*, 58–61.
68. Zhao, Y.; Qin, G.; Sun, Z.; Che, D.; Bao, N.; Zhang, X. Effects of soybean agglutinin on intestinal barrier permeability and tight junction protein expression in weaned piglets. *Int. J. Mol. Sci.* **2011**, *12*, 8502–8512. [CrossRef]
69. Rahimnejad, S.; Zhang, J.; Wang, L.; Sun, Y.; Zhang, C. Evaluation of *Bacillus pumillus* SE5 fermented soybean meal as a fish meal replacer in spotted seabass (*Lateolabrax maculatus*) feed. *Aquaculture* **2021**, *531*, 735975. [CrossRef]
70. He, Y.; Yang, Y.; Dong, Y.; Yan, C.; Zhang, B. The effects of flavomycin and colistin sulfate pre-treatment on ileal bacterial community composition, the response to *salmonella typhimurium* and host gene expression in broiler chickens. *Microorganisms* **2019**, *7*, 574. [CrossRef]
71. Zhou, Z.; Yao, W.; Ye, B.; Wu, X.; Li, X.; Dong, Y. Effects of replacing fishmeal protein with poultry by-product meal protein and soybean meal protein on growth, feed intake, feed utilization, gut and liver histology of hybrid grouper (*Epinephelus fuscoguttatus* ♀× *Epinephelus lanceolatus* ♂) juveniles. *Aquaculture* **2020**, *516*, 734503. [CrossRef]
72. Blais, A.; Fan, C.; Voisin, T.; Aattouri, N.; Dubarry, M.; Blachier, F.; Tomé, D. Effects of lactoferrin on intestinal epithelial cell growth and differentiation: An in vivo and in vitro study. *Biometals* **2014**, *27*, 857–874. [CrossRef]
73. Shimizu, H. Development of an enteric-coated lactoferrin tablet and its application. *Biometals* **2004**, *17*, 343–347. [CrossRef] [PubMed]
74. Wang, Y.; Shan, T.; Xu, Z.; Liu, J.; Feng, J. Effect of lactoferrin on the growth performance, intestinal morphology, and expression of PR-39 and protegrin-1 genes in weaned piglets. *J. Anim. Sci.* **2006**, *84*, 2636–2641. [CrossRef] [PubMed]
75. Fu, R.; Li, Z.; Zhou, R.; Li, C.; Shao, S.; Li, J. The mechanism of intestinal flora dysregulation mediated by intestinal bacterial biofilm to induce constipation. *Bioengineered* **2021**, *12*, 6484–6498. [CrossRef] [PubMed]
76. Xu, J.; Li, Y.; Yang, Z.; Li, C.; Liang, H.; Wu, Z.; Pu, W. Yeast probiotics shape the gut microbiome and improve the health of early-weaned piglets. *Front. Microbiol.* **2018**, *9*, 2011. [CrossRef] [PubMed]
77. Li, C.; Zhang, B.; Liu, C.; Zhou, H.; Wang, X.; Mai, K.; He, G. Effects of dietary raw or *Enterococcus faecium* fermented soybean meal on growth, antioxidant status, intestinal microbiota, morphology, and inflammatory responses in turbot (*Scophthalmus maximus* L.). *Fish Shellfish Immunol.* **2020**, *100*, 261–271. [CrossRef] [PubMed]
78. Estruch, G.; Collado, M.C.; Penaranda, D.S.; Tomas, V.A.; Jover, C.M.; Perez, M.G.; Martinez-Llorens, S. Impact of fishmeal replacement in diets for gilthead sea bream (*Sparus aurata*) on the gastrointestinal microbiota determined by pyrosequencing the 16S rRNA Gene. *PLoS ONE* **2015**, *10*, e0136389. [CrossRef]
79. Wang, P.; Zhou, Q.; Feng, J.; He, J.; Lou, Y.; Zhu, J. Effect of dietary fermented soybean meal on growth, intestinal morphology and microbiota in juvenile large yellow croaker, *Larimichthys crocea*. *Aquac. Res.* **2019**, *50*, 748–757. [CrossRef]
80. Ruhee, R.T.; Ma, S.; Suzuki, K. Sulforaphane protects cells against lipopolysaccharide-stimulated inflammation in murine macrophages. *Antioxidants* **2019**, *8*, 577. [CrossRef]
81. Liu, J.; Li, B.; Lee, C.; Zhu, H.; Zheng, S.; Pierro, A. Protective effects of lactoferrin on injured intestinal epithelial cells. *J. Pediatr. Surg.* **2019**, *54*, 2509–2513. [CrossRef]
82. Prgomet, C.; Prenner, M.L.; Schwarz, F.J.; Pfaffl, M.W. Effect of lactoferrin on selected immune system parameters and the gastrointestinal morphology in growing calves. *J. Anim. Physiol. Anim. Nutr.* **2007**, *91*, 109–119. [CrossRef]
83. Valenti, P.; Berlutti, F.; Conte, M.P.; Longhi, C.; Seganti, L. Lactoferrin functions: Current status and perspectives. *J. Clin. Gastroenterol.* **2004**, *38*, S127–S129. [CrossRef] [PubMed]
84. Wakabayashi, H.; Takakura, N.; Yamauchi, K.; Tamura, Y. Modulation of immunity-related gene expression in small intestines of mice by oral administration of lactoferrin. *Clin. Vaccine Immunol.* **2006**, *13*, 239–245. [CrossRef] [PubMed]
85. Lin, S.; Zhou, X.; Zhou, Y.; Kuang, W.; Chen, Y.; Luo, L.; Dai, F. Intestinal morphology, immunity and microbiota response to dietary fibers in largemouth bass, *Micropterus salmoide*. *Fish Shellfish Immunol.* **2020**, *103*, 135–142. [CrossRef] [PubMed]
86. Yang, Y.; Wang, Y.; Lu, Y.; Li, Q. Effect of replacing fish meal with soybean meal on growth, feed utilization and nitrogen and phosphorus excretion on rainbow trout (*Oncorhynchus mykiss*). *Aquac. Int.* **2010**, *19*, 405–419. [CrossRef]
87. Gu, M.; Bai, N.; Zhang, Y.; Krogdahl, Å. Soybean meal induces enteritis in turbot *Scophthalmus maximus* at high supplementation levels. *Aquaculture* **2016**, *464*, 286–295. [CrossRef]
88. Urán, P.A.; Gonçalves, A.A.; Taverne-Thiele, J.J.; Schrama, J.W.; Verreth, J.A.J.; Rombout, J.H.W.M. Soybean meal induces intestinal inflammation in common carp (*Cyprinus carpio* L.). *Fish Shellfish Immunol.* **2008**, *25*, 751–760. [CrossRef]
89. Tusi, S.K.; Manesh, T.E.; Fathollahi, M.S.; Bagherian, A. Can tert-butylhydroquinone improve the healing of extracted tooth socket in rats? *J. Dent. Res.* **2017**, *14*, 8–12.
90. Van der Oost, R.; Beyer, J.; Vermeulen, N.P.E. Fish bioaccumulation and biomarkers in environmental risk assessment: A review. *Environ. Toxicol. Pharmacol.* **2003**, *13*, 57–149. [CrossRef]

91. Wan, J.; Zhang, J.; Chen, D.; Yu, B.; Huang, Z.; Mao, X.; Zheng, P.; Yu, J.; He, J. Alginate oligosaccharide enhances intestinal integrity of weaned pigs through altering intestinal inflammatory responses and antioxidant status. *RSC Adv.* **2018**, *8*, 13482–13492. [CrossRef]
92. Aguiló, A.; Tauler, P.; Fuentespina, E.; Tur, J.A.; Córdova, A.; Pons, A. Antioxidant response to oxidative stress induced by exhaustive exercise. *Physiol. Behav.* **2005**, *84*, 1–7. [CrossRef]
93. Lin, Y.H.; Mui, J.J. Comparison of dietary inclusion of commercial and fermented soybean meal on oxidative status and non-specific immune responses in white shrimp, *Litopenaeus vannamei*. *Fish Shellfish Immunol.* **2017**, *63*, 208–212. [CrossRef]
94. Baeverfjord, G.T.; Krogdahl, A. Development and regression of soybean meal induced enteritis in Atlantic salmon, *Salmo salar* L., distal intestine: A comparison with the intestines of fasted fish. *J. Fish Dis.* **1996**, *19*, 375–387. [CrossRef]
95. Olsvik, P.A.; Torstensen, B.E.; Hemre, G.I.; Sanden, M.; Waagbø, R. Hepatic oxidative stress in Atlantic salmon (*Salmo salar* L.) transferred from a diet based on marine feed ingredients to a diet based on plant ingredients. *Aquac. Nutr.* **2011**, *17*, e424–e436. [CrossRef]
96. Hashem, N.M.A.; El Son, M.A.M.; Ateya, A.I.; Saleh, R.M. Impact of lactoferrin supplementation on oxidative stress, gene expression and immunity dysfunction induced by *Aeromonas veronii* in Nile tilapia (*Oreochromis niloticus*). *Aquac. Res.* **2022**, *53*, 2392–2407. [CrossRef]
97. Zhuang, Y.; Huang, H.; Liu, X.; Wang, N.; Zhong, G. Effect of bovine lactoferricin on the growth performance, digestive capacity, immune responses and disease resistance in Pacific white shrimp, *Penaeus vannamei*. *Fish Shellfish Immunol.* **2022**, *123*, 282–289. [CrossRef] [PubMed]
98. Jiang, M.; Wen, H.; Gou, G.W.; Liu, T.L.; Lu, X.; Deng, D.F. Preliminary study to evaluate the effects of dietary bile acids on growth performance and lipid metabolism of juvenile genetically improved farmed tilapia (*Oreochromis niloticus*) fed plant ingredient-based diets. *Aquac. Nutr.* **2018**, *24*, 1175–1183. [CrossRef]
99. Dossou, S.; Koshio, S.; Ishikawa, M.; Yokoyama, S.; Dawood, M.A.O.; El Basuini, M.F.; El-Hais, A.M.; Olivier, A. Effect of partial replacement of fish meal by fermented rapeseed meal on growth, immune response and oxidative condition of red sea bream juvenile, *Pagrus major*. *Aquaculture* **2018**, *490*, 228–235. [CrossRef]
100. Yamamoto, T.; Iwashita, Y.; Matsunari, H.; Sugita, T.; Furuita, H.; Akimoto, A.; Okamatsu, K.; Suzuki, N. Influence of fermentation conditions for soybean meal in a non-fish meal diet on the growth performance and physiological condition of rainbow trout *Oncorhynchus mykiss*. *Aquaculture* **2010**, *309*, 173–180. [CrossRef]
101. Ye, H.; Xu, M.; Chen, L.; Tan, X.; Chen, S.; Zou, C.; Sun, Z.; Liu, Q.; Ye, C.; Wang, A. Effects of dietary plant protein sources influencing hepatic lipid metabolism and hepatocyte apoptosis in hybrid grouper (*Epinephelus lanceolatus* ♂ × *Epinephelus fuscoguttatus* ♀). *Aquaculture* **2019**, *506*, 437–444. [CrossRef]

metabolites

Article

Response of Intestinal Microbiota to the Variation in Diets in Grass Carp (*Ctenopharyngodon idella*)

Gang Yang [1], Yuhan Xiang [1], Shanshan Wang [1], Yujie Tao [1], Lichen Xie [1], Lixin Bao [1], Kaikai Shen [1], Jiamin Li [2], Baoqing Hu [1], Chungen Wen [1], Vikas Kumar [3] and Mo Peng [2,*]

1 Department of Fisheries Science, School of Life Science, Nanchang University, Nanchang 330031, China
2 School of Animal Science and Technology, Jiangxi Agricultural University, Nanchang 330045, China
3 Aquaculture Research Institute, Department of Animal, Veterinary and Food Sciences, University of Idaho, Moscow, ID 83844, USA
* Correspondence: pengmowell@jxau.edu.cn

Citation: Yang, G.; Xiang, Y.; Wang, S.; Tao, Y.; Xie, L.; Bao, L.; Shen, K.; Li, J.; Hu, B.; Wen, C.; et al. Response of Intestinal Microbiota to the Variation in Diets in Grass Carp (*Ctenopharyngodon idella*). *Metabolites* **2022**, *12*, 1115. https://doi.org/10.3390/metabo12111115

Academic Editors: Qingchao Wang, Yan He, Qingsong Tan and Baichuan Deng

Received: 19 October 2022
Accepted: 13 November 2022
Published: 15 November 2022

Abstract: The intestinal microbiota is important for the nutrient metabolism of fish and is significantly influenced by the host's diet. The effect of ryegrass and commercial diets on the intestinal microbiota of grass carp was compared in this study. In comparison to ryegrass, artificial feed significantly reduced the microbial diversity in the intestine, which was measured by a decrease in the observed OTUs, ACE, Shannon, and the InvSimpson index. Although grass carp fed with ryegrass and artificial feed shared a dominant phyla Firmicutes and Proteobacteria, the microbial composition was clearly distinguishable between the two groups. In grass carp fed with ryegrass, Alphaproteobacteria, Gammaproteobacteria, and Actinobacteria predominated, whereas Bacilli was significantly higher in the artificial feed group due to an increase in Weissella and an unassigned Bacillales bacteria, as well as a significant increase in a potential pathogen: *Aeromonas australiensis*. Grass carp fed with ryegrass exhibited a more complex ecological network performed by the intestinal bacterial community, which was dominated by cooperative interactions; this was also observed in grass carp fed with artificial feed. Despite this, the increase in *A. australiensis* increased the competitive interaction within this ecological network, which contributed to the vulnerable perturbation of the intestinal microbiota. The alteration of the microbial composition through diet can further affect microbial function. The intestinal microbial function in grass carp fed with ryegrass was rich in amino acids and exhibited an increased energy metabolism in order to compensate for a low-nutrient diet intake, while the artificial feed elevated the microbial lipid metabolism through the promotion of its synthesis in the primary and secondary bile acids, together with a notable enhancement of fatty acid biosynthesis. These results indicated that diet can affect the homeostasis of the intestinal microbiota by altering the microbial composition and the interspecific interactions, whilst microbial function can respond to a variation in diet.

Keywords: *Ctenopharyngodon idella*; ecological network; homeostasis; microbial function

1. Introduction

The intestine is the most important organ of digestion and absorption in fish, and it is also a complex ecosystem as it harbors an extremely diverse and complex microbial community [1–3]. It is widely acknowledged that the intestinal microbiota performs critical functions for the host, such as the production of digestion-related enzymes, vitamin synthesis, pathogen protection, and immune maturation [4–6]. The microbiota in the intestine provides a large number of fermented metabolites for the host, particularly herbivores [7–9]. The intestinal bacterial community is a complex micro-ecological system that is significantly influenced by several factors, including the nutritional components of the host's food [6,10]. Since fishmeal is scarce, plant protein is frequently utilized in aquafeed; however, this may lead to intestinal inflammation and microbial dysbiosis in fish [11–13]. The number of species, their abundance, and their intricate microbial interactions play a critical role in the homeostasis of the intestinal microbiota [14]. The microbial interactions in the intestine are widely acknowledged to be dynamic in order to connect trillions of bacteria into a sophisticated ecological network [9,15,16]. Intestinal microbiota includes more than 100 times as many genes as the host, enabling it to encode a diverse range of enzymes with a variety of different metabolic capabilities [17]. Numerous studies have shown that the intestinal microbiota can act as an additional metabolic organ, contributing significantly to the host's amino acid, glucose, energy, and lipid metabolism by producing fermentation by-products [18–21].

The herbivorous grass carp (*Ctenopharyngodon idella*) is China's most productive freshwater fish. Numerous studies have characterized the grass carp's intestinal microbiota [22–25] and cellulase-producing bacteria, which assists in the digestion of fish that are fed a high-fiber diet [7,8,26]. However, the existing knowledge about the bacterial community in the intestine is about the composition and function of microbial communities in grass carp; however, little is known about species-species interactions within the bacterial community. Our previous research looked at the microbiota in different parts of the intestine of grass carp, and we discovered that intricate interspecific interactions could boost the efficiency of the bacterial community fermentation [16]. It is widely acknowledged that the overuse of plant protein can disturb the interspecific interactions and result in a disorder of the intestinal bacterial community in fish.

Diets not only provide nutrients to fish and the intestinal microbiota, but they also influence fish health by modulating the intestinal microbiome. The objective of this study was to investigate the response of grass carp's intestinal microbiota to ryegrass and a commercial diet that is high in plant protein by evaluating the variation in composition, interspecific interactions, and metabolic functions of the intestinal bacterial community. Thus, the findings of this study may provide an in-depth understanding of the alteration of intestinal microbiota in grass carp in response to dietary variations, providing a theoretical basis for intervening in the intestinal microbiota to maintain grass carp health.

2. Materials and Methods

2.1. Sample Collection

The grass carp were raised in artificial ponds (area, 200 m^2; depth mean, 1.5 m) in the Institute of Special Aquaculture, Yichun, China. The grass carp, in two ponds, were fed with ryegrass (GF group: 12.73% protein, 1.38% lipid, and 26.52% fiber of total dry matter) and a commercial diet, provided by Tongwei Co., Ltd., Chendu, China (CF group: 30% protein (mainly plant protein from soybean meal, cottonseed meal, and rapeseed meal), 5% lipid, and 10.13% fiber). The fish were fed to apparent satiation twice a day (8:30 and 16:00) for one month without antibiotics. Grass carp fed with ryegrass (n = 8, weight mean, 316.45 ± 32.62 g) and a commercial diet (n = 8, weight mean, 357.81 ± 68.31 g) were collected after 3 h of feeding in the morning. In each group, twelve fish in one pond were chosen randomly from among more than 20 caught fish, and the other fish were returned. Before dissection, the skin surface of the grass carp was sterilized with 70% ethanol to reduce contamination, and then the digesta from the middle intestine was collected and stored in sterile freezing tubes under −80 °C. The experimental protocols of grass carp handling and sampling have been approved by the department of Laboratory Animal Science, Nanchang University (Approval Number: NCU-208-2021).

2.2. Illumina Sequencing of Bacterial 16S rRNA Gene

PowerFecal™ DNA Isolation Kit (MoBio Laboratories, Inc., Carlsbad, CA, USA) was used for DNA extraction of digesta samples. Amplification of the 16S rRNA V3-V4 region was performed as described previously with barcoded fusion primers of 341F and 805R [27]. High-throughput sequencing was performed using the Illumina HiSeq platform at Novogen Co., Ltd., Beijing, China. All the sequencing data can be found in the Sequence Read Archive (SRA) database at NCBI under accession number PRJNA880788.

2.3. Bioinformatics and Statistical Analysis

The raw sequences were sorted into different samples according to the barcodes by using the BIPES pipeline, followed by a quality-control step to remove any low-quality amplicon sequences by VSEARCH [28]. The clean sequences were then clustered into operational taxonomic units (OTUs) with a 99% sequence similarity and annotated using the Ribosomal Database (rdp_16s_v16_sp). A total of 1,369,277 effective sequences and 1201 OTUs were generated from all samples. Alpha diversity and the relative abundance of taxa analyses were calculated by R software v 4.1.3. The Wilcoxon test was used to test the α-diversity index, and the relative abundance of taxa using R software. Treemap was used to visualize the significantly abundant OTU's, the annotated taxonomy, the P-value, and the relative abundance, in which the size of the bubbles indicated the relative abundance of the raw read counts [29]. Principal coordinates analysis (PCoA), based on the Bray–Curtis dissimilarity analyses, were employed to visualize the bacterial community structure and the differences in the bacterial community was calculated by Permutational analysis of variance (PERMANOVA) based on the Bray–Curtis distance [30].

Using abundance profiles of the individual OTUs, a molecular ecological network analysis was performed to evaluate bacterial species-to-species interactions within a community (http://ieg2.ou.edu/MENA (accessed on 15 July 2022)). A Random Matrix Theory (RMT)-based approach was used for an ecological network construction and topological role identification [31]. The network was visualized using Circos and Cytoscape 3.9.0. Based on a modularity property, each network was separated into modules by the fast greedy modularity optimization. According to values of within-module connectivity (Zi) and among module connectivity (Pi), the topological roles of different nodes can be categorized into four types: peripherals (Zi $\leq$ 2.5, Pi $\leq$ 0.62), connectors (Zi $\leq$ 2.5, Pi > 0.62), module hubs (Zi > 2.5, Pi $\leq$ 0.62) and network hubs (Zi > 2.5, Pi > 0.62).

Functional gene and Kyoto Encyclopedia of Genes and Genomes (KEGG) pathways were predicted using PICRUSt2 software [32] against a Greengenes reference database (Greengenes 13.5). The non-metric multidimensional scaling (NMDS) and analysis of similarity (ANOSIM) were used to evaluate the overall differences in predicted bacterial functional compositions based on the Bray–Curtis distance at KEGG orthology (KO) level [33]. A two-sided Welch's t test was used to identify significant different metabolic pathways in the two groups by software STAMP, with $p < 0.05$ considered significant.

3. Results

3.1. Diversity and Composition of the Bacterial Community

The bacterial community of grass carp in the GF group displayed a significantly higher value in the number of observed OTUs, ACE, Shannon, and InvSimpson index when compared to grass carp fed with the commercial diet ($p < 0.05$), whereas no significant differences were observed in chao1 (Table 1). The most observed phylum of the bacterial communities included Firmicutes (GF: 20.49%; CF: 61.59%), Proteobacteria (GF: 56.59%; CF: 25.42%), Actinobacteria (GF: 10.77%; CF: 2.97%), Fusobacteria (GF: 0.24%; CF: 5.62%), and Chloroflexi (GF: 3.93%; CF: 1.02%) were detected in the mid-intestine (Figure 1 and Table S1). Specifically, Alphaproteobacteria (43.84%), Bacilli (16.11%), Actinobacteria (10.77%), and Gammaproteobacteria (9.51%) took dominance in the middle intestine of grass carp fed with ryegrass, while the grass carp in the CF group were enriched with classes of Bacilli (55.58%), Alphaproteobacteria (16.63%), Gammaproteobacteria (7.51%), Fusobacteria (5.62%), and Clostridia (5.24%) in the intestine (Figure 2 and Table S2).

Table 1. Alpha diversity estimates of the bacterial communities (mean ± S.E.; n = 8) [1].

Groups	Observed OTUs	Chao1	ACE	Shannon	InvSimpson
GF	518.88 ± 21.96 ^{b}	618.72 ± 16.8	605.7 ± 17.79 ^{b}	3.78 ± 0.23 ^{b}	16.8 ± 3.49 ^{b}
CF	304.13 ± 41.13 ^{a}	351.08 ± 51.7	344.53 ± 50.91 ^{a}	2.76 ± 0.2 ^{a}	5.86 ± 1.35 ^{a}

[1] Values in the same row with the different superscript are significantly different ($p < 0.05$).

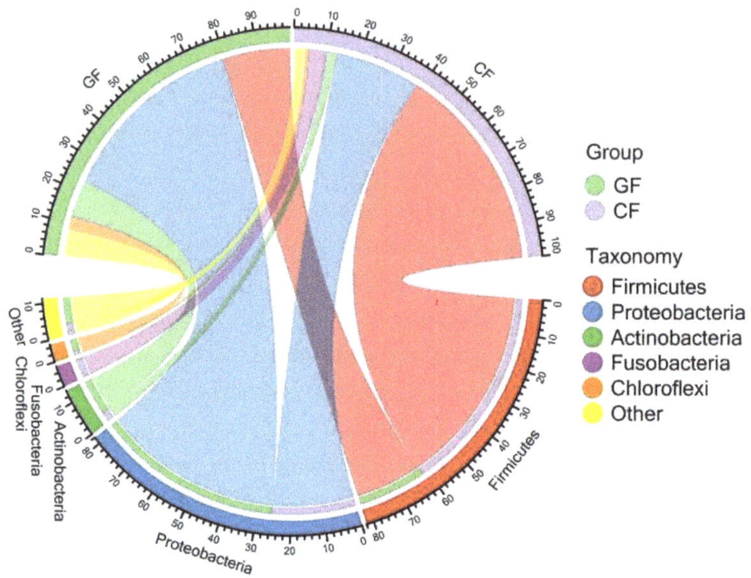

Figure 1. Chord diagram exhibited the relative abundance of bacterial phyla above ≥ a cutoff value of 2%.

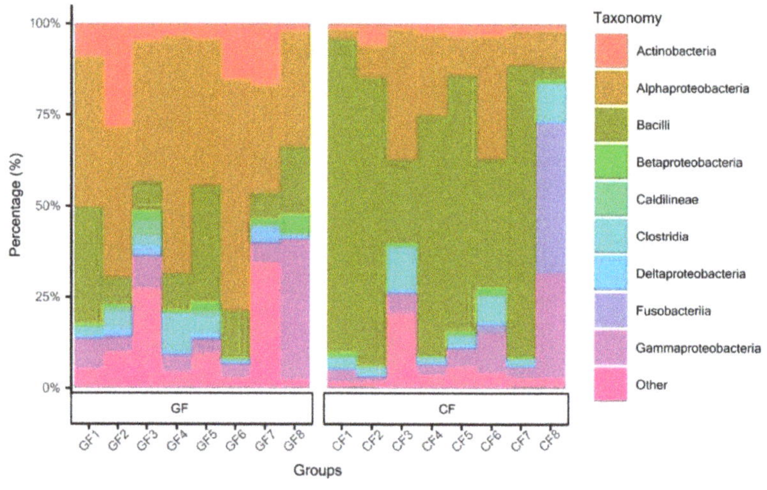

Figure 2. Relative abundances of the top 10 bacterial classes.

As shown in Figure 3, grass carp in the GF group were significantly rich in Rhizobiales (mainly Rhizobiaceae, Rhizobiaceae, and Methylobacteriaceae), Microbacteriaceae from Actinobacteria, and Bacillaceae_1 from Bacilli ($p < 0.05$), whereas Weissella from Lactobacillales (Bacilli), *Cetobacterium_somerae* from Fusobacteriia, *Aeromonas_australiensis* from Gammaproteobacteria, and unassigned bacteria from Bacillales (Bacilli) dominated the bacterial community of grass carp from the CF group ($p < 0.05$). Additionally, the PCoA analysis exhibited a clear separation in the bacterial communities between the GF and the CF groups at the OTU level, and a significant difference was further revealed using PERMANOVA ($p = 0.003$, Figure 4).

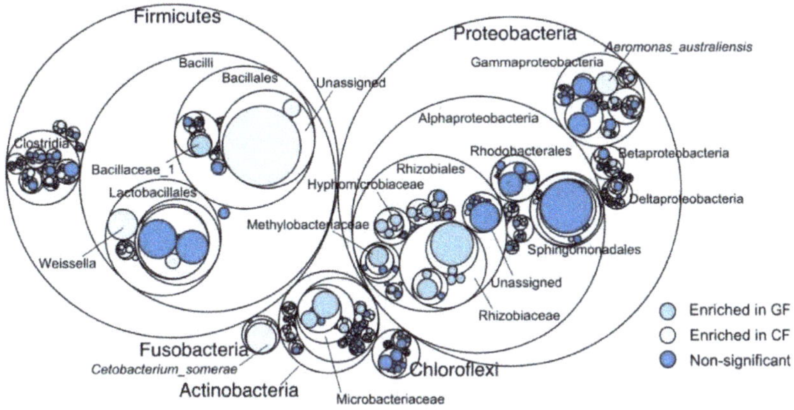

Figure 3. Maptree plot descriptions of the taxonomic differences based on 16S rRNA sequences. The largest circles represent phyla level, the inner circles represent class, order, family, and genus for panel.

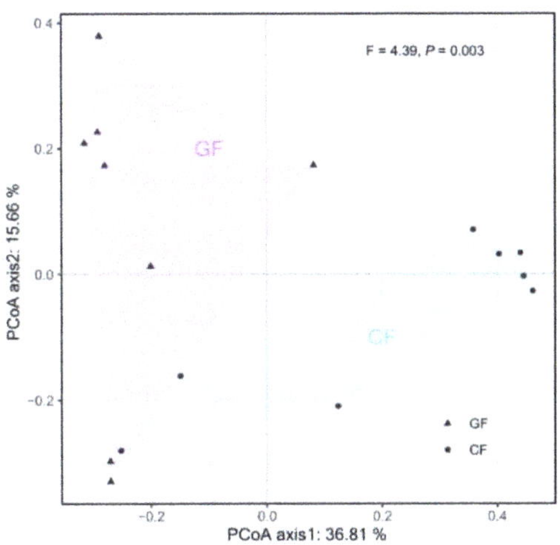

Figure 4. Principal coordinates analysis (PCoA) plot based on Bray–Curtis dissimilarity visualizing dissimilarities in the intestinal bacterial community of grass carp fed with ryegrass and a commercial diet.

3.2. Ecological Network Analysis

A circos plot displayed a classified composition and species-species interactions within the bacterial community (Figure 5A), which consisted of different OTUs from 27 bacterial classes (Table 2). The GF network represented 441 OTUs and 2946 edges (gray edges: 2099; red edges: 847), and the CF network displayed 252 OTUs and 856 edges (gray edges: 592; red edges: 264). The gray and red edges indicated the positive and negative interactions between two OTUs. The GF network recorded major OTUs ($\geq$20) from Actinobacteria (69 OTUs), Alphaproteobacteria (66 OTUs), Clostridia (41 OTUs), Gammaproteobacteria (40 OTUs), Acidobacteria (29 OTUs), Deltaproteobacteria (25 OTUs), and Bacilli (23 OTUs), whereas the major OTUs from Alphaproteobacteria (47 OTUs), Bacilli (46 OTUs), Actinobacteria (40 OTUs), and Gammaproteobacteria (35 OTUs) were observed in the CF network.

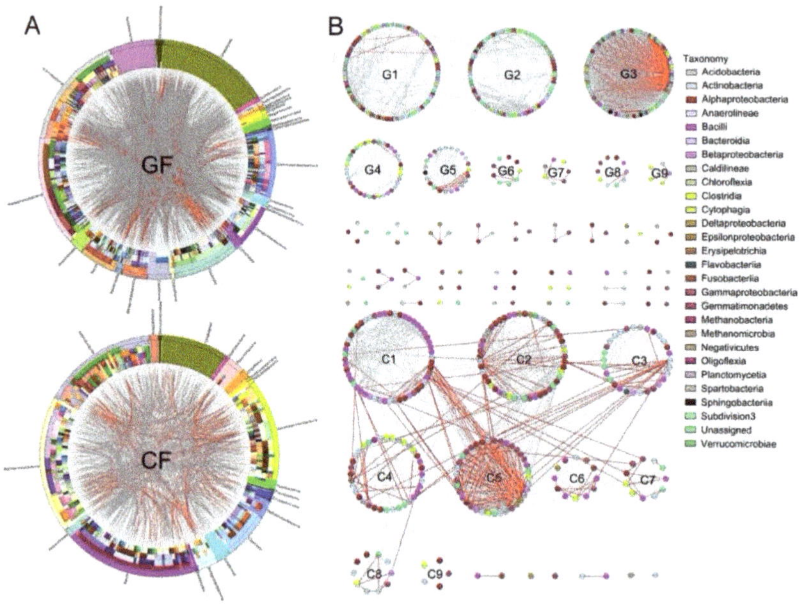

Figure 5. Circular plot (**A**) and ecological network (**B**) descriptions of the interaction between species within intestinal bacterial community. The width of the bars represents the abundance of each taxon. The bands with different colors demonstrate the source of different genera. The taxonomic levels were class, order, family, genera, and species from the outside to the inside of the circle. Each node in the network graph indicates one OTU. Colors of the nodes indicate different major classes. The edges (gray edge = positive interaction and red edge = negative interaction) inside the circle and ecological network represent the interactions between species.

In the ecological network, an RMT-based approach was employed to delineate separate modules. Strikingly, in Figure 5B, the ecological network within the bacterial community of the grass carp fed with ryegrass consisted of 9 modules ($\geq$5 nodes), and 4 larger sub-modules with $\geq$ 30 nodes were G1 (105 OTUs), G2 (86 OTUs), G3 (87 OTUs), and G4 (36 OTUs). Similarly, the CF network also had 9 modules with more than 5 nodes, and 5 sub-modules with $\geq$ 30 nodes were observed including C1 (53 OTUs), C2 (50 OTUs), C3 (30 OTUs), C4 (36 OTUs), and C5 (37 OTUs). Negative interactions were observed dominantly within the G3 and C5 sub-modules, whilst many negative edges were recorded between C1 and C5, or C3 and C5 sub-modules. Each specie performed different topological roles in the ecological network, in which most of the nodes were peripherals and several nodes performed as module hubs or connectors (Figure 6). As shown in Table 3, the GF network recorded 7 module hubs in G1 (3 OTUs), G2 (2 OTUs) and G4 (2 OTUs) sub-modules, among which these nodes were from Alphaproteobacteria (OTU_891, OTU_901 and OTU_911), Actinobacteria (OTU_568), Chloroflexi (OTU_756), Erysipelotrichia (OTU_376), and Gammaproteobacteria (OTU_153). In CF network, 4 module hubs (OTU_149, OTU_549, OTU_892 and OTU_1171) and 3 connectors (OTU_376, OTU_522 and OTU_894) were observed in C2 (3 OTUs), C3 (1 OTU) and C4 (3 OTUs) sub-modules, among which these nodes were from Alphaproteobacteria (OTU_892 and OTU_894), Actinobacteria (OTU_549), Clostridia (OTU_1171), Erysipelotrichia (OTU_376), Gammaproteobacteria (OTU_149), and Planctomycetia (OTU_522).

Table 2. The composition of the ecological network.

Index	GF	CF
Acidobacteria	29	3
Actinobacteria	69	40
Alphaproteobacteria	66	47
Anaerolineae	9	3
Bacilli	23	46
Bacteroidia	4	1
Betaproteobacteria	16	7
Caldilineae	14	5
Chloroflexia	2	0
Clostridia	41	16
Cytophagia	4	0
Deltaproteobacteria	25	6
Epsilonproteobacteria	1	0
Erysipelotrichia	3	3
Flavobacteriia	0	1
Fusobacteriia	3	3
Gammaproteobacteria	40	35
Gemmatimonadetes	1	0
Methanomicrobia	4	0
Methanobacteria	2	2
Negativicutes	3	
Oligoflexia	2	3
Planctomycetia	1	6
Spartobacteria	2	1
Sphingobacteriia	3	0
Subdivision3	2	0
Verrucomicrobiae	4	1
Unassigned	68	23
Total number of OTUs	441	252
The number of modules (≥5 OTUs)	9	9
The number of module hubs	7	4
The number of connectors	0	3
The number of gray edges	2099	592
The number of red edges	847	264
Total number of edges	2946	856

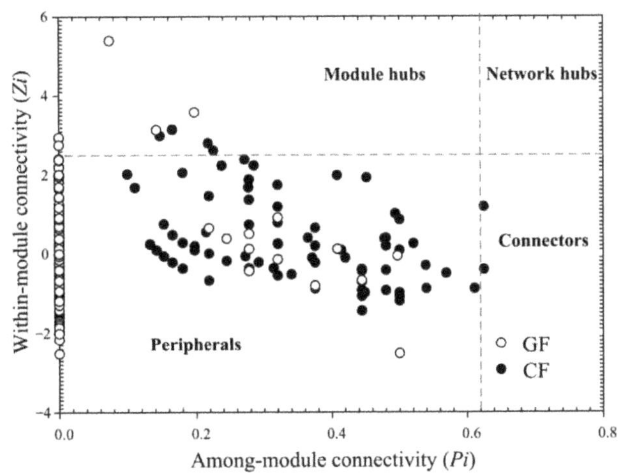

Figure 6. *Z-P* plot showing the distribution of OTUs based on their topological roles.

Table 3. Topological roles of intestinal microbiota in grass carp.

Treatment	Topological Roles	OTUs	Module Number	Phylogenetic Associations
GF	Module hubs	OTU_376	1	Erysipelotrichia
	Module hubs	OTU_568	4	Actinobacteria
	Module hubs	OTU_891	2	Alphaproteobacteria
	Module hubs	OTU_756	2	Chloroflexi
	Module hubs	OTU_153	4	Gammaproteobacteria
	Module hubs	OTU_901	1	Alphaproteobacteria
	Module hubs	OTU_911	1	Alphaproteobacteria
CF	Module hubs	OTU_549	3	Actinobacteria
	Module hubs	OTU_1171	2	Clostridia
	Module hubs	OTU_892	2	Alphaproteobacteria
	Module hubs	OTU_149	2	Gammaproteobacteria
	Connectors	OTU_522	4	Planctomycetia
	Connectors	OTU_894	4	Alphaproteobacteria
	Connectors	OTU_376	4	Erysipelotrichia

3.3. Functional Predictions of Intestinal Microbiota with PICRUSt2

The NMDS analysis exhibited a clear distinction in the functional composition of the intestinal microbiota between the GF and the CF groups at a KO level, and an analysis of similarity (ANOSIM) further confirmed the remarkable differences in the bacterial functional composition between the GF and the CF groups ($p = 0.007$, Figure 7). To study the microbial metabolic function, the KEGG functional categories were analyzed, including amino acid metabolism, carbohydrate metabolism, energy metabolism, lipid metabolism, and protein families metabolism (Figure 8). A significant difference in 47 pathways and 6 protein families, which are involved in nutrient metabolism, were observed between the GF and the CF groups ($p < 0.05$). In particular, the bacterial communities in the grass carp fed with ryegrass was significantly rich in 13 pathways in amino acid metabolism, 6 pathways in carbohydrate metabolism, 3 pathways in energy metabolism, 5 pathways in lipid metabolism, and 2 protein families related to metabolism ($p < 0.05$). The dietary commercial diet notably promoted bacterial metabolic function including 7 pathways in amino acid metabolism, 6 pathways in carbohydrate metabolism, 2 pathways in energy metabolism, 6 pathways in lipid metabolism, and 4 protein families related to metabolism in grass carp ($p < 0.05$).

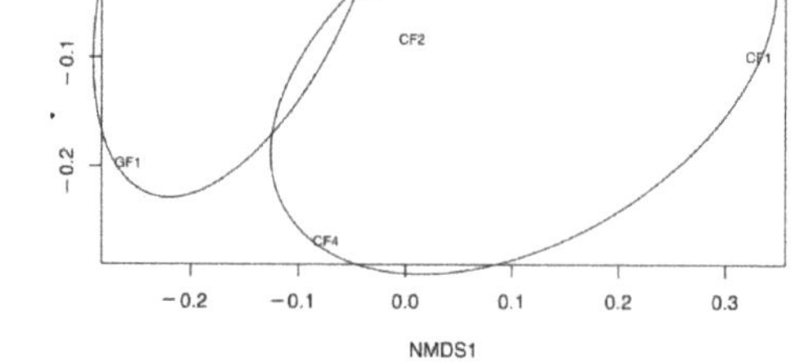

Figure 7. Non-metric multidimensional scaling (NMDS) plot visualizing bacterial functional community dissimilarities using Bray–Curtis distance.

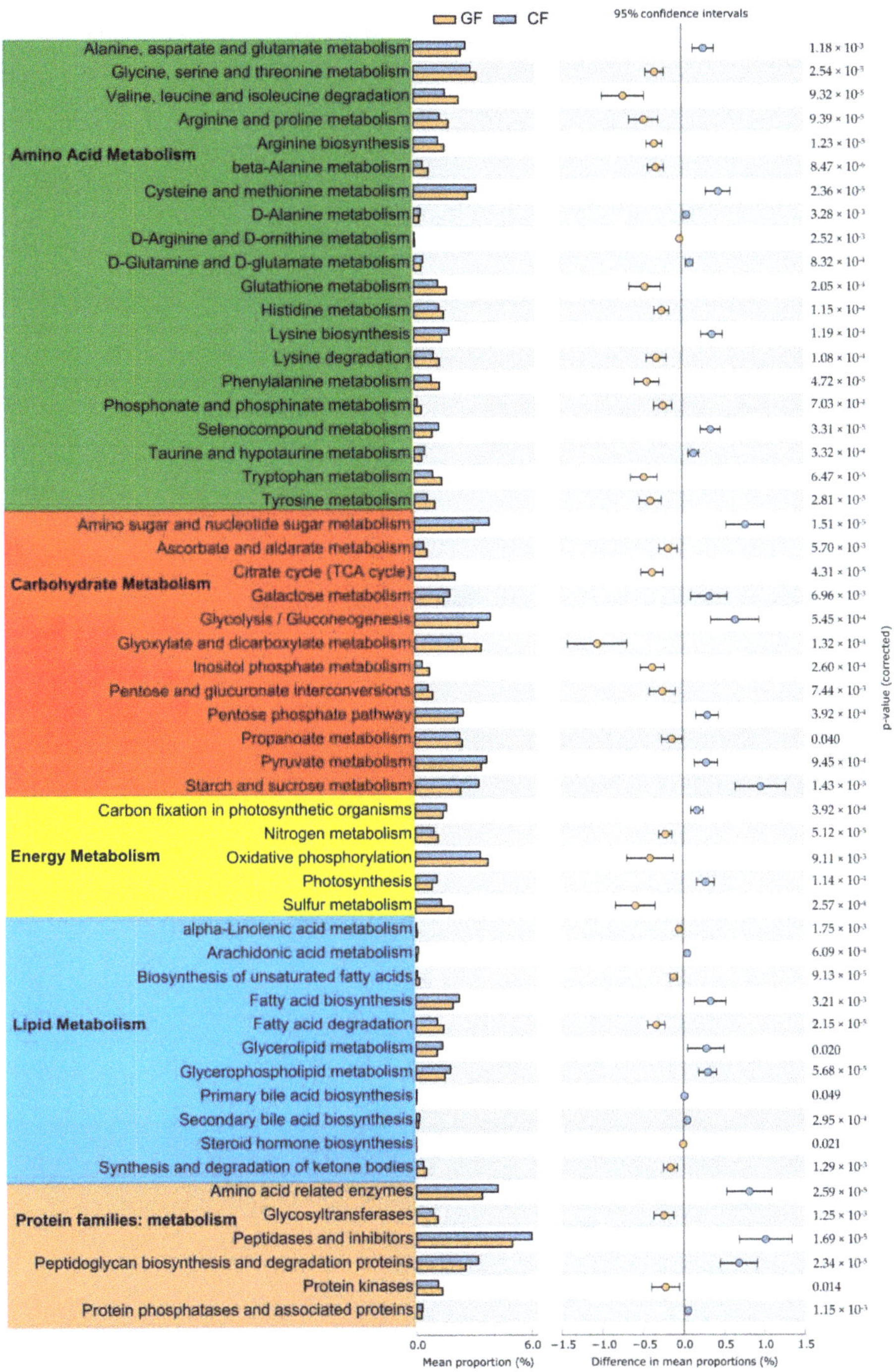

Figure 8. Significant changes in intestinal bacterial Kyoto Encyclopedia of Genes and Genomes (KEGG) pathways between ryegrass diet-fed and commercial diet-fed grass carp using the response ratio method at a 95% confidence interval (CI).

4. Discussion

The importance of intestinal microbiota in host health has been highlighted in recent decades, owing to its benefits to nutrient metabolism and immune maturation [4,34,35]. It is widely accepted that diets can influence the composition of the intestinal bacterial community because microbes require nutrients and energy from the food consumed by hosts [36]. Grass carp feed on aquatic weeds in the wild, and previous research found that grass carp fed a Sudan grass diet were rich in Firmicutes, Proteobacteria, Fusobacteria, and Actinobacteria in the intestine [37]. In the current study, Proteobacteria, Firmicutes, and Actinobacteria were found to be the most prevalent phyla in the intestines of grass carp fed with ryegrass. Grass carp, the most important aquaculture species in China, are primarily fed an artificial diet rich in plant protein. Previous research has shown that commercially formulated feed can significantly alter the structure of the intestinal bacterial community in grass carp, resulting in a decrease in microbial biodiversity [37]. Indeed, the current findings show that the artificial feed reduced diversity, and the proportion of Alphaproteobacteria and Actinobacteria, while increasing Bacilli. Furthermore, in this study, grass carp fed a high plant protein diet had a significantly higher proportion of *Aeromonas australiensis* from Gammaproteobacteria. *Aeromonas* are Gram-negative microbes that live in aquatic environments and have been identified as an opportunistic pathogen for fish [38,39]. Previous research has suggested that *Aeromonas* sp. infection can cause intestinal inflammation and mucosal barrier function damage in grass carp [40,41], crucian carp [42], and Nile Tilapia (*Oreochromis niloticus*) [43]. As a result, the significant increase in *A. australiensis*, caused by the artificial feed, increased the potential threat to the health of grass carp in this study.

As species perform similar or complementary functions, competitive and cooperative interactions occur between different species in the same habitat, for example competition to occupy more space and to obtain more nutrients [44]. The intestinal bacterial community can form a complex ecological network based on these interspecific interactions, allowing the intestinal microbiota to maintain a dynamic homeostasis in the host [14]. The microbial interactions were clearly degraded in the grass carp fed with an artificial diet due to a decrease in the microbial diversity, which can have a negative effect on the stability of a bacterial community because the intestinal microbiota is more easily disturbed by external factors. A cooperation dominated community is thought to be more stable because cooperative interactions are more resistant to population perturbations in spatial conditions, whereas competitive interactions are more vulnerable to disruption [45,46]. Indeed, the current findings suggested that cooperation dominated the intestinal bacterial community of grass carp, whereas the increased percentage of competitive interactions in the group fed with the artificial diet can be attributed to the significant increase in the opportunistic pathogen *A. australiensis*. Despite the fact that the grass carp, in two groups, showed distinctive submodules in the ecological networks, the networks in this study were primarily made up of the dominating microbial flora of each group. From an ecological perspective, connections, and module hubs generalists act as structural and functional keystones and play a crucial role in sustaining the ecological network [47]. According to the latest findings, the network's generalist population was unaffected by the artificial diet. By weakening interspecific relationships and reducing cooperation in the ecological network, the artificial diet may collectively have a detrimental influence on the gut microbiota homeostasis.

Since the genes of microbes encode several enzymes involved in protein, carbohydrate, lipid, and energy metabolism, microbial fermentation plays a critical role in the host's nutritional metabolism [35,48,49]. The structure and operation of the gut bacterial community can adjust to changes in the dietary environment in a similar manner [36]. Due to differences in gene expression across the different microbes in the current study, dietary ryegrass, and artificial feed, as expected, shaped two distinctly different intestinal microbiota structures in grass carp. These differences were accompanied by a notable variation in the metabolic activities of the microbial community. Fish were frequently affected with the issue of aberrant fat deposition during breeding due to the use of a high-protein and a high-fat artificial feed. [50,51]. Evidence in both humans and animals has shown a robust correlation between obesity and a high Firmicutes ratio [52,53]. Here, compared to a ryegrass diet, the artificial feed did increase the proportion of Firmicutes in grass carp. This was followed by an increase in lipid metabolism, specifically an improvement in the biosynthesis of fatty acids, glycerols, and glycerospholipids, as well as a promotion of the production of primary and secondary bile acids. However, when grass carp were fed with ryegrass to supplement their low-nutrient diet, the microbial function related to amino acid and energy metabolism was more active. Furthermore, it has been demonstrated that microbial fermentation of complex, non-digestible dietary carbohydrates aids the host in obtaining useable forms of energy from a plant-based diet [8]. Grass carp are generally considered herbivorous, but no gene in the grass carp genome encodes an enzyme to digest cellulose, which comprises the principal component of a plant-based diet [54]. Evidence has confirmed that the microbial function of metabolizing cellulose plays an important role in the nutrient metabolism of an herbivore, and the abundant enzymes involved in the digestion of complex carbohydrates were also observed in the intestinal microbiota of grass carp [8]. Here, grass carp in the ryegrass diet group displayed a higher proportion of Actinobacteria, which proved to harbor an increased level of carbohydrate enzyme genes for the degradation of cellulose [55]. Nevertheless, it was shown that grass carp fed with an artificial feed, where starch was the predominant source of carbohydrate, had an increased microbial starch and sucrose metabolism. Additionally, a group of microorganisms that are interdependent on the effectiveness of fermentation are necessary for the microbial process of fermentation [9,56]. Therefore, the increased cooperative interaction may encourage the microbial fermentation of a ryegrass diet, assisting the grass carp to obtain more nutrients and energy. However, the microbial function in grass carp fed with the artificial diet was more susceptible to diseases in this study because of the per durability in the bacterial population.

The modification of intestinal microbiota is an effective method for keeping fish healthy because it plays a significant role in the metabolism of nutrients and the host's health. In fact, several studies have shown that by managing the homeostasis of fish's intestinal microbiota with functional feed additives improves fish health [57,58]. The current study evaluated the intestinal bacterial communities of grass carp fed with an artificial diet and ryegrass, and it would serve as a theoretical guide for the control of grass carp intestinal microbiota.

5. Conclusions

Intestinal bacterial community characteristics in grass carp were altered by dietary ryegrass and artificial feed. An opportunistic pathogen, *A. australiensis*, was found to drastically increase in the artificial feed diet. Additionally, artificial feed may impact the lipid metabolism of grass carp by raising the ratio of Firmicutes and decreasing the disturbance resistance of intestinal microbiota.

Supplementary Materials: The following supporting information can be downloaded at: https://www.mdpi.com/article/10.3390/metabo12111115/s1, Table S1: Relative abundance of different bacterial phyla in grass carp, Table S2: Relative abundance of different bacterial classes in grass carp.

Author Contributions: M.P. designed the experiments and supervised the manuscript; G.Y. carried out the sample collection, data analysis, visualization and Writing—original draft with the help of Y.X., S.W., Y.T., L.B., L.X., K.S., J.L. and B.H.; C.W. and V.K. revised the manuscript. All authors have read and agreed to the published version of the manuscript.

Funding: This study was supported by Double Thousand Program of Jiangxi Province (2019), Excellent Youth Cultivation Project of National Natural Science Foundation of China (20202ZDB01010), and Jiangxi Agriculture Research System (JXARS-03).

Institutional Review Board Statement: Grass carp used in this trial were supplied by the Institute of Special Aquaculture (Yichun, China) with the approval of the department of Laboratory Animal Science, Nanchang University (Approval Number: NCU-208-2021).

Informed Consent Statement: Not applicable.

Data Availability Statement: The data presented in this study are available in the main article and the Supplementary Materials.

Conflicts of Interest: The authors declare that they have no competing interest.

References

1. Tarnecki, A.M.; Burgos, F.A.; Ray, C.L.; Arias, C.R. Fish intestinal microbiome: Diversity and symbiosis unravelled by metagenomics. *J. Appl. Microbiol.* **2017**, *123*, 2–17. [CrossRef] [PubMed]
2. Talwar, C.; Nagar, S.; Lal, R.; Negi, R.K. Fish gut microbiome: Current approaches and future perspectives. *Indian J. Microbiol.* **2018**, *58*, 397–414. [CrossRef] [PubMed]
3. Kim, P.S.; Shin, N.-R.; Lee, J.-B.; Kim, M.-S.; Whon, T.W.; Hyun, D.-W.; Yun, J.-H.; Jung, M.-J.; Kim, J.Y.; Bae, J.-W. Host habitat is the major determinant of the gut microbiome of fish. *Microbiome* **2021**, *9*, 166. [CrossRef] [PubMed]
4. Costa, M.C.; Weese, J.S. Understanding the intestinal microbiome in health and disease. *Vet. Clin. N. Am. Equine Pract.* **2018**, *34*, 1–12. [CrossRef] [PubMed]
5. Dupont, H.L.; Jiang, Z.-D.; Dupont, A.W.; Utay, N.S. The intestinal microbiome in human health and disease. *Trans. Am. Clin. Climatol. Assoc.* **2020**, *131*, 178. [PubMed]
6. Power, S.E.; O'Toole, P.W.; Stanton, C.; Ross, R.P.; Fitzgerald, G.F. Intestinal microbiota, diet and health. *Br. J. Nutr.* **2014**, *111*, 387–402. [CrossRef]
7. Liu, H.; Guo, X.; Gooneratne, R.; Lai, R.; Zeng, C.; Zhan, F.; Wang, W. The gut microbiome and degradation enzyme activity of wild freshwater fishes influenced by their trophic levels. *Sci. Rep.* **2016**, *6*, 24340. [CrossRef]
8. Wu, S.; Ren, Y.; Peng, C.; Hao, Y.; Xiong, F.; Wang, G.; Li, W.; Zou, H.; Angert, E.R. Metatranscriptomic discovery of plant biomass-degrading capacity from grass carp intestinal microbiomes. *FEMS Microbiol. Ecol.* **2015**, *91*, fiv107. [CrossRef]
9. Dearing, M.D.; Kohl, K.D. Beyond fermentation: Other important services provided to endothermic herbivores by their gut microbiota. *Integr. Comp. Biol.* **2017**, *57*, 723–731. [CrossRef]
10. Wilson, A.S.; Koller, K.R.; Ramaboli, M.C.; Nesengani, L.T.; Ocvirk, S.; Chen, C.; Flanagan, C.A.; Sapp, F.R.; Merritt, Z.T.; Bhatti, F. Diet and the human gut microbiome: An international review. *Digest. Dis. Sci.* **2020**, *65*, 723–740. [CrossRef]
11. Agboola, J.O.; Chikwati, E.M.; Hansen, J.Ø.; Kortner, T.M.; Mydland, L.T.; Krogdahl, Å.; Djordjevic, B.; Schrama, J.W.; Øverland, M. A meta-analysis to determine factors associated with the severity of enteritis in Atlantic salmon (*Salmo salar*) fed soybean meal-based diets. *Aquaculture* **2022**, *555*, 738214. [CrossRef]
12. Zhang, W.; Tan, B.; Deng, J.; Haitao, Z. Multiomics analysis of soybean meal induced marine fish enteritis in juvenile pearl gentian grouper, *Epinephelus fuscoguttatus*♀× *Epinephelus lanceolatus*♂. *Sci. Rep.* **2021**, *11*, 23319. [CrossRef] [PubMed]
13. Infante-Villamil, S.; Huerlimann, R.; Jerry, D.R. Microbiome diversity and dysbiosis in aquaculture. *Rev. Aquac.* **2021**, *13*, 1077–1096. [CrossRef]
14. Coyte, K.Z.; Schluter, J.; Foster, K.R. The ecology of the microbiome: Networks, competition, and stability. *Science* **2015**, *350*, 663–666. [CrossRef] [PubMed]
15. Montoya, J.M.; Pimm, S.L.; Solé, R.V. Ecological networks and their fragility. *Nature* **2006**, *442*, 259–264. [CrossRef]
16. Yang, G.; Jian, S.Q.; Cao, H.; Wen, C.; Hu, B.; Peng, M.; Peng, L.; Yuan, J.; Liang, L. Changes in microbiota along the intestine of grass carp (*Ctenopharyngodon idella*): Community, interspecific interactions, and functions. *Aquaculture* **2019**, *498*, 151–161. [CrossRef]
17. Zhu, B.; Wang, X.; Li, L. Human gut microbiome: The second genome of human body. *Protein Cell* **2010**, *1*, 718–725. [CrossRef]
18. Aron-Wisnewsky, J.; Warmbrunn, M.V.; Nieuwdorp, M.; Clément, K. Metabolism and metabolic disorders and the microbiome: The intestinal microbiota associated with obesity, lipid metabolism, and metabolic health—Pathophysiology and therapeutic strategies. *Gastroenterology* **2021**, *160*, 573–599. [CrossRef]
19. Portincasa, P.; Bonfrate, L.; Vacca, M.; De Angelis, M.; Farella, I.; Lanza, E.; Khalil, M.; Wang, D.Q.-H.; Sperandio, M.; Di Ciaula, A. Gut Microbiota and Short Chain Fatty Acids: Implications in Glucose Homeostasis. *Int. J. Mol. Sci.* **2022**, *23*, 1105. [CrossRef]
20. Lin, R.; Liu, W.; Piao, M.; Zhu, H. A review of the relationship between the gut microbiota and amino acid metabolism. *Amino Acids* **2017**, *49*, 2083–2090. [CrossRef]
21. Nieuwdorp, M.; Gilijamse, P.W.; Pai, N.; Kaplan, L.M. Role of the microbiome in energy regulation and metabolism. *Gastroenterology* **2014**, *146*, 1525–1533. [CrossRef] [PubMed]
22. Tran, N.T.; Zhang, J.; Xiong, F.; Wang, G.-T.; Li, W.-X.; Wu, S.-G. Altered gut microbiota associated with intestinal disease in grass carp (*Ctenopharyngodon idellus*). *World J. Microb. Biot.* **2018**, *34*, 71. [CrossRef] [PubMed]
23. Sun, B.-Y.; Yang, H.-X.; He, W.; Tian, D.-Y.; Kou, H.-Y.; Wu, K.; Yang, C.-G.; Cheng, Z.-Q.; Song, X.-H. A grass carp model with an antibiotic-disrupted intestinal microbiota. *Aquaculture* **2021**, *541*, 736790. [CrossRef]
24. Zhang, J.; Xiong, F.; Wang, G.T.; Li, W.X.; Li, M.; Zou, H.; Wu, S.G. The influence of diet on the grass carp intestinal microbiota and bile acids. *Aquac. Res.* **2017**, *48*, 4934–4944. [CrossRef]
25. Wu, S.; Wang, G.; Angert, E.R.; Wang, W.; Li, W.; Zou, H. Composition, diversity, and origin of the bacterial community in grass carp intestine. *PLoS ONE* **2012**, *7*, e30440. [CrossRef] [PubMed]

26. Tran, N.T.; Wang, G.T.; Wu, S.G. A review of intestinal microbes in grass carp *Ctenopharyngodon idellus* (Valenciennes). *Aquac. Res.* **2017**, *48*, 3287–3297. [CrossRef]
27. Yang, G.; Xu, Z.; Tian, X.; Dong, S.; Peng, M. Intestinal microbiota and immune related genes in sea cucumber (*Apostichopus japonicus*) response to dietary β-glucan supplementation. *Biochem. Biophys. Res. Commun.* **2015**, *458*, 98–103. [CrossRef]
28. Rognes, T.; Flouri, T.; Nichols, B.; Quince, C.; Mahé, F. VSEARCH: A versatile open source tool for metagenomics. *PeerJ* **2016**, *4*, e2584. [CrossRef]
29. Carrión, V.J.; Perez-Jaramillo, J.; Cordovez, V.; Tracanna, V.; De Hollander, M.; Ruiz-Buck, D.; Mendes, L.W.; van Ijcken, W.F.; Gomez-Exposito, R.; Elsayed, S.S. Pathogen-induced activation of disease-suppressive functions in the endophytic root microbiome. *Science* **2019**, *366*, 606–612. [CrossRef]
30. Anderson, M.J. Permutational multivariate analysis of variance. *Dep. Stat. Univ. Auckl. Auckl.* **2015**, *26*, 32–46.
31. Deng, Y.; Jiang, Y.-H.; Yang, Y.; He, Z.; Luo, F.; Zhou, J. Molecular ecological network analyses. *BMC Bioinf.* **2012**, *13*, 113. [CrossRef]
32. Douglas, G.M.; Maffei, V.J.; Zaneveld, J.R.; Yurgel, S.N.; Brown, J.R.; Taylor, C.M.; Huttenhower, C.; Langille, M.G. PICRUSt2 for prediction of metagenome functions. *Nat. Biotechnol.* **2020**, *38*, 685–688. [CrossRef] [PubMed]
33. Clarke, K.R. Non-parametric multivariate analyses of changes in community structure. *Aust. Ecol.* **1993**, *18*, 117–143. [CrossRef]
34. de Vos, W.M.; Tilg, H.; Van Hul, M.; Cani, P.D. Gut microbiome and health: Mechanistic insights. *Gut* **2022**, *71*, 1020–1032. [CrossRef] [PubMed]
35. Mallott, E.K.; Amato, K.R. Host specificity of the gut microbiome. *Nat. Rev. Microbiol.* **2021**, *19*, 639–653. [CrossRef]
36. David, L.A.; Maurice, C.F.; Carmody, R.N.; Gootenberg, D.B.; Button, J.E.; Wolfe, B.E.; Ling, A.V.; Devlin, A.S.; Varma, Y.; Fischbach, M.A. Diet rapidly and reproducibly alters the human gut microbiome. *Nature* **2014**, *505*, 559–563. [CrossRef]
37. Feng, W.; Zhang, J.; Jakovlić, I.; Xiong, F.; Wu, S.; Zou, H.; Li, W.; Li, M.; Wang, G. Gut segments outweigh the diet in shaping the intestinal microbiota composition in grass carp *Ctenopharyngodon idellus*. *AMB Express* **2019**, *9*, 44. [CrossRef]
38. Fečkaninová, A.; Koščová, J.; Mudroňová, D.; Popelka, P.; Toropilova, J. The use of probiotic bacteria against Aeromonas infections in salmonid aquaculture. *Aquaculture* **2017**, *469*, 1–8. [CrossRef]
39. Igbinosa, I.H.; Beshiru, A.; Odjadjare, E.E.; Ateba, C.N.; Igbinosa, E.O. Pathogenic potentials of *Aeromonas* species isolated from aquaculture and abattoir environments. *Microb. Pathogen.* **2017**, *107*, 185–192. [CrossRef]
40. Song, X.; Zhao, J.; Bo, Y.; Liu, Z.; Wu, K.; Gong, C. *Aeromonas hydrophila* induces intestinal inflammation in grass carp (*Ctenopharyngodon idella*): An experimental model. *Aquaculture* **2014**, *434*, 171–178. [CrossRef]
41. Song, X.; Hu, X.; Sun, B.; Bo, Y.; Wu, K.; Xiao, L.; Gong, C. A transcriptome analysis focusing on inflammation-related genes of grass carp intestines following infection with *Aeromonas hydrophila*. *Sci. Rep.* **2017**, *7*, 40777. [CrossRef] [PubMed]
42. Dong, Y.; Yang, Y.; Liu, J.; Awan, F.; Lu, C.; Liu, Y. Inhibition of *Aeromonas hydrophila*-induced intestinal inflammation and mucosal barrier function damage in crucian carp by oral administration of *Lactococcus lactis*. *Fish Shellfish Immunol.* **2018**, *83*, 359–367. [CrossRef] [PubMed]
43. Gewaily, M.S.; Shukry, M.; Abdel-Kader, M.F.; Alkafafy, M.; Farrag, F.A.; Moustafa, E.M.; Doan, H.V.; Abd-Elghany, M.F.; Abdelhamid, A.F.; Eltanahy, A. Dietary *Lactobacillus plantarum* relieves Nile tilapia (*Oreochromis niloticus*) juvenile from oxidative stress, immunosuppression, and inflammation induced by deltamethrin and *Aeromonas hydrophila*. *Front. Mar. Sci.* **2021**, *8*, 621558.
44. Zhou, J.; Deng, Y.; Luo, F.; He, Z.; Yang, Y. Phylogenetic molecular ecological network of soil microbial communities in response to elevated CO_2. *mBio* **2011**, *2*, e00122-11. [CrossRef] [PubMed]
45. Pande, S.; Kaftan, F.; Lang, S.; Svatoš, A.; Germerodt, S.; Kost, C. Privatization of cooperative benefits stabilizes mutualistic cross-feeding interactions in spatially structured environments. *ISME J.* **2016**, *10*, 1413–1423. [CrossRef]
46. Ren, X.; Murray, R.M. Cooperation enhances robustness of coexistence in spatially structured consortia. In Proceedings of the 2019 18th European Control Conference (ECC), Naples, Italy, 25–28 June 2019; pp. 2651–2656.
47. Olesen, J.M.; Bascompte, J.; Dupont, Y.L.; Jordano, P. The modularity of pollination networks. *Proc. Natl. Acad. Sci. USA* **2007**, *104*, 19891–19896. [CrossRef]
48. Hill, D.; Sugrue, I.; Arendt, E.; Hill, C.; Stanton, C.; Ross, R.P. Recent advances in microbial fermentation for dairy and health. *F1000Research* **2017**, *6*, 751. [CrossRef]
49. Tremaroli, V.; Bäckhed, F. Functional interactions between the gut microbiota and host metabolism. *Nature* **2012**, *489*, 242–249. [CrossRef]
50. Shefat, S.H.T.; Karim, M.A. Nutritional diseases of fish in aquaculture and their management: A review. *Acta Sci. Pharm. Sci.* **2018**, *2*, 50–58.
51. Yao, T.; Gu, X.; Liang, X.; Fall, F.N.; Cao, A.; Zhang, S.; Guan, Y.; Sun, B.; Xue, M. Tolerance assessment of dietary bile acids in common carp (*Cyprinus carpio* L.) fed a high plant protein diet. *Aquaculture* **2021**, *543*, 737012. [CrossRef]
52. Million, M.; Lagier, J.-C.; Yahav, D.; Paul, M. Gut bacterial microbiota and obesity. *Clin. Microbiol. Infect.* **2013**, *19*, 305–313. [CrossRef] [PubMed]
53. Aoun, A.; Darwish, F.; Hamod, N. The influence of the gut microbiome on obesity in adults and the role of probiotics, prebiotics, and synbiotics for weight loss. *Prev. Nutr. Food Sci.* **2020**, *25*, 113. [CrossRef] [PubMed]
54. Ni, J.; Yan, Q.; Yu, Y.; Zhang, T. Factors influencing the grass carp gut microbiome and its effect on metabolism. *FEMS Microbiol. Ecol.* **2014**, *87*, 704–714. [CrossRef] [PubMed]

55. Leo, V.V.; Asem, D.; Singh, B.P. Actinobacteria: A highly potent source for holocellulose degrading enzymes. In *Actinobacteria, New and Future Developments in Microbial Biotechnology and Bioengineering*; Singh, J.S., Singh, D.P., Eds.; Elsevier: Amsterdam, The Netherlands, 2018; pp. 191–205. ISBN 978-0-444-63994-3.
56. Smid, E.; Lacroix, C. Microbe–microbe interactions in mixed culture food fermentations. *Curr. Opin. Biotechnol.* **2013**, *24*, 148–154. [CrossRef]
57. Beltrán, J.; Esteban, M. Nature-identical compounds as feed additives in aquaculture. *Fish Shellfish Immunol.* **2022**, *123*, 409–416. [CrossRef]
58. Vargas-Albores, F.; Martínez-Córdova, L.; Hernández-Mendoza, A.; Cicala, F.; Lago-Lestón, A.; Martínez-Porchas, M. Therapeutic modulation of fish gut microbiota, a feasible strategy for aquaculture? *Aquaculture* **2021**, *544*, 737050. [CrossRef]

metabolites

MDPI

Article

Profiling Phospholipids within Atlantic Salmon *Salmo salar* with Regards to a Novel Terrestrial Omega-3 Oil Source

Richard Broughton [1,*], Douglas R. Tocher [1,2], Johnathan A. Napier [3] and Mónica B. Betancor [1]

1 Institute of Aquaculture, Faculty of Natural Sciences, University of Stirling, Stirling FK9 4LA, UK
2 Guangdong Provincial Key Laboratory of Marine Biotechnology, Institute of Marine Sciences, Shantou University, Shantou 515063, China
3 Department of Plant Sciences, Rothamsted Research, Harpenden AL5 2JQ, UK
* Correspondence: richard.broughton@stir.ac.uk

Abstract: The development and inclusion of novel oils derived from genetically modified (GM) oilseeds into aquafeeds, to supplement and supplant current terrestrial oilseeds, as well as fish oils, warrants a more thorough investigation into lipid biochemical alterations within finfish species, such as Atlantic salmon. Five tissues were examined across two harvesting timepoints to establish whether lipid isomeric alterations could be detected between a standard commercial diet versus a diet that incorporated the long-chain polyunsaturated fatty acids (LC-PUFA), EPA (eicosapentaenoic acid), and DHA (docosahexaenoic acid), derived from the GM oilseed *Camelina sativa*. Tissue-dependent trends were detected, indicating that certain organs, such as the brain, have a basal limit to LC-PUFA incorporation, though enrichment of these fatty acids is possible. Lipid acyl alterations, as well as putative stereospecific numbering (sn) isomer alterations, were also detected, providing evidence that GM oils may modify lipid structure, with lipids of interest providing a set of targeted markers by which lipid alterations can be monitored across various novel diets.

Keywords: lipidomics; phospholipids; salmon; aquaculture; GM; novel feeds

Citation: Broughton, R.; Tocher, D.R.; Napier, J.A.; Betancor, M.B. Profiling Phospholipids within Atlantic Salmon *Salmo salar* with Regards to a Novel Terrestrial Omega-3 Oil Source. *Metabolites* **2022**, *12*, 851. https://doi.org/10.3390/metabo12090851

Academic Editors: Qingchao Wang, Yan He and Qingsong Tan

Received: 4 August 2022
Accepted: 8 September 2022
Published: 10 September 2022

1. Introduction

Aquaculture is tasked with fulfilling the demands of a growing population, owing to constraints on the capacity of traditional wild capture fisheries [1]. In the future, the role of fisheries, as well as many other food production systems, will likely be impacted by climate change, though the impact is likely to not be distributed equally across countries and communities [2]. Therefore, to meet the demands of a growing population, aquaculture will need to provide the additional capacity by which sustainable and nutritious seafood is produced. Regarding human nutrition, fish consumption is advised on a weekly, or bi-weekly basis, based on the health benefits provided by the high levels of protein, micronutrients, and omega-3 long-chain polyunsaturated fatty acids (LC-PUFA) [3,4]. It is the latter, however, the omega-3 fatty acids that are present in significant amounts, which consumers particularly associate with seafood. Owing to the biotrophic accumulation of these fatty acids, aquaculture feeds need to contain either marine-derived fishmeal and/or fish oil to supply finfish with omega-3 LC-PUFA. Vegetable oils, such as rapeseed, are typically used to supply and increase the energy content of feeds and can provide fatty acids including 18:3n-3, however terrestrial plant oils lack LC-PUFA. Presently other alternative oil sources rich in marine fatty acids are being investigated (e.g., oils derived from heterotrophic microalgae), but scalability and price are still a constraint, though are improving and are already used within the industry.

Therefore, nutritional quality for consumers and issues related to fish health are potential issues in aquaculture [5], owing to the growing demand to produce fish rich in omega-3 LC-PUFA with constrained lipid supplies. Omega-3 LC-PUFA containing lipids can be obtained from lower trophic levels, including krill, copepods, and algae.

However, depleting the natural stocks of these organisms could alter the ecosystem balance, resulting in potentially disruptive ecosystem alteration [6,7]. Therefore, novel solutions are required to address the need for these high-value lipids. One solution is the adoption of genetically modified (GM) oil seed crops, capable of elongating and desaturating fatty acids beyond the standard terrestrial lipids [8–10], which has been undertaken within *Brassica napus* and *Camelina sativa* [11], though only *Brassica napus* varieties are currently commercially offered. Within the present study, an oil derived from *Camelina sativa*, rich in both 20:5n-3 (eicosapentaenoic acid, EPA) and 22:6n-3 (docosahexaenoic acid, DHA) was trialled to assess its impact on the lipid profiles of several tissues of Atlantic salmon (*Salmo salar*) in comparison to a standard commercial-like vegetable oil/fish oil blend (control). The assessment of lipid composition at two different timepoints was used to ascertain the plasticity of lipids, with the first timepoint 14 weeks prior to the final sampling. Little work has been conducted applying lipidomic techniques to the assessment of novel feeds, and whilst lipid composition may be impacted by factors such as season and fecundity, assessment of diets within commercially relevant conditions was of initial interest and provides a basis for further exploration of more fundamental alterations to lipid biochemistry.

The use of oils derived from GM sources may allow for the replacement of fish oil, and traditional vegetable oils, however, owing to the synthetic biology approach, the structure of the lipids in GM-derived oils may differ from those produced natively, through both de novo and dietary incorporation. Alterations may stem from the combinations of desaturases and elongases used within the host plants [11,12], as well as the potential for the host organism itself to impact lipid biosynthesis. This approach has resulted, in certain instances, in acyl rearrangement within triacylglycerols (TAG). Both NMR and lipase studies have generally concluded that DHA preferentially occupies the sn-2 position in fish oils [13–16], with EPA usually being equally split or being more likely in the outer sn-1/3 positions, while it has been shown that DHA resides preferentially in the outer sn-1/3 positions in TAGs in engineered systems [9,17], while EPA positioning seems less affected. The relative proportions of DHA in single cell oils such as *Schizochytrium* sp. and *Crypthecodinium* appear to be enriched within the outer sn-1/3 positions [18], though these are calculated values rather than directly observed. However, few comparisons of other fatty acids in fish oils and GM oils have been conducted, possibly because ^{13}C NMR, the predominant method for acyl sn determination has no distinct carboxylic functional group peak for certain fatty acids, or peaks represent a mixture of fatty acids [19], therefore warranting a lipase-based approach. This regiospecificity has also now been modified within TAGs for specific fatty acids, for the desired market [20]. Such structural differences are generally not evident using standard lipid analysis, such as GC, whereby the intact lipid structure is usually destroyed, or LC-MS with standard reverse phase conditions that are unlikely to resolve sn positional isomers or D/L stereoisomers. However, the synthetic construction of TAG, the main dietary lipid within aquafeeds, inherently involves multiple elongation and desaturation steps, potentially with several lipid substrates, and may contain subtle variations not commonly found within traditional lipid sources. This may impact both the digestion and subsequent utilisation of the fatty acids by various fish tissues, resulting in a unique signal, detectable as alterations within the isomer ratios. Detection of isomeric variations in relation to GM oils has received little attention and potentially may impact fish health through sn specific lipase action for eicosanoid production, and incorporation into LC-PUFA-dependent tissues, such as the brain and eye.

To that end, a preliminary screening protocol based on supercritical fluid chromatography SFC-MS/MS was utilised to determine and characterise lipids, which distinguished fish that had been fed control and GM oil-supplemented feeds, with candidate compounds further explored using non-standard, targeted LC-MS/MS methodology, which was designed to resolve phospholipid isomers.

2. Materials and Methods

2.1. Materials

Acetonitrile, methanol, and isopropanol were supplied by Rathburn Chemicals (Walkerburn, UK), and formic acid, water, and ammonium formate were purchased from Merck (Dorset, UK), with all solvents being LC-MS grade. CO_2 (99.8% industrial grade) was purchased from BOC (Grangemouth, UK). HPLC grade methyl-tert-butyl ether (MTBE) and chloroform were purchased from Fisher Scientific, and LC-MS grade sample vials were purchased from Waters (Waters, Milford, MA, USA).

2.2. Fish Growth Conditions and Sampling

All procedures were conducted in compliance with the Animals Scientific Procedures Act 1986 (Home Office Code of Practice) and in accordance with the regulations set forward by the Directive 2010/63/EU of the European Parliament and of the Council of 22 September 2010 on the protection of animals used for scientific purposes. Additionally, the experimental protocol was reviewed and approved by the Animal Welfare and Ethical Review Board at the University of Stirling (AWERB(16-17)83).

A total of 900 post-smolt, mixed sex, Atlantic salmon (*Salmo salar* L.) with an average initial body weight of 187.2 ± 1.9 g (mean ± SD) were distributed into six 5 m seawater floating pens (150 fish per pen) and fed one of two experimental diets in triplicate from week 25, 2018 to week 10, 2019 with the average fish weight at the intermediate point being 1941.8 ± 156.3 g, and the final average weight being 3510.8 g ± 158.3 (mean ± SD). The intermediate timepoint recorded an average weight of 2071 g ± 97.3 for the control diet, 1812 g ± 46.3 for the GM diet, and for the final timepoint, 3601.9 g ± 307.6 for the control diet and 3108.3 g ± 87.0 for the GM diet, with no significant difference between dietary treatments. The diets were isolipidic (36%) and isoproteic (36%) and formulated to contain either a 2:1 (*v*/*v*) blend of vegetable (rapeseed) and fish oil (control diet), or an oil derived from the genetically modified oilseed *Camelina sativa*, rich in both EPA and DHA (GM diet) (Table 1). Six fish per pen were humanely euthanised by anaesthetic overdose of metacaine sulphonate (>150 mg/L) at two timepoints, week 48 2018 (intermediate), and week 10 2019 (final). Five tissues including the brain, eye, gill, intestine, and liver were collected at each timepoint and immediately frozen at −70 °C prior to lipid extraction.

2.3. Lipid Extraction

Lipids were extracted from pooled tissues, with three pooled fish per biological replicate, with two pools taken from each pen and were treated as six biological replicates per diet (n = 6). Lipids were extracted using the MTBE method [21]. Samples were kept on ice during extraction, with gill and intestine samples manually chopped prior to homogenisation. Pre-cut or intact tissues were weighed into 50 mL test tubes, 3 mL of 90% *v*/*v* ice-cold methanol per gram of tissue was added and samples were homogenised using an Ultra-Turrax tissue disrupter (T25, IKA, Darmstadt, Germany). Once homogenous, the total sample volume was measured, and a volume equating to 200 mg of sample was taken into a glass test tube and 5 mL of MTBE added, samples vortexed, and left on ice for 5 min. Then, 1.25 mL of water was added, samples vortexed, and centrifuged at 800 g for 5 min at room temperature to allow the solvent system to separate. The upper phase was taken into a clean glass tube, and a second extraction was carried out on the lower phase using 2 mL of MTBE/methanol/water (10:3:2.5, *v*/*v*/*v*). After centrifugation, the second upper phase extract was pooled with the first, and the total extract was dried under nitrogen. Total lipid extracts were then weighed, and a stock solution of 10 mg/mL in chloroform/methanol (2:1, *v*/*v*) was made. QC samples were made by mixing equal volumes of both control and GM extracts together, to create a mixed sample that could be reinjected several times throughout a run. The stock concentration was used for positive mode, whilst a 2.5 times concentration was used for negative mode injections.

Table 1. Diet formulation and fatty acid composition of the control and GM diets. The provenances of the ingredients used were [a] Biomar AS, Brande, Denmark, [b] Rothamsted Research, Harpenden, UK, [c] DSM, Basel, Switzerland.

	Control	GM
Feed ingredients (%)		
Fishmeal [a]	7.5	7.5
Vegetable protein [a]	36.1	36.1
Land animal protein [a]	10.0	10.0
Wheat [a]	11.2	11.2
Fish oil [a]	10.8	-
Rapeseed oil [a]	20.8	-
Camelina oil (transgenic) [b]	-	31.6
Premix [c]	3.6	3.6
Yttrium oxide [c]	0.005	0.005
Fatty acid profile (%)		
Total saturated[1]	18.1	14.7
Total monoenes[2]	49.1	22.2
18:2n-6	15.2	19.6
18:3n-6	0.1	1.5
20:4n-6	0.4	1.7
Total n-6 PUFA	16.0	25.2
18:3n-3	5.8	19.1
20:5n-3	4.8	5.7
22:5n-3	0.7	3.8
22:6n-3	2.4	5.3
Total n-3 PUFA	15.1	37.7
MOPA		
Moisture %	6	6
Energy–crude (MJ/Kg)	26	26
Protein–crude (%)	36	36
Fat–crude (%)	36	36
Ash (%)	4.2	4.2

2.4. Untargeted Supercritical Fluid Chromatography (SFC)-MS/MS

Untargeted SFC-MS/MS was carried out as described previously [22], with polar lipids studied separately from neutral lipids. SFC separation was carried out on an Acquity UPC^2 system (Waters, Milford, MA, USA), which used CO_2 as solvent A, and methanol/acetonitrile/water (75:20:5, *v/v/v*) containing 0.1% (*v/v*) formic acid and 0.15% ammonium formate (*w/v*) as co-solvent B. The makeup solvent comprised of methanol/ isopropanol (80:20, *v/v*) containing 0.1% (*w/v*) ammonium formate. A 5 cm Viridis C18 HSS column (50 × 2.1 mm, 1.8 µm particle size; Waters, Milford, MA, USA) was used with neutral lipid analysis decoupled from polar lipid analysis by diverting column eluent to waste prior to the elution of polar lipids. SFC gradient elution comprised of solvent B being held at 2% for 2.5 min, then increased to 64% at 18.7 min, held at 64% solvent B until 24 min, then decreased to 2% at 27.5 min, and held at 2% for a final run time of 29.5 min. The back pressure regulator was set to 1500 psi, the flow rate was set at 1.2 mL/min, and the makeup solvent flow rate set to 0.25 mL/min, with the column temperature set at 52.5 °C. Two µL of the sample was injected on-column.

The UPC2 system was attached to a Xevo G2-XS Q-TOF mass spectrometer (Waters, Milford, MA, USA), which was operated in both positive and negative ESI mode. The capillary voltages were set to 3 kV, with the sampling cone voltage set at 28 V, the source offset set to 80, and the source and desolvation temperatures set at 120 and 300 °C, respectively. The cone and desolvation gas flows were set to 50 and 1000 L/h, respectively. The system was operated in MS^e mode, with the scan rate set between 200 and 1600 Da at 0.25 sec/scan, with a low energy collision energy of 2 V and a ramped voltage of 28–40 V for the high energy scan. The lock mass compound was leu-enkephalin, which was infused at a rate of 10 µL/min during the run to correct for mass deviations.

2.5. Semi-targeted Reverse Phase LC-MS/MS

Semi-targeted analysis was conducted on a Waters I class UPLC, attached to a Xevo TQ-S, based on a method by Nakanishi et al. [23]. The reverse phase method utilised two 10 cm Aquity BEH C18 columns (100 × 2.1 mm, 1.7 µm) in tandem, and four solvents. Solvent A comprised acetonitrile/water (90:10, v/v) with 0.1% ammonium formate (*w/v*), solvent B comprised of acetonitrile/methanol/isopropyl alcohol/water (47.5:45:2.5:5, *v/v/v/v*) with 0.1% ammonium formate (*w/v*), solvent C comprised of acetonitrile/methanol/isopropyl alcohol (40:45:15, *v/v/v*) containing 0.1% ammonium formate (*w/v*), and solvent D comprised 100% isopropyl alcohol. The flow rate was set at 0.33 mL/min, with a column temperature of 65 °C. The system gradient used is shown in Table 2.

Table 2. Reverse phase gradient conditions.

	Solvent			
Time	**A**	**B**	**C**	**D**
0	100	0	0	0
5	100	0	0	0
5.1	0	100	0	0
18.5	0	100	0	0
35	0	50.2	49.8	0
42	0	20	80	0
42.1	35	0	0	65
52	35	0	0	65
56	100	0	0	0
61	100	0	0	0

The Xevo TQ-S was used in data-dependent acquisition mode, with an inclusion list for fragmentation based upon the lipids identified using untargeted QTOF methodology. The system was run in negative mode with a capillary voltage of 3 kV, a cone voltage of 28 V, a source offset of 80, a source and desolvation temperature of 130 and 280 °C respectively, and a cone and desolvation gas flow of 150 and 1000 L/h, respectively. The survey scan was adjusted based on the target lipids and was set to 680–970 Da with a scan time of 0.35 s. Fragmentation of lipids was carried out through the use of an inclusion list, which utilised a ScanWave DS fragmentation of precursor ions. The fragmentation mass range was set from 100–950 Da and scanned at 5000 amu/sec.

2.6. Data Analysis

The UPC2 and Xevo G2-XS QTOF were operated using Masslynx v4.2, whereas the Xevo TQ-S was operated using Masslynx V4.1. Xevo G2-XS QTOF data were analysed using Progenesis QI v 3.0 (Nonlinear Dynamics, London, UK), with multivariate OPLS-DA analysis run using Simca-P v12.0 (Umetrics, Umeå, Sweden). Each tissue type and timepoint was analysed separately, comparing the GM diet against the control diet, with data normalised using the all-ions approach. Progenesis parameters were set as; ANOVA score < 0.05, a fold change of >1.2, and multiple testing values (Q values) of <0.05 were used to filter results. Multivariate analysis was then used to discriminate between the control and GM groups, using OPLS-DA to determine compounds of significance, with a w (1) score higher than ± 0.04, and a p (corr) score greater than ± 0.6. QC samples were assessed to establish whether these clustered centrally between the two treatments. These compounds were reimported back into Progenesis for lipid identification, based on retention time, accurate mass, and fragmentation. Data were then exported to Excel, with the positive and negative mode data then merged to create a unified dataset.

The selected compounds from each tissue type and timepoint were converted into their negative ion form, with these compounds forming the basis of the m/z range to be chosen for the LC-MS/MS approach, as well as the data-dependent acquisition fragmentation inclusion list. Those lipids which were identified within the tissue or specific timepoint by QTOF were integrated using Masslynx, with the fragmentation profiles interpreted to deduce the lipid acyl composition. Integrated profiles were exported to Excel for further analysis, which included percentage calculations of higher-level lipid structure, e.g., PC 42:10, and ratio calculations of putative sn isomers. Data relating to lipid structure and statistical values are given in Supplementary Tables S1–S20. Data were multiple test corrected using the Benjamini-Hochberg procedure, and those below the cut-off point are highlighted in red, as shown in the supplementary data.

3. Results

3.1. Tissue Responses to GM and Control Diets

Five tissues were examined in the present study, two of which, the brain and eye, were found to be more conservative in terms of fold change when the average fold change across polar lipid classes was considered (Figure 1). The largest number of lipid alterations were observed within phosphatidylcholine (PC) and phosphatidylethanolamine (PE), but these were generally limited to around, on average, 2-fold in both these tissues. Average fold changes in PE were generally much greater for the other three tissues, ranging from 3 to 5-fold on average. On the other hand, PC was more conservative, only extending up to a change of 3-fold on average. Similarly, phosphatidylglycerol (PG) generally showed higher fold variances in the intestine and liver compared with the brain as did phosphatidylinositol (PI), which was tightly controlled in the brain, showing similar fold changes across timepoints, whereas the other tissues, especially liver, generally showed both greater variance and higher average fold changes. Ceramides and the more tissue-specific glycosylated sphingolipids and galactolipids showed little in the way of overall tissue trends, with average fold changes close to 2, and tightly controlled variances. A few cardiolipins were also detected, although the majority appeared to respond to the control diet, with the greatest diversity detected within the eye at the final timepoint. Cardiolipins were not summarised further owing to the small number detected.

With regards to differential accumulation of lipid class contents between diets, whilst various individual lipids were observed within the different tissues, a few general trends were apparent. The sum of the lipids comprising the main phospholipid classes, PC and PE, demonstrated quite marked alterations in the gill at both timepoints (Figure 2), with these two classes generally being elevated when fed the control diet. The intestine and the earlier liver timepoint generally showed increased PC in fish fed the control diet, whereas the final brain timepoint, on average, showed PC was increased in fish fed the GM diet. However, PE displayed a relatively equal split for all tissues except the gill. The hexose ceramides, which also included the galactolipids monogalactosyl diacylglycerol (MGDG) to increase the span of the group, were found to be predominantly increased by the GM diet within the brain, as were the limited number of ceramides. This appeared in contrast to the eye where, in general, fish fed the control diet showed increases in tissue-specific glycosylated lipids, distinct from the brain (Figure 3). Diet had little impact on ceramide contents in the final gill and intermediate intestine samples, though both the intermediate gill and liver showed increased ceramides in fish fed the GM diet.

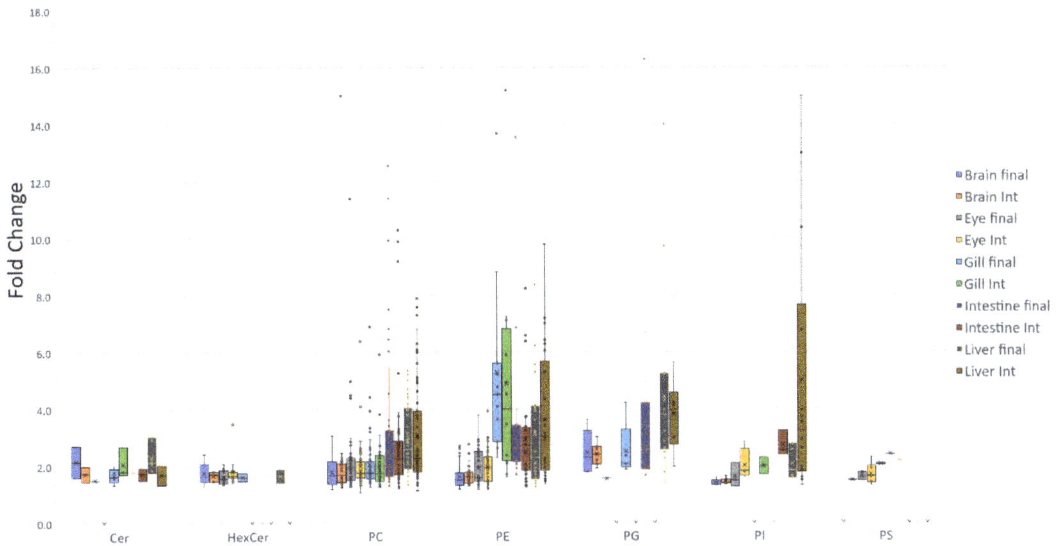

Figure 1. Phospholipid class fold distribution for untargeted QTOF data. The average fold distribution of lipids determined to be significant showed that, in general, the brain and eye exhibited smaller fold changes across the majority of lipid classes. The liver, however, generally exhibited the greatest variance in lipid alteration in response to diet.

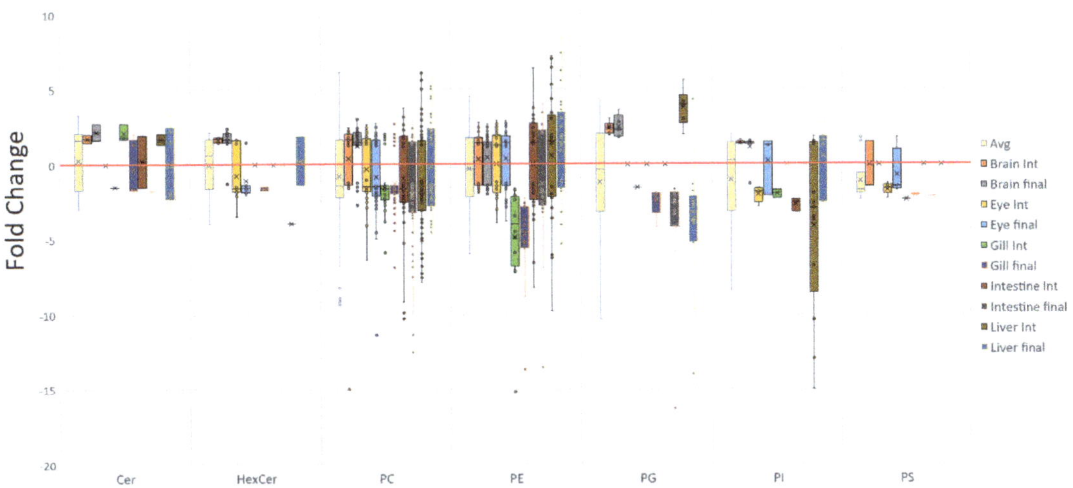

Figure 2. Phospholipid bias toward GM or control diets. Positive values indicate a general trend of being increased in fish fed the GM diet, whereas negative values indicate a general trend of being reduced by the GM diet, compared to fish fed the control diet. The brain in general was found to be positively affected by the GM diet, whereas the opposite was the case in the gill with the major lipid phospholipid class PC being primarily increased in fish fed the control diet.

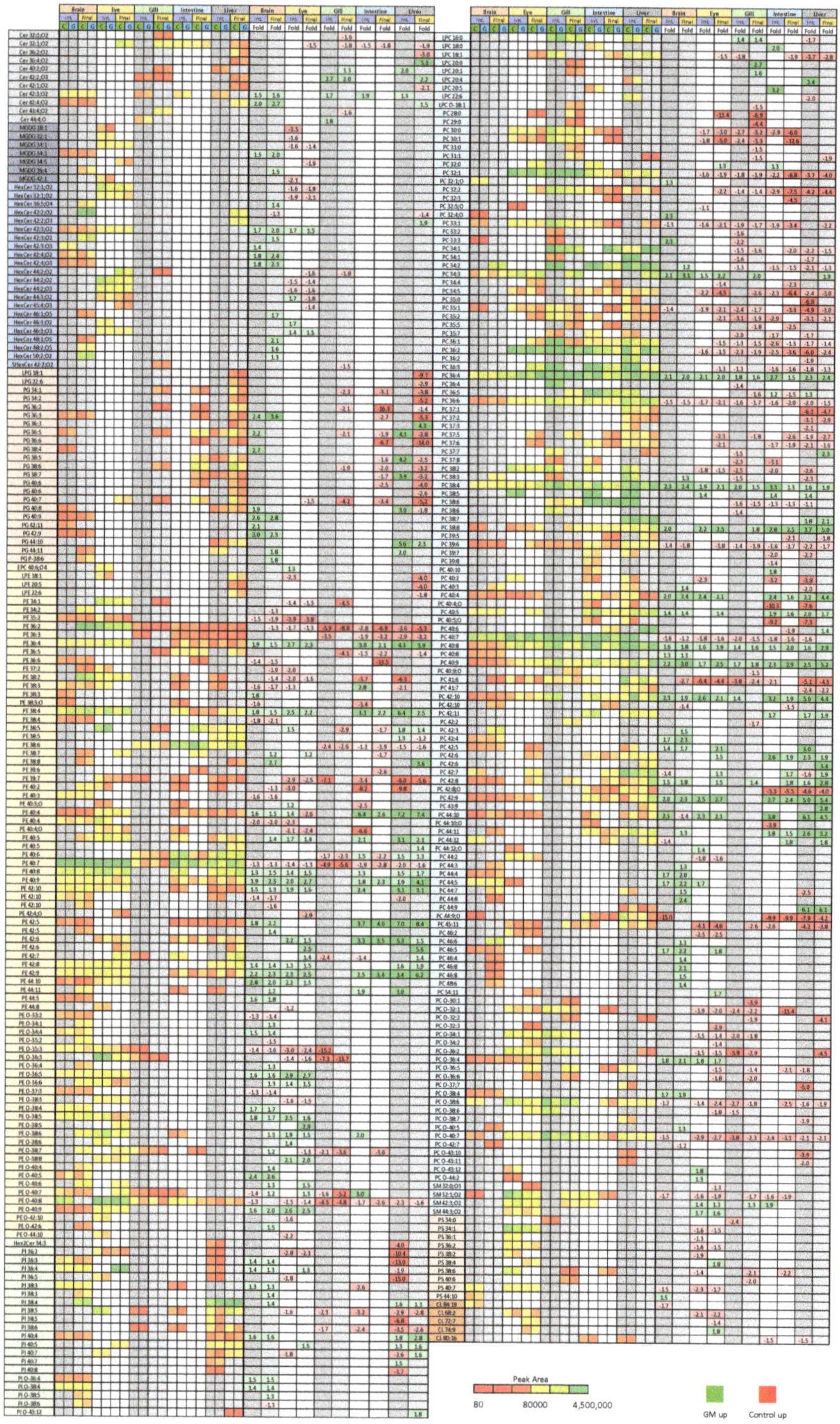

Figure 3. Lipids found to be significantly elevated by either the control or GM diet using the untargeted SFC-MS/MS methodology. Lipids were determined via ANOVA and OPLS-DA analysis

and formed the basis for the semi-targeted reverse phase LC-MS/MS analysis. Data are given as all ion normalised peak areas (with colour scale for reference), and fold change data, with green representing elevated levels in the GM diet vs. control, and red representing elevated levels in the control diet vs. GM.

As above, PG again seemed to be selectively increased within brain tissue, as well as the intermediate liver sampling point, however, this was reversed in the final liver sample, which showed that PG tended to be increased in fish fed the control diet. This was also found to be the case within the final gill and intestine samples. Finally, for both PI and phosphatidylserine (PS), the brain seemed to be most affected when fed the GM diet, as again, PI was almost exclusively increased within this tissue. In contrast, PI in the eye appeared to be increased in fish fed the control diet, at least at the intermediate timepoint. Similar trends in PI were observed in the gill, intestine, and liver at the same timepoint, though the liver demonstrated a much wider variation, possibly owing to the relatively low abundances detected within the tissue. Significant levels of PS were only detected in the eye and appeared to increase in fish fed the control diet. It should be noted, however, that individual lipids were not always detected or studied in each tissue, although general trends present within both the tissue and at different timepoints appear to be evident.

When the individual lipids are deconstructed into both their carbon and double bond numbers (Figure 4), several trends become evident. With regards to carbon number, brain tissue illustrates that there appears to be a lower limit of lipid modification, which tightens with brain maturity. On average, the brain appears to defy the trends of the other tissues, with all even carbon number lipids being generally elevated in response to the GM diet, which is especially evident at the final timepoint. The gill, and to some extent the earlier intestine timepoint, bucked this trend somewhat, being more likely to show increased fold change with a higher carbon number in fish fed the control diet. Trends that appeared relatively consistently included the majority of odd chain lipids being increased in fish fed the control diet and, with the exception of brain tissue, 30, 32, 34, and, to some extent, 36 carbon lipids were also elevated in fish fed the control diet. Excluding gill, the larger 42 carbon lipids appeared to be elevated with the GM diet, with 44 carbon lipids also favoured within the liver.

Regarding the number of double bonds, the control diet appeared to result in increased mono- and di-unsaturated lipids and, to a lesser extent, saturated lipids. The brain once again showed the opposite trend for monounsaturated lipids, most notably by the final timepoint, showing elevated levels in fish fed the GM diet. The brain generally showed a tight spread of fold changes, as shown with the earlier 2-fold change on average, and usually in a predominant direction. At the intermediate sampling point, the brain conformed to the general pattern shown in the other tissues, with 6 and 7 double bond-containing lipids increased in fish fed the control diet. In fish fed the GM diet, the brain generally showed increased levels of the majority of other double bond numbers, whereas the eye showed increased levels of 8 and 9 double bond-containing lipids. Regarding overall trends for more unsaturated lipids, 4, 9 and to a lesser extent 10 double bond lipids on average appeared to be favoured in fish fed the GM diet, though not as consistently as with mono- and di-unsaturated lipids in salmon fed the control diet. Gill and intestine appeared to show much tighter control in terms of fold changes and often were the tissues most reliably modulated by the control diet, with lipids containing 3, 5, 6, and 7 double bonds more consistently increased by the control diet.

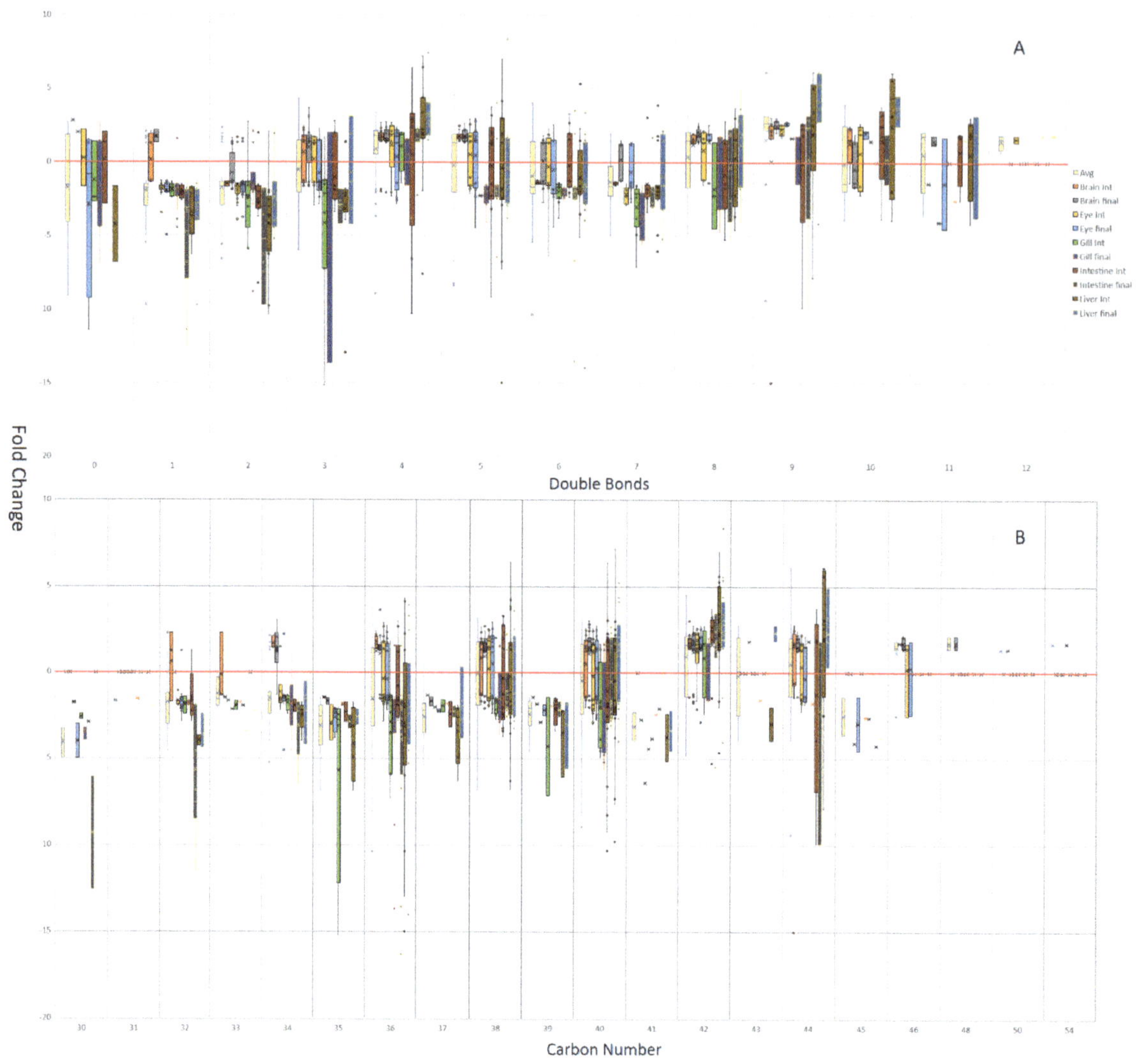

Figure 4. Lipid double bond (**A**) and carbon number (**B**) bias toward the GM or control diets, with positive values indicating a general trend of being increased by the GM diet, whereas negative values indicate a general trend of being decreased by the GM diet compared to fish fed the control diet. The brain was consistently found to have the majority of lipids increased by the GM diet, in contrast to gill, which was biased by being more likely to show lipids increased by the control diet. On average, lipids containing 1, 2, and 7 double bonds, and comprised of 32, 34, and 36 carbons, were more likely to be increased by the control diet, whereas lipids containing 4, 9, and 10 double bonds and comprised of 42 carbons, were more likely to be increased by the GM diet.

This consensus between both tissue and timepoints was present for a large number of individual lipids, with lipid responses consistent across a diet (Figure 3). Where no response was observed for a specific tissue or timepoint it could be assumed that either the trend would be in the same direction or was not observed due to the change being too small and/or not significant. These consistent patterns were most evident across PC and PE, with increased levels in fish fed the GM diet being seen in PE 42:9, 38:4, and 36:4, as well as PC 36:4 and 38:4, for example. Counter trends existed in PE 40:7 and ether lipids such as PE O-40:8, as well as PC 36:6 and 40:7, which saw consistent elevation by the control diet. It also appears that PC and choline-containing lipids had more individual species modulated by the control diet, approximately 78 versus 53 for the GM diet. Of note, the brain in general tended not to show much response to the control diet, as evidenced by a greater proportion of positive fold changes, representing proportionally higher lipid elevation brought about by the GM diet. This can be seen in Figure 3, especially within the PC class, where lipid species elevated by the control diet are not seen to be modulated within the brain, with inter-tissue trends only seen in lipid species that were elevated by the GM diet, indicating some potential form of lipid compositional baseline or threshold, which can only be increased, but not decreased. Gill on the other hand tended to lack the wealth of individual lipids which were modified in other tissues, although was generally starkly contrasted to the brain, being more sensitive to the control diet and rarely showing individual lipids upregulated by the GM diet.

These trends across tissues also highlight the possibility that certain trends may exist across entire lipid classes, even in the absence of comparable data across both tissue and time. For example, with PG there appeared to be a diet-related overall trend shared across individual tissues or timepoints, for example, the final timepoints of gill, intestine, and liver, even though the same PGs are not always detected across different tissues. There are also contra directional changes for both individual and entire lipid species indicating that, in fact, a whole lipid class may be modulated within a tissue in the same manner by dietary conditions. These contra directional changes occur to a limited extent, being evident in the liver across time, predominantly in the PG lipid class, although occasionally occurring in PC, for example, PC 42:7. Inter tissue variations also exist, for example, PE O-40:7, PE 40:6, PG 36:5 and PC 42:7.

3.2. Isomeric Analysis

Those lipids which demonstrated significant change within the QTOF dataset (Figure 3) were selectively targeted using reverse phase LC-MS/MS methodology. Owing to the selective nature of the method, fully comprehensive profiling of each lipid class was not possible. Therefore, predominantly PC and PE were studied, owing to their amenable fragmentation profiles in negative mode, though only those lipids detected in the initial analysis above could be analysed in further detail. It was demonstrated, however, that the reverse phase method was capable of resolving potential sn positional isomers (Figure 5). As illustrated in Figure 5, acyl ion fragments were clearly evident in negative mode, however, their use in determining the sn configuration was hampered by the non-reproducible ion ratios that derive from the acyl fragments. Therefore, the isomeric configurations are currently uncertain and were designated as putative sn isomers, with these lipid isomers given a/b/c suffixes.

Once lipids were separated into individual acyl isomers, those with two or more isomers were taken forward, with several acyl trends determined (Figure 6). On average across all tissues, and within the PC lipid pool, and most evident for the fold change data, lipids containing 14 and 16 acyl groups were more likely to be associated with the control diet, that is, complex lipids were more likely to contain one of these acyl groups in one of the two acyl positions (Figure 6A). Conversely, lipids containing acyl groups of 20 and 24 carbons in length were more likely to be associated with the GM diet, though 18 and 22 carbon acyl-containing lipids, on average, also tended to be more likely to be increased by the GM diet. No obvious discernible pattern was determined within the PE lipid pool (Figure 6B). With regards to double bond number, across the PC and PE (Figure 6C/D) lipid pools within the selected lipids and across all tissues, monounsaturated acyl groups were more likely to be attached to complex lipids and increased in salmon fed the control diet. Contrary to this, 3 and 4 double bond-containing acyl groups incorporated into complex lipids were more likely to be influenced by the GM diet, though only the PC lipid pool saw a general increase of 5 double bond-containing acyl groups in fish fed the GM diet.

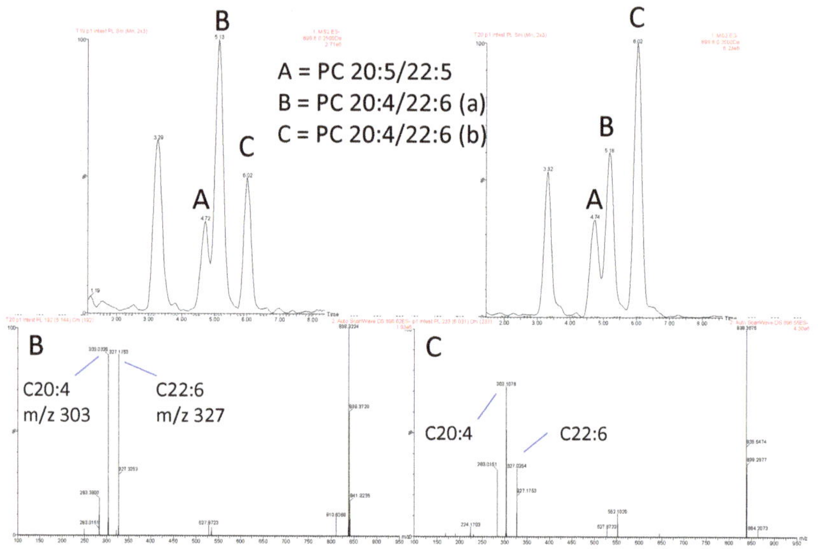

Figure 5. Acyl and putative sn isomer separation based on BEH C18 reverse phase analysis. Putative identification of sn isomers was made based on unit mass and CID fragmentation of precursor ions in negative ion mode. In this example, PC 42:10 is comprised of 3 isomers, with a shift in the relative proportion of the two putative sn isomers in response to diet. (a) and (b) nomenclature is used to designate the putative sn isomers.

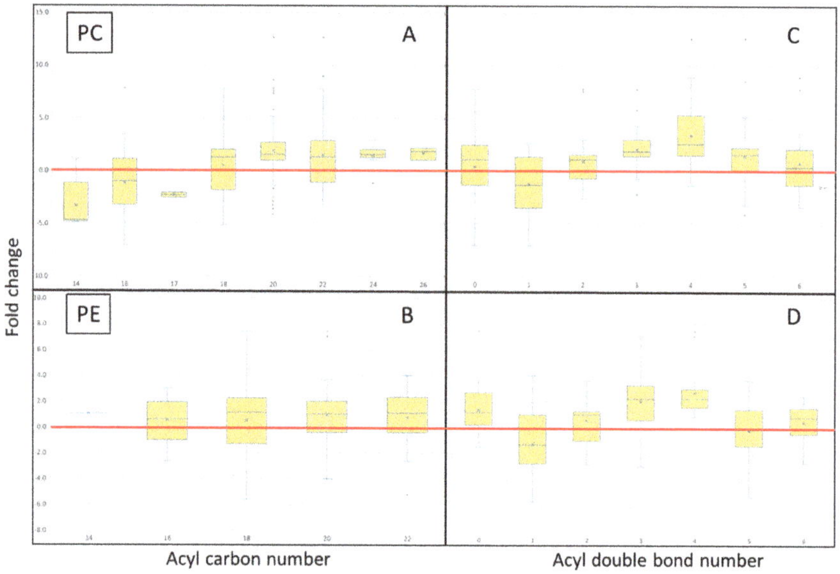

Figure 6. Averaged individual PC and PE bias plots for all tissues and timepoints for both acyl carbon number and number of double bonds, expressed in fold change. (**A**) acyl carbon number in PC, (**B**) acyl carbon number in PE, (**C**) double bond number in PC, and (**D**) double bond number in PE. The GM diet preferentially increased lipids containing acyl chains with 3 and 4 double bonds, comprising 20 and 22 carbons. The control diet preferentially increased lipids containing acyl chains containing 1 double bond, comprising 14 and 16 carbons.

When the PC and PE pools were combined but split on the basis of tissue and timepoint (Figure 7), only those that contained a reliable number of lipids were analysed, with these constituting acyl groups containing 16 to 22 carbons. Sixteen carbon acyl groups were found to be marginally increased in fish fed the GM diet within the liver, though only when looking at percentage data. Acyl groups comprising 18 carbons tended to be preferentially incorporated within complex lipids in fish fed the control diet, though only evident in percentage data, with gill exhibiting the greatest response. Fold data clearly showed a preference for 20 carbon acyl chains being increased by the GM diet, with the strongest effects observed in the brain and eye, while gill showed no preferential regulation. Acyl groups containing 22 carbons were less clear cut, although fold data demonstrated that intestine and liver, as well as the final eye timepoint, responded with marginally increased inclusion in complex lipids when fed the GM diet. In contrast, gill generally presented increased incorporation of these acyl groups when fed the control diet.

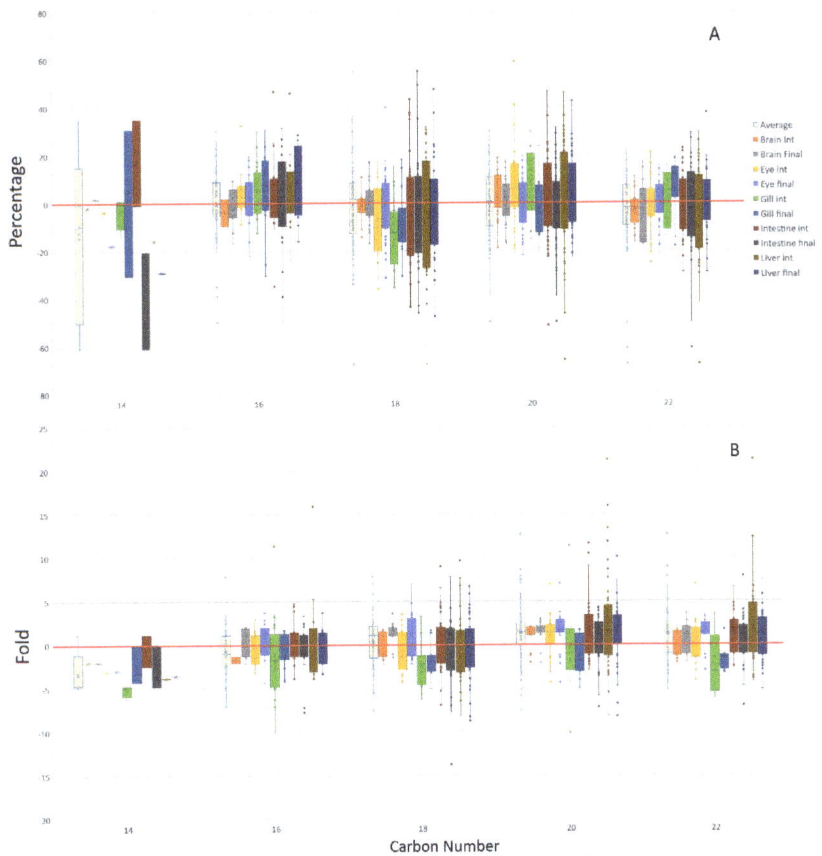

Figure 7. Acyl lipid carbon number bias toward the GM or control diets for combined PC and PE lipid pools represented as a percentage (**A**) or fold data (**B**), with positive values GM biased, and negative values control biased.

Regarding the levels of unsaturation for the combined PC and PE pools (Figure 8), fold change data tended to give a clearer indication of patterns within tissues and timepoints. Saturated acyl groups appeared to partially be more likely to be incorporated into complex lipids, and were increased by the GM diet, whilst the final timepoints for both brain and eye, containing di-unsaturated acyl groups, also experienced similarly increased levels when expressed as fold change in response to the GM diet. Acyl groups containing

3 and 4 double bonds were more consistently increased by the GM diet, with acyl groups containing 5 and 6 double bonds also marginally elevated. As evidenced by the fold change data, a countertrend was evident within the gill at both timepoints, indicating that the inverse was true; that the control diet increased the production of lipids containing acyl groups containing 5 or 6 double bonds. As seen previously, mono-unsaturated acyl groups tended to be increased by the control diet.

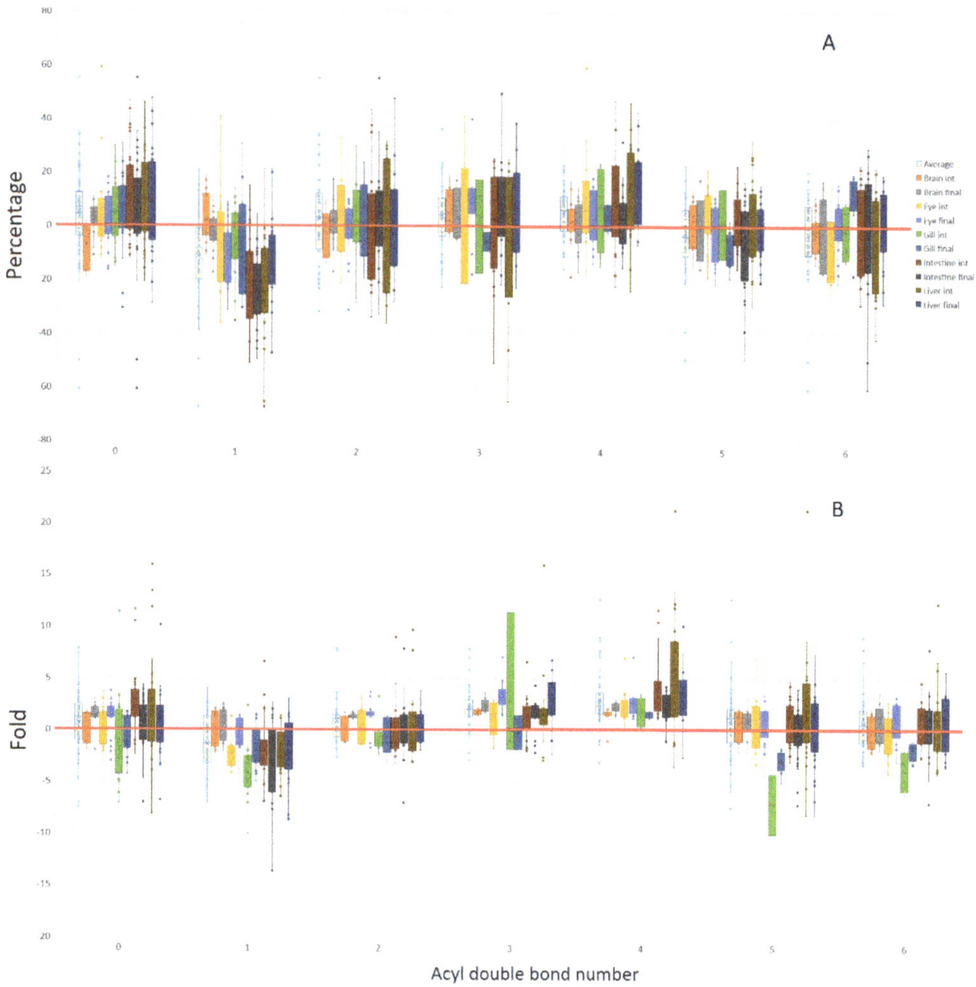

Figure 8. Acyl lipid double bond number bias toward the GM or control diets for combined PC and PE lipid pools, represented as a percentage (**A**) or fold data (**B**), with positive values GM biased, and negative values control biased. On average, acyl groups containing 1 double bond were increased by the control diet, whereas those containing 3–4 double bonds, tended to be increased by the GM diet.

The limited number of significantly modified PC and PE lipids in the brain stemmed from the initial QTOF analysis, and again, individual acyl isomers were muted in terms of fold change, and more so in terms of percentage compositional shifts within the carbon and double bond classes, when compared with the other tissues (Supplemental Figures S1 and S2). On the other end of the spectrum, unsurprisingly, the liver and intestine showed the greatest variation with respect to percentage composition, although the eye at the intermediate timepoint also demonstrated a surprisingly large variation.

The liver intermediate timepoint demonstrated the greatest fold variance, with the gill intermediate timepoint showing the second highest variance. The number and distribution of acyl isomers which constitute a lipid class structure, e.g., PC 36:2, tend to be relatively consistent across tissues and timepoints, with the caveat that a total data set encompassing all observed acyl isomers is not available for all tissues and timepoints, owing to the technical limitations of the instrumentation. However, that noted, several tissue or timepoint acyl alterations are evident, with noticeable lipids being PE 40:7, PE 40:4, PC 36:6, PC 36:3, PC 36:2, PC 38:4, PC 40:7. The initial evidence suggests that tissues might be classifiable based on the complex lipid acyl makeup, although missing data currently hampers the building of such a model.

Organ wide acyl modification again showed a relatively consistent consensus with regards to the response to the two diets (Supplemental Figures S1 and S2), although tissue-specific dietary effects may also be alluded to by, for example, differences in the magnitude of the lipid response, and the counter-trend of certain lipids in specific tissues. Those effects, which run contrary to the organ wide consensus, or differ markedly in magnitude in either percentage or fold terms can be observed within PC 42:10, PC 36:4, PC 36:2, PC 36:6, PC 32:1, PE 42:6, PE 38:6, and PE 36:2. Whilst percentage composition data and fold data generally tend to align, there are deviations whereby counter-trends are more readily identified. On the other hand, correlations that are similar and strongly reflected in both organs and timepoints are most clearly visible in lipids such as PC 36:3, PC 36:2, PC 40:8, and PC 40:7, although even these lipids demonstrate outliers with regards to acyl composition within a certain tissue, for example. With regards to temporal patterns, which may demonstrate a strengthening or weakening of a fold or percentile change over time, consistent patterns are difficult to determine. However, within the initial QTOF data, several consistent temporal patterns, either strengthening or weakening of fold changes, could be observed across time, as illustrated by PE 36:4, PE 38:4, PE 40:9, PE 42:9, PC 32:1, PC 40:7, PC 40:8, PC 40:9 and PC 42:10. Acyl data demonstrated fewer of these temporal trends, with the majority being more evident within percentage composition data within lipids, such as sn positional isomers PC 16:1/20:5(b), PC 20:4/20:5(a), PC 18:3/22:6, and all those acyl isomers comprising PC 42:10. It was found that there was some overlap between the datasets, for example, PC 40:9 and PC 42:10, indicating that there may be both proportional and absolute temporal trends in effect within a specific complex lipid, i.e., PC 42:10.

A key requirement of the reverse phase method was to resolve what are thought to be sn positional acyl isomers. Positional isomers were denoted as such based on their fragmentation spectra, however, their sn designation cannot be confidently attributed and therefore isomers were designated as either a or b, or c if more than two were detected, and are taken as a ratio of one isomer over the other. Surprisingly, several putative positional isomers were detected (Figure 9) and, in one lipid, PC 32:1, four isomers were detected. Where representative numbers of tissues were found to contain these isomers, certain isomeric trends appear to be influenced by diet, albeit not all statistically valid. This was highlighted most prominently within PC 16:0/20:2, PC 16:0/20:3, PC 16:0/20:4, and PC 20:4/22:6, where one isomer appears favoured in salmon fed either the control or GM diets. Again, there appeared to be tissue-specific variations with regards to either magnitude, for example, eye, when compared with intestine and liver for PC 20:4/22:6, and direction, as seen with the brain (intermediate timepoint) in PC 18:1/22:6, which may be characteristic of the organ. In many instances, only fatty acyl isomers were detected as opposed to putative sn isomers, again indicating a potential tissue or temporal discriminating factor. However, owing to the uncertainty of the isomeric identification, further elucidation of lipid biosynthesis and its potential alteration by diet was difficult to elaborate on.

	Brain Int.			Brain Final			Eye Int.			Eye Final			Gill Int.			Gill Final			Intest. Int.			Intest. Final			Liver Int.			Liver Final		
Lipid Isomer	Co	GM	p	Co	GM	p	Co	GM	p	Co	GM	p	Co	GM	p	Co	GM	p	Co	GM	p	Co	GM	p	Co	GM	p	Co	GM	p
PE P-18:0/20:4 b/a	2.1	1.9																												
PE 18:0/18:2 a/b																						1.5	0.2	**						
PE 18:1/22:6 a/b	30.5	23.8																	65.5	10.5	*									
PE 18:0/22:5 a/b																			3.0	5.9	*							5.4	18.9	**
PE 18:1/22:5 a/b																												2.1	5.4	
PC 14:0/16:1 a/b													0.9	0.8		2.4	0.0	*												
PC 16:0/17:1 a/b													2.6	2.1																
PC 16:0/18:2 a/b				1.7	1.9								17.0	3.9	**				47.4	6.8	**									
PE 20:2/22:6 a/b	2.0	1.9								6.7	6.2														14.1	4.6	**			
PE 20:4/22:6 a/b	1.6	1.4		1.2	1.5		1.5	1.3		1.7	1.7								4.8	1.8	**									
PC 16:0/20:2 b/a							1.4	3.2	**	1.9	2.6	*	1.8	3.0	*	2.6	2.9											2.1	2.8	
PC 16:0/20:4 b/a							6.0	12.2	**	5.5	11.2	**				7.5	12.3	**	3.0	11.3	**	1.3	4.3	**	1.1	6.6	**	1.3	5.1	**
PC 16:0/20:3 c/a										4.6	2.2	*																		
PC 16:1/20:5 a/b							2.0	1.4																						
PC 16:1/22:6 a/b																									2.1	1.0	**	2.5	1.1	*
PC 16:0/20:3 a/b													20.7	1.7	**				3.4	2.8		8.2	4.7		49.4	3.0	**			
PC 16:0/22:5 a/b																			4.4	5.9	**									
PC 18:0/20:4 b/a							7.7	4.8	*																			1.7	3.1	*
PC 18:1/20:4 b/a																									6.7	12.9	*			
PC 18:0/22:5 a/b																												7.5	26.9	**
PC 18:1/22:6 b/a	0.6	0.4	**				1.7	0.9	**	2.2	1.2	*	1.8	1.2		2.4	1.0													
PC 18:2/22:6 a/b																						90.5	10.8	*						
PC 20:1/20:3 a/b																			3.5	0.6	**									
PC 20:4/20:5 a/b																									6.6	2.0				
PC 20:4/22:6 a/b	1.1	0.8					1.1	0.5	**	1.1	0.8	**							2.2	0.7	**	3.5	1.3		3.7	1.2	*	2.2	1.9	
PC 20:0/22:6 b/a																			0.9	2.3	*				2.1	3.5	*	1.5	4.0	*
PI 18:0/20:4 a/b				2.7	1.7																									
PC 20:4/24:1 a/b	1.4	0.8	*	0.5	0.5																									
PC 20:2/C24:1 b/a				1.8	1.6																									
PC 20:4/26:1 a/b				1.2	1.8	*																								
PG 22:5/22:6 a/b																									36.8	10.3	**			

Figure 9. Potential sn isomers identified using the reverse phase LC-MS/MS methodology. Data are presented as the ratio of one isomer over the second isomer, with ratios determined using peak area data. Blue squares indicate that sn positional isomers were not detected, however fatty acyl isomers were detected. * p = 0.01–0.05, ** $p < 0.01$, red boxes indicate p values fell below the Benjamini-Hochberg critical value.

4. Discussion

4.1. Lipid Class Trends

Previous studies have found that tissue fatty acid compositions generally reflect the diet, and diets containing elevated levels of LC-PUFA such as EPA and DHA generally tend to promote LC-PUFA enrichment within tissues [24–26], and the subsequent elevation of complex lipids containing these fatty acids [22] with the opposite being true for diets containing mostly mono- and di-unsaturated fatty acids derived from terrestrial sources. In the present study, several tissues at two different timepoints were investigated to establish whether a universal trend exists in fish fed different diets, or whether tissue and temporal variations exist. For the five tissues studied, general trends were apparent across all tissues and timepoints, although exceptions were common and illustrated that certain tissues appear more plastic than others. Overall, a general trend was established, that a diet enriched in the LC-PUFA, EPA, and DHA, as well as 18:3n-3, was more likely to contain elevated levels of lipids containing larger carbon numbers, such as 42 and 44 carbons, with the inverse being the case for an 18:1-rich diet, which showed an elevation in 32, 34 and 36 carbon-containing lipids, corresponding to lipids such as PC 32:1, 32:2. 34:1 and 34:2. With regards to the unsaturation level, the LC-PUFA enriched diet resulted in increased levels of highly unsaturated complex lipids, containing 8, 9, and 10 double bonds, though lipids containing 4 double bonds were also more prevalent, owing to the inclusion of acyl chains likely derived from 18:3n-3, as illustrated by acyl chains containing 3 and 4 double bonds being incorporated into complex lipids. The control diet generally resulted in mono- and di-unsaturated complex lipids being enhanced within tissues, although lipids containing 6 or 7 desaturations also appeared to increase on average, with examples such as PC 36:6 and PC 40:7. Regarding acyl chains, the LC-PUFA enriched diet appeared to favour acyl configurations of 20–22 carbons, containing 3–4 double bonds, with 5–6 double bonds being marginally increased by the diet. Acyl configurations which were increased with the control diet generally contained 16 carbons, and 1 double bond, though 18 carbon acyl chain inclusion preference could not be determined conclusively owing to their high prevalence in both diets as 18:1 and 18:3n-3.

4.2. Tissue Plasticity

With regards to the amplitude with which lipid modulation occurred, the tissue with the lowest plasticity was the brain, followed by the eye to some extent, which generally corroborates previous work [24–26]. The brain generally exhibited a limit on the fold change induced by diet, being the most limited in range compared with other tissues. This limitation in response was highlighted by the lack of meaningful modulation by the control diet, suggesting a potential homeostatic process by which certain fatty acids cannot be included beyond a certain threshold, as evidenced by the responsiveness of other tissues to the control diet while allowing enhancement of longer and more unsaturated lipids if these lipids are elevated within the diet. Previously, the response of the brain to diet, particularly to vegetable oil and fish oil, was contrary to this, with DHA within the brain appearing unresponsive to dietary DHA level, though this was only examined for the total pool of DHA and 18:2n-6, with the latter only comprising less than 2% of total fatty acids, though appeared to be modulated by a diet rich in 18:2n-6 [27]. Within rats, however, a small increase in total DHA within brain tissue was observed with increased dietary DHA, though 18:1n-9 and 18:2n-6 within the brain showed no obvious response to dietary levels of these fatty acids [28]. A similar, yet more pronounced effect was found within PE in rat brains, with increased DHA in the diet benefiting both the total DHA and multiple diacyl and ether lipids, with 18:1n-9 and 18:2n-6 remaining relatively unchanged [29]. It appears then, that increasing dietary LC-PUFA has a small, but beneficial role in the brain, resulting consistently in increased levels of lipids containing 8 or 9 double bonds, with 4 and 5 double bond-containing lipids also elevated. This increase appeared to occur across the entire range of carbon chain lengths, and for the majority of lipid classes. Glycosylated lipids also appeared to be responsive to higher levels of polyunsaturated lipids within the diet, with a countertrend being observed in the eye. These trends may be driven by 18:3n-3 and its attachment to the long-chain base of the glycosylated sphingolipid, whereas eye glycolipids appear driven by mono- and di-unsaturated fatty acids. Interestingly there appeared to be little crossover between the compounds identified within the two tissues, indicating that both the brain and eye may benefit from tailored unsaturation profiles that are unique to their function. In humans, the functionality of the lens appeared to be due to the unique membrane structure, high degree of saturation, and high levels of dihydrosphingomyelin [30], despite glycosylated sphingolipids accounting for less than 1% of the total retinal lipids, with cerebrosides constituting a greater proportion in brain tissue, predominantly within the myelin sheath [31].

Gill, on the other hand, appeared in direct contrast to the brain, being predominantly modulated by the control diet. This appeared to occur across the majority of carbon numbers, within the two major lipid classes PC and PE, and, in those lipids containing 8 or fewer double bonds (other than those with 4), although PE showed a relatively small number of lipid modulations as demonstrated by the lower number of fold changes detected. However, gill was found to respond in line for key lipids, which was especially noticeable within PC. Certain highly unsaturated lipids such as PC 40:8 and 40:9 were increased by the increase in total polyunsaturates found within the GM diet, as did intermediate lipids such as PC 36:4 and 38:4, driven by 20:4 and 22:4. The most plastic and variable organs observed were the intestine and liver, which was unsurprising owing to their roles in both digestion and lipid modification, and their diverse enzymatic activity and general role in lipid modification and export to other tissues. Therefore, the greater fold change variance, compared with relatively static tissues, such as the brain and eye, appeared to be a consequence of their function, as well as the propensity of tissues such as the liver and intestine to autolyse and continue to modify metabolites post-harvest.

4.3. Acyl Complexity and Counter Trends

The primary goal was to establish whether isomeric effects could be discriminated against and whether this additional level of detail would help to further our understanding of the impact of diet and its complex effects on lipid metabolism. What was illustrated was the diverse acyl variety present in tissues, which generally reflected the diet, although significant tissue and temporal variations existed that may provide an opportunity to both classify and understand the alterations in lipid metabolism in finer detail. With the general tendency for the majority of tissues to follow a certain trend, as illustrated with the two diets tested, those trends which run counter to the rest may provide insight into the tissue or temporal lipid regulation. The most obvious counterpoints arose within the brain and gill, though smaller deviations from the consensus were also present. Some notable exceptions include variations in abundance, whilst still maintaining comparable fold changes, distinguishing brain, eye, and gill, for example. These were evident in PE 36:2, PE 40:7, PE O-40:8, PC 36:4, PC 36:6, PC 40:8 and PC 40:9 as examples. The altered abundance profiles, as measured by peak intensity, provided a measure of the prevalence of these lipids, and that abundance or deficits of specific lipids may be markers of certain tissues and timepoints. Certain lipid types also effectively discriminate both brain and eye by their presence and are also altered by diet. These lipids constituted the ether lipids (plasmalogens) and the glycosylated lipids, bound both as sphingolipids and MGDG. Ideally, tissues and timepoints may be distinguished through multivariate analysis. However, owing to the initial untargeted approach and subsequent refinement of these detected lipids, the large number of missing values between different tissues and timepoints makes such modelling difficult, and likely to be swayed by the absent data. Understanding the lipid flux within individual tissues, and its alteration by diet is likely a complicated endeavour, more so than in metabolomics, where specific pathways for individual metabolites have been established. Within lipids, although pathways such as the Kennedy, Lands cycle, fatty acid biosynthesis, and lipid class synthesis have been established, the fact that these pathways act, or have an impact, upon multiple substrates, be they biogenic or diet-derived, generally results in lipids being grouped into their classes, rather than individual metabolites, potentially obfuscating key links between lipids. Such finer grain lipid metabolism, however, can be deconvoluted through flux analysis, usually through stable isotope analysis [32]. This may go some way to understanding the complex interplay between the myriad of individual lipids, rather than the current broad lipid class models.

On studying the effect of growth/age, and whether differences could be determined which might indicate the age of the fish and/or whether lipid effects are strengthened or weakened over time, there were instances, predominantly in the initial screening data, indicating that fold or proportional strengthening and weakening effects might occur over time. How these trends relate, and whether they are the result of lipid flux as the fish increases fat storage and reduces bodily growth, for example, is currently difficult to determine due to the limited number of observable trends. The correlation between timepoint data, specifically the magnitude shift of the fold change or percentage composition over time, might enable individual pools to be connected that, in turn, might help to delineate whether lipids are biogenic in origin or dietarily derived. It is also suspected that different timepoints may be distinguishable and therefore classifiable, owing to differences in individual lipids in either fold or percentage composition terms, and that these different timepoints may eventually be linked to differences in growth stage, sexual maturity, and water temperature, though attributing lipid alterations to these factors is currently not possible and requires further investigation. This classification, however, would require further validation through the targeted acquisition of lipids across all tissues and timepoints.

The detection of acyl isomeric differences in relation to diet is primarily attributable to the fatty acids found within the diet and are likely due to the digestion of fatty acids from TAG, the main lipid class in aquafeeds. However, acyl isomeric and putative sn isomers could also be due to the configuration of TAG molecules. Whilst it is difficult to directly determine an association, owing to the lack of definitive sn positional data, the fact that isomeric trends exist indicates that further structural analysis is warranted. The mechanism by which acyl configuration is maintained is likely through monoacylglycerols (MAG) after lipase digestion within the gut, with 2-MAG the predominant pancreatic lipase product [33] taken up across the enterocyte membrane. It is these 2-MAGs that possess a structural fingerprint that can be built upon, with various acyl-CoAs derived from free fatty acids then used by monoacyl- and diacylglycerol-acyltransferases to reconstruct TAG, or through phospholipid-specific pathways. While free fatty acids cannot transfer a structural signal, if deficient in LC-PUFA, elongation of dietary and de novo fatty acids may carry with it a specific structural signal, for example, by biosynthesis of DHA through Sprecher's shunt, which may be distinct from that obtained from direct incorporation of LC-PUFA into complex lipids. The former, that isomeric signals derive from structured TAG, is predicated on the fact that structural alterations exist within lipids used for aquafeeds. It has been found that the positional distribution of specific acyl groups can vary, with both genetically engineered *C. sativa* and *B. napus* demonstrating a strong preference for DHA incorporation into the sn-1/3 position in TAG [9,17], while EPA was either incorporated to a greater extent into the sn-1/3 position, or was equally distributed between the positions [34], but was enriched in the sn-2 position in PC. The distribution of DHA appears similar to that found within certain single-cell oils [18]. Fish body oil, however, was found to have DHA preferentially esterified (80%) to the sn-2 position, whereas EPA was split equally between sn-2 and sn-1/3 positions in TAG [13]. In contrast, PC was found to contain EPA and DHA preferentially bound (77%) to the sn-2 position. DHA incorporation in TAG has also been corroborated in anchovy oil, with DHA 1.7 times as likely to be found in the sn-2 position, though EPA was found to be 4.3 times as likely to be found in either the sn-1 or -3 positions [14]. In general, fish oils appear to favour DHA in the sn-2 position, whereas EPA is generally found in the sn-1/3 position [15,16]. Differences in positioning might also be observed in 18:2n-6 and 18:3n-3 and their subsequent elongation products, as both these fatty acids are present in both fish and terrestrial plant oils, albeit 18:3n-3 is difficult to resolve in ^{13}C NMR and, depending on the plant and cultivar, might be limited in concentration. This relative positioning of EPA and DHA within TAG has also been found to be beneficial to serum and liver cholesterol and TAG concentrations [35], though more so in the sn-2 position for DHA, and in the sn-1/3 position for EPA, with EPA in this position resulting in modulation of the eicosanoids PGI_2 and TXA_2. This indicates that certain structural forms of lipids, specifically TAG, which are predominantly found within feed formulations, may have some impact on fish health, or have implications for lipid biosynthesis within the fish itself, with these biochemical modulations initially appearing to be attuned to the regiospecific configuration of fish oils. The initial evidence that modified plant oils do in fact deviate from standard fish oils, specifically in DHA positioning within TAG, lends credence to the fact that isomeric signals may be transmitted from the diet to organs, with this signal being found within the putative sn isomeric ratios. The detection of these acyl isomers provides a valuable resource for the development of more targeted approaches, with the aim of filling in data gaps resultant from an untargeted approach and exploring those lipids that display alternative acyl configurations, as well as altered fold and compositional responses to diet.

With regards to assessing these isomeric alterations routinely, targeted approaches utilizing multiple reaction monitoring (MRM) are potentially the methods of choice, although traditional HILIC and reverse phase methodologies cannot reliably separate the myriad of acyl isomers that constitute the lipidome. Whilst certain columns such as charged surface hybrid (CSH) [36] are known to separate isomers based on double bond configuration, the thorough assessment of isomers utilising isocratically derived methods is slow and time-consuming, usually requiring separate methods for both polar and neutral lipids. This is generally impractical, though likely to become easier with the increased adoption of ion mobility. The determination of what are putatively identified as sn positional isomers, and the identification of trends associated with one isomer over another, for example, PC 20:4/22:6 and PC 16:0/20:4, as well as the detection of four potential isomers of PC 16:0/16:1, indicates that there may be a signal which can be transferred from the diet or derives from the biogenesis of specific lipids under certain conditions, such as LC-PUFA scarcity. It is also worth noting that not all tissues and timepoints were found to contain these putative sn isomers, only being found to contain fatty acyl rather than positional isomers. This may be due to low abundances and the inability to consistently resolve closely eluting isomers, although it may also be a characteristic feature of the tissue or timepoint.

5. Conclusions

Overall, it was found that phospholipid molecular species in salmon tissues generally reflected the given diet, though variations were found in the response to the diets, specifically in the brain and gill, indicating some need for homeostatic regulation of certain fatty acids. Tissue-specific variations also existed within certain lipid classes, such as the glycosylated lipids, including alterations in abundance, fold amplitude, lipid acyl configurations as well as putative sn isomers, across tissues and timepoints, indicating the existence of unique organ and temporal lipid signatures, which both defined the tissue and suggested differences in their lipid biosynthetic pathways. The presence of putative sn isomers highlighted the need to explore the interplay between novel diets and organ biochemistry in more detail, with structurally different lipids derived from synthetic biology potentially capable of influencing the structure of organ lipids that, in turn, may have consequences for processes such as eicosanoid formation.

Supplementary Materials: The following supporting information can be downloaded at: https://www.mdpi.com/article/10.3390/metabo12090851/s1, Figure S1: Individual lipid isomers detected using LC-MS/MS triple quad mass spectrometry, across 5 tissues and two timepoints (absolute and fold); Figure S2: Individual lipid isomers detected using LC-MS/MS triple quad mass spectrometry, across 5 tissues and two timepoints (%).; Table S1: SFC-QTOF analysis of brain tissue at the final timepoint; Table S2: SFC-QTOF analysis of brain tissue at the intermediate timepoint; Table S3: SFC-QTOF analysis of eye tissue at the final timepoint; Table S4: SFC-QTOF analysis of eye tissue at the intermediate timepoint; Table S5: SFC-QTOF analysis of gill tissue at the final timepoint; Table S6: SFC-QTOF analysis of gill tissue at the intermediate timepoint; Table S7: SFC-QTOF analysis of intestine tissue at the final timepoint; Table S8: SFC-QTOF analysis of intestine tissue at the intermediate timepoint; Table S9: SFC-QTOF analysis of liver tissue at the final timepoint; Table S10: SFC-QTOF analysis of liver tissue at the intermediate timepoint; Table S11: LC-MS/MS data for brain tissue at the final timepoint; Table S12: LC-MS/MS data for brain tissue at the intermediate timepoint; Table S13: LC-MS/MS data for eye tissue at the final timepoint; Table S14: LC-MS/MS data for eye tissue at the intermediate timepoint; Table S15: LC-MS/MS data for gill tissue at the final timepoint; Table S16: LC-MS/MS data for gill tissue at the intermediate timepoint; Table S17: LC-MS/MS data for intestine tissue at the final timepoint; Table S18: LC-MS/MS data for intestine tissue at the intermediate timepoint; Table S19: LC-MS/MS data for liver tissue at the final timepoint; Table S20: LC-MS/MS data for liver tissue at the intermediate timepoint.

Author Contributions: Conceptualization, D.R.T. and J.A.N.; Formal analysis, R.B.; Funding acquisition, D.R.T., J.A.N. and M.B.B.; Investigation, R.B.; Methodology, R.B.; Project administration, D.R.T., J.A.N. and M.B.B.; Supervision, M.B.B.; Writing–original draft, R.B.; Writing–review & editing, D.R.T., J.A.N. and M.B.B. All authors have read and agreed to the published version of the manuscript.

Funding: This research was funded by BBSRC, grant numbers BB/N022157/1, BB/S005919/1 and BB/S006400/1, as was the APC.

Institutional Review Board Statement: The animal study protocol was approved by the Institutional Review Board of The University of Stirling (AWERB(16-17)83).

Informed Consent Statement: Not applicable.

Data Availability Statement: Data is contained within the article and supplementary material, as listed previously.

Conflicts of Interest: The authors declare no conflict of interest, and the funders had no role in the design of the study; in the collection, analyses, or interpretation of data; in the writing of the manuscript, or in the decision to publish the results.

References

1. FAO. The State of World Fisheries and Aquaculture 2020. In *Sustainability in Action*; FAO: Rome, Italy, 2020; Volume 32, p. 244. [CrossRef]
2. Barange, M.; Bahri, T.; Beveridge, M.; Cochrane, K.; Funge-Smith, S.; Poulain, F. *Impacts of Climate Change on Fisheries and Aquaculture: Synthesis of Current Knowledge, Adaptation and Mitigation Options*; FAO: Rome, Italy, 2018; pp. 1–628.
3. Sprague, M.; Chau, T.C.; Givens, D.I. Iodine Content of Wild and Farmed Seafood and Its Estimated Contribution to UK Dietary Iodine Intake. *Nutrients* **2021**, *14*, 195. [CrossRef] [PubMed]
4. Sprague, M.; Fawcett, S.; Betancor, M.; Struthers, W.; Tocher, D. Variation in the nutritional composition of farmed Atlantic salmon (*Salmo salar* L.) fillets with emphasis on EPA and DHA contents. *J. Food Compos. Anal.* **2020**, *94*, 103618. [CrossRef]
5. de Roos, B.; Sneddon, A.A.; Sprague, M.; Horgan, G.W.; Brouwer, I.A. The potential impact of compositional changes in farmed fish on its health-giving properties: Is it time to reconsider current dietary recommendations? *Public Health Nutr.* **2017**, *20*, 2042–2049. [CrossRef] [PubMed]
6. Jónasdóttir, S.H.; Visser, A.W.; Richardson, K.; Heath, M.R. Seasonal copepod lipid pump promotes carbon sequestration in the deep North Atlantic. *Proc. Natl. Acad. Sci. USA* **2015**, *112*, 12122–12126. [CrossRef] [PubMed]
7. Cavan, E.L.; Belcher, A.; Atkinson, A.; Hill, S.L.; Kawaguchi, S.; McCormack, S.; Meyer, B.; Nicol, S.; Ratnarajah, L.; Schmidt, K.; et al. The importance of Antarctic krill in biogeochemical cycles. *Nat. Commun.* **2019**, *10*, 4742. [CrossRef] [PubMed]
8. Walsh, T.A.; Bevan, S.A.; Gachotte, D.J.; Larsen, C.M.; Moskal, W.A.; Merlo, P.A.O.; Sidorenko, L.V.; Hampton, R.E.; Stoltz, V.; Pareddy, D.; et al. Canola engineered with a microalgal polyketide synthase-like system produces oil enriched in docosahexaenoic acid. *Nat. Biotechnol.* **2016**, *34*, 881–887. [CrossRef] [PubMed]
9. Petrie, J.R.; Zhou, X.-R.; Leonforte, A.; McAllister, J.; Shrestha, P.; Kennedy, Y.; Belide, S.; Buzza, G.; Gororo, N.; Gao, W.; et al. Development of a Brassica napus (Canola) Crop Containing Fish Oil-Like Levels of DHA in the Seed Oil. *Front. Plant Sci.* **2020**, *11*, 727. [CrossRef]
10. Napier, J.A.; Olsen, R.; Tocher, D. Update on GM canola crops as novel sources of omega-3 fish oils. *Plant Biotechnol. J.* **2018**, *17*, 703–705. [CrossRef]
11. Han, L.; Usher, S.; Sandgrind, S.; Hassall, K.L.; Sayanova, O.; Michaelson, L.V.; Haslam, R.; Napier, J.A. High level accumulation of EPA and DHA in field-grown transgenic Camelina—A multi-territory evaluation of TAG accumulation and heterogeneity. *Plant Biotechnol. J.* **2020**, *18*, 2280–2291. [CrossRef]
12. Usher, S.; Han, L.; Haslam, R.; Michaelson, L.V.; Sturtevant, D.; Aziz, M.; Chapman, K.D.; Sayanova, O.; Napier, J.A. Tailoring seed oil composition in the real world: Optimising omega-3 long chain polyunsaturated fatty acid accumulation in transgenic Camelina sativa. *Sci. Rep.* **2017**, *7*, 6570. [CrossRef]
13. Ruiz-Lopez, N.; Stubhaug, I.; Igapharraguerre, I.; Rimbach, G.; Menoyo, D. Positional Distribution of Fatty Acids in Triacylglycerols and Phospholipids from Fillets of Atlantic Salmon (*Salmo Salar*) Fed Vegetable and Fish Oil Blends. *Mar. Drugs* **2015**, *13*, 4255–4269. [CrossRef] [PubMed]
14. Moharana, T.R.; Byreddy, A.R.; Puri, M.; Barrow, C.; Rao, N.M. Selective Enrichment of Omega-3 Fatty Acids in Oils by Phospholipase A1. *PLoS ONE* **2016**, *11*, e0151370. [CrossRef]
15. Ando, Y.; Satake, M.; Takahashi, Y. Reinvestigation of positional distribution of fatty acids in docosahexaenoic acid-rich fish oil triacyl-sn-glycerols. *Lipids* **2000**, *35*, 579–582. [CrossRef] [PubMed]
16. Akanbi, T.O.; Adcock, J.; Barrow, C.J. Selective concentration of EPA and DHA using Thermomyces lanuginosus lipase is due to fatty acid selectivity and not regioselectivity. *Food Chem.* **2013**, *138*, 615–620. [CrossRef]
17. Petrie, J.R.; Shrestha, P.; Belide, S.; Kennedy, Y.; Lester, G.; Liu, Q.; Divi, U.K.; Mulder, R.; Mansour, M.P.; Nichols, P.D.; et al. Metabolic Engineering Camelina sativa with Fish Oil-Like Levels of DHA. *PLoS ONE* **2014**, *9*, e85061. [CrossRef]
18. Jin, J.; Jin, Q.; Wang, X.; Akoh, C.C. High Sn-2 Docosahexaenoic Acid Lipids for Brain Benefits, and Their Enzymatic Syntheses: A Review. *Engineering* **2020**, *6*, 424–431. [CrossRef]
19. Lankhorst, P.P.; Chang, A.-N. The Application of NMR in Compositional and Quantitative Analysis of Oils and Lipids. In *Modern Magnetic Resonance*; Springer: Berlin/Heidelberg, Germany, 2018; pp. 1743–1764. [CrossRef]
20. van Erp, H.; Bryant, F.M.; Martin-Moreno, J.; Michaelson, L.V.; Bhutada, G.; Eastmond, P.J. Engineering the stereoisomeric structure of seed oil to mimic human milk fat. *Proc. Natl. Acad. Sci. USA* **2019**, *116*, 20947–20952. [CrossRef]
21. Matyash, V.; Liebisch, G.; Kurzchalia, T.V.; Shevchenko, A.; Schwudke, D. Lipid extraction by methyl-tert-butyl ether for high-throughput lipidomics. *J. Lipid Res.* **2008**, *49*, 1137–1146. [CrossRef]
22. Broughton, R.; Tocher, D.R.; Betancor, M.B. Development of a C18 Supercritical Fluid Chromatography-Tandem Mass Spectrometry Methodology for the Analysis of Very-Long-Chain Polyunsaturated Fatty Acid Lipid Matrices and Its Application to Fish Oil Substitutes Derived from Genetically Modified Oilseeds in the Aquaculture Sector. *ACS Omega* **2020**, *5*, 22289–22298. [CrossRef]

23. Nakanishi, H.; Iida, Y.; Shimizu, T.; Taguchi, R. Separation and quantification of sn-1 and sn-2 fatty acid positional isomers in phosphatidylcholine by RPLC-ESIMS/MS. *J. Biochem.* **2010**, *147*, 245–256. [CrossRef]
24. Betancor, M.; Sprague, M.; Sayanova, O.; Usher, S.; Campbell, P.; Napier, J.; Caballero, M.; Tocher, D. Evaluation of a high-EPA oil from transgenic Camelina sativa in feeds for Atlantic salmon (*Salmo salar* L.): Effects on tissue fatty acid composition, histology and gene expression. *Aquaculture* **2015**, *444*, 1–12. [CrossRef] [PubMed]
25. Betancor, M.B.; Howarth, F.J.; Glencross, B.D.; Tocher, D.R. Influence of dietary docosahexaenoic acid in combination with other long-chain polyunsaturated fatty acids on expression of biosynthesis genes and phospholipid fatty acid compositions in tissues of post-smolt Atlantic salmon (*Salmo salar*). *Comp. Biochem. Physiol. Part B Biochem. Mol. Biol.* **2014**, *172-173*, 74–89. [CrossRef] [PubMed]
26. Zhang, M.; Chen, C.; You, C.; Chen, B.; Wang, S.; Li, Y. Effects of different dietary ratios of docosahexaenoic to eicosapentaenoic acid (DHA/EPA) on the growth, non-specific immune indices, tissue fatty acid compositions and expression of genes related to LC-PUFA biosynthesis in juvenile golden pompano Trachinotus ovatus. *Aquaculture* **2019**, *505*, 488–495. [CrossRef]
27. Olsvik, P.A.; Amlund, H.; Torstensen, B.E. Dietary lipids modulate methylmercury toxicity in Atlantic salmon. *Food Chem. Toxicol.* **2011**, *49*, 3258–3271. [CrossRef] [PubMed]
28. Saito, M.; Ueno, M.; Kubo, K.; Yamaguchi, M. Dose-Response Effect of Dietary Docosahexaenoic Acid on Fatty Acid Profiles of Serum and Tissue Lipids in Rats. *J. Agric. Food Chem.* **1998**, *46*, 184–193. [CrossRef] [PubMed]
29. Kitajka, K.; Puskás, L.G.; Zvara, Á.; Hackler, L.; Barceló-Coblijn, G.; Yeo, Y.K.; Farkas, T. The role of n-3 polyunsaturated fatty acids in brain: Modulation of rat brain gene expression by dietary n-3 fatty acids. *Proc. Natl. Acad. Sci. USA* **2002**, *99*, 2619–2624. [CrossRef]
30. Borchman, D.; Yappert, M.C. Lipids and the ocular lens. *J. Lipid Res.* **2010**, *51*, 2473–2488. [CrossRef] [PubMed]
31. Brush, R.S.; Tran, J.-T.A.; Henry, K.R.; McClellan, M.E.; Elliott, M.H.; Mandal, N.A. Retinal Sphingolipids and Their Very-Long-Chain Fatty Acid–Containing Species. *Investig. Ophthalmol. Vis. Sci.* **2010**, *51*, 4422–4431. [CrossRef]
32. Schlame, M.; Xu, Y.; Erdjument-Bromage, H.; Neubert, T.A.; Ren, M. Lipidome-wide 13C flux analysis: A novel tool to estimate the turnover of lipids in organisms and cultures. *J. Lipid Res.* **2020**, *61*, 95–104. [CrossRef]
33. Iqbal, J.; Hussain, M.M. Intestinal lipid absorption. *Am. J. Physiol. Endocrinol. Metab.* **2009**, *296*, 1183–1194. [CrossRef]
34. Ruiz-Lopez, N.; Haslam, R.P.; Usher, S.; Napier, J.A.; Sayanova, O. An alternative pathway for the effective production of the omega-3 long-chain polyunsaturates EPA and ETA in transgenic oilseeds. *Plant Biotechnol. J.* **2015**, *13*, 1264–1275. [CrossRef] [PubMed]
35. Yoshinaga, K.; Sasaki, K.; Watanabe, H.; Nagao, K.; Inoue, N.; Shirouchi, B.; Yanagita, T.; Nagai, T.; Mizobe, H.; Kojima, K.; et al. Differential effects of triacylglycerol positional isomers containing n-3 series highly unsaturated fatty acids on lipid metabolism in C57BL/6J mice. *J. Nutr. Biochem.* **2015**, *26*, 57–63. [CrossRef] [PubMed]
36. Damen, C.W.N.; Isaac, G.; Langridge, J.; Hankemeier, T.; Vreeken, R.J. Enhanced lipid isomer separation in human plasma using reversed-phase UPLC with ion-mobility/high-resolution MS detection. *J. Lipid Res.* **2014**, *55*, 1772–1783. [CrossRef] [PubMed]

metabolites MDPI

Article

miR-1/AMPK-Mediated Glucose and Lipid Metabolism under Chronic Hypothermia in the Liver of Freshwater Drum, *Aplodinotus grunniens*

Jianxiang Chen [1,2], Changyou Song [1,2], Haibo Wen [1,2], Guangxiang Liu [1], Ningyuan Wu [1], Hongxia Li [1,2,*], Miaomiao Xue [1] and Pao Xu [1,2,*]

1 Wuxi Fisheries College, Nanjing Agricultural University, Wuxi 214081, China; 2020113020@stu.njau.edu.cn (J.C.); songchangyou@ffrc.cn (C.S.); wenhb@ffrc.cn (H.W.); 2019113019@stu.njau.edu.cn (G.L.); 2020813059@stu.njau.edu.cn (N.W.); 2021113016@stu.njau.edu.cn (M.X.)

2 Key Laboratory of Freshwater Fisheries and Germplasm Resources Utilization, Ministry of Agriculture and Rural Affairs, Freshwater Fisheries Research Center, Chinese Academy of Fishery Sciences, Wuxi 214081, China

* Correspondence: lihx@ffrc.cn (H.L.); xup@ffrc.cn (P.X.)

Abstract: Our previous study demonstrated that low temperature could induce hepatic inflammation and suppress the immune and oxidation resistance of freshwater drum. However, the metabolism, especially the glucose and lipid metabolism involved, is poorly studied. To further explore the chronic hypothermia response of freshwater drum, an 8-day hypothermia experiment was conducted at 10 °C to investigate the effect of chronic hypothermia on glucose and lipid metabolism via biochemical and physiological indexes, and metabolic enzyme activities, miRNAs and mRNA-miRNA integrate analysis in the liver. Plasma and hepatic biochemical parameters reveal chronic hypothermia-promoted energy expenditure. Metabolic enzyme levels uncover that glycolysis was enhanced but lipid metabolism was suppressed. Differentially expressed miRNAs induced by hypothermia were mainly involved in glucose and lipid metabolism, programmed cell death, disease, and cancerization. Specifically, KEGG enrichment indicates that AMPK signaling was dysregulated. mRNA-miRNA integrated analysis manifests miR-1 and AMPK, which were actively co-related in the regulatory network. Furthermore, transcriptional expression of key genes demonstrates hypothermia-activated AMPK signaling by miR-1 and subsequently inhibited the downstream glucogenic and glycogenic gene expression and gene expression of fatty acid synthesis. However, glycogenesis was alleviated to the control level while fatty acid synthesis was still suppressed at 8 d. Meanwhile, the gene expressions of glycolysis and fatty acid oxidation were augmented under hypothermia. In conclusion, these results suggest that miR-1/AMPK is an important target for chronic hypothermia control. It provides a theoretical basis for hypothermia resistance on freshwater drum.

Keywords: hypothermia; glucose and lipid metabolism; miR-1; AMPK; freshwater drum

Citation: Chen, J.; Song, C.; Wen, H.; Liu, G.; Wu, N.; Li, H.; Xue, M.; Xu, P. miR-1/AMPK-Mediated Glucose and Lipid Metabolism under Chronic Hypothermia in the Liver of Freshwater Drum, *Aplodinotus grunniens*. *Metabolites* **2022**, *12*, 697. https://doi.org/10.3390/metabo12080697

Academic Editors: Qingchao Wang, Yan He and Qingsong Tan

Received: 24 June 2022
Accepted: 25 July 2022
Published: 27 July 2022

1. Introduction

Water temperature is one of the most essential external factors for aquatic poikilotherms [1]. Temperature shifts outside of the optimal temperature range have negative impacts on the physiology and metabolism of the animal [2]. As a result, chronic exposure to suboptimal temperatures can also compromise the overall health of the animal [3]. Seasonal water temperature variances are major contributors to suboptimal temperatures. In China, water temperatures within a whole year vary enormously as the seasons change. In Wuxi, for example, the water temperature can reach 30 °C in summer while close to 0 °C in winter. Additionally, changes in latitude, topography, and precipitation contribute to the large temperature difference between the north and south of China. These have led to tremendous economic loss in the aquaculture industry every year. Therefore, the prevention and control of low temperatures are crucial to farming. In aquaculture, physical methods, crossbreeding, molecular markers, and nutritional enhancement are commonly used to improve fish resistance to low temperatures.

Low temperature stress induces a decrease in enzyme activity and cell membrane fluidity, resulting in cell functional damages, and ultimately impacts fish health and even individual death [4]. When cells suffer cold stimulation, the signal can be transmitted to nuclei via various stress pathways to initiate the stress response and establish new homeostasis [5]. To cope with hypothermia, energy expenditures are inevitable. Therefore, it is necessary to mobilize more endogenous energy materials to augment energy generation. There are a variety of studies suggesting that carbohydrate and lipid metabolism play important roles in the resistance to hypothermia. Glucose is generally considered the primary energy substance. Under low temperature stress, raising plasma glucose and enhanced glycolysis supply energy for life activities [6]. However, with the prolonging of time, plasma glucose content gradually drops below the normal level [7]. Meanwhile, lipid metabolism in fish is sensitive to temperature changes [8]. Cold stress promotes the synthesis of unsaturated fatty acids to strengthen fish adaptability to low temperatures [9].

AMPK, AMP-activated protein kinase, is common in eukaryotic organisms. AMPK is widely regarded as a switch that regulates cellular energy metabolism. AMPK can perceive variations in cellular energy metabolism and maintain the balance between energy supply and demand by multiple processes in cellular metabolism [10]. When intracellular adenosine triphosphate (ATP) is reduced, AMPK prompts glucose uptake and inhibits glycogen synthesis to promote glucose conversion to glycolysis [11,12]. Additionally, it has been suggested that AMPK inhibits acetyl-CoA carboxylase (ACC) activity through phosphorylation [13], which reduces malonyl-CoA synthesis to weaken the inhibition of carnitine o-palmitoyltransferase 1 (CPT1) [14,15], thereby enhancing fatty acid oxidation to generate more ATP. Furthermore, ACC is the rate-limiting enzyme in fatty acid synthesis, so AMPK is also involved in suppressing fatty acid synthesis [16].

More recently, omics approaches have become a research focus aiming at understanding the functions, mechanisms, and interactions as well as relationships between genes, transcripts, proteins, lipids, and other biomolecules [17]. Transcriptomics refers to a discipline on gene transcription and regulation in cells, which analyzes gene expression on the RNA level. RNA can be divided into messenger RNA (mRNA) and non-coding RNA (ncRNA) according to their structure and function. Among several types of ncRNAs, microRNAs (miRNAs) are a group of small RNAs (approximately 22-nt-long), which regulate gene expression mainly at the post-transcriptional level through translational repression or mRNA degradation [18]. Recent studies have highlighted that miRNAs play important roles in different stress adaptations [19,20]. Nevertheless, few authors reported on how low temperature stress affects fish miRNAs. It is, therefore, necessary to investigate the miRNAs and their target genes under low temperature stress.

Freshwater drum (*Aplodinotus grunniens*) is a kind of fish that is endemic to North and Central America. It is the only species in the genus of *Aplodinotus* that perpetually inhabits freshwater [21]. With respect to edibility, the freshwater drum is featured to possess a higher edible proportion, with delicious and nutritious flesh rich in proteins, amino acids, and fatty acids. Moreover, the freshwater drum has no intermuscular bones, which improves the fish quality and processing of the aquatic product [22]. With these prospects, we imported the larvae from the USA in 2016 and have achieved a milestone in artificial breeding and cultivation in 2019, providing a breakthrough and prospect for aquaculture. However, the mechanisms of hypothermia response in freshwater drum are still poorly understood. Summer water temperature in central and eastern China ranges from 25 to 35 °C, while the water temperature in the winter is generally below 10 °C. According to the previous study, the freshwater drum could adapt to the water temperature ranging from 7 to 30 °C [23]. Meanwhile, juvenile freshwater drum suffered increased mortality as the water temperature dropped to 1 °C and below [24]. Combined with our pre-experiment, as lukewarm water fish, the optimum temperature of freshwater drum should range from 18–26 °C. Therefore, it is necessary to study the low temperature response mechanism of freshwater drum for the development of aquaculture. This study from the perspective of metabolism was conducted to evaluate the effect of chronic hypothermia on glucose and lipid metabolism in livers of freshwater drum. In the present study, miRNA sequencing was applied to identify the key regulators under chronic hypothermia exposure. The epigenetic regulation between miRNA and encoding genes was also investigated. These results will suggest that glucolipid metabolism plays an important role in the low temperature response of freshwater drum. Meanwhile, these will provide a theoretical foundation for the regulation of miRNAs on glucose and lipid metabolism in freshwater drum under hypothermia exposure, which may provide a new method for the prevention and control of low temperature.

2. Results

2.1. Effects of Chronic Hypothermia on Plasma Biochemical Parameters of Freshwater Drum

To explore the effects of chronic hypothermia on freshwater drum, plasma biochemical parameters were tested (shown in Figure 1). The levels of glucose (Glu), total cholesterol (TC), and triglyceride (TG) are elevated at the beginning of hypothermia. However, these significantly decreased at 2 d and 8 d (Figure 1A–C, $p < 0.05$). Unexpectedly, TP exhibited no statistical differences among all groups, though there was a downward trend at 2 d (Figure 1D, $p > 0.05$).

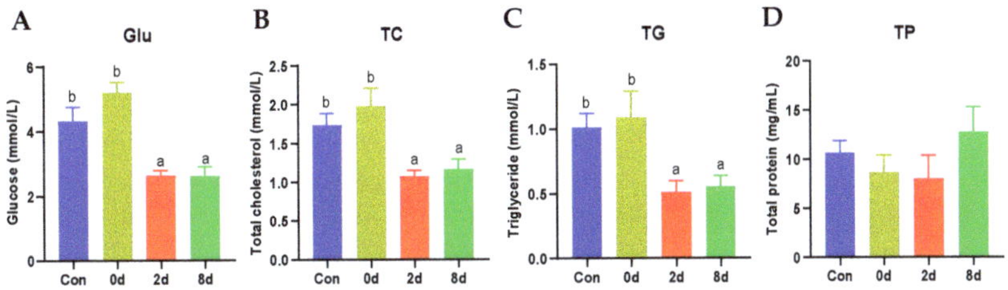

Figure 1. Effects of chronic hypothermia on plasma biochemical parameters of freshwater drum. (**A**) Glucose, Glu; (**B**) Total cholesterol, TC; (**C**) Triglyceride, TG; (**D**) Total protein, TP. Data were calculated by one-way ANOVA analysis with SPSS 23.0 (IBM SPSS Statistics, Version 23.0, Armonk, NY, USA). The significant differences are considered to exist when $p < 0.05$, and marked with different letters. Results were expressed as mean ± SEM, $n = 6$.

2.2. Effects of Chronic Hypothermia on Hepatic Biochemical Parameters of Freshwater Drum

Hepatic biochemical parameters were next evaluated and shown in Figure 2. Hypothermia had no impact on TP level (Figure 2A, $p > 0.05$), which was in line with the result in Figure 1D. The levels of hepatic glycogen (HG) and Na^+/K^+ ATPase declined rapidly and reached a minimum at 2 d (Figure 2B,D, $p < 0.05$), while it approached to control level at 8 d ($p > 0.05$). Meanwhile, the regulation of adenosine triphosphate (ATP) was significantly inhibited in both groups (Figure 2C, $p < 0.05$). However, the activities of aspartate aminotransferase (AST) and alanine aminotransferase (ALT) were not impacted by hypothermia (Figure 2E,F, $p > 0.05$).

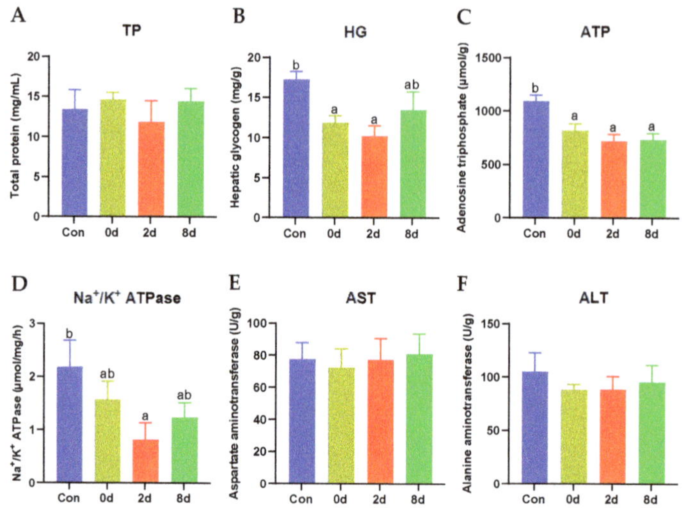

Figure 2. Effects of chronic hypothermia on hepatic biochemical parameters of freshwater drum. (**A**) Total protein, TP; (**B**) Hepatic glycogen, HG; (**C**) Adenosine triphosphate, ATP; (**D**) Na^+/K^+ ATPase. (**E**) Aspartate aminotransferase, AST; (**F**) Alanine aminotransferase, ALT. Data were calculated by one-way ANOVA analysis with SPSS 23.0. The significant differences are considered to exist when $p < 0.05$, and marked with different letters. Results were expressed as mean ± SEM, $n = 6$.

2.3. Effects of Chronic Hypothermia on Glucometabolism of Freshwater Drum

To further explore the effects of hypothermia, enzymes on glucometabolism were tested (shown in Figure 3). The activities of enzymes on glycolysis including glucokinase (GK), hexokinase (HK), phosphofructokinase (PFK), and pyruvate (PK) rose remarkably and peaked at 2 d (Figure 3A–D, $p < 0.05$); thereafter, these enzymes decreased to levels that did not differ significantly in comparison to the control group with continued hypothermia. Meanwhile, the activity of lactate dehydrogenase (LDH), a key enzyme in anaerobic glycolysis, displayed no significant difference under low temperature (Figure 3E, $p > 0.05$).

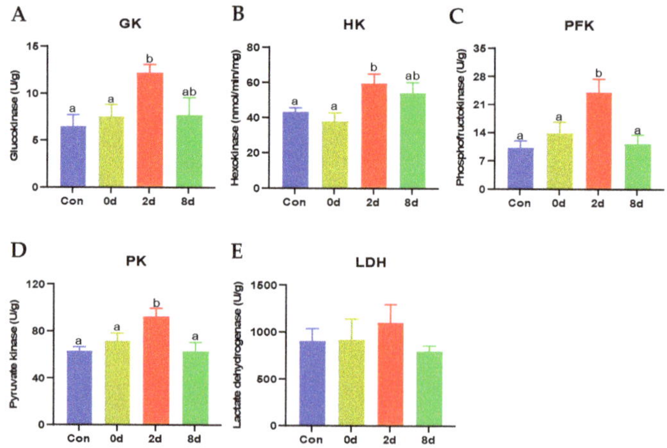

Figure 3. Effects of chronic hypothermia on glucometabolism of freshwater drum. (**A**) Glucokinase, GK; (**B**) Hexokinase, HK; (**C**) Phosphofructokinase, PFK; (**D**) Pyruvate kinase, PK; (**E**) Lactate dehydrogenase, LDH. Data were calculated by one-way ANOVA analysis with SPSS 23.0. The significant differences are considered to exist when $p < 0.05$, and marked with different letters. Results were expressed as mean ± SEM, $n = 6$.

2.4. Effects of Chronic Hypothermia on Lipid Metabolism of Freshwater Drum

Enzymes on lipid metabolism were next evaluated (shown in Figure 4). The activities of lipoprotein lipase (LPL), total esterase (TE), and fatty acid synthetase (FAS) were inhibited significantly with prolonged hypothermia exposure (Figure 4A,C,D, $p < 0.05$). However, hepatic lipase (HL) displayed no significant differences in contrast to the control group (Figure 4B, $p > 0.05$).

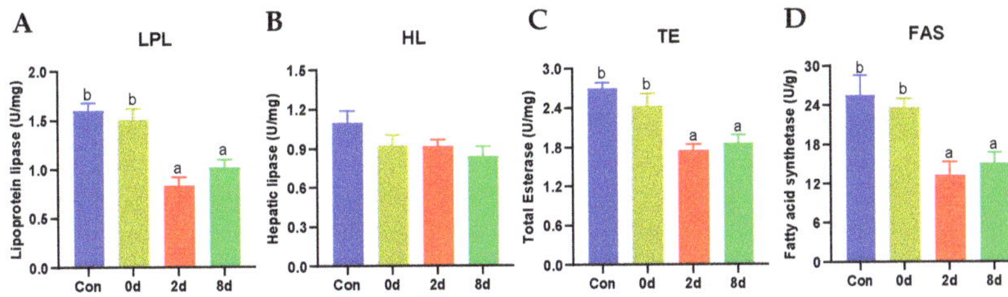

Figure 4. Effects of chronic hypothermia on lipid metabolism of freshwater drum. (**A**) Lipoprotein lipase, LPL; (**B**) Hepatic lipase, HL; (**C**) Total Esterase, TE; (**D**) Fatty acid synthetase, FAS. Data were calculated by one-way ANOVA analysis with SPSS 23.0. The significant differences are considered to exist when $p < 0.05$, and marked with different letters. Results were expressed as mean ± SEM, $n = 6$.

2.5. miRNA Analysis Reveals AMPK Signaling Was Active to Glucose and Lipid Metabolism under Hypothermia Exposure

To reveal the underlying mechanism of glucose and lipid metabolism under hypothermia, high-throughput sequencing was conducted to reveal the dynamically regulated miRNAs. According to the above results, the treatment group of 2 d was selected for research. We first used univariate statistical analyses, including fold change (FC) analysis and Students' *t*-test. In comparison with the control group, a total of 813 differentially expressed miRNAs (DEMs) at 2 d was identified with the threshold of $|\log_2 FC| > 1$ and $p < 0.05$, including 321 up-regulated and 492 down-regulated DEMs (Figure 5A). Meanwhile, these 813 DEMs were clustered into different subclusters according to their expression levels in the heatmap (Figure 5B).

To uncover the underlying regulation of these miRNAs, target prediction was conducted with the background of RNA-seq results. The obtained target differentially expressed RNAs (DEGs) were subjected to the Gene Ontology (GO) and Kyoto Encyclopedia of Genes and Genomes (KEGG) analysis. GO enrichment indicated the target genes were mainly involved in the biological process of acute-phase response (GO:0006953), acute inflammatory response (GO:0002526), iron ion transport (GO:0006826), cellular component of keratin filament (GO:0045095), molecular function of oxidoreductase activity (GO:0016722), ferroxidase activity (GO:0004322), and ATP binding (GO:0005524) (Figure 5C, Supplementary Material Table S1). Meanwhile, KEGG enrichment demonstrated that these target DEGs were enriched in carbohydrate and lipid metabolism (AMPK signaling pathway, Insulin signaling pathway, Glycolysis/Gluconeogenesis, Fat digestion and absorption, Glycerophospholipid metabolism), programmed cell death (Ferroptosis and Apoptosis), disease (Vibrio cholerae infection, Type II diabetes mellitus, Epstein-Barr virus infection and Graft-versus-host disease), and cancerization (Proteoglycans in cancer, Endometrial cancer, and Non-small cell lung cancer) (Figure 5D, Table S2).

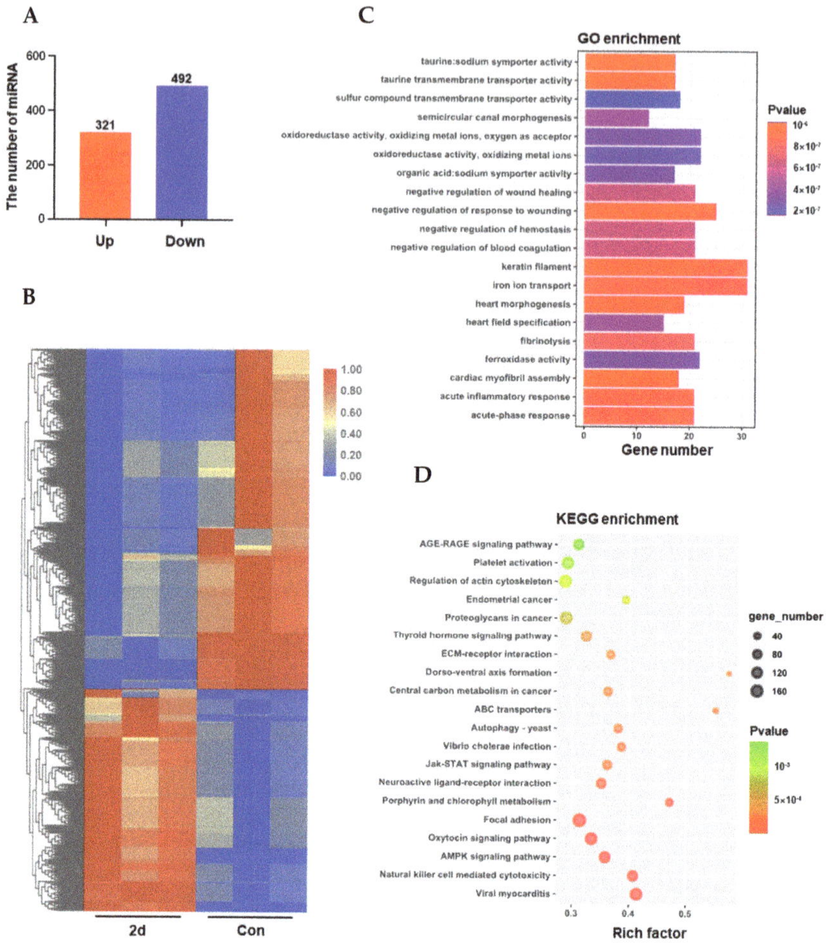

Figure 5. miRNA analysis reveals that AMPK signaling was active in glucose and lipid metabolism under hypothermia exposure. (**A**) up-and down-regulated differentially expressed miRNAs; (**B**) Hierarchical clustering of differentially expressed miRNAs among the 6 libraries; (**C**) GO enrichment of differentially expressed miRNAs 2 d vs. Con; (**D**) KEGG enrichment of differentially expressed miRNAs 2 d vs. Con.

2.6. *mRNA-miRNA Integrate Analysis Reveals miR-1 and AMPK Were Involved in Hypothermia Exposure*

To further explore the potential regulatory mechanism, mRNA-miRNA-integrated network analysis was performed. As shown in Figure 6, seven DEMs play important roles in the network with their target genes. Concomitantly, miR-1, miR-133a-5p, miR-133a-2-5p, miR-1388-5p, miR-2187-5p, miR-724, and miR-155 were mutually regulated by targeting to specific encoding genes. In particular, we found that miR-1/AMPK was involved in hypothermia response.

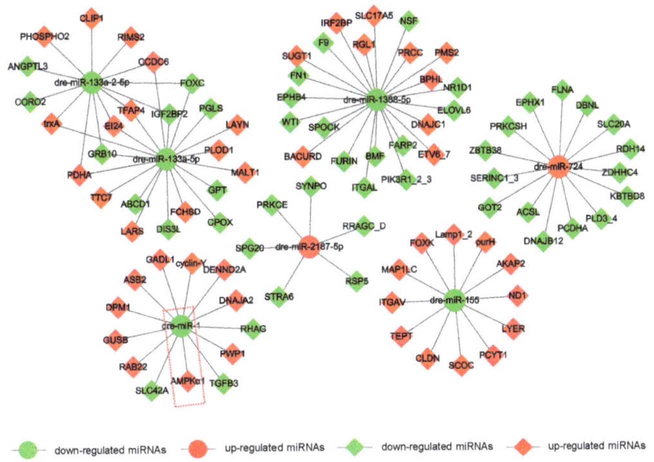

Figure 6. mRNA-miRNA integrate analysis reveals miR-1 and AMPK were involved in hypothermia exposure. Sig-diff miRNAs were selected to conduct mRNA-miRNA integrate analysis, and DEGs were targeted as the background. Data retrieved from high-throughput sequencing were applied to the integrated network by Cytoscape 3.7.2. The red dotted box represents there was a targeting relationship between miR-1 and AMPKα1.

2.7. AMPK Signaling Was Dysregulated under Hypothermia Based on Transcriptomic Analysis

To uncover whether AMPK signaling was active in hypothermia, we next quantified the relative expression of the key genes in AMPK signaling from the RNA-seq databases that were constructed previously. From the heatmap cluster (Figure 7A) and relative expression levels (Figure 7B), AMPK, PFK-2 and CPT1 were activated ($p < 0.05$), while G6Pase, PEPCK, TORC2, GS, SREBP1, FAS, ACC1, SCD1 and eEF-2 were significantly inhibited ($p < 0.05$).

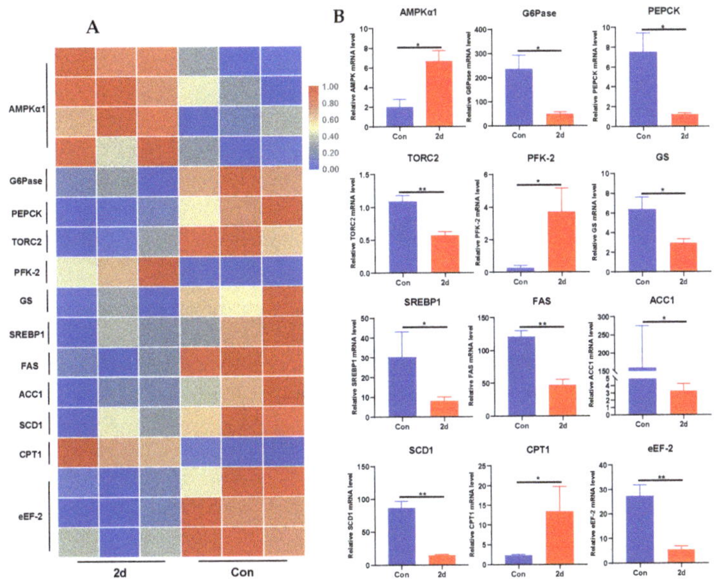

Figure 7. AMPK signaling was dysregulated under hypothermia based on transcriptomic analysis. (**A**) heatmap; (**B**) transcriptional expression of key genes in AMPK signaling pathway retrieved from the transcriptome. Data were analyzed by Student's t-test with SPSS 23.0, asterisk represents the statistical difference (*, $p < 0.05$; **, $p < 0.01$). Results were expressed as mean $\pm$ SEM, $n = 3$.

2.8. miR-1/AMPK Signaling Was Involved in Glucose and Fatty Acid Metabolism under Chronic Hypothermia

Target predictions indicate that AMPKα1 was the target gene of miR-1 (Figure 8A). To further evaluate whether the AMPK signaling pathway was involved in glucose and lipid metabolism under chronic hypothermia, we investigated the transcriptional expression of key genes. From the results, the regulation between miR-1 and AMPK under chronic hypothermia was confirmed, indicating that the expression of miR-1 was inhibited by hypothermia and persisted up to 8 days ($p < 0.05$, Figure 8A). Synchronously, the expression of AMPK was also activated at 2 d and 8 d ($p < 0.05$, Figure 8B).

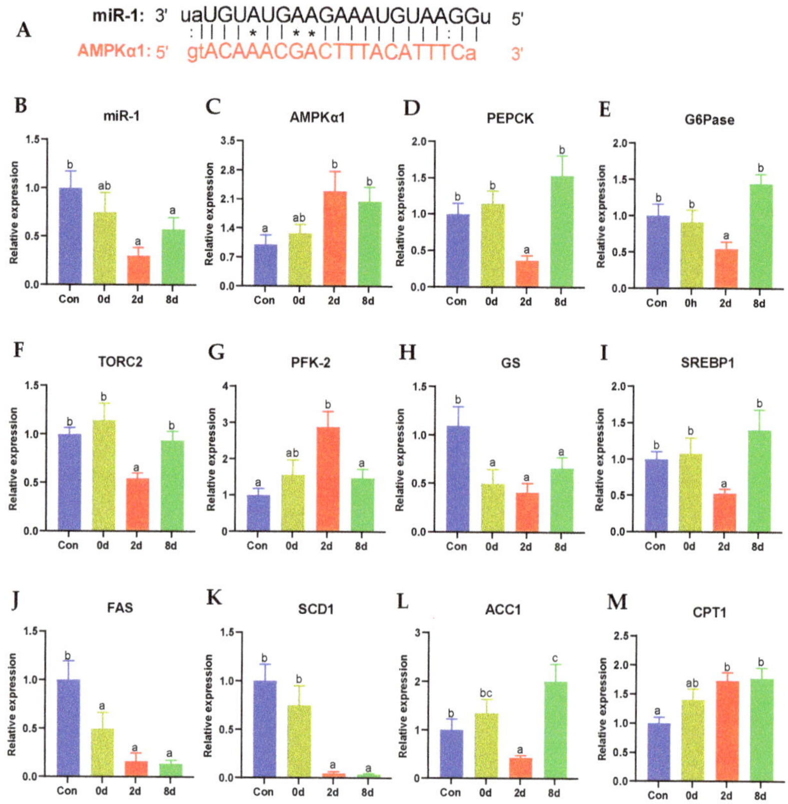

Figure 8. miR-1/AMPK signaling was involved in glucose and fatty acid metabolism under chronic hypothermia. (**A**) target prediction between miR-1 and AMPKα1; (**B**) miR-1; (**C**) AMP-activated protein kinase, AMPK; (**D**) Phosphoenolpyruvate carboxykinase, PEPCK; (**E**) Glucose-6-phosphatase, G6Pase; (**F**) CREB-regulated transcription coactivator 2, TORC2; (**G**) 6-phosphofructo-2-kinase-2, PFK-2; (**H**) Glycogen synthase, GS; (**I**) Sterol regulatory element-binding transcription factor 1, SREBP1; (**J**) Fatty acid synthase, FAS; (**K**) Stearoyl-CoA desaturase 1, SCD1; (**L**) Acetyl-CoA carboxylase 1, ACC1; (**M**) Carnitine O-palmitoyltransferase 1, CPT1. Data were calculated by one-way ANOVA analysis with SPSS 23.0. The significant differences are considered to exist when $p < 0.05$, and marked with different letters. Results were expressed as mean ± SEM, $n = 9$.

Apart from miR-1 and AMPK, gene expressions of PEPCK, G6Pase, and TORC2 on gluconeogenesis and GS on glycogen synthesis were significantly inhibited at 2 d (Figure 8C–E,G, $p < 0.05$), while the expression of PFK-2 on glycolysis was activated (Figure 8F, $p < 0.05$). Importantly, the expression of these genes at 8 d showed no significant changes apart from GS (Figure 8C–G, $p > 0.05$). Additionally, the mRNA expressions of SREBP1 and ACC1 demonstrated remarkable decreases at 2 d (Figure 8H,K, $p < 0.05$), while it notably increased after hypothermia for 8 days (Figure 8H,K, $p < 0.01$). Simultaneously, the gene expressions of FAS and SCD1 declined significantly throughout the chronic hypothermia (Figure 8I,J, $p < 0.05$). However, the mRNA expression of CPT1, a crucial gene for fatty acid oxidation, promptly elevated in both groups (Figure 8L, $p < 0.05$).

2.9. Epigenetic Schematic of AMPK Signaling under Hypothermia Induction in Freshwater Drum

In the present study, we validated the relationship between miRNAs and target genes with high throughput sequencing, target prediction, and RT-PCR. Based on the above results, we raise the hypothetical regulation schematic of miR-1/AMPK signaling on glucose and lipid metabolism under hypothermia (Figure 9). Hypothermia inhibited miR-1 and subsequently activated AMPK signaling and downstream modulators to affect glucose and lipid metabolism on freshwater drum.

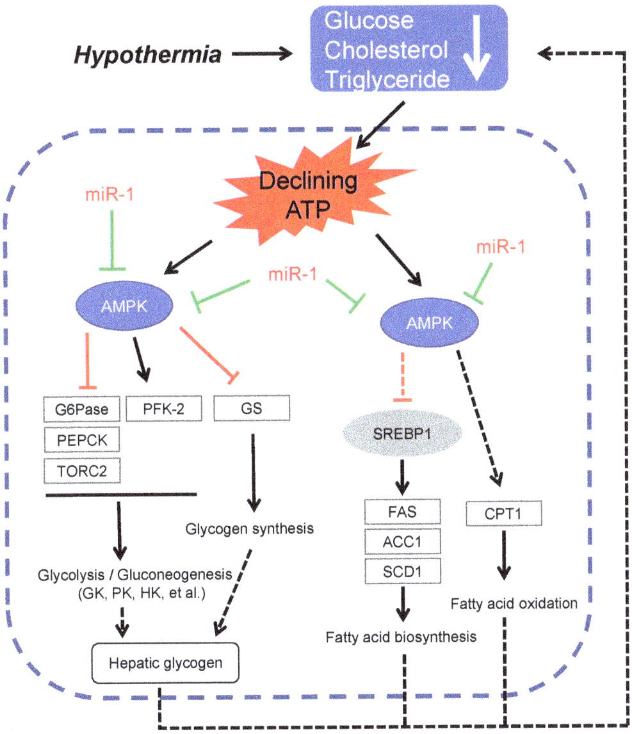

Figure 9. Epigenetic schematic of AMPK signaling under hypothermia induction in freshwater drum. miR-1 negatively regulates AMPK. This epigenetic network proves miR-1/AMPK is the important target for glucose and lipid metabolism under hypothermia response for freshwater drum.

3. Discussion

Freezing in winter takes a heavy toll on the Chinese aquaculture industry every year [25]. This study was conducted to investigate the effects of chronic low temperature on carbohydrate and lipid metabolism in freshwater drum. Some studies have demonstrated that water temperature is an important factor affecting fish physiology [26]. In the present study, the content of ATP and Na^+/K^+ ATPase in the liver decreased remarkably under chronic hypothermia exposure. The freshwater drum resists chronic hypothermia by a large amount of energy expenditure to maintain normal physiological activities.

Glucose is one of the main energy sources. Glucometabolism is divided into catabolism and anabolism. Glycolysis and gluconeogenesis are two opposite metabolic pathways, which could be mutually regulated, and so the activation or inhibition of key enzymes in the two metabolic pathways can cooperate [27,28]. In the present study, glucose and hepatic glycogen levels further corroborated that freshwater drum expended lots of energy at chronic low temperatures. Glucometabolism was involved in hypothermia exposure but slowly returned to a normal level over time. Glucose can provide energy for stress response in freshwater drum at the early stage of hypothermia exposure. Chronic hypothermia-promoted glycolysis thereinto aerobic glycolysis was predominant. Inversely, gluconeogenesis was suppressed. These reactions are the self-protection for freshwater drum under stress environment by increasing energy production and reducing energy consumption to adapt to chronic hypothermia [29].

Lipids are important energy storage substances in animals [30]. Lipids play an important role in fish health; lipid metabolic disorders can directly affect fish growth, development, and physiological activities, including its stress resistance [31]. Among them, triglyceride is an important form of energy storage, and it is oxidized for energy supply. Triglycerides are hydrolyzed to fatty acids and glycerol by lipase such as LPL [32], and then fatty acids are β-oxidized to acetyl-CoA, using the energy supply in the tricarboxylic acid cycle [33]. Simultaneously, acetyl-CoA is also the raw material for fatty acids and cholesterol synthesis [34,35]. Under the catalysis of FAS and other synthases, plenty of ATPs are consumed to synthesize fatty acids in the liver [35]. In this present study, chronic hypothermia decreased plasma triglyceride and cholesterol levels. The freshwater drum was kept in a state of energy deficiency under chronic hypothermia. Glucometabolism failed to meet enough energy supply under stress. Therefore, the decomposition of triglycerides and other lipids was supposed to participate in the regulation of energy homeostasis [36]. Meanwhile, cholesterol is a sterol type of lipid that serves as an essential structural component of animal cell membranes. Chronic hypothermia exposure inevitably suppressed cholesterol synthesis and could even deplete cholesterol to disrupt cell membrane function. Lipolysis-related enzymes and fatty acid synthase in the liver were inhibited, and these were not alleviated over time. Hypothermia caused effects on the rapid reduction of perivisceral lipids and non-polar lipid deposition in liver [37]. As a result, the strong reduction, particularly in LPL activity, may correspond to a form of liver protection, limiting lipid uptake from plasma lipoproteins [38]. Synchronously, chronic hypothermia tends to restrain the activity of fatty acid synthase to maintain energy balance. Therefore, lipid metabolism persisting up to the end of the experiment may be more necessary than glucometabolism at chronic low temperatures.

Protein is an important nutrient in the organism. All the structures and metabolism in an organism require the participation of protein. Under stress, aquatic animals can provide energy by proteolysis in different organs [39,40]. AST and ALT are the most important aminotransferases in the liver, which are related to protein catabolism [41,42]. In this experiment, TP in plasma and liver both remained unchanged. Similarly, there were no changes in ALT and AST. These results indicate that chronic hypothermia did not cause functional damage to the liver in freshwater drum. Proteolysis is not the main way to supply energy to maintain physiological activity in freshwater drum during chronic low temperature exposure.

Determining the miRNA profiles under stress is extremely valuable to the characterization of the regulation of biological functions [43]. Previous studies have demonstrated that miRNAs are closely related to temperature stress [44,45]. To further reveal the underlying epigenetic mechanism, the integrated analysis between miRNAs and target mRNAs was conducted with throughput sequencing in the present study. The results reveal that AMPK signaling was activated to cope with chronic hypothermia stress. Specifically, miR-1 targeted AMPK signaling was involved in the regulatory network. Transcriptomic analysis reveals that the downstream target genes were dysregulated under hypothermia stress. These data confirm the prospect that AMPK is the target in response to hypothermia.

It is reported that miR-1 is closely related to glucose and lipid metabolism. Sun et al. [46] identified that there was a significant decrease in the levels of miR-1 involved in glycolysis under temperature stress. In the present study, miR-1 was inhibited under hypothermia, and the mRNA-miRNA integrate analysis reveals that miR-1 and AMPK were involved in chronic hypothermia stress, and the contrasting expression of miR-1 and AMPK indicate that miR-1 negatively regulates AMPK under hypothermia. AMPK is a crucial molecule in the regulation of biological energy metabolism. Jia et al. [47] reported a miR-1-mediated AMPK pathway to inhibit rat cardiac fibroblasts fibrosis induced by high glucose. However, in another study, HIF-1α was demonstrated to be a direct functional target of miR-1. The downregulation of miR-1 significantly increased HIF-1α expression, resulting in enhanced tumor glycolysis [48]. Taken together, these findings suggest that miR-1 is a critical regulator of fundamental biological processes in freshwater drum under chronic hypothermia.

To further discuss whether miR-1/AMPK medicated the carbohydrate and lipid metabolism under chronic hypothermia stress, the transcriptional expression of downstream genes in the AMPK signaling pathway was evaluated with RT-PCR. The results indicate that the AMPK signaling pathway was activated. Hypothermia promoted the regulation of glycolysis, and fatty acid oxidation while inhibiting glycogen synthesis, gluconeogenesis, and fatty acid synthesis. Over time, glucometabolism was alleviated, while lipid metabolism was still affected. Notably, the gene expressions of SERBP1 and ACC1 were up-regulated after hypothermia for 8 days, meaning chronic hypothermia potentially promoted the synthesis of unsaturated fatty acids to strengthen fish adaptability. Collectively, these results further demonstrate that chronic hypothermia impacts the regulation of carbohydrate and lipid metabolism depending on the AMPK mediated by miR-1. Moreover, the lipid metabolism has a significant implication in the chronic hypothermia response of freshwater drum, and even lipid consumption can last through the whole overwintering period [49], which is consistent with the results of metabolic enzyme activities. Together, these results will provide a reference for the prevention and control of low temperature on freshwater drum.

4. Materials and Methods

4.1. Ethics Statement

This study was approved by the Animal Care and Use Committee of Nanjing Agricultural University (Nanjing, China). All animal procedures were performed according to the Guideline for the Care and Use of Laboratory Animals in China.

4.2. Experimental Animals and Rearing Conditions

The hypothermia experiment was conducted at Wuxi Fisheries College of Nanjing Agricultural University. Laboratory fish were the first-generation larvae of freshwater drum introduced from the United States by the Freshwater Fisheries Research Center, Chinese Academy of Fishery Sciences. Freshwater drum were reared in a temperature-adjustable circulating water system (specifications for φ 820 $\times$ 700 mm) consisting of 12 tanks (300 L each). Prior to the experiment, fish were acclimated in the tank fed with fresh bait at 25 °C for 14 days. After acclimation, fish averaging 20.88 $\pm$ 2.75 g were randomly assigned into 9 tanks (3 tanks per group, 20 fish per tank) and were fed with fresh bait (3–5% of their body weight) twice a day (8:00 and 16:00). During the 8-day experiment, we cleaned up food scraps and feces daily. During the experiment, the temperature gradually decreased from 25 °C to 10 °C in 15 h at a rate of 1 °C/h and was then maintained at 10 °C for 0 d, 2 d and 8 d. A temperature of 25 °C was set as the control group (Con). Throughout the experiment, dissolved oxygen was kept as >6 mg L^{-1}, pH 7.2–7.8, NO_2^- < 0.02 mg L^{-1}, and NH_3 < 0.05 mg L^{-1}.

4.3. Sample Collection

Experimental samples were collected at 0 d, 2 d and 8 d. Fish were starved for 24 h to evacuate the alimentary tract contents before sampling. Fifteen fish from each tank were randomly sampled and anesthetized with MS-222 (100 mg L^{-1}) at each time point. Blood samples were obtained from the caudal vein and injected into anticoagulant tubes. These samples were centrifuged at 5000 rpm at 4 °C for 10 min to extract the plasma. The plasma was stored at −80 °C for the measurement of biochemical indicators. Meanwhile, the sampled fish were dissected to collect the liver on ice, froze in liquid nitrogen immediately, and stored at −80 °C for subsequent analysis.

4.4. Plasma and Liver Biochemical Indicators Analysis

Plasma samples of two fish from each tank were used to measure parameters including glucose (Glu), total cholesterol (TC), triglyceride (TG), and total protein (TP). Additionally, liver samples of two fish from each tank were used to measure parameters including total protein (TP), hepatic glycogen (HG), adenosine triphosphate (ATP), Na^+/K^+ ATPase, aspartate aminotransferase (AST), alanine aminotransferase (ALT), glucokinase (GK), hexokinase (HK), phosphofructokinase (PFK), pyruvate kinase (PK), lactate dehydrogenase (LDH), lipoprotein lipase (LPL), hepatic lipase (HL), total esterase (TE) and fatty acid synthetase (FAS) according to the manufacturer's instructions. All the assay kits were purchased from Nanjing Jiancheng Bioengineering Institute, China. In detail, Glu was detected by the hexokinase method (Category No: F006-1-1), TC was determined by the colorimetric method (Category No: A111-2), TG was determined by the colorimetric method (Category No: A110-2), TP was determined by Coomassie brilliant blue (Category No: A045-2), HG was determined by colorimetric method (Category No: A043-1-1), ATP was determined by colorimetric method (Category No: A095-1-1), Na^+/K^+ ATPase was determined by colorimetric method (Category No: A016-1), AST was determined by Reit′s method (Category No: C010-1-1), ALT was determined by Reit′s method (Category No: C009-1-1), GK was determined by enzyme linked immunosorbent assay (ELISA) (Category No: H439-1), HK was determined by UV-spectrophotometric method (Category No: A077-3-1), PFK was determined by ultraviolet colorimetry method (Category No: A129-1-1), PK was determined by ultraviolet colorimetry method (Category No: A076-1-1), LDH was determined by colorimetry method (Category No: A020-1-2), LPL, HL and TE were determined by colorimetry method (Category No: A067-1-2), and FAS was determined by ELISA (Category No: H231-1-1).

4.5. RNA Extraction and De Novo High-Throughput Sequencing

4.5.1. RNA Extraction, cDNA Library Construction, and RNA-Seq

The total liver RNA of each group was extracted using TRIzol Reagent according to protocols (Takara, Dalian, China). In each group, nine liver tissues were selected to conduct the high-throughput sequence, wherein three fish in each group were randomly mixed and three biological replicates were finally applied for RNA-seq. After the qualified three micrograms of total RNA from each sample, ployA by magnetic beads with Oligo (dT) can be used to isolate mRNA from total RNA. The mRNA could be randomly fractured by fragmentation buffer, and small fragments of about 300 bp could be separated by magnetic bead screening. RNA fragments were converted to cDNA using random primers, followed by second-strand cDNA synthesis and end repair. End Repair Mix was added to fill the sticky end of the double-stranded cDNA into the flat end, and then base A was added to the 3′ end to connect the Y-shaped joint. Adaptor-tagged cDNA fragments were enriched using the manufacturer's cocktail and 15-cycle PCR. The target bands were recovered from 2% agarose gel and PCR amplification to obtain the final sequencing libraries. Illumina Hiseq6000 was used for sequencing after the library was qualified by quality inspection. Sequencing reads are paired-end 2 × 150 bp (PE150).

4.5.2. De Novo Assembly, Functional Annotation and the Differentially Expressed Genes (DEGs) Analysis

Unqualified raw data was filtered into clean data using cut adapt software before being assembled, including removing the joint sequences in reads and the reads without inserted fragments due to self-connection of the joint, removing low qualities reads (bases of Q $\leq$ 10 ranks over 20% of the whole read), removing reads with over 10% N, cutting the adapters and the less than 30 bp sequences. All sequences after quality control generated contig and singleton through de novo assembly, which were finally connected to obtain transcripts. All transcripts were compared with six databases (NR, Swiss-PROT, Pfam, COG, GO, and KEGG databases) to obtain functional annotation information. The transcriptome was quantified by RSEM. The DEGs were identified based on Fragments Per Kilobase of exon model per Million mapped reads (FPKM). As the sequencing depth of samples differs from each other, the absolute gene expression was normalized to the FPKM value, which made FKPM the expression quantity of genes. Deseq2 was used to analyze the variation on DEGs (the significant difference threshold was $|\log_2 FC| > 1$, $p < 0.05$). GO term and KEGG pathway enrichment were conducted to analyze DEGs and pathways.

4.6. miRNA Sequencing, Identification, and Target Gene Prediction

4.6.1. miRNA Sequencing

The Con and 2 d groups were applied for miRNA analysis. In each group, nine liver tissues were selected to conduct the high-throughput sequence, wherein three fish in each group were randomly mixed and three biological replicates were finally applied. After total RNA was extracted, 3′ and 5′ adaptors were added to synthesize the transcript into cDNA. After PCR amplification, fragments between 140 and 150 bp were selected for miRNA sequencing with Illumina Hiseq6000 platform. Fastx-Toolkit (Majorbio Bio-pharm Technology Co., Ltd, Shanghai, China) was applied to remove adapters, lower quality bases (mass value less than 20), and too long or too short reads (more than 32 nt and less than 18 nt). The clean readings were mapped to NR, Rfam, and miRbase for annotation (less than 2 mismatches).

4.6.2. Differentially Expressed miRNAs (DEMs) Identification and Target Gene Prediction

Clean reads were compared with miRBase and Rfam databases to obtain known miRNA annotation information. New miRNAs were predicted from reads without annotated information by miRDeep2 software. Reads with $|\log_2 FC| > 1$ and $p < 0.05$ were identified as DEMs. The database miRanda, targetscan, and RNAhybrid were applied for target gene prediction. Targeted genes with corrected $p < 0.05$ (corrected with Bonferroni) were identified as targeted DEGs, and target DEGs were also subjected to Gene Ontology (GO) and KEGG enrichment analysis with Blast2GO.

4.7. Validation of Differentially Expressed Genes Obtained from RNA-Seq

Real-time quantitative PCR (RT-PCR) was conducted according to our previously established methods [50] to validate the expressions of key genes involved in glucose and lipid metabolism ($n = 9$). All the primers were synthesized in Sangon Biotech Co., Ltd. (Shanghai, China). Details of primers are listed in Table 1. RT-PCR was performed with SYBR Green (Takara, Dalian, China) on Takara 800 Fast Real-Time PCR System according to the manufacturer's protocol.

Table 1. Primers and sequences referred in the experiments.

Gene	Primer	Sequence (5′-3′)
miRNA primer		
miR-1	F	CGGGAATGTAAAGAAGTATG
U6	F	CTCGCTTCGGCAGCACA
	R	AACGCTTCACGAATTTGCGT
mRNA primer		
AMPK	F	TCCCTCCTACAGCAACAAC
	R	GACGCCAGGTAGAAATCC
PEPCK	F	TCGTCTCATTCGGTAGTGGT
	R	CGTAGGTTGCCTTGGTTGT
G6Pase	F	GCGGACCTGAGGAACACCTT
	R	AAACACCTGCGGCTCCCATA
TORC2	F	AAGTTTAGCGAGAAGATAGCG
	R	TCAAGAGTGGATGGAAGG
PFK-2	F	CCAGATAATGAGGAGGGTTT
	R	TCTTGTTGTATTTGTGGCATC
GS	F	GAGCCTCCGCACATCGTA
	R	CGCCTGCTTCCTTATCCA
SREBP1	F	TTCCTCTCCCTCAACCCTCT
	R	TTACGGGCTCTCCATACACC
FAS	F	TGGCATCGAGTACAACAAGC
	R	TTGGCACGAAGTAGCATCAC
SCD1	F	CGGGGCTTCTTCTTCTCTCA
	R	GAAGCACATGACTAGCACGG
ACC1	F	CTGGAGGAGACGGTGAAAAG
	R	TGCGTATCTGCTTGAGGATG
CPT1	F	GCACCAGAACCTTTACCGA
	R	TCCGCCCAGTGATATGAAC
eEF-2	F	TCAGCCCTGTGGTGAGAAT
	R	TGTGCTTGTTGGGTGACTTT
β-actin	F	AGGCTGTGCTGTCCCTGTAT
	R	GCTGTGGTGGTGAAGGAGTAG

Note: The mRNA sequences for each gene were obtained from freshwater drum liver transcriptome sequencing database of the Gene Center of Freshwater Fisheries Research Center, Chinese Academy of Fishery Sciences. Primers for RT-PCR were designed using Primer premier 5.0.

4.8. Statistical Analysis

All data in the study were represented as mean ± standard error mean (SEM), which was calculated using SPSS software (version 23.0). The data of plasma, liver biochemical indicators, and metabolic enzyme activity were analyzed with one-way analysis of variance (ANOVA) followed by a Duncan multiple-range test. Independent samples *t*-test was conducted to analyze the transcriptional expression of detected genes based on the transcriptome. Relative RNA expression was calculated using the $2^{-\Delta\Delta CT}$ comparative CT method and analyzed with one-way analysis of variance (ANOVA) followed by a Duncan multiple-range test.

5. Conclusions

In this study, chronic hypothermia stress-induced alteration in glucose and lipid metabolism in freshwater drum. Lipid metabolism plays a critical role in the chronic hypothermia response of freshwater drum. miRNA sequencing and integrated analysis with mRNAs conjointly reveal that miR-1/AMPK was involved in hypothermia stress. Transcriptional expression uncovers the miR-1/AMPK contributions to the regulation of carbohydrate and lipid metabolism to maintain normal physiological activities in freshwater drum under chronic hypothermia stress. In conclusion, miR-1/AMPK could be an important target for chronic hypothermia responses in freshwater drum.

Supplementary Materials: The following supporting information can be downloaded at: https://www.mdpi.com/article/10.3390/metabo12080697/s1, Table S1: GO enrichment analysis of differentially targeted genes on DEMs; Table S2: KEGG enrichment analysis of differentially targeted genes on DEMs.

Author Contributions: J.C., conceptualization, methodology, validation, investigation and writing original—draft; C.S., methodology, writing—review and editing, supervision and funding acquisition; H.W., resources; G.L., validation, investigation and data curation; N.W., software and investigation; H.L., supervision and funding acquisition; M.X., formal analysis and validation; P.X., supervision and project administration. All authors have read and agreed to the published version of the manuscript.

Funding: This work was supported by the National Nonprofit Institute Research Grant of Freshwater Fisheries Research Center, Chinese Academy of Fishery Sciences (2020JBFR03, 2021JBFM13, 2020JBMF02), Jiangsu Agriculture Science and Technology Innovation Fund (CX(20)2025).

Institutional Review Board Statement: This study was approved by the Animal Care and Use Committee of Nanjing Agricultural University (Nanjing, China) WXFC 2022-0006, approved 23 March. All animal procedures were carried out in accordance with the China Laboratory Animal Care and Use Guidelines. The ethics in this experiment are the same as the ethics previously published [51].

Informed Consent Statement: Not applicable.

Data Availability Statement: The data that support the findings of this study are available from the corresponding author upon reasonable request.

Acknowledgments: The authors gratefully acknowledge Xueyan Ma, Guohua Lv, and Liang Zhang from the Freshwater Fisheries Research Center, Chinese Academy of Fishery Sciences for their assistance during the experimental period.

Conflicts of Interest: The authors declare no conflict of interest.

References

1. Song, C.; Cui, Y.; Liu, B.; Xie, J.; Ge, X.; Xu, P.; Lin, Y. HSP60 and HSP90β from blunt snout bream, *Megalobrama amblycephala*: Molecular cloning, characterization, and comparative response to intermittent thermal stress and Aeromonas hydrophila infection. *Fish Shellfish Immunol.* **2017**, *74*, 119–132. [CrossRef]
2. Zhou, C.; Zhang, Z.; Zhang, L.; Liu, Y.; Liu, P.F. Effects of temperature on growth performance and metabolism of juvenile sea bass (*Dicentrarchus labrax*). *Aquaculture* **2021**, *537*, 736458. [CrossRef]
3. Abram, Q.H.; Dixon, B.; Katzenback, B.A. Impacts of low temperature on the teleost immune system. *Biology* **2017**, *6*, 39. [CrossRef]
4. Snyder, R.J.; Hennessey, T.M. Cold tolerance and homeoviscous adaptation in freshwater alewives (*Alosa pseudoharengus*). *Fish Physiol. Biochem.* **2003**, *29*, 117–126. [CrossRef]
5. Donaldson, M.R.; Cooke, S.J.; Patterson, D.A.; Macdonald, J.S. Cold shock and fish. *J. Fish Biol.* **2008**, *73*, 1491–1530. [CrossRef]
6. Kuo, C.M.; Hsieh, S.L. Comparisons of physiological and biochemical responses between milkfish (*Chanos chanos*) and grass carp (*Ctenopharyngodon idella*) to cold shock. *Aquaculture* **2006**, *251*, 525–536. [CrossRef]
7. Qiang, J.; Yang, H.; Wang, H.; Kpundeh, M.D.; He, J.; Xu, P. Physiological responses and HSP70 mRNA expression in GIFT tilapia juveniles, *Oreochromis niloticus* under short-term crowding. *Aquac. Res.* **2015**, *46*, 335–345. [CrossRef]
8. Farkas, T.; Csengeri, I.; Majoros, F.; Oláh, J. Metabolism of fatty acids in fish: III. Combined effect of environmental temperature and diet on formation and deposition of fatty acids in the carp, Cyprinus carpio Linnaeus 1758. *Aquaculture* **2015**, *20*, 29–40. [CrossRef]
9. Sellner, P.A.; Hazel, J.R. Incorporation of polyunsaturated fatty acids into lipids of rainbow trout hepatocytes. *Am. J. Physiol. Regul. Integr. Comp. Physiol.* **1982**, *243*, 223–228. [CrossRef] [PubMed]
10. Dzamko, N.L.; Steinberg, G.R. AMPK-dependent hormonal regulation of whole-body energy metabolism. *Acta Physiol.* **2009**, *196*, 115–127. [CrossRef] [PubMed]
11. Merrill, G.F.; Kurth, E.J.; Hardie, D.G.; Winder, W.W. AICA riboside increases AMP-activated protein kinase, fatty acid oxidation, and glucose uptake in rat muscle. *Am. J. Physiol.* **1997**, *273*, 1107–1112. [CrossRef] [PubMed]
12. Lim, C.T.; Kola, B.; Korbonits, M. AMPK as a mediator of hormonal signalling. *J. Mol. Endocrinol.* **2010**, *44*, 87–97. [CrossRef] [PubMed]
13. Healy, J.E.; Gearhart, C.N.; Bateman, J.L.; Handa, R.J.; Florant, G.L. AMPK and ACC change with fasting and physiological condition in euthermic and hibernating golden-mantled ground squirrels (*Callospermophilus lateralis*). *Comp. Biochem. Physiol. Part A* **2011**, *159*, 322–331. [CrossRef]

14. McGarry, J.D.; Mills, S.E.; Long, C.S.; Foster, D.W. Observations on the affinity for carnitine, and malonyl-CoA sensitivity, of carnitine palmitoyltransferase I in animal and human tissues. Demonstration of the presence of malonyl-CoA in non-hepatic tissues of the rat. *Biochem. J.* **1983**, *214*, 21–28. [CrossRef] [PubMed]
15. Abu-Elheiga, L.; Matzuk, M.M.; Abo-hashema, K.A.H.; Wakil, S.J. Continuous Fatty Acid Oxidation and Reduced Fat Storage in Mice Lacking Acetyl-CoA Carboxylase 2. *Science* **2001**, *291*, 2613–2616. [CrossRef] [PubMed]
16. Carlson, C.L.; Winder, W.W. Liver AMP-activated protein kinase and acetyl-CoA carboxylase during and after exercise. *J. Appl. Physiol.* **1999**, *86*, 669–674. [CrossRef] [PubMed]
17. Horgan, R.P.; Kenny, L.C. 'Omic' technologies: Genomics, transcriptomics, proteomics and metabolomics. *Obstet. Gynaecol.* **2011**, *13*, 189–195. [CrossRef]
18. Bartel, D.P. MicroRNAs: Genomics, biogenesis, mechanism, and function. *Cell* **2004**, *116*, 281–297. [CrossRef]
19. Song, C.; Liu, B.; Ge, X.; Li, H.; Liu, B.; Xu, P. miR-34a/Notch1b mediated autophagy and apoptosis contributes to oxidative stress amelioration by emodin in the intestine of teleost *Megalobrama amblycephala*. *Aquaculture* **2021**, *547*, 737441. [CrossRef]
20. Song, C.; Liu, B.; Xu, P.; Ge, X.; Tang, Y.; Li, H.; Su, S. mir-144 is the epigenetic target for emodin to ameliorate oxidative stress induced by dietary oxidized fish oil via Nrf2 signaling in Wuchang bream, *Megalobrama amblycephala*. *Aquaculture* **2021**, *534*, 736357. [CrossRef]
21. Hernández-Gómez, R.E.; Contreras-Sánchez, W.M.; Hernández-Franyutti, A.; Perera-García, M.A.; Torres-Martínez, A. Testicular structure and development of the male germinal epithelium in the freshwater drum *Aplodinotus grunniens* (Perciformes: *Sciaenidae*) from the Usumacinta River, Southern Mexico. *Acta Zool.* **2021**, 1–19. [CrossRef]
22. Song, C.; Wen, H.; Liu, G.; Ma, X.; Lv, G.; Wu, N.; Chen, J.; Xue, M.; Li, H.; Xu, P. Gut Microbes Reveal Pseudomonas Medicates Ingestion Preference via Protein Utilization and Cellular Homeostasis Under Feed Domestication in Freshwater Drum, *Aplodinotus grunniens*. *Front. Microbiol.* **2022**, 861705. [CrossRef] [PubMed]
23. William, B.W. Life history aspects of smallmouth buffalo and freshwater drum in wheeler reservoir, Alabama. *Proc. Annu. Conf. Southeast. Assoc. Game Fish Comm.* **1968**, *22*, 479–495.
24. Bodensteiner, L.S.; Lewis, W.M. Role of Temperature, Dissolved Oxygen, and Backwaters in the Winter Survival of Freshwater Drum (*Aplodinotus grunniens*) in the Mississippi River. *Can. J. Fish. Aquat. Sci.* **1992**, *49*, 173–184. [CrossRef]
25. Ou, Y.J. Impact on fishery production by frost disaster in southern China. *Chin. Fish. Econ.* **2008**, *26*, 89–93.
26. Lahnsteiner, F.; Caberlotto, S. Motility of gilthead seabream sparus aurata spermatozoa and its relation to temperature, energy metabolism and oxidative stress. *Aquaculture* **2012**, *370–371*, 76–83. [CrossRef]
27. Pilkis, S.J.; Granner, D.K. Molecular Physiology of the Regulation of Hepatic Gluconeogenesis and Glycolysis. *Annu. Rev. Physiol.* **1992**, *54*, 885–909. [CrossRef]
28. Metón, I.; Fernández, F.; Baanante, I.V. Short- and long-term effects of refeeding on key enzyme activities in glycolysis-gluconeogenesis in the liver of gilthead seabream (*Sparus aurata*). *Aquaculture* **2003**, *225*, 99–107. [CrossRef]
29. Shaklee, J.B.; Christiansen, J.A.; Sidell, B.D.; Prosser, C.L.; Whitt, G.S. Molecular aspects of temperature acclimation in fish: Contributions of changes in enzyme activities and isozyme patterns to metabolic reorganization in the green sunfish. *J. Exp. Zool. Part A Ecol. Genet. Physiol.* **1977**, *201*, 1–20. [CrossRef]
30. Mark, A.S. Lipid dynamics in fish: Aspects of absorption, transportation, deposition and mobilization. *Comp. Biochem. Physiol. Part B Comp. Biochem.* **1988**, *90*, 679–690. [CrossRef]
31. Martins, D.A.; Rocha, F.; Castanheira, F.; Mendes, A.; Pousão-Ferreira, P.; Bandarra, N.; Coutinho, J.; Morais, S.; Yúfera, M.; Conceição, L.E.C.; et al. Effects of dietary arachidonic acid on cortisol production and gene expression in stress response in Senegalese sole (*Solea senegalensis*) post-larvae. *Fish Physiol. Biochem.* **2013**, *39*, 1223–1238. [CrossRef]
32. Choi, S.S.; Diehl, A.M. Hepatic triglyceride synthesis and nonalcoholic fatty liver disease. *Curr. Opin. Lipidol.* **2008**, *19*, 295–300. [CrossRef] [PubMed]
33. Ma, Y.; Temkin, S.M.; Hawkridge, A.M.; Guo, C.; Wang, W.; Wang, X.Y.; Fang, X.J. Fatty acid oxidation: An emerging facet of metabolic transformation in cancer. *Cancer Lett.* **2018**, *435*, 92–100. [CrossRef] [PubMed]
34. Munday, M.R. Regulation of mammalian acetyl-coa carboxylase. *Biochem. Soc. Trans.* **2002**, *30*, 1059–1064. [CrossRef] [PubMed]
35. Ke, R.; Xu, Q.; Li, C.; Luo, L.; Huang, D. Mechanisms of AMPK in the maintenance of ATP balance during energy metabolism. *Cell Biol. Int.* **2017**, *42*, 384–392. [CrossRef] [PubMed]
36. Cheng, C.; Ye, C.; Guo, Z.; Wang, A. Immune and physiological responses of pufferfish (*Takifugu obscurus*) under cold stress. *Fish Shellfish Immunol.* **2017**, *64*, 137–145. [CrossRef]
37. Ibarz, A.; Blasco, J.; Beltrán, M.; Gallardo, M.A.; Sánchez, J.; Sala, R.; Fernández-Borràs, J. Cold-induced alterations on proximate composition and fatty acid profiles of several tissues in gilthead sea bream (*Sparus aurata*). *Aquaculture* **2005**, *249*, 477–486. [CrossRef]
38. Ibarz, A.; Beltrán, M.; Fernández-Borràs, J.; Gallardo, M.A.; Sánchez, J.; Blasco, J. Alterations in lipid metabolism and use of energy depots of gilthead sea bream (*Sparus aurata*) at low temperatures. *Aquaculture* **2007**, *262*, 470–480. [CrossRef]
39. Galt, N.J.; Froehlich, J.M.; McCormick, S.D.; Biga, P.R. A comparative evaluation of crowding stress on muscle HSP90 and myostatin expression in salmonids. *Aquaculture* **2018**, *483*, 141–148. [CrossRef]
40. Marqueze, A.; Garbino, C.F.; Trapp, M.; Kucharski, L.C.; Fagundes, M.; Ferreira, D.; Koakoski, G.; Rosa, J.G.S. Protein and lipid metabolism adjustments in silver catfish (*Rhamdia quelen*) during different periods of fasting and refeeding. *Braz. J. Biol.* **2018**, *78*, 464–471. [CrossRef]

41. Fynn-Aikins, K.; Hughes, S.G.; Vandenberg, G.W. Protein retention and liver aminotransferase activities in Atlantic salmon fed diets containing different energy sources. *Comp. Biochem. Physiol. Part A Physiol.* **1995**, *111*, 163–170. [CrossRef]
42. Li, S.; Ji, H.; Zhang, B.; Zhou, J.; Yu, H. Defatted black soldier fly (*Hermetia illucens*) larvae meal in diets for juvenile jian carp (cyprinus carpio var. jian): Growth performance, antioxidant enzyme activities, digestive enzyme activities, intestine and hepatopancreas histological structure. *Aquaculture* **2017**, *477*, 62–70. [CrossRef]
43. Ambros, V. The functions of animal micrornas. *Nature* **2004**, *431*, 350–355. [CrossRef]
44. Sun, J.; Zhao, L.; Wu, H.; Lian, W.; Cui, C.; Du, Z.; Luo, W.; Li, M.; Yang, S. Analysis of miRNA-seq in the liver of common carp (*Cyprinus carpio* L.) in response to different environmental temperatures. *Funct. Integr. Genom.* **2019**, *19*, 265–280. [CrossRef] [PubMed]
45. Yang, R.; Dai, Z.; Chen, S.; Chen, L. MicroRNA-mediated gene regulation plays a minor role in the transcriptomic plasticity of cold-acclimated zebrafish brain tissue. *BMC Genom.* **2011**, *12*, 605. [CrossRef] [PubMed]
46. Sun, J.; Liu, Q.; Zhao, L.; Cui, C.; Wu, H.; Liao, L.; Tang, G.; Yang, S. Potential regulation by miRNAs on glucose metabolism in liver of common carp (*Cyprinus carpio*) at different temperatures. *Comp. Biochem. Physiol. Part D Genom. Proteom.* **2019**, *32*, 100628. [CrossRef] [PubMed]
47. Jia, Q.; An, W.; Xu, Y.; Qiao, S.; Chen, W. Role of microrna-1-mediated amp-activated protein kinase pathway in cardiac fibroblasts induced by high glucose in rats. *Zhonghua Wei Zhong Bing Ji Jiu Yi Xue* **2018**, *30*, 145–150. [CrossRef]
48. Xu, W.; Zhang, Z.; Zou, K.; Cheng, Y.; Yang, M.; Chen, H.; Wang, H.; Zhao, J.; Chen, P.; He, L. MiR-1 suppresses tumor cell proliferation in colorectal cancer by inhibition of Smad3-mediated tumor glycolysis. *Cell Death Dis.* **2017**, *8*, e2761. [CrossRef] [PubMed]
49. Gong, Y.; Chen, W.; Han, D.; Zhu, X.; Yang, Y.; Jin, J.; Liu, H.; Xie, S. Effects of food restriction on growth, body composition and gene expression related in regulation of lipid metabolism and food intake in grass carp. *Aquaculture* **2017**, *469*, 28–35. [CrossRef]
50. Li, H.; Qiang, J.; Song, C.; Xu, P. Transcriptome profiling reveal Acanthopanax senticosus improves growth performance, immunity and antioxidant capacity by regulating lipid metabolism in GIFT (*Oreochromis niloticus*). *Comp. Biochem. Physiol. Part D Genom. Proteom.* **2020**, *37*, 100784. [CrossRef] [PubMed]
51. Song, C.; Liu, B.; Xu, P.; Xie, J.; Ge, X.; Zhou, Q.; Sun, C.; Zhang, H.; Shan, F.; Yang, Z. Oxidized Fish Oil Injury Stress in *Megalobrama Amblycephala*: Evaluated by Growth, Intestinal Physiology, and Transcriptome-Based PI3K-Akt/NF-κB/TCR Inflammatory Signaling. *Fish Shellfish Immunol.* **2018**, *81*, 446–455. [CrossRef]

metabolites

Article

Time-Restricted Feeding Could Not Reduce Rainbow Trout Lipid Deposition Induced by Artificial Night Light

Hanying Xu [1,2], Ce Shi [1,2,3,*], Yangfang Ye [1,3], Changbin Song [4], Changkao Mu [1,3] and Chunlin Wang [1,3,*]

1 Key Laboratory of Aquacultural Biotechnology, Ministry of Education, Ningbo University, 818 Fenghua Road, Ningbo 315211, China
2 Marine Economic Research Center, Dong Hai Strategic Research Institute, Ningbo University, 818 Fenghua Road, Ningbo 315211, China
3 Collaborative Innovation Center for Zhejiang Marine High-Efficiency and Healthy Aquaculture, 818 Fenghua Road, Ningbo 315211, China
4 Institute of Semiconductors, Chinese Academy of Sciences, Beijing 100083, China
* Correspondence: shice@nbu.edu.cn (C.S.); wangchunlin@nbu.edu.cn (C.W.)

Abstract: Artificial night light (ALAN) could lead to circadian rhythm disorders and disrupt normal lipid metabolism, while time-restricted feeding (TRF) could maintain metabolic homeostasis. In mammals, TRF has been demonstrated to have extraordinary effects on the metabolic regulation caused by circadian rhythm disorders, but studies in lower vertebrates such as fish are still scarce. In this study, the impacts of ALAN on the body composition and lipid metabolism of juvenile rainbow trout were investigated by continuous light (LL) exposure as well as whether TRF could alleviate the negative effects of LL. The results showed that LL upregulated the expression of lipid synthesis (*fas* and *srebp-1c*) genes and suppressed the expression of lipid lipolysis (*pparβ*, *cpt-1a*, and *lpl*) genes in the liver, finally promoting lipid accumulation in juvenile rainbow trout. However, LL downregulated the expression of genes (Δ*6-fad*, Δ*9-fad*, *elovl2*, and *elovl5*) related to long-chain polyunsaturated fatty acid (LC-PUFA) synthesis, resulting in a significant decrease in the proportion of LC-PUFA in the dorsal muscle. In serum, LL led to a decrease in glucose (Glu) levels and an increase in triglyceride (TG) and high-density lipoprotein cholesterol (H-DLC) levels. On the other hand, TRF (mid-dark stage feeding (D)) and mid-light stage feeding (L)) upregulated the expression of both the lipid synthesis (*srebp-1c* and *pparγ*), lipolysis (*pparα*, *pparβ*, and *cpt-1a*), and lipid transport (*cd36/fat* and *fatp-1*) genes, finally increasing the whole-body lipid, liver protein, and lipid content. Meanwhile, TRF (D and L groups) increased the proportion of polyunsaturated fatty acid (PUFA) and LC-PUFA in serum. In contrast, random feeding (R group) increased the serum Glu levels and decreased TG, total cholesterol (T-CHO), and H-DLC levels, suggesting stress and poor nutritional status. In conclusion, ALAN led to lipid accumulation and a significant decrease in muscle LC-PUFA proportion, and TRF failed to rescue these negative effects.

Keywords: artificial night light; feeding regime; lipid metabolism; serum metabolites; rainbow trout

Citation: Xu, H.; Shi, C.; Ye, Y.; Song, C.; Mu, C.; Wang, C. Time-Restricted Feeding Could Not Reduce Rainbow Trout Lipid Deposition Induced by Artificial Night Light. *Metabolites* **2022**, *12*, 904. https://doi.org/10.3390/metabo12100904

Academic Editors: Yan He, Qingsong Tan and Qingchao Wang

Received: 2 September 2022
Accepted: 22 September 2022
Published: 26 September 2022

1. Introduction

Light is an essential environmental factor that affects the various life activities of organisms on Earth. With the advent and large-scale application of artificial light sources, artificial light at night (ALAN) has been recognized as one of the fastest-growing factors altering the natural environment and is considered as a pollutant called light pollution [1]. Light pollution has been shown to negatively affect human society and natural ecology [1–3]. For living organisms, the light–dark cycle (photoperiod) is the dominant zeitgeber that guides and synchronizes circadian rhythms [4]. Circadian rhythms play an important role in maintaining energy homeostasis, and many genes and enzymes related to nutrient metabolism exhibit strong circadian rhythms [5]. ALAN breaks the normal photoperiod and disrupts the circadian rhythm of the organism. In animal models and human studies, nocturnal light exposure has been associated with metabolic disorders, leading to an increased risk of metabolic diseases such as obesity [6] and type 2 diabetes [7]. In addition to directly affecting metabolic processes through biological clock rhythms, nocturnal light exposure also affects metabolic function by suppressing melatonin, altering glucocorticoids, and changing sleep architecture [8,9]. Metabolic disruptions caused by light pollution are of increasing concern, not only with regard to human health concerns, but also for the growth, survival, and welfare of farmed animals.

Lipids are crucial nutrients that can provide energy, essential fatty acids, phospholipids, sterols, and other substances for the life activities of organisms [10,11]. For carnivorous fish, lipids are preferred energy providers due to the limited ability to utilize carbohydrates [12,13]. Maintaining normal lipid metabolic processes is the basis for the growth and development of carnivorous fish. Briefly, lipid metabolism is the process of lipid uptake and transport, synthesis, and catabolism, in which various transcription factors and enzymes are involved [14,15]. The de novo synthesis of fatty acids begins with the synthesis of acetyl coenzyme A into palmitic acid (C16:0) and stearic acid (C18:0) in the presence of fatty acid synthase (FAS), followed by the further synthesis of polyunsaturated fatty acids (PUFA) in the presence of desaturases (e.g., Δ6-fatty acid desaturase (Δ6-fad) and Δ9-fatty acid desaturase (Δ9-fad)) and elongases (e.g., elongation of very long-chain fatty acid protein 2 (elovl2) and elongation of very long-chain fatty acid protein 5 (elovl5)) [15]. The *fas* gene expression is regulated by the upstream transcription factor sterol regulatory element binding protein 1c (SREBP-1c) [16,17]. The process of lipid oxidative catabolism begins with the hydrolysis of triglycerides to fatty acids and monoacylglycerols by the action of lipoprotein lipase (LPL) [18,19]. Further oxidation of fatty acids for energy supply involves the transport of long-chain fatty acids in mitochondria, and in the liver, carnitine palmitoyltransferase 1a (cpt-1a) is the key rate-limiting enzyme [15,20]. During lipid uptake and transport, cluster of differentiation 36/fatty acid translocase (CD36/FAT) [21] and fatty acid transport protein 1 (FATP-1) [22] play essential roles. In addition, peroxisome proliferator-activated receptors (PPARs, e.g., PPARα, PPARβ, PPARγ) play a crucial role in regulating lipid metabolic processes [23,24].

Feeding/meal time can directly activate nutrient-sensing pathways to regulate metabolic processes in organisms and is independent of photoperiod [25,26]. Time-restricted feeding (TRF) is thought to prevent and treat metabolic diseases by maintaining optimal nutrient utilization [25]. In rodents, TRF was found to reduce the total cholesterol, triglyceride, glucose, and insulin levels and improve glucose control and insulin sensitivity [27]. Furthermore, in some metabolic disorder animal models and populations suffering from metabolic diseases such as in mice fed high-fat diets, TRF reduced the body fat accumulation, improved glucose tolerance, and stabilized the circadian rhythm of the central biological clock compared to ad libitum feeding [28–30]. In a biological clock-deficient mouse model, TRF prevented the development of obesity and metabolic syndrome [31]; TRF restored muscle function in a *Drosophila* model of obesity and rhythm disorders [32]. In overweight patients with type 2 diabetes, TRF improved glucose and insulin sensitivity [33]. In addition, in mice, TRF repaired the attenuation of biological clock rhythms in peripheral tissues (liver and white adipose tissue) due to continuous light (LL) [34]. In recent years, several review publications have summarized the positive role of TRF in regulating metabolic homeostasis [26,35,36].

Although TRF has shown extraordinary effects on metabolic regulation in humans and some model organisms, studies in lower vertebrates such as fish are still scarce [37]. Aquaculture is currently the fastest-growing form of food production globally, providing more than 82 million tons of high-quality protein for humans in 2018 [38]. Although there have been several reports focusing on the impact of light pollution on aquatic organisms [39–41], the negative impacts of light pollution on aquatic animals have not received much attention compared to terrestrial animals [42]. In fish, light pollution research has focused on behavior [43–45], community structure [46], physiology [47,48], and fitness [49], with very limited information on metabolism [50,51]. The vast majority of studies have reported that fish melatonin levels were suppressed by ALAN [42,48,52]. Melatonin is an important mediator of the conversion of exogenous temporal signals into endogenous biological rhythms [53], and the disruption of melatonin rhythms in fish by ALAN can also have an impact on lipid metabolism. Previous studies have reported that LL exposure led to an increased lipid content in some fish such as Atlantic salmon (*Salmo salar*) [54,55] and gibel carp (*Carassius auratus*) [50], which may be attributed to disturbed lipid metabolism under LL.

Rainbow trout (*Oncorhynchus mykiss*) is an economically important cold-water fish. As an important representative species of salmonids, it occupies an essential position in aquaculture production, with global aquaculture production reaching 848,051 tons and a value of 3.88 billion USD in 2018 [56]. Rainbow trout is also an important model fish, and there have been several studies on the effects of photoperiod and restricted feeding on metabolism [57,58]. As a representative species of teleost fish, it is necessary to determine the impact of light pollution on its lipid metabolism, and whether TRF can mitigate this state. Thus, this study investigated (1) the effects of ALAN on the body composition and lipid metabolism of juvenile rainbow trout by simulating light pollution with continuous light (LL) and (2) whether TRF could mitigate the negative effects of light pollution through three feeding regimes.

2. Materials and Methods

2.1. Fish

This study was approved by the Animal Care and Use Committee of Ningbo University and carried out in the recirculating aquaculture system (RAS) (HISHING, Qingdao, China) at the School of Marine Sciences pilot plant, Ningbo University from July to November 2020. The juvenile rainbow trout were purchased from a commercial nursery (Shandong, China). Fish were acclimatized for one month in the RAS and randomly fed with a commercial diet (about 2% body weight, Tech-Bank, Ningbo, China, Supplementary Table S1). The feeding schedule was provided by random number generator software (RAND function of Microsoft Excel) according to Nisembaum et al. (2012) [59]. The environmental conditions were maintained as follows: the photoperiod was 12L:12D (lights-on at 6:00, the light intensity on the water surface is 100–200 lx), the water temperature at 16.5 °C $\pm$ 1 °C, the dissolved oxygen was higher than 9 mg/L, and the ammonia nitrogen was lower than 0.05 mg/L.

2.2. Experiment Design

After the acclimation, 840 healthy fish (18.98 $\pm$ 1.69 g/fish) were weighed and randomly assigned to 42 culture tanks (volume 600 L, Supplementary Figure S1), with 20 fish in each tank. The experiment included two photoperiods: 24L:0D (LL, constant light with 100–200 lx light intensity on the water surface) and 12L:12D (LD, lights-on at 6:00, light-off at 18:00, the light intensity on the water surface is 100–200 lx); three feeding regimes: random feeding (R), mid-dark stage feeding (D), and mid-light stage feeding (L). In this study, automatic feeding machines (YF-9258, Fish Baby, Sichuan, China) were used, and the feeding time and quantity were set in advance. A total of six experimental treatments were combined: R-LL, D-LL, L-LL, R-LD, D-LD, and L-LD, and each treatment included seven culture tanks (each treatment included seven replicates). The growth experiment lasted three months, and the cultural management and environmental conditions were consistent with the acclimation.

2.3. Sample Collection

At the end of the three-month growth experiment, all fish were deprived of food for 24 h. To reduce the effect of food anticipatory activity (FAA) on the blood metabolite levels, sampling was started 4 h after the feeding point. That is, all fish were starved for 28 h. MS-222 was used to anesthetize the fish, and 12 fish were randomly selected from each tank for sample collection. Two fish were randomly selected and frozen at −20 °C to determine the body composition. The blood was immediately collected from the remaining 10 fish by the tail vein method. The blood was stored in a 1.5 mL EP tube, left at 4 °C overnight, and then centrifuged at 2500 rpm at 4 °C for 10 min. The supernatant (serum) was stored at −80 °C until analysis of the hematological parameters. The fish was immediately dissected after the blood was taken; the liver, intestine, and dorsal muscle were separated and quick-frozen in liquid nitrogen, then stored at −80 °C until analysis.

2.4. Biochemical Analysis

The biochemical composition of the whole fish, dorsal muscle, and liver was determined according to AOAC (1995) [60]. First, the sample was dried to constant weight by a freeze dryer (LL1500, Thermo Scientific, Waltham, MA, USA), and the reduced mass was the moisture; then, the crude protein content in the sample was determined by a Kjeldahl analyzer (K355/K437, Buchi, Flawil, Switzerland) and the crude lipid in the sample was determined by a Soxhlet extractor (E816, Buchi, Flawil, Switzerland). The ash content of the whole fish was determined by a muffle furnace at 550 °C for 12 h. Total protein (TP), glucose (Glu), triglyceride (TG), lactic acid (LA), total cholesterol (T-CHO), high-density lipoprotein cholesterol (H-DLC), and low-density lipoprotein cholesterol (L-DLC) in serum and glycogen in the muscle and liver were all measured using commercial kits (Nanjing Jiancheng Bioengineering Institute, Nanjing, China) according to the instructions. Fatty acids in the liver, dorsal muscle, and serum were determined by gas chromatography (GC7890B, Agilent Technologies, Santa Clara, CA, USA) as described in Liu et al. (2021) [61].

2.5. Gene Expression

The total RNA of the liver was extracted with a commercial kit RNA isolator (R401-01, Vazyme, Nanjing, China). The quality of total RNA was checked by an ultramicro-spectrophotometer (Nanodrop 2000, Thermo Scientific, Waltham, MA, USA) and 1% gel electrophoresis. The RNA was reverse transcribed into cDNA using a HiFiScript cDNA Synthesis Kit (CW2569M, CWBIO, Beijing, China).

Real-time PCR was used to analyze the relative expression of the lipid metabolism-related genes (fatty acid synthase (*fas*), Δ6-fatty acid desaturase (*Δ6-fad*), Δ9-fatty acid desaturase (*Δ9-fad*), elongation of very long-chain fatty acid protein 2 (elovl2), elongation of very long-chain fatty acid protein 5 (*elovl5*), sterol regulatory element binding protein 1c (*srebp1c*), peroxisome proliferators-activated receptor α (*pparα*), peroxisome proliferators-activated receptor β (*pparβ*), peroxisome proliferators-activated receptor γ (*pparγ*), carnitine palmitoyl transferase 1a (*cpt1a*), lipoprotein lipase (*lpl*), cluster of differentiation 36/fatty acid translocase (*cd36/fat*), and fatty acid transport protein 1 (*fatp1*)) in the liver. The total reaction volume was 20 μL including 10 μL of 2 × MagicSYBR Mixture (CW3008H, CWBIO, Beijing, China), 2 μL of cDNA, 0.4 μL of each primer (10 μM), and 7.2 μL of ddH_2O. A Real-Time PCR System (QuantStudio™ 6 Flex, Life Technologies, Carlsbad, CA, USA) was used with the program as follows: 95 °C for 30 s; 45 cycles at 95 °C for 5 s, 60 °C for 30 s; and 95 °C for 15 s. The specific primers used in this study are shown in Supplementary Table S2 and were synthesized by a commercial company (Youkang Biological Technology Co., Ltd., Hangzhou, China). The relative expression level of the target genes was normalized by *β-actin* and elongation factor-1α (*ef1α*) and calculated by the comparative CT method ($2^{-\Delta\Delta CT}$ method) [62].

2.6. Statistical Analysis

All statistical analyses were performed on SPSS 22.0 and R 4.1.2 software. First, all data were checked for homogeneity and normal distribution through Levene's test and the Kolmogorov–Smirnov test, respectively. Then, a two-way ANOVA was performed with the photoperiod and feeding regime. Meanwhile, t-tests were performed for the photoperiod, and one-way ANOVA followed by Duncan's multiple range test were performed for the feeding regime. $p < 0.05$ and $p < 0.01$ were considered as significant differences and extremely significant differences, respectively. In addition, the principal component analysis (PCA) was used to analyze the liver, dorsal muscle, and serum fatty acid profile. Finally, the structural equation model (SEM), based on partial least squares path modeling (PLS-PM) [63], demonstrates the relationships between the photoperiod, feeding regime, lipid metabolism genes, serum metabolites, and body composition. Data related to growth performance and feed utilization of juvenile rainbow trout were also obtained in this study and have been presented in a separate unpublished paper.

3. Results

3.1. Body Composition

The juvenile rainbow trout whole-body composition was influenced by the photoperiod and feeding regime (Figure 1a and Supplementary Table S3). The LL environment significantly increased whole-body lipid content and decreased whole-body ash content ($p < 0.05$). The whole-body moisture content in the R group was significantly higher than that in the L and D groups, and the whole-body lipid content in the D group was significantly higher than that in the R and L groups ($p < 0.05$). There was no interaction between the photoperiod and feeding regime on the whole-body composition ($p > 0.05$).

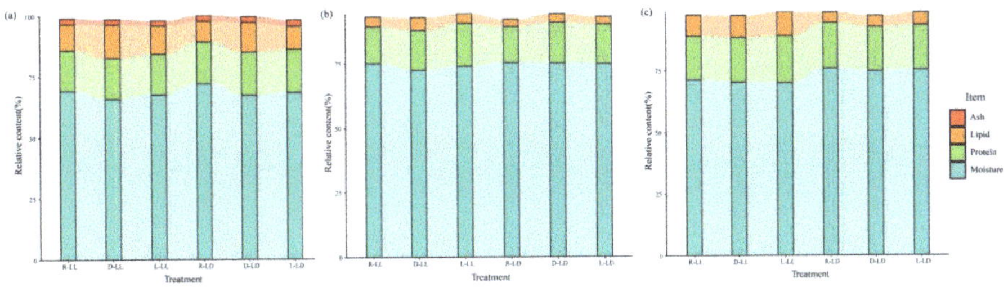

Figure 1. The body composition of juvenile *Oncorhynchus mykiss* under different experimental treatment. (**a**) Whole fish. (**b**) Liver. (**c**) Dorsal muscle.

The juvenile rainbow trout liver composition was influenced by the photoperiod and feeding regime (Figure 1b and Supplementary Table S3). The LL environment significantly reduced the liver moisture content and increased the liver lipid content ($p < 0.05$). The liver protein content in the R group was significantly lower than those in the L and D groups ($p < 0.05$). The liver glycogen content showed R > D > L ($p < 0.05$) (Figure 4h and Supplementary Table S3). Photoperiod and feeding regime had an interactive effect on the liver glycogen content ($p < 0.05$). In the LL environment, the R group liver glycogen content was significantly higher than the D and L groups ($p < 0.05$). In the LD environment, the L group liver glycogen content was significantly lower than the R and D groups ($p < 0.05$). Under the D treatment, the liver glycogen content of the LD group was significantly higher than that of the LL group.

The juvenile rainbow trout dorsal muscle composition was influenced by the photoperiod (Figure 1c and Supplementary Table S3). The LL environment significantly reduced the dorsal muscle moisture content and elevated the dorsal muscle lipid content ($p < 0.05$) and the dorsal muscle glycogen was significantly higher in the L group than in the R and D groups; in the D treatment, the dorsal muscle glycogen was markedly lower in the LL group than in the LD group ($p < 0.05$) (Figure 4i and Supplementary Table S3). There was no interaction between the photoperiod and feeding regime on the dorsal muscle composition ($p > 0.05$).

The liver fatty acid profile was influenced by the photoperiod and feeding regime (Figure 2a and Supplementary Table S4). The LL environment significantly reduced the saturated fatty acid (SFA) proportion; the D group significantly increased the monounsaturated fatty acid (MUFA) proportion and significantly reduced the polyunsaturated fatty acid (PUFA) and long-chain polyunsaturated fatty acid (LC-PUFA) proportions compared to the R and L groups ($p < 0.05$). Photoperiod and feeding regime had an interactive effect on the liver SFA and PUFA proportions ($p < 0.05$). The results of the principal component analysis (PCA) revealed that the liver fatty acid profile in group D (D-LL and D-LD) was distinguished from the other groups (Figure 2b,c).

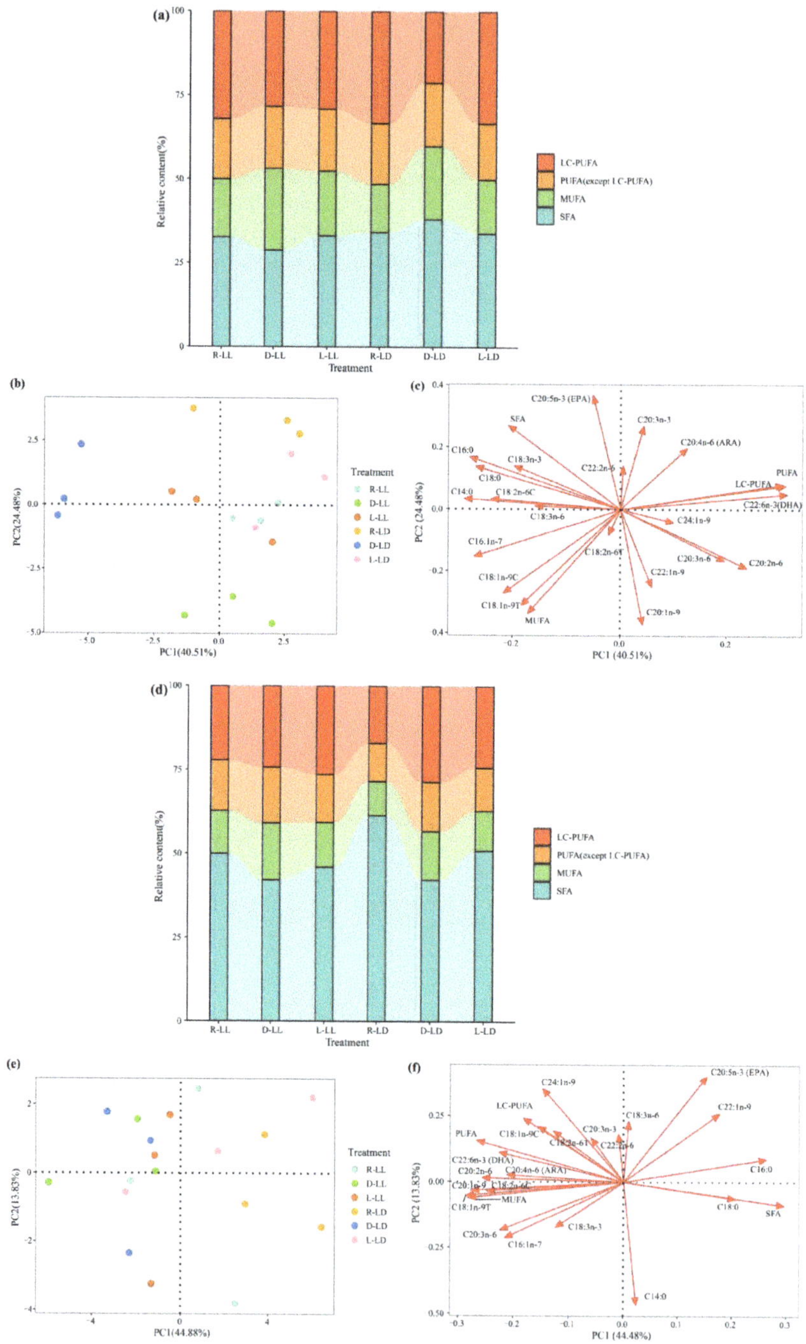

Figure 2. *Cont.*

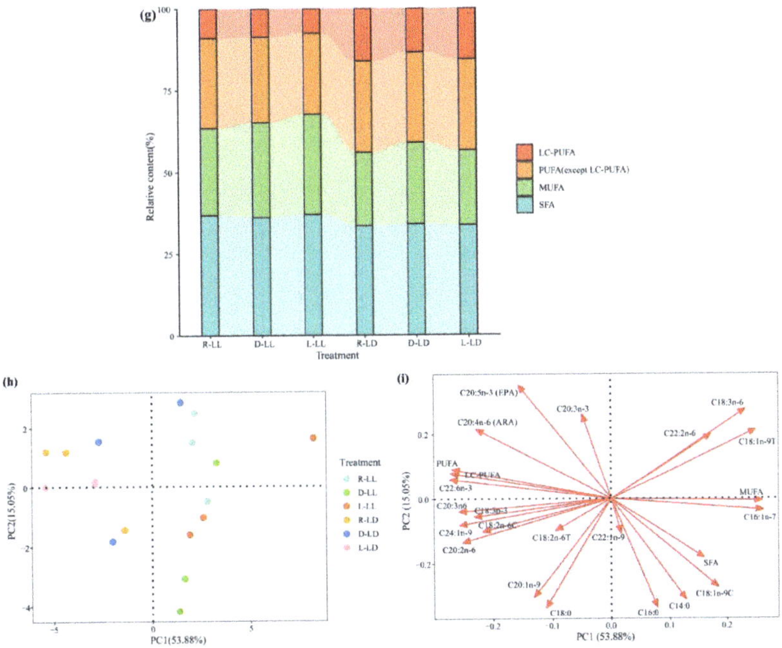

Figure 2. Fatty acids profile of juvenile *Oncorhynchus mykiss* under different experimental treatments. (**a**) Proportion of liver fatty acids; (**b**) PCA score plot of liver fatty acids; (**c**) PCA loading plot of liver fatty acids. (**d**) Proportion of serum fatty acids; (**e**) PCA score plot of serum fatty acids; (**f**) PCA loading plot of serum fatty acids. (**g**) Proportion of dorsal muscle fatty acids; (**h**) PCA score plot of dorsal muscle fatty acids; (**i**) PCA loading plot of dorsal muscle fatty acids.

The serum fatty acids profile was mainly influenced by the feeding regime (Figure 2d and Supplementary Table S5). In the R group, the SFA proportion was significantly higher than that in the D group, the MUFA proportion was significantly lower than that in the D group, and the PUFA and LC-PUFA proportions were markedly lower than those in the D and L groups ($p < 0.05$). The PCA results revealed that the serum fatty acid profiles of the different treatments were not significantly separated (Figure 2e,f).

The dorsal muscle fatty acid profile was mainly influenced by the photoperiod (Figure 2g and Supplementary Table S6). The LL environment significantly increased the SFA and MUFA proportions and decreased the PUFA and LC-PUFA proportions ($p < 0.05$). The PCA results revealed that different photoperiods (LL vs. LD) were able to separate the fatty acid profiles of the dorsal muscle (Figure 2h,i).

3.2. Lipid Metabolism Genes

In the present study, lipid metabolism-related genes were influenced by the photoperiod and feeding regime (Figure 3 and Supplementary Table S7). Lipid metabolism genes were simply classified into three major classes according to their functions: lipid synthesis and deposition, lipolysis and oxidation, and lipid transport.

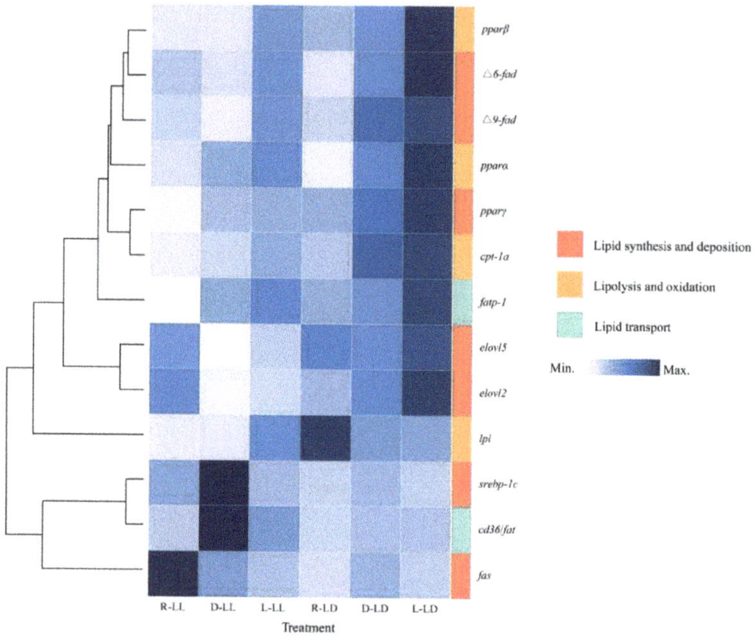

Figure 3. Gene expression of juvenile *Oncorhynchus mykiss* under different experimental treatments. *fas*: Fatty acid synthase, Δ6-*fad*: Δ6-fatty acid desaturase, Δ9-*fad*: Δ9-fatty acid desaturase, *elovl2*: Elongation of very long-chain fatty acid protein 2, *elovl5*: Elongation of very long-chain fatty acid protein 5, *srebp-1c*: Sterol regulatory element binding protein 1c, *pparα*: Peroxisome proliferators-activated receptor α, *pparβ*: Peroxisome proliferators-activated receptor β, *pparγ*: Peroxisome proliferators-activated receptor γ, *cpt-1a*: Carnitine palmitoyl transferase 1a, *lpl*: Lipoprotein lipase, *cd36/fat*: Cluster of differentiation 36/Fatty acid translocase, *fatp-1*: Fatty acid transport protein 1.

For lipid synthesis and deposition, the *fas*, Δ9-*fad*, *elovl2*, *elovl5*, *srebp-1c*, and *pparγ* genes' expression were affected by the photoperiod (Figure 3 and Supplementary Table S7). The LL environment resulted in the upregulation of *fas* and *srebp-1c* gene expression and the downregulation of the expression of the Δ9-*fad*, *elovl2*, *elovl5*, and *pparγ* genes. The feeding regime affected the expression of the *fas*, Δ6-*fad*, Δ9-*fad*, *srebp-1c*, and *pparγ* genes. The L group upregulated Δ6-*fad* and Δ9-*fad* gene expression, the D group upregulated *srebp-1c* gene expression, and the R group upregulated *pparγ* gene expression. There was an interactive effect of photoperiod and feeding strategy on *fas*, Δ6-*fad*, Δ9-*fad*, and *elovl2* gene expression.

For lipolysis and oxidation, *pparβ*, *cpt-1a*, and *lpl* gene expression was affected by the photoperiod (Figure 3 and Supplementary Table S7). The LL environment downregulated the expression of these three genes. The feeding regime affected the expression of the *pparα*, *pparβ*, and *cpt-1a* genes. The R group downregulated *pparα* gene expression, the L group upregulated *pparβ* gene expression, and the R group downregulated *cpt-1a* gene expression. There was an interactive effect of photoperiod and feeding regime on *cpt-1a* and *lpl* gene expression.

For lipid transport, *cd36/fat* and *fatp-1* gene expression was influenced by the photoperiod (Figure 3 and Supplementary Table S7). The LL environment upregulated *cd36/fat* gene expression and downregulated *fatp-1* gene expression. The feeding regime affected the expression of the *cd36/fat* and *fatp-1* genes. The expression of the *cd36/fat* gene was upregulated in group D. The trend of *fatp-1* gene expression can be shown as R < D < L.

3.3. Serum Metabolites

Serum glucose (Glu), triglyceride (TG), total cholesterol (T-CHO), lactic acid (LA), and high-density lipoprotein cholesterol (H-DLC) were affected by the photoperiod and feeding regime (Figure 4a–e and Supplementary Table S3). The LL environment significantly reduced the LA levels ($p < 0.05$). The R group had a significantly higher Glu than the D and L groups, and the R group had significantly lower TG, LA, and H-DLC than the D group ($p < 0.05$). Individually, under the LD environment, TG was significantly lower in the R group than in the D and L groups. Under the LL environment, T-CHO showed R < D < L while under the L treatment, the LL environment significantly increased the H-DLC levels ($p < 0.05$).

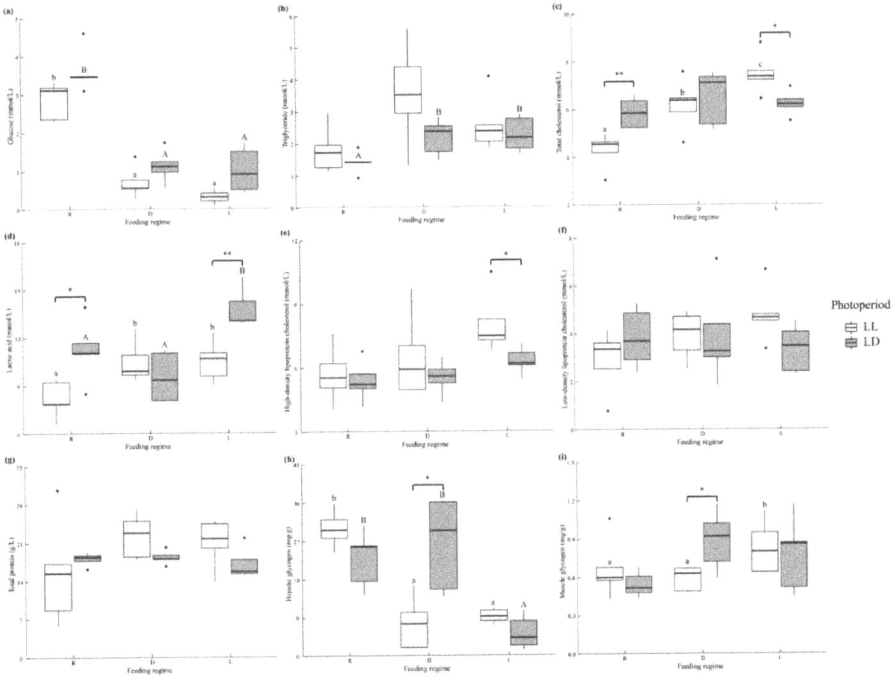

Figure 4. Serum metabolites and glycogen of juvenile *Oncorhynchus mykiss* under different experimental treatments (mean ± SD, $n = 6$). (**a**) Glucose; (**b**) triglyceride; (**c**) total cholesterol (**d**) lactic acid; (**e**) high-density lipoprotein cholesterol; (**f**) low-density lipoprotein cholesterol; (**g**) total protein; (**h**) hepatic glycogen; (**i**) muscle glycogen. Different lowercase letters and capital letters indicate a significant difference among the different feeding strategy at the LL (constant light) and LD (12L: 12D, lights-on at 6:00), respectively ($p < 0.05$). Asterisks denote significant differences between photoperiods at the same feeding strategy (* $p < 0.05$; ** $p < 0.01$).

3.4. Structural Equation Model

The structural equation model (SEM) based on partial least squares path modeling (PLS-PM) demonstrates the relationships between the photoperiod, feeding regime, lipid metabolism genes, serum metabolites, and body composition (Figure 5). The observed variables in this study were classified into the above five latent variables, and those with a loading value < 0.7 were excluded (Supplementary Figure S2) [63]. Photoperiod was significantly negatively correlated with lipid metabolism genes (path coefficients (PC) = −0.6027, $p < 0.01$) and serum metabolites (PC = −0.3925, $p < 0.05$), and was positively correlated with body composition (PC = 0.9750, $p < 0.01$). Feeding regime was significantly positively correlated with lipid metabolism genes (PC = 0.7363, $p < 0.01$) and serum metabolites

(PC = 1.7234, $p < 0.01$), and negatively correlated with body composition (PC = −0.7069, $p = 0.215$). Lipid metabolism genes were significantly negatively correlated with serum metabolites (PC = −1.2794, $p < 0.01$) and lipid metabolism genes (PC = 0.5055, $p = 0.294$) and serum metabolites (PC = 0.6361, $p < 0.05$) were positively correlated with body composition. The goodness of fit (GOF) of the model was 0.8368.

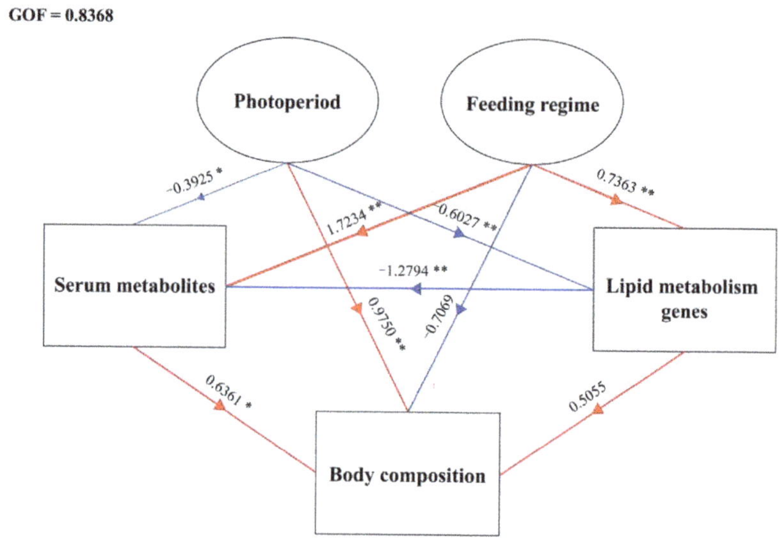

Figure 5. The structural equation model (SEM) based on partial least squares path modeling (PLS-PM) demonstrates the relationships between the photoperiod, feeding regime, digestive enzyme, lipid metabolism genes, serum hormones, serum metabolites, and body composition. Numbers on arrows are path coefficients. * means $p < 0.05$, ** means $p < 0.01$. Arrow widths show the strength of the causal relationship. Red arrows indicate positive correlation, blue arrows indicate negative correlation.

4. Discussion

In the present study, LL decreased the moisture content in the liver and dorsal muscle and increased the lipid content in the whole-body, liver, and dorsal muscle. Similarly, LL increased the whole-body lipid content of Atlantic salmon (*Salmo salar*) [55]. In gibel carp (*Carassius auratus*), the lipid content in the whole-body, liver, and muscle gradually increased with prolonged light exposure [50]. An interesting phenomenon was observed that LL led to a decrease in the ash content of the whole-body. Some studies in Atlantic salmon have found that LL decreases the bone mineral content [64], affects mineralization, and delays osteoid incorporation [65], and even causes vertebrae malformations [66]. Similarly, LL caused a higher lower jaw malformation in European sea bass (*Dicentrarchus labrax*) [67]. In gilthead seabream (*Sparus aurata*), Mhalhel et al. (2020) [68] found that exogenous melatonin supplementation affected normal skeletogenesis and caused bone deformities by regulating the expression of genes related to bone formation. Fish receive external light signals and convert environmental time cues into endogenous biological signals through melatonin secretion [53]. Therefore, the increased bone deformity caused by LL may be related to the loss of coordination between the skeletal muscle and bone function caused by the disturbance of the circadian clock [68]. Although no bone deformities were observed in juvenile rainbow trout in the LL group, the significantly low whole-body ash and higher lipid deposition may be a warning sign of concern in the metabolic disturbance.

Aside from light, the feeding regime also affected the rainbow trout's body composition. The R group fish had a higher whole-body moisture, lower whole-body lipid, and lower liver protein, which may reflect the poor nutritional status and feed utilization. This result is not surprising because random feeding could not induce food anticipatory activity (FAA) [69–71], thus failing to increase locomotor activity and optimize digestive and metabolic processes before feeding [72–74]. In addition, the serum Glu level in the R group was significantly higher than in the TRF group. Similar results have also been reported in gilthead seabream [74,75]. Sánchez et al. (2009) [75] concluded that random feeding stressed the fish, increasing the plasma Glu levels. The animal can store excess glucose in the form of glycogen in the liver and muscle [76]. In fish, hepatic glycogen synthesis is one of the metabolic pathways of blood glucose, and elevated blood glucose levels often cause a simultaneous increase in hepatic glycogen content [77]. Similarly, significantly higher liver glycogen was also observed in the R group. High hepatic glycogen levels reflect the passive adaptation of juvenile rainbow trout to high blood glucose. However, as a carnivorous fish, rainbow trout have a minimal ability to utilize glucose, so high glycogen levels may become a metabolic burden. On the other hand, random feeding may encourage fish to store more glycogen in response to possible food deficiencies.

However, contrary to expectations, TRF did not alleviate the whole-body, liver, and muscle lipid deposition caused by the LL. Fish body lipid content results from the balance between lipid synthesis and catabolism. Previous studies have attributed the elevated body lipid content to the following four conditions: (1) stable lipolysis and increased lipogenesis; (2) decreased lipolysis and increased lipogenesis; (3) decreased lipolysis and stable lipogenesis; and (4) slightly increased lipolysis and vastly increased lipogenesis [78]. The liver is the central organ of lipid metabolism, and the expression of lipid metabolism-related genes in the liver reflects the lipid metabolism of juvenile rainbow trout. In the present study, LL upregulated the expression of *fas* and *srebp-1c* genes and downregulated the expression of *pparα*, *pparβ*, *cpt-1a*, and *lpl* genes. Based on the above results, the rise in the lipid content of juvenile rainbow trout caused by LL may be due to (2) or (3). In addition, *cd36/fat* gene expression was significantly upregulated, and *fatp-1* gene expression was significantly downregulated in the LL group. Studies in mice have found that CD36/FAT protein promotes the intestinal absorption of fatty acids [21], and the upregulation of *cd36/fat* expression in the liver is associated with hepatic TG accumulation, elevated serum TG, and obesity [79]. In this study, the serum TG and H-DLC levels were also increased by LL. The *fatp-1* gene is involved in the uptake and oxidation of long-chain fatty acids [22] and its expression is upregulated in the liver by the upstream gene *pparα* [80]. The present study also observed the same expression trend of the *pparα* and *fatp-1* genes.

Notably, under the LL environment, TRF (especially in the L group) downregulated the mRNA abundance of *fas* and *srebp-1c* genes and upregulated the mRNA abundance of the *pparα*, *pparβ*, *cpt-1a*, *lpl*, and *fatp-1* genes. Similarly in mice livers, TRF was found to downregulate the expression of srebp-1c genes and upregulate the expression of *cpt-1a* genes [81]. Theoretically, lipid synthesis was inhibited in the TRF group, while oxidative catabolism was enhanced, which should lead to a decrease in lipid content. However, TRF also induced FAA (unpublished data), which led to an enhanced ability of rainbow trout to obtain lipids from the diet. In serum, the TG, T-CHO, and H-DLC levels in the TRF group were higher than those in the R group, reflecting a more active lipid metabolism and better nutritional status [82], thus supporting that the increased lipid absorption and transport level contributed to the body lipid deposition. Thus, in the RTF group, the uptake and utilization of exogenous lipids were probably prioritized over the synthesis and catabolism of endogenous lipids, which ultimately did not alleviate the lipid deposition caused by LL. On the other hand, although TRF corrected the attenuation of peripheral biological clock oscillations caused by LL [34] and alleviated the development of metabolic disease in a biological clock defect/metabolic disorder model in mammals [30,31], the relationship between the central and peripheral biological clocks in fish is more complex, and there does not seem to be a strict hierarchy between the central and peripheral biological clocks [83], which may be one of the reasons why TRF failed to alter the lipid deposition in the present study.

Though TRF did not alter the lipid deposition caused by LL, the resting period feeding still had a significantly different effect on the fish. The whole-body and the liver lipids were significantly higher in the D group than in the R and L groups in the LD photoperiod. In the wild, food is not continuously available, and fish exhibit different feeding rhythms to balance food availability and the occurrence of predators [84]. Fish metabolism is regulated by central and peripheral biological clocks (synchronized by light and food, respectively) [85]. Rainbow trout are diurnal feeders [86–88], and midnight feeding decouples the synchronization of the central and peripheral biological clocks, disrupts the metabolic process, and leads to lipid deposition and obesity [89]. Interestingly, it is noteworthy that the midnight feeding did not have the same effect in LL. This result suggests that the light–dark cycle may be a prerequisite for resting period feeding to induce lipid deposition, or that strict active and resting periods may not exist when light zeitgeber is absent. In the present study, compared with the L group, the D group had a higher expression of the *fas*, *srebp-1c*, and cd36/fat genes and a lower expression of the *pparα*, *pparβ*, and *cpt-1a* genes. These results suggest that lipid accumulation in the D group may be attributable to the upregulation of lipid synthesis and uptake. On the other hand, the upregulation of lipid oxidative catabolism in the L group may be associated with higher energy requirements for more frequent activity. The enhanced locomotor activity by TRF has also been widely reported in fishes [90,91]. In addition, higher muscle glycogen was observed in the L group. Rainbow trout muscles have been shown to have the ability to synthesize glycogen in situ [92], so high muscle glycogen may be an adaptation to the high energy demand of frequent locomotion in the L group. Muscle glycogen is not directly catabolized for energy but needs to be catabolized into LA and transported through the blood circulation to the liver for metabolism. A high serum LA level was detected in the L group.

Furthermore, this study also analyzed the fatty acid composition in the liver, serum, and muscle of juvenile rainbow trout. As the central organ of lipid metabolism, the liver reflects the metabolic process of fatty acid synthesis and decomposition. Fatty acids are the favored source of metabolic energy in fish [93]. Tocher (2003) [94] concluded that the priority of fatty acid oxidation for energy supply in most fish was SFA > MUFA > PUFA > LC-PUFA. In this study, the liver's SFA (mainly C16:0) proportion was lower in the LL group than in the LD group, indicating an enhanced energy metabolism due to ALAN [95], as C16:0 is preferentially used for energy consumption when subjected to LL stress [96,97]. Furthermore, interestingly, the MUFA proportion in the LL group was higher than that in the LD group, but the PUFA and LC-PUFA proportions did not differ between the two photoperiods, suggesting that the excess MUFA in the LL group did not further synthesize PUFA and LC-PUFA. MUFA can be further synthesized into PUFA and LC-PUFA by elongase and desaturase [93]. The relative expression levels of the *Δ6-fad*, *Δ9-fad*, *elovl2*, and *elovl5* genes in the LL group were lower than those in the LD group. These results imply that LL may hinder the synthesis of PUFA and LC-PUFA by inhibiting the expression of elongase and desaturase genes. In Atlantic salmon, Nemova (2021) [54] found that LL increased the LC-PUFA (EPA and DHA) levels and 16:0/18:1n9 ratio, which was considered as preparation for smoltification. However, this phenomenon was not observed in the present study, probably because rainbow trout are landlocked salmonids.

Regarding the feeding regime, the TRF groups had higher MUFA and lower PUFA and LC-PUFA (especially in the D group) compared with the R group. The gene expression analysis of the PUFA-related synthases showed that the expression levels of the *Δ6-fad* and *Δ9-fad* genes in the TRF groups were higher than in the R group. In contrast, the feeding regime did not affect the relative expression of the *elovl2* and *elovl5* genes, implying that the expression patterns of the desaturase and elongase genes were different in rainbow trout. Furthermore, an interaction between the feeding regime and photoperiod on the fatty acid profile of rainbow trout liver was observed. Under the LD photoperiod, the proportion of PUFA was significantly lower in the D group than in the R and L groups. However, there was no statistical difference in the proportion of PUFA under the LL, and the proportion of PUFA in the D-LD group was also significantly lower than that in the D-LL group. First, compared with the L group, the expression levels of the *Δ6-fad*, *Δ9-fad*, *elovl2*, and *elovl5* genes were lower in the D group, indicating that mismatched feeding may inhibit the synthesis of PUFA in rainbow trout. Similarly, mismatched feeding resulted in the downregulation of the expression of the PUFA synthesis gene (*elovl6*) in darkbarbel catfish (*Pelteobagrus vachellii*) [98]. In addition, the cd36/fat gene expression was significantly higher in the D-LL group than in the D-LD group, indicating that LL promoted fatty acid uptake in the D-LL group. These phenomena eventually led to a significantly lower PUFA proportion observed only in the D-LD group. The results of liver fatty acid PCA analysis also clearly distinguished the D-LD group from the other groups.

The fatty acid profile in serum reflects the transport of fatty acids, and unsurprisingly, feeding regime caused extensive changes, with little contribution from the photoperiod. Briefly, the TRF groups had lower SFA and relatively higher MUFA, PUFA, and LC-PUFA than the R group. Additionally, a higher expression of the *fatp1* gene was detected in the TRF group. Compared with SFA, PUFA and LC-PUFA play a more important role in fish [93]. The present study results suggest that TRF facilitated the absorption and transport of these critical fatty acids and that FAA induced by TRF may play an important role in this process. In contrast to the serum, the fatty acid composition in muscle was mainly affected by the photoperiod and was almost independent of the feeding regime. Compared to the LD group, the LL group had a higher SFA, MUFA, and lower PUFA. A previous study suggested that highly unsaturated fatty acids (HUFA) (mainly EPA and DHA) have anti-stress effects in fish [99]. In the present study, LL exposure may have caused stress in rainbow trout [97], leading to a decrease in the proportion of HUFA in muscle. On the other hand, it was found that rainbow trout [100] and Mesopotamian catfish (*Silurus triostegus*) [101] had a higher MUFA and lower PUFA in summer (long daylight) than in winter (short daylight). Xie et al. (2013) [99] suggested that seasons affect fatty acids mainly related to changes in the photoperiod and temperature. The differential effects of photoperiod and feeding regime on the fatty acid profiles of different tissues reflect the functional variability between tissues. It has also been shown that the target tissues of light and food (as zeitgeber) are different in fish [90]. Further studies are needed to elucidate the molecular mechanisms underlying the differential effects of light and food on the fatty acid profiles of different tissues.

5. Conclusions

The present study investigated (1) the effects of ALAN on the body composition and lipid metabolism of juvenile rainbow trout (*Oncorhynchus mykiss*) through a LL environment and (2) whether the negative effects could be alleviated through TRF. The results showed that ALAN strongly impacted the lipid metabolism in juvenile rainbow trout, increased lipid synthesis, and decreased lipid oxidative catabolism, leading to lipid accumulation and a significant decrease in LC-PUFA proportion in the muscle. Unexpectedly, TRF could not alleviate rainbow trout lipid deposition caused by ALAN. Even subjective nocturnal feeding (D-LL) tended to exacerbate lipid deposition. Unlike the strict subordination in mammals, the independent relationship between the central and peripheral biological clocks in fish could be an important reason as to why TRF does not work. Further studies are needed to elucidate the mechanism behind this.

Supplementary Materials: The following supporting information can be downloaded at: https://www.mdpi.com/article/10.3390/metabo12100904/s1, Figure S1: Schematic diagram of the culture tank, the left side is the side view, and the right side is the top view with the top cover removed (1. Feeding port, 2. LED full-spectrum light, 3. Shading cloth, 4. Top cover, 5. Water outlet, 6. water outlet pipe, 7. water inlet pipe, 8. water inlet, 9. blue tank wall, 10. beam for fixing LED full spectrum light); Figure S2: Correlation between latent and observed variables. Numbers on arrows are loading values (The loading value < 0.7 are excluded). *fatp-1*: Fatty acid transport protein 1, *cpt-1a*: Carnitine palmitoyl transferase 1a, *pparβ*: Peroxisome proliferators-activated receptor β, *pparα*: Peroxisome proliferators-activated receptor α, *pparγ*: Peroxisome proliferators-activated receptor γ, *elovl2*: Elongation of very long-chain fatty acid protein 2, Δ*9-fad*: Δ9-fatty acid desaturase, Δ*6-fad*: Δ6-fatty acid desaturase, TP: Total protein, L-DLC: Low-density lipoprotein cholesterol, H-DLC: High-density lipoprotein cholesterol, T-CHO: Total cholesterol, TG: Triglyceride, Glu: Glucose, M-LCPUFA: Dorsal muscle long-chain polyunsaturated fatty acid, M-PUFA: Dorsal muscle polyunsaturated fatty acid, M-MUFA: Dorsal muscle monounsaturated fatty acid, M-Lipid: Dorsal muscle lipid, M-Moisture: Dorsal muscle moisture, L-Lipid, Liver lipid. Table S1: The main nutrients of the commercial diet used in this experiment; Table S2: The specific primers used for real-time PCR in this study; Table S3: Proximate composition of juvenile Oncorhynchus mykiss under different experimental treatment; Table S4: Liver fatty acids composition of juvenile Oncorhynchus mykiss under different experimental treatment; Table S5: Serum fatty acids composition of juvenile Oncorhynchus mykiss under different experimental treatment; Table S6: Dorsal muscle fatty acids composition of juvenile Oncorhynchus mykiss under different experimental treatment; Table S7: Gene expression of juvenile Oncorhynchus mykiss under different experimental treatment. References [102–109] cited in the supplementary materials.

Author Contributions: Conceptualization, C.S. (Ce Shi) and H.X.; methodology, H.X.; formal analysis, H.X.; resources, C.S. (Ce Shi); in-vestigation, H.X.; writing—original draft preparation, H.X.; writing—review and editing, H.X., C.S. (Ce Shi), Y.Y., C.S. (Changbin Song), C.M. and C.W.; supervision, C.S. and C.W.; project administration, C.W.; funding acquisition, C.S. (Ce Shi). All authors have read and agreed to the published version of the manuscript.

Funding: This study was supported by the National Key Research and Development Program of China (Project No. 2019YFD0901000), K.C. Wong Magna Fund in Ningbo University.

Institutional Review Board Statement: The committee on the Ethics of Animal Experiments of Ningbo University (No. SYXK20190005) was established in 2019, but the committee authorities were only for rabbits, mice, and rats, not including aquatic animals. In this study, juvenile rainbow trout were purchased from commercial farms, and all experimental operations involving animals complied with the requirements of the governing regulation for the use of experimental animals in Zhejiang Province (Zhejiang provincial government order No. 263, released on 17 August 2009, effective from 1 October 2010), were performed according to the Experimental Animal Management Law of China, and supervised by the Animal Ethics Committee of the School of Marine Science, Ningbo University.

Informed Consent Statement: Not applicable.

Data Availability Statement: The data presented in this study are not publicly available but are available upon request from the corresponding author.

Acknowledgments: The authors also thank Herve Migaud for his constructive comments on this research.

Conflicts of Interest: The authors declare no conflict of interest.

References

1. Cinzano, P.; Falchi, F.; Elvidge, C.D. The first World Atlas of the artificial night sky brightness. *Mon. Not. R. Astron. Soc.* **2001**, *328*, 689–707. [CrossRef]
2. Longcore, T.; Rich, C. Ecological light pollution. *Front. Ecol. Environ.* **2004**, *2*, 191–198. [CrossRef]
3. Tancredi, S.; Urbano, T.; Vinceti, M.; Filippini, T. Artificial light at night and risk of mental disorders: A systematic review. *Sci. Total Environ.* **2022**, *833*, 155185. [CrossRef] [PubMed]
4. Foster, R.G.; Helfrich-Förster, C. The regulation of circadian clocks by light in fruitflies and mice. *Philos. Trans. R. Soc. B-Biol. Sci.* **2001**, *356*, 1779–1789. [CrossRef] [PubMed]
5. Ma, D.; Li, S.M.; Molusky, M.M.; Lin, J.D. Circadian autophagy rhythm: A link between clock and metabolism? *Trends Endocrinol. Metab.* **2012**, *23*, 319–325. [CrossRef] [PubMed]
6. Lai, K.Y.; Sarkar, C.; Ni, M.Y.; Gallacher, J.; Webster, C. Exposure to light at night (LAN) and risk of obesity: A systematic review and meta-analysis of observational studies. *Environ. Res.* **2020**, *187*, 109637. [CrossRef]
7. Obayashi, K.; Yamagami, Y.; Kurumatani, N.; Saeki, K. Bedroom lighting environment and incident diabetes mellitus: A longitudinal study of the HEIJO-KYO cohort. *Sleep Med.* **2020**, *65*, 1–3. [CrossRef] [PubMed]
8. Huang, W.Y.; Ramsey, K.M.; Marcheva, B.; Bass, J. Circadian rhythms, sleep, and metabolism. *J. Clin. Investig.* **2011**, *121*, 2133–2141. [CrossRef] [PubMed]
9. Fleury, G.; Masis-Vargas, A.; Kalsbeek, A. Metabolic Implications of Exposure to Light at Night: Lessons from Animal and Human Studies. *Obesity* **2020**, *28*, S18–S28. [CrossRef]
10. Watanabe, T. Lipid nutrition in fish. *Comp. Biochem. Physiol. Part B Comp. Biochem.* **1982**, *73*, 3–15. [CrossRef]
11. Xu, H.Y.; Han, T.; Li, X.Y.; Wang, J.T.; Zheng, P.Q.; Yin, F.; Wang, C.L. Effects of dietary lipid levels on survival, growth performance, and antioxidant ability of the early juvenile *Scylla paramamosain*. *Aquaculture* **2020**, *528*, 735559. [CrossRef]
12. Wilson, R.P. Utilization of dietary carbohydrate by fish. *Aquaculture* **1994**, *124*, 67–80. [CrossRef]
13. Stone, D.A.J. Dietary carbohydrate utilization by fish. *Rev. Fish. Sci.* **2003**, *11*, 337–369. [CrossRef]
14. Greene, D.H.; Selivonchick, D.P. Lipid metabolism in fish. *Prog. Lipid Res.* **1987**, *26*, 53–85. [CrossRef]

15. Turchini, G.M.; Francis, D.S.; Du, Z.Y.; Olsen, R.E.; Ringø, E.; Tocher, D.R. The lipids. In *Fish Nutrition*, 4th ed.; Hardy, R.W., Kaushik, S.J., Eds.; Academic Press: Cambridge, MA, USA, 2022; pp. 303–467. [CrossRef]
16. Shimano, H. Sterol Regulatory Element-binding Protein-1 as a Dominant Transcription Factor for Gene Regulation of Lipogenic Enzymes in the Liver. *Trends Cardiovasc. Med.* **2000**, *10*, 275–278. [CrossRef]
17. Pai, W.Y.; Hsu, C.C.; Lai, C.Y.; Chang, T.Z.; Tsai, Y.L.; Her, G.M. Cannabinoid receptor 1 promotes hepatic lipid accumulation and lipotoxicity through the induction of SREBP-1c expression in zebrafish. *Transgenic Res.* **2013**, *22*, 823–838. [CrossRef] [PubMed]
18. Nilsson-Ehle, P.; Garfinkel, A.S.; Schotz, M.C. Lipolytic enzymes and plasma lipoprotein metabolism. *Annu. Rev. Biochem.* **1980**, *49*, 667–693. [CrossRef]
19. Wang, L.; Kaneko, G.; Takahashi, S.I.; Watabe, S.; Ushio, H. Identification and gene expression profile analysis of a major type of lipoprotein lipase in adult medaka *Oryzias latipes*. *Fish. Sci.* **2015**, *81*, 163–173. [CrossRef]
20. Wang, C.C.; Si, L.F.; Li, W.Y.; Zheng, J.L. A functional gene encoding carnitine palmitoyltransferase 1 and its transcriptional and kinetic regulation during fasting in large yellow croaker. *Comp. Biochem. Physiol. B-Biochem. Mol. Biol.* **2019**, *231*, 26–33. [CrossRef] [PubMed]
21. Nassir, F.; Wilson, B.; Han, X.L.; Gross, R.W.; Abumrad, N.A. CD36 is important for fatty acid and cholesterol uptake by the proximal but not distal intestine. *J. Biol. Chem.* **2007**, *282*, 19493–19501. [CrossRef]
22. Stahl, A. A current review of fatty acid transport proteins (SLC27). *Pflug. Arch.-Eur. J. Physiol.* **2004**, *447*, 722–727. [CrossRef] [PubMed]
23. Schoonjans, K.; Staels, B.; Auwerx, J. Role of the peroxisome proliferator-activated receptor (PPAR) in mediating the effects of fibrates and fatty acids on gene expression. *J. Lipid Res.* **1996**, *37*, 907–925. [CrossRef]
24. Tsai, M.L.; Chen, H.Y.; Tseng, M.C.; Chang, R.C. Cloning of peroxisorne proliferators activated receptors in the cobia (*Rachycentron canadum*) and their expression at different life-cycle stages under cage aquaculture. *Gene* **2008**, *425*, 69–78. [CrossRef] [PubMed]
25. Panda, S. Circadian physiology of metabolism. *Science* **2016**, *354*, 1008–1015. [CrossRef] [PubMed]
26. Zeb, F.; Wu, X.Y.; Fatima, S.; Zaman, M.H.; Khan, S.A.; Safdar, M.; Alam, I.; Feng, Q. Time-restricted feeding regulates molecular mechanisms with involvement of circadian rhythm to prevent metabolic diseases. *Nutrition* **2021**, *89*, 111244. [CrossRef] [PubMed]
27. Chaix, A.; Manoogian, E.N.C.; Melkani, G.C.; Panda, S. Time-restricted eating to prevent and manage chronic metabolic diseases. *Annu. Rev. Nutr.* **2019**, *39*, 291–315. [CrossRef]
28. Hatori, M.; Vollmers, C.; Zarrinpar, A.; DiTacchio, L.; Bushong, E.A.; Gill, S.; Leblanc, M.; Chaix, A.; Joens, M.; Fitzpatrick, J.A.J.; et al. Time-restricted feeding without reducing caloric intake prevents metabolic diseases in mice fed a high-fat diet. *Cell Metab.* **2012**, *15*, 848–860. [CrossRef] [PubMed]
29. Chaix, A.; Zarrinpar, A.; Miu, P.; Panda, S. Time-restricted feeding is a preventative and therapeutic intervention against diverse nutritional challenges. *Cell Metab.* **2014**, *20*, 991–1005. [CrossRef] [PubMed]
30. Ye, Y.Q.; Xu, H.P.; Xie, Z.B.; Wang, L.; Sun, Y.N.; Yang, H.Y.; Hu, D.D.; Mao, Y.L. Time-restricted feeding reduces the detrimental effects of a high-fat diet, possibly by modulating the circadian rhythm of hepatic lipid metabolism and gut microbiota. *Front. Nutr.* **2020**, *7*, 596285. [CrossRef] [PubMed]
31. Chaix, A.; Lin, T.; Le, H.D.; Chang, M.W.; Panda, S. Time-restricted feeding prevents obesity and metabolic syndrome in mice lacking a circadian clock. *Cell Metab.* **2019**, *29*, 303–319. [CrossRef] [PubMed]
32. Villanueva, J.E.; Livelo, C.; Trujillo, A.S.; Chandran, S.; Woodworth, B.; Andrade, L.; Le, H.D.; Manor, U.; Panda, S.; Melkani, G.C. Time-restricted feeding restores muscle function in Drosophila models of obesity and circadian-rhythm disruption. *Nat. Commun.* **2019**, *10*, 2700. [CrossRef] [PubMed]
33. Che, T.T.; Yan, C.; Tian, D.Y.; Zhang, X.; Liu, X.J.; Wu, Z.M. Time-restricted feeding improves blood glucose and insulin sensitivity in overweight patients with type 2 diabetes: A randomised controlled trial. *Nutr. Metab.* **2021**, *18*, 88. [CrossRef]
34. Yamamuro, D.; Takahashi, M.; Nagashima, S.; Wakabayashi, T.; Yamazaki, H.; Takei, A.; Takei, S.; Sakai, K.; Ebihara, K.; Iwasaki, Y.; et al. Peripheral circadian rhythms in the liver and white adipose tissue of mice are attenuated by constant light and restored by time-restricted feeding. *PLoS ONE* **2020**, *15*, e0234439. [CrossRef]
35. Flanagan, A.; Bechtold, D.A.; Pot, G.K.; Johnston, J.D. Chrono-nutrition: From molecular and neuronal mechanisms to human epidemiology and timed feeding patterns. *J. Neurochem.* **2021**, *157*, 53–72. [CrossRef] [PubMed]
36. Kang, J.; Ratamess, N.A.; Faigenbaum, A.D.; Bush, J.A.; Beller, N.; Vargas, A.; Fardman, B.; Andriopoulos, T. Effect of Time-Restricted Feeding on Anthropometric, Metabolic, and Fitness Parameters: A Systematic Review. *J. Am. Coll. Nutr.* **2021**, 1–16. [CrossRef] [PubMed]
37. Paredes, J.F.; López-Olmeda, J.F.; Martínez, F.J.; Sánchez-Vázquez, F.J. Daily rhythms of lipid metabolic gene expression in zebra fish liver: Response to light/dark and feeding cycles. *Chronobiol. Int.* **2015**, *32*, 1438–1448. [CrossRef]
38. FAO. *The State of World Fisheries and Aquaculture 2020*; Sustainability in Action; FAO: Rome, Italy, 2020. [CrossRef]
39. Davies, T.W.; Duffy, J.P.; Bennie, J.; Gaston, K.J. The nature, extent, and ecological implications of marine light pollution. *Front. Ecol. Environ.* **2014**, *12*, 347–355. [CrossRef]
40. Bolton, D.; Mayer-Pinto, M.; Clark, G.F.; Dafforn, K.A.; Brassil, W.A.; Becker, A.; Johnston, E.L. Coastal urban lighting has ecological consequences for multiple trophic levels under the sea. *Sci. Total Environ.* **2017**, *576*, 1–9. [CrossRef]
41. Li, D.; Huang, J.; Zhou, Q.M.; Gu, L.; Sun, Y.F.; Zhang, L.; Yang, Z. Artificial light pollution with different wavelengths at night interferes with development, reproduction, and antipredator defenses of *Daphnia magna*. *Environ. Sci. Technol.* **2022**, *56*, 1702–1712. [CrossRef]

42. Bassi, A.; Love, O.P.; Cooke, S.J.; Warriner, T.R.; Harris, C.M.; Madliger, C.L. Effects of artificial light at night on fishes: A synthesis with future research priorities. *Fish Fish.* **2022**, *23*, 631–647. [CrossRef]
43. Becker, A.; Whitfield, A.K.; Cowley, P.D.; Järnegren, J.; Naesje, T.F. Potential effects of artificial light associated with anthropogenic infrastructure on the abundance and foraging behaviour of estuary-associated fishes. *J. Appl. Ecol.* **2013**, *50*, 43–50. [CrossRef]
44. Pulgar, J.; Zeballos, D.; Vargas, J.; Aldana, M.; Manriquez, P.H.; Manriquez, K.; Quijón, P.A.; Widdicombe, S.; Anguita, C.; Quintanilla, D.; et al. Endogenous cycles, activity patterns and energy expenditure of an intertidal fish is modified by artificial light pollution at night (ALAN). *Environ. Pollut.* **2019**, *244*, 361–366. [CrossRef] [PubMed]
45. Nelson, T.R.; Michel, C.J.; Gary, M.P.; Lehman, B.M.; Demetras, N.J.; Hammen, J.J.; Horn, M.J. Effects of artificial lighting at night on predator density and salmonid predation. *Trans. Am. Fish. Soc.* **2021**, *150*, 147–159. [CrossRef]
46. Zapata, M.J.; Sullivan SM, P.; Gray, S.M. Artificial lighting at night in estuaries-Implications from individuals to ecosystems. *Estuaries Coasts* **2019**, *42*, 309–330. [CrossRef]
47. Brüning, A.; Hölker, F.; Franke, S.; Kleiner, W.; Kloas, W. Influence of light intensity and spectral composition of artificial light at night on melatonin rhythm and mRNA expression of gonadotropins in roach Rutilus rutilus. *Fish Physiol. Biochem.* **2018**, *44*, 1–12. [CrossRef] [PubMed]
48. Khan, Z.A.; Labala, R.K.; Yumnamcha, T.; Devi, S.D.; Mondal, G.; Devi, H.S.; Rajiv, C.; Bharali, R.; Chattoraj, A. Artificial Light at Night (ALAN), an alarm to ovarian physiology: A study of possible chronodisruption on zebrafish (*Danio rerio*). *Sci. Total Environ.* **2018**, *628*, 1407–1421. [CrossRef]
49. O'Connor, J.J.; Fobert, E.K.; Besson, M.; Jacob, H.; Lecchini, D. Live fast, die young: Behavioural and physiological impacts of light pollution on a marine fish during larval recruitment. *Mar. Pollut. Bull.* **2019**, *146*, 908–914. [CrossRef]
50. Wei, H.; Cai, W.J.; Liu, H.K.; Han, D.; Zhu, X.M.; Yang, Y.X.; Jin, J.Y.; Xie, S.Q. Effects of photoperiod on growth, lipid metabolism and oxidative stress of juvenile gibel carp (*Carassius auratus*). *J. Photochem. Photobiol. B Biol.* **2019**, *198*, 111552. [CrossRef]
51. Basili, D.; Lutfi, E.; Falcinelli, S.; Balbuena-Pecino, S.; Navarro, I.; Bertolucci, C.; Capilla, E.; Carnevali, O. Photoperiod Manipulation Affects Transcriptional Profile of Genes Related to Lipid Metabolism and Apoptosis in Zebrafish (*Danio rerio*) Larvae: Potential Roles of Gut Microbiota. *Microb. Ecol.* **2020**, *79*, 933–946. [CrossRef]
52. Kupprat, F.; Hölker, F.; Kloas, W. Can skyglow reduce nocturnal melatonin concentrations in Eurasian perch? *Environ. Pollut.* **2020**, *262*, 114324. [CrossRef]
53. Mondal, G.; Devi, S.D.; Khan, Z.A.; Yumnamcha, T.; Rajiv, C.; Devi, H.S.; Chattoraj, A. The influence of feeding on the daily rhythm of mRNA expression on melatonin bio-synthesizing enzyme genes and clock associated genes in the zebrafish (*Danio rerio*) gut. *Biol. Rhythm. Res.* **2021**, *53*, 1073–1090. [CrossRef]
54. Nemova, N. Different photoperiod regimes affect the fatty acid profile of juvenile Atlantic salmon *Salmo salar* L.: Aquaculture and conservation approach to restoration of wild populations. *J. Am. Oil Chem. Soc.* **2021**, *98*, 236–237.
55. Ytrestøyl, T.; Hjelle, E.; Kolarevic, J.; Takle, H.; Rebl, A.; Afanasyev, S.; Krasnov, A.; Brunsvik, P.; Terjesen, B.F. Photoperiod in recirculation aquaculture systems and timing of seawater transfer affect seawater growth performance of Atlantic salmon (*Salmo salar*). *J. World Aquac. Soc.* **2022**, 1–23. [CrossRef]
56. FAO. *FAO Yearbook. Fishery and Aquaculture Statistics 2018*; FAO: Rome, Italy, 2020. [CrossRef]
57. Polakof, S.; Míguez, J.M.; Soengas, J.L. Daily changes in parameters of energy metabolism in liver, white muscle, and gills of rainbow trout: Dependence on feeding. *Comp. Biochem. Physiol. A-Mol. Integr. Physiol.* **2007**, *147*, 363–374. [CrossRef] [PubMed]
58. Hernández-Pérez, J.; Míguez, J.M.; Librán-Pérez, M.; Otero-Rodiño, C.; Naderi, F.; Soengas, J.L.; López-Patiño, M.A. Daily rhythms in activity and mRNA abundance of enzymes involved in glucose and lipid metabolism in liver of rainbow trout, *Oncorhynchus mykiss. Influ. Light Food Availab. Chronobiol. Int.* **2015**, *32*, 1391–1408. [CrossRef] [PubMed]
59. Nisembaum, L.G.; Velarde, E.; Tinoco, A.B.; Azpeleta, C.; de Pedro, N.; Alonso-Gómez, A.L.; Delgado, M.J.; Isorna, E. Light-dark cycle and feeding time differentially entrains the gut molecular clock of the goldfish (*Carassius auratus*). *Chronobiol. Int.* **2012**, *29*, 665–673. [CrossRef]
60. AOAC. Official methods of analysis of AOAC international. In *Official Analytical Chemists*, 16th ed.; Cunniff, P., Ed.; AOAC International: Arlington, VA, USA, 1995; pp. 1141–1154.
61. Liu, T.; Xu, H.Y.; Han, T.; Wang, J.T.; Yin, F.; Wang, C.L. Effect of dietary egg yolk lecithin levels on survival, growth, lipid metabolism, and antioxidant capacity of early juvenile green mud crab *Scylla paramamosain*. *Aquaculture* **2021**, *540*, 736706. [CrossRef]
62. Livak, K.J.; Schmittgen, T.D. Analysis of relative gene expression data using real-time quantitative PCR and the $2^{-\Delta\Delta CT}$ method. *Methods* **2001**, *25*, 402–408. [CrossRef]
63. Sanchez, G. PLS Path Modeling with R. Trowchez Editions. Berkeley. 2013. Available online: http://www.gastonsanchez.com/PLSPathModelingwithR.pdf (accessed on 25 April 2022).
64. Fjelldal, P.G.; Nordgarden, U.; Berg, A.; Grotmol, S.; Totland, G.K.; Wargelius, A.; Hansen, T. Vertebrae of the trunk and tail display different growth rates in response to photoperiod in Atlantic salmon, *Salmo salar* L., post-smolts. *Aquaculture* **2005**, *250*, 516–524. [CrossRef]
65. Wargelius, A.; Fjelldal, P.G.; Nordgarden, U.; Hansen, T. Continuous light affects mineralization and delays osteoid incorporation in vertebral bone of Atlantic salmon (*Salmo salar* L.). *J. Exp. Biol.* **2009**, *212*, 656–661. [CrossRef]

66. Fjelldal, P.G.; Lock, E.J.; Hansen, T.; Waagbo, R.; Wargelius, A.; Martens, L.G.; El-Mowafi, A.; Onsrud, R. Continuous light induces bone resorption and affects vertebral morphology in Atlantic salmon (*Salmo salar* L.) fed a phosphorous deficient diet. *Aquac. Nutr.* **2012**, *18*, 610–619. [CrossRef]
67. Villamizar, N.; Garciá-Alcazar, A.; Sánchez-Vázquez, F.J. Effect of light spectrum and photoperiod on the growth, development and survival of European sea bass (*Dicentrarchus labrax*) larvae. *Aquaculture* **2009**, *292*, 80–86. [CrossRef]
68. Mhalhel, K.; Germanà, A.; Abbate, F.; Guerrera, M.C.; Levanti, M.; Laurà, R.; Montalbano, G. The Effect of Orally Supplemented Melatonin on Larval Performance and Skeletal Deformities in Farmed Gilthead Seabream (*Sparus aurata*). *Int. J. Mol. Sci.* **2020**, *21*, 9597. [CrossRef]
69. Mistlberger, R.E. Circadian food-anticipatory activity: Formal models and physiological mechanisms. *Neurosci. Biobehav. Rev.* **1994**, *18*, 171–195. [CrossRef]
70. Reebs, S.G.; Gallant, B.Y. Food-anticipatory activity as a cue for local enhancement in golden shiners (Pisces: Cyprinidae, *Notemigonus crysoleucas*). *Ethology* **1997**, *103*, 1060–1069. [CrossRef]
71. Chen, W.M.; Purser, G.J. The effect of feeding regime on growth, locomotor activity pattern and the development of food anticipatory activity in greenback flounder. *J. Fish Biol.* **2001**, *58*, 177–187. [CrossRef]
72. Stephan, F.K. The "other" circadian system: Food as zeitgeber. *J. Biol. Rhythm.* **2002**, *17*, 284–292. [CrossRef]
73. Vera, L.M.; De Pedro, N.; Gomez-Milán, E.; Delgado, M.J.; Sánchez-Muros, M.J.; Madrid, J.A.; Sánchez-Vázquez, F.J. Feeding entrainment of locomotor activity rhythms, digestive enzymes and neuroendocrine factors in goldfish. *Physiol. Behav.* **2007**, *90*, 518–524. [CrossRef]
74. Montoya, A.; López-Olmeda, J.F.; Yúfera, M.; Sánchez-Muros, M.J.; Sánchez-Vázquez, F.J. Feeding time synchronises daily rhythms of behaviour and digestive physiology in gilthead seabream (*Sparus aurata*). *Aquaculture* **2010**, *306*, 315–321. [CrossRef]
75. Sánchez, J.A.; López-Olmeda, J.F.; Blanco-Vives, B.; Sánchez-Vázquez, F.J. Effects of feeding schedule on locomotor activity rhythms and stress response in sea bream. *Physiol. Behav.* **2009**, *98*, 125–129. [CrossRef]
76. Polakof, S.; Panserat, S.; Soengas, J.L.; Moon, T.W. Glucose metabolism in fish: A review. *J. Comp. Physiol. B-Biochem. Syst. Environ. Physiol.* **2012**, *182*, 1015–1045. [CrossRef] [PubMed]
77. Capilla, E.; Medale, F.; Navarro, I.; Panserat, S.; Vachot, C.; Kaushik, S.; Gutierrez, J. Muscle insulin binding and plasma levels in relation to liver glucokinase activity, glucose metabolism and dietary carbohydrates in rainbow trout. *Regul. Pept.* **2003**, *110*, 123–132. [CrossRef]
78. Gong, Y.; Chen, W.; Dong, H.; Zhu, X.; Yang, Y.; Jin, J.; Liu, H.; Xie, S. Effects of food restriction on growth, body composition and gene expression related in regulation of lipid metabolism and food intake in grass carp. *Aquaculture* **2017**, *469*, 28–35. [CrossRef]
79. Koonen, D.P.Y.; Jacobs, R.L.; Febbraio, M.; Young, M.E.; Soltys, C.L.M.; Ong, H.; Vance, D.E.; Dyck, J.R.B. Increased hepatic CD36 expression contributes to dyslipidemia associated with diet-induced obesity. *Diabetes* **2007**, *56*, 2863–2871. [CrossRef] [PubMed]
80. Frohnert, B.I.; Hui, T.Y.; Bernlohr, D.A. Identification of a functional peroxisome proliferator-responsive element in the murine fatty acid transport protein gene. *J. Biol. Chem.* **1999**, *274*, 3970–3977. [CrossRef] [PubMed]
81. Vieira, R.F.L.; Munoz, V.R.; Junqueira, R.L.; de Oliveira, F.; Gaspar, R.C.; Nakandakari, S.C.B.R.; Costa, S.D.; Torsoni, M.A.; da Silva, A.S.R.; Cintra, D.E.; et al. Time-restricted feeding combined with aerobic exercise training can prevent weight gain and improve metabolic disorders in mice fed a high-fat diet. *J. Physiol.* **2021**, *600*, 797–813. [CrossRef]
82. Wagner, T.; Congleton, J.L. Blood chemistry correlates of nutritional condition, tissue damage, and stress in migrating juvenile chinook salmon (*Oncorhynchus tshawytscha*). *Can. J. Fish. Aquat. Sci.* **2004**, *61*, 1066–1074. [CrossRef]
83. Sánchez-Bretaño, A.; Gueguen, M.M.; Cano-Nicolau, J.; Kah, O.; Alonso-Gomez, A.L.; Delgado, M.J.; Isorna, E. Anatomical distribution and daily profile of gper1b gene expression in brain and peripheral structures of goldfish (*Carassius auratus*). *Chronobiol. Int.* **2015**, *32*, 889–902. [CrossRef]
84. López-Olmeda, J.F.; Sánchez-Vázquez, F.J. Feeding Rhythms in Fish: From Behavioural to Molecular Approach. In *Biological Clock in Fish*; Kulczykowska, E., Popek, W., Kapoor, B.G., Eds.; Science Publishers: Enfield, CT, USA, 2010; pp. 155–184.
85. Vera, L.M.; Negrini, P.; Zagatti, C.; Frigato, E.; Sanchez-Vazquez, F.J.; Bertolucci, C. Light and feeding entrainment of the molecular circadian clock in a marine teleost (*Sparus aurata*). *Chronobiol. Int.* **2013**, *30*, 649–661. [CrossRef]
86. Iigo, M.; Tabata, M. Circadian Rhythms of Locomotor Activity in the Rainbow Trout *Oncorhynchus mykiss*. *Fish. Sci.* **1997**, *63*, 77–80. [CrossRef]
87. Sánchez-Vázquez, F.J.; Tabata, M. Circadian rhythms of demand-feeding and locomotor activity in rainbow trout. *J. Fish Biol.* **1998**, *52*, 255–267. [CrossRef]
88. Bolliet, V.; Aranda, A.; Boujard, T. Demand-feeding rhythm in rainbow trout and European catfish: Synchronisation by photoperiod and food availability. *Physiol. Behav.* **2001**, *73*, 625–633. [CrossRef]
89. Salgado-Delgado, R.C.; Saderi, N.; Basualdo, M.D.; Guerrero-Vargas, N.N.; Escobar, C.; Buijs, R.M. Shift Work or Food Intake during the Rest Phase Promotes Metabolic Disruption and Desynchrony of Liver Genes in Male Rats. *PLoS ONE* **2013**, *8*, e60052. [CrossRef]
90. López-Olmeda, J.F. Nonphotic entrainment in fish. *Comp. Biochem. Physiol.* **2017**, *203 Pt A*, 133–143. [CrossRef]
91. Fortes-Silva, R.; Do Valle, S.V.; Lopez-Olmeda, J.F. Daily rhythms of swimming activity, synchronization to different feeding times and effects on anesthesia practice in an Amazon fish species (*Colossoma macropomum*). *Chronobiol. Int.* **2018**, *35*, 1713–1722. [CrossRef] [PubMed]

92. Frolow, J.; Milligan, C.L. Hormonal regulation of glycogen metabolism in white muscle slices from rainbow trout (*Oncorhynchus mykiss* Walbaum). *Am. J. Physiol.-Regul. Integr. Comp. Physiol.* **2004**, *287*, R1344–R1353. [CrossRef] [PubMed]
93. Sargent, J.R.; Tocher, D.R.; Bell, J.G. The Lipid. In *Fish Nutrition*; Halver, J.E., Hardy, R.W., Eds.; Academic Press: Cambridge, MA, USA, 2002; pp. 181–257.
94. Tocher, D.R. Metabolism and functions of lipids and fatty acids in teleost fish. *Rev. Fish. Sci.* **2003**, *11*, 107–184. [CrossRef]
95. Hillyer, K.E.; Beale, D.J.; Shima, J.S. Artificial light at night interacts with predatory threat to alter reef fish metabolite profiles. *Sci. Total Environ.* **2021**, *769*, 144482. [CrossRef]
96. Henderson, R.J.; Sargent, J.R.; Hopkins, C.C.E. Changes in the content and fatty acid composition of lipid in an isolated population of the capelin *Mallotus villosus* during sexual maturation and spawning. *Mar. Biol.* **1984**, *78*, 255–263. [CrossRef]
97. Valenzuela, A.; Rodríguez, I.; Schulz, B.; Cortés, R.; Acosta, J.; Campos, V.; Escobar-Aguirre, S. Effects of continuous light (LD24:0) modulate the expression of lysozyme, mucin and peripheral blood cells in rainbow trout. *Fishes* **2022**, *7*, 28. [CrossRef]
98. Qin, C.J.; Gong, Q.; Wen, Z.Y.; Zou, Y.C.; Yuan, D.Y.; Shao, T.; Li, H.T. Comparative analysis of the liver transcriptome of *Pelteobagrus vachellii* with an alternative feeding time. *Comp. Biochem. Physiol. D-Genom. Proteom.* **2017**, *22*, 131–138. [CrossRef] [PubMed]
99. Xie, D.Z.; Wang, S.Q.; You, C.H.; Chen, F.; Zhang, Q.H.; Li, Y.Y. Influencing factors and mechanisms on HUFA biosynthesis in teleosts. *J. Fish. Sci. China* **2013**, *20*, 456–466. [CrossRef]
100. Calabretti, A.; Cateni, F.; Procida, G.; Favretto, L.G. Influence of environmental temperature on composition of lipids in edible flesh of rainbow trout (*Oncorhynchus mykiss*). *J. Sci. Food Agric.* **2003**, *83*, 1493–1498. [CrossRef]
101. Cengiz, E.I.; Unlu, E.; Bashan, M.; Satar, A.; Uysal, E. Effect of seasonal variations on the fatty acid composition of total lipid phospholipid and triacylglicerol in the dorsal muscle of Mesopotamian catfish (*Silurus triostegus* Heckel, 1843) in Tigris River (Turkey). *Turk. J. Fish. Aquat. Sci.* **2012**, *2012*, 31–37. [CrossRef]
102. Skiba-Cassy, S.; Geurden, I.; Panserat, S.; Seiliez, I. Dietary methionine imbalance alters the transcriptional regulation of genes involved in glucose, lipid and amino acid metabolism in the liver of rainbow trout (*Oncorhynchus mykiss*). *Aquaculture* **2016**, *454*, 56–65. [CrossRef]
103. Song, X.R.; Marandel, L.; Skiba-Cassy, S.; Corraze, G.; Dupont-Nivet, M.; Quillet, E.; Geurden, I.; Panserat, S. Regulation by dietary carbohydrates of intermediary metabolism in liver and muscle of two isogenic lines of rainbow trout. *Front. Physiol.* **2018**, *9*, 1579. [CrossRef]
104. Kamalam, B.S.; Médale, F.; Larroquet, L.; Corraze, G.; Panserat, S. Metabolism and fatty acid profile in fat and lean rainbow trout lines fed with vegetable oil: Effect of carbohydrates. *PLoS ONE* **2013**, *8*, e76570. [CrossRef] [PubMed]
105. Velasco, C.; Comesaña, S.; Conde-Sieira, M.; Míguez, J.M.; Soengas, J.L. Effects of CCK-8 and GLP-1 on fatty acid sensing and food intake regulation in trout. *J. Mol. Endocrinol.* **2019**, *62*, 101–116. [CrossRef] [PubMed]
106. Conde-Sieira, M.; Capelli, V.; Álvarez-Otero, R.; Díaz-Rúa, A.; Velasco, C.; Comesaña, S.; López, M.; Soengas, J.L. Hypothalamic AMPKα2 regulates liver energy metabolism in rainbow trout through vagal innervation. *Am. J. Physiol-Reg. I.* **2020**, *318*, R122–R134. [CrossRef]
107. Sánchez-Gurmaches, J.; Cruz-Garcia, L.; Gutiérrez, J.; Navarro, I. Adiponectin effects and gene expression in rainbow trout: An in vivo and in vitro approach. *J. Exp. Biol.* **2012**, *215*, 1373–1383. [CrossRef] [PubMed]
108. Naderi, F.; Míguez, J.M.; Soengas, J.L.; López-Patiño, M.A. SIRT1 mediates the effect of stress on hypothalamic clock genes and food intake regulators in rainbow trout, *Oncorhynchus mykiss*. *Comp. Biochem. Phys. A* **2019**, *235*, 102–111. [CrossRef] [PubMed]
109. Panserat, S.; Plagnes-Juan, E.; Gazzola, E.; Palma, M.; Magnoni, L.J.; Marandel, L.; Viegas, I. Hepatic glycerol metabolism-related genes in carnivorous rainbow trout (*Oncorhynchus mykiss*): Insights into molecular characteristics, ontogenesis, and nutritional regulation. *Front. Physiol.* **2020**, *11*, 882. [CrossRef] [PubMed]

MDPI
St. Alban-Anlage 66
4052 Basel
Switzerland
www.mdpi.com

Metabolites Editorial Office
E-mail: metabolites@mdpi.com
www.mdpi.com/journal/metabolites

www.ingramcontent.com/pod-product-compliance
Lightning Source LLC
Chambersburg PA
CBHW061832210926
51850CB00006B/864
* 9 7 8 3 0 3 6 5 9 9 8 5 4 *